The NEW SELF-SUFFICIENT GARDENER

The NEW SELF-SUFFICIENT GARDENER

JOHN SEYMOUR

London, New York, Munich, Melbourne, and Delhi

Produced for Dorling Kindersley by
Editor Pip Morgan
Designer Edward Kinsey
Digital color enhancement Simon Roulstone

For Dorling Kindersley Ltd
Editor Ariane Durkin
Designer Kathryn Wilding
Managing Art Editor Heather McCarry
Managing Editor Dawn Henderson
Production Editor Jenny Woodcock

Original edition published by Faber and Faber in 1978
Editorial director Christopher Davis
Art director Roger Bristow
Managing editor Jackie Douglas
Editor David Reynolds
Art editor Chris Meehan
Text editors Sybil del Strother Peter Leech
Designer Sue Rawkins
Editorial assistant David Lamb
Illustrators Peter Morter Eric Thomas Robert Micklewright
David Bryant Jim Robins

First American Edition, 2008

Published in the United States by
DK Publishing
375 Hudson Street, New York, New York 10014

09 10 11 10 9 8 7 6 5

ND107–April 2008

Published in Great Britain by Dorling Kindersley Limited.

A catalog record for this book is available from the Library of Congress.

ISBN 978-0-7566-2898-7

Color reproduction by MDP, UK
Printed and bound in China by Toppan

WARNING
This book reproduces the original 1978 text of this classic by John Seymour. Some of the pesticides he recommends are no longer legally available, and the publisher has made every effort to identify these by inserting asterisks throughout the text, and has set out a list at the end of the book identifying products that have now been banned, and suggesting some alternatives where available. The publisher does not accept any liability for any damage or injuries resulting from the use of any pesticides.

Foreword

The growing, harvesting, preserving, and, most of all, the eating of food were a major part of the Seymour family's life.

Everyone was involved with the production of food. As children we'd help to plant the seeds, weed, harvest potatoes, shell peas, and collect eggs. We would milk the cow and scoop the froth from the top of the milk bucket to drink. There was thick Jersey cream to pour over our oatmeal.

John and Sally had a down-to-earth practical philosophy for growing vegetables. If you dug enough good compost or manure into the soil, almost anything would grow—nourish the soil to nourish the plants, which in turn will nourish you.

I have used this book constantly throughout my gardening life. In every page I hear John's voice—practical, inspiring, and encouraging. John was a great believer in "working" the soil and in using one's "common sense." Not for him the phases of the moon.

A few years before his death I visited him, having just lovingly planted my onion sets, placing them tenderly in the soil and crooning to them encouragingly. I found John in his garden shoving his onions into the ground with force—"Get in there, you little brutes," he said. Needless to say, they grew regardless and were as good a crop of onions as I have seen.

JANE SEYMOUR AUGUST 2007

Contents

Introduction

When I was a boy in the English countryside—fifty years ago and more—country people really did garden for self-sufficiency. They had to. Early each morning the lovely sound of cockerels crowing was heard everywhere, because nearly all country people kept hens. Country children would be seen coming home along the lanes from school, their arms laden with wild plants they had gathered for their rabbits. Nearly every cottager had a pig: "a pig in the sty and one in the larder," they used to say. Their pigs and chickens and rabbits gave enormous fertility to their gardens, as did the emptyings of the lavatory bucket—there was no waterborne sewage in the countryside in those days. Most farmers delivered a cartload or two of farmyard manure to their farm workers once a year, and this was put on the garden.

These country gardens were enormously productive. They were bursting with fertility. No countryman in those days would have dreamed of buying any vegetable and there was absolutely no need for him to do so. As people grew richer and food became cheaper, and as farming became increasingly mechanized, sending more and more countrymen into the cities to find their livings, so these magnificent food-producing gardens all but ceased to exist.

Gardens acquired a new role, that of a status symbol in the game of keeping up with the neighbors. The vegetable yield ceased to be the vital factor; instead, such matters as the greenness of the lawn and the display of flowers were all-important. The country wife might still grow a few lettuces and cabbages, but they would be down at the end of the garden, well out of sight.

But now, once again, the pendulum is swinging the other way. As food, and everything else, is becoming more expensive, there is coming about a great renaissance of gardening for self-sufficiency. People find that they are saving a significant part of their salaries by doing it, that their food tastes better and does them more good, and that their children are healthier. They themselves benefit from some hard work in the fresh air, and from being involved with the benign cycle of the seasons and with the satisfying process of helping nature create beautiful and nourishing food out of what is apparently nothing.

A few years ago, in Europe, there were community gardens with nobody working them. Now there's a long waiting list everywhere. In the United States, plots of land for food-growing are sought with great eagerness. Everywhere gardening is losing its image as a retired person's sport; young people are learning

to do it and lively minds are getting to work on new techniques.

Organic gardening

Alongside this reemergence of the desire to grow food has come an awareness that the earth's supply of mineral-derived energy is limited. Simplistic gardening with chemical fertilizers is becoming a thing of the past as the new generation of gardeners is learning to do without the petrochemicals on which gardening has become dependent, and to rely solely on nature's very own methods. People are rediscovering the value of waste organic matter animal and vegetable. Methods of making compost and of keeping animals in the garden are again being treated as matters of crucial importance.

The whole question of chemical versus organic gardening is still, at the present time, a major controversy. In my view, the best proof is in the eating. My own garden is far from perfect. I sow more than I can reap, like most serious gardeners, and end up with too many weeds, not enough time to hoe them, and occasionally crops that I am ashamed of. But on the whole my garden is lush and fertile, and I grow a great deal of produce in it. Orthodox gardeners, who use chemicals on their gardens, come and look at my crops sometimes and refuse to believe that I put no inorganic fertilizer on them. But I don't: not an ounce of artificial nitrogen has gone on my land for thirteen years now. And even when the weeds do overwhelm me, I am often surprised that there seems to be fertility enough for them and my crops. This year my onions, carrots, and parsnips, which were all interplanted with each other, were neglected and overrun with weeds. And yet out of the mess, I have dug large and beautiful carrots and parsnips, and the onions, all hanging in strings now, are twice as big as my fist, luscious and firm.

I am not advocating the cultivation of weeds. But rather than douse them with some "selective weed spray" (which is only selective up to a point, of course—if a chemical damages one form of life, you can be sure that it will do at least some damage to many other forms of life), I would rather have a few weeds and maybe a slightly smaller crop.

I am surprised when I read the advice of many gardening writers that their readers should apply this or that amount of "complete fertilizer," or of some commercial high-nitrogen chemical, or that the soil should be doused with herbicides, pesticides, or fungicides ("-cide" means poison and no gardener should forget that fact). I wonder whether these writers have tried gardening without chemicals. Applied nitrates definitely harm the soil in the long term, although it must be admitted that they have a dramatic effect on the growth of crops in soil that is already hooked on them—that is to say, soil that has no nitrogen-fixing capacity of its own. However, the main point that needs to be made is that some of the best garden—and farm—crops in the world are grown without any of these expensive and dangerous aids at all. The highest yield possible with any crop can be, and very often is, achieved without chemicals.

But here there is a trap that many a would-be organic gardener has fallen into. "I am 'organic'; I don't put any artificial fertilizers on my land," they say. But neither do they put anything else! Nothing will come out of nothing, and if you continue to take crops out of your soil and put nothing back, you will eventually grow, well, next to nothing. If you see a garden, which is held to be organic, filled with nothing but miserable insect-eaten crops surrounded by weeds, it is probable that you have encountered this negative approach. Some advocates of organic gardening also put forward eccentric notions like planting crops according to the phases of the moon, sprinkling tiny amounts of obscure substances on the soil, and so on. Plant seeds will germinate and plants will grow when temperature, humidity, and nutrients are right for them. The organic philosophy has no need of these irrational and superstitious notions. The organic approach is based on sound fact and science, and its practice can be seen to be effective and correct.

The organic philosophy is well summed up as adherence to the following six laws: first, the gardener must work with nature and not against it; second, nature is diverse and therefore the gardener must practice diversity; third, the gardener must husband other forms of life—animal or vegetable—in environments as close as possible to those for which they evolved; fourth, the gardener must return to the soil as much, or nearly as much, as he takes from it; fifth, the gardener must feed the soil and not the plant; sixth, the gardener must study nature as a whole and never any part of it in isolation.

Garden animals

It is, of course, possible to live on vegetables alone. In fact, it is theoretically possible to live on peas, beans, and potatoes, if you have enough of them. However, those gardeners who are neither vegans nor vegetarians will find that on even the smallest pieces of land, they can keep rabbits, chickens, or both. Any animals you have will make a major contribution to the fertility of your garden, and if you add rabbit and chicken meat, and eggs, to your vegetable and fruit produce, you can certainly be a self-sufficient gardener in the true sense, and you will be coming close to providing yourself with the varied diet that is now fairly commonplace in the West. Dairy products and the meat of large animals will be almost the only things that you need to buy.

Many people feel that animals will tie them down: they won't be able to take a vacation, or even go away for the weekend. But what if they team up with neighbors, and either keep birds and animals cooperatively, or else agree to take turns looking after each other's livestock? To feed a few rabbits and a few hens, and collect the eggs, is the work of minutes a day.

The advantage of keeping birds and animals is immense. As well as the food they provide, almost more important is the fertility they give to the garden. Detailed instructions for keeping garden animals are given on pp. 230 to 239.

I have organized the chapters of this book in which I discuss the cultivation of vegetables and fruit (see pp. 113–190) on the basis of the natural families and orders into which plants have been grouped by botanists. I have done this because there are very close relationships between plants of the same family, and you will find that if you think of plants in terms of their relations, you will eventually get a very real feeling for the characters of the different plants.

For example, you grow potatoes and tomatoes for quite different reasons: the former for its underground tubers and the latter for its fruit. You could, of course, grow them successfully for years without realizing that they are very closely related. But once you have them classified in your head as members of the *Solanaceae*, or nightshade family, you immediately see and feel their close relationship and take an interest in it. And, of course, you soon find that potatoes and tomatoes suffer from all the same diseases and respond well or poorly to much the same sort of treatment.

It is also fun to spot the similarities that make, for example, apples, strawberries, and roses close relations. It is useful to know that pears are closely related to hawthorns, and plums to blackthorns, for it is well worth grafting scions, or fruiting spurs, of fruit trees on to the rootstocks of their wild, and therefore more hardy, relations. A sense of the specialness, similarities, and differences of different kinds of plants adds enormously to the pleasure of gardening; it gives an insight into plants and engenders in the gardener the sort of sympathy that good husbanders should have—the sympathy that is best described by the phrase "a green thumb."

Finally, let me stress that self-sufficient gardening needs commitment. It requires the determination to produce as much good food as is humanly possible on whatever area of land is at your disposal. Every patio, every flat roof, every windowsill, should be looked upon as a possible food-producing area. New methods of organic gardening that produce high yields in small spaces should be considered. More than this, self-sufficient gardening requires a commitment to understanding your plants, your soil, and the workings of nature: its seasons and its cycles. I have given over a chapter of this book to explain the workings of the seasons. Other chapters discuss the growing of the individual varieties of vegetables, fruit, and herbs in detail. In the remainder of this introduction, I will try to explain about the soil and the cycles of nature as I understand them. But before I move on to these topics, I would like to introduce you to the deep-bed method of gardening.

THE DEEP-BED METHOD

Unless you have a very large area of land at your disposal, the key to success as a self-sufficient gardener is to be able to grow a lot in a small space. Of the new techniques for growing more vegetables in smaller areas, the most important in my view is what I call the deep-bed method, which is being developed in California by several

Americans, several Chinese immigrants to the US, and an Englishman named Alan Chadwick. The method is derived from age-old techniques that have been practiced in France and China, but which have never been widely adopted in the West.

The essence of the method is to dig deeply and then never step on the bed. This means your plants are growing in very loose, deeply dug soil; their roots will go down instead of sideways. You therefore get bigger plants, and can grow them closer together. The deep-bed method is discussed in detail on pp. 106 to 112. Throughout the chapters on the cultivation of vegetables and fruit (pp. 113-190), I have included instructions for growing each crop by the deep-bed method, wherever this differs from conventional practice.

A deep bed should produce about four times the yield by weight that a conventional bed will produce. A deep bed of 100 square feet (9 sq m) can produce from 200 to 400 lb (90–180 kg) of vegetables a year. According to the US Department of Agriculture's statistics, the average American eats 322 lb (145 kg) of vegetables a year. Thus one tiny bed—just pace out 20 feet by 5 feet (6 x 1.5 m) on the floor to get an idea of the size—can keep one adult supplied with vegetables. From what I have seen of deep-bed gardens in the United States, and from my first hand experience on my own land, I can say that the claims made for this method are by no means exaggerated. I think it highly likely that more and more serious vegetable gardeners will adopt this method. If your aim is to grow as many vegetables as you can in the space available to you, then I urge you to study the technique and try it.

NATURE'S CYCLES

Early in the evolution of living things, the animal and vegetable kingdoms diverged along different paths. Since then they have evolved in coexistence, each dependent on the other. Plants are essential to animals, for only plants can store the energy of the sun by photosynthesis, and fix the free nitrogen in the air into compounds from which both they and the animals can draw energy.

A DEEP BED

You can create a deep bed by digging to a spade's depth and loosening the soil to a further spade's depth. You must incorporate a large amount of manure and never step on the bed. The very loose soil will allow the roots of your crops to penetrate deep down, instead of spreading sideways as they do in conventional beds. You will get bigger vegetables and you will be able to grow them closer together.

At the same time, plants could not exist without animals. Nearly all flowering plants—which means nearly all the higher plants—are dependent on animals, especially birds and insects, for pollination. Without animals they could not reproduce. But beyond this, there is a benign cycle between plants and animals at all stages of their lives. Plants inhale carbon dioxide, which would kill animals in too great a quantity, and exhale oxygen, without which animals cannot live. Animals inhale the oxygen and exhale the carbon dioxide that is needed by plants.

Animals consume plants and, quite simply, transform plant tissue into more complex matter. Simple plant protein is converted into more complex animal protein. Animals void such food as they cannot directly absorb, and this falls on the earth, making instantly-available food that is taken up again by plants. Plants feed animals, and animals feed plants. This is the basic cycle of nature, without which life on this planet would cease to exist.

As well as the animal–plant cycle, there are other natural cycles that define the ceaseless circulation of the elements within Earth's atmosphere. Two of these are of the utmost importance to gardeners: the water cycle and the nitrogen cycle.

The water cycle

The water cycle is the simpler of the two. Put briefly, water is evaporated from the sea, the lakes, the rivers, and the soil by the sun, and it is also transpired by plants and animals. It is carried around the atmosphere by winds, and precipitated in the form of rain or snow, some of which falls back into the sea, but much of which falls on the land.

If it falls on good soil, with plenty of humus in it, it soaks in. Some of it remains in the soil, held like water in a sponge. The rest sinks down deeper, until it reaches impervious rock. It then makes its way down any slope it can until, perhaps, it emerges on the surface lower down a hill, runs down to a stream, and eventually it gets to the sea.

Now such water as remains in the soil may possibly reach the surface and be evaporated again, or it may be taken up by a plant, in which case it will probably enter the roots of the plant. It will ascend the plant, carrying whatever soluble chemicals there are dissolved in it, for it has taken these things up from the soil. Some of this water will make part of the plant's tissues, and deliver the nutrients it has in solution to the various cells of the plant. The remaining water, which is not taken into the plant tissue, will be transpired through the stomata of the leaves, the small apertures in the leaves' skin.

Without this movement of water from the soil below it to the sky above it, a plant could not eat, nor could it grow. Plants depend entirely on water to bring them their food. This does not mean that you have to swamp your plants with water. Most land plants need moist soil, not waterlogged soil, to keep them healthy and growing. If their roots are immersed completely in water for any length of time, ultimately they will die.

The nitrogen cycle

Nitrogen is an essential ingredient of all plants and animals. The air is a mixture of oxygen and nitrogen; but the two elements are not in compound, simply in mixture. This means that the nitrogen is what is called "free nitrogen," meaning that it is still free to combine with another element, or elements, to form a compound. But the higher plants cannot use free nitrogen: they must have it combined as a compound with at least one other element. For example, one part of nitrogen combined with three parts of hydrogen produces ammonia, which, when it has undergone further changes, can be used by plants.

Fortunately, certain bacteria, and certain algae, are capable of "fixing" nitrogen—in other words, making it available to higher forms of life in the form of a compound. Also, the tremendous power in a flash of lightning can fix nitrogen. (It has been conjectured that nitrogen fixed by lightning made possible the first forms of life on Earth.) And nitrogen can also be fixed artificially by the same process as lightning uses. Nitrogen can be combined artificially with hydrogen to form ammonia. The ammonia can then be combined with oxygen and other chemicals to make such substances as ammonium sulfate, urea, ammonium nitrate, sodium nitrate, and calcium nitrate, all of which are used as artificial nitrogen fertilizers.

Nitrogen fixation by artificial means requires one constant: an enormous expenditure of power. Therefore, as power becomes more expensive, and more difficult and dangerous to produce, the nitrogen fixed free by bacteria becomes more and more valuable. Fortunately, by employing perfectly simple and well-tried methods of gardening, you can encourage nitrogen fixation by natural means and grow crops as good as any that ever have been grown with artificial nitrogen, with nitrogen fixed by natural processes alone.

If you study the illustration and caption below, you will see that much fixed nitrogen simply goes around in a short circle—plants, microorganisms, plants, microorganisms, and so on—and in good soil conditions, little nitrogen is released into the air. But any that is released to the air comes back again, eventually, fixed by nitrogen-fixing bacteria.

Other fixed nitrogen goes in a rather longer circle—plants, animals, microorganisms, plants, animals, and so on. No chemist or biologist has ever been able to explain this, but animals are capable of transforming plant matter with a very low nitrogen content into manure with a fairly high nitrogen content in a matter of hours. Keeping animals in your garden is extremely good for your soil and your vegetables. If you have animals, or if you can import animal manure from elsewhere, you will never be short of fixed nitrogen to feed your crops.

There are some important facts about the nitrogen cycle that gardeners need to know. First, all dead animal or vegetable tissue put in or on the soil will eventually release its nitrogen for the use of plants. But—and this is of great importance—it may do so very slowly, because of what is called the nitrogen–carbon ratio. If there is not enough nitrogen to balance the carbon that forms a large part of the body of every living organism, the putrefactive bacteria, which break down organic matter and release nitrates for the higher plants to use, will have

THE NITROGEN CYCLE
The power in a flash of lightning can fix nitrogen—take it from the air and leave it in the soil in a compound form that can be absorbed by plants. More often, nitrogen is fixed by bacteria in the soil, some of which live in nodules on the roots of leguminous plants. Plants turn nitrogen into protein. Animals eat plants and produce more complex protein. Animal waste and the dead bodies of plants and animals return the protein to the soil. Bacteria work on this protein, once again producing nitrogen compounds that will feed plants, and nitrogen that returns to the air.

to borrow nitrogen that is already in the soil. This means that, temporarily, they will rob the soil of nitrogen. However, eventually they will complete their work and release into the soil the nitrogen they have borrowed, and also the nitrogen they have gotten from the organic matter they have eaten.

In practice, this means that if you dig or plow in material that is low in nitrogen, such as straw, sawdust, or plants that have already gone to seed, you should either put some highly nitrogenous substance in with it to feed the bacteria that are to break them down, or else be prepared to wait a long time before that soil is again very fertile. Leguminous plants like alfalfa or clover, turned into the soil before they have flowered, are sufficiently high in nitrogen to break down almost immediately and release their nitrogen in a matter of a very few weeks; the warmer the weather, the quicker the process. Straw from sweet corn, and old plant residues that have seeded, will take a year or even two.

One lesson to be learned from this is: only dig in green manure crops if they are young and succulent. If any plant has already seeded, which means that most of its nitrogen has gone into its seed, put it on your compost pile. This is the value of your compost pile: it rots down organic matter, so that the matter can give its fixed nitrogen back to the soil, without robbing the soil in the process. This is why extra nitrogen, ideally in the form of animal manure, should be added to the material in your compost pile to help the rotting process. Organic matter will rot down in your compost pile even if you don't give it more nitrogen, but the process will take a long time.

The other lesson that the nitrogen cycle teaches is that a gardener should grow as many plants that belong to the family *Fabaceae*, the pea and bean family, as he can. As I have explained, the members of this family have nodules on their roots containing bacteria that fix nitrogen. Leguminous plants are also extremely nutritious and very high in protein. In fact, it would be difficult on an all-vegetable diet to live without them: the seeds of peas and beans are called, are the best form of vegetable protein you can get.

THE SOIL

A spadeful of soil may look like a simple, innocuous substance. But it is, in fact, of such enormous complexity that it is doubtful if humankind will ever fully understand it. First of all, if it is good soil, it is filled with life. In every teaspoonful of soil there are millions of organisms—bacteria of numerous species as well as algae, microscopic animals, the mycelium of fungi, and viruses. In larger quantities of good soil, you are sure to find worms and the larvae of numerous beetles and other insects. It has been calculated that there are from five to ten tons of living matter in every acre of soil.

The interrelationships of these various animals and plants are of great complexity. There are long and involved food chains, and subtle mutually beneficial arrangements. There are chemical processes of such complexity that no scientist has ever been able to duplicate them in his laboratory. For example, there are five species of bacteria that we

know of that can fix nitrogen from the air and convert it into the type of amino acid that can make protein for plants and ultimately people. Two other species of bacteria have the baleful effect of turning useful nitrites and nitrates, which could have been used by plants, into free nitrogen gas again; three species can turn ammonia into nitrites; another can turn nitrites into nitrates, which plants can use; and a huge array of bacteria, fungi, and actinomycetes turn protein and other dead organic matter into ammonia. That simple spadeful of soil is a chemical factory of a sophistication that no human chemist has ever been able even to approach.

The origins of soil

Fundamentally, soil is rock that has been pulverized by such agents as heat and cold, water, and wind, and, very importantly, has been subjected to the eroding effect of lichens, bacteria, algae, and other living creatures. The hardest rock in the world, as long as it is exposed to light, is being gradually gnawed away by plant life.

For the purposes of the gardener, although a geologist might disapprove, it is enough to say that most of the land surface of Earth consists of a layer of soil lying on top of rock. Between

The Ecology of the Soil

All terrestrial life comes from the soil and returns to it. And all terrestrial death comes to life again through the soil, because decomposing organic matter contains nearly all the nutrients that plants require. The good gardener will respect this natural cycle and thereby ensure that the soil in his garden is always living and life-giving.

SOIL LAYERS
Every soil can be divided into three distinct layers: topsoil, which in fertile soil is rich in humus (decayed organic matter); subsoil, which is composed mostly of rock particles; and rock, from which the basis of all soil is formed. Minerals are found in all layers.

PLANTS
The roots of different plants and trees push outward and downward to varying extents. Where many species grow together, nourishment is drawn from all layers of the soil through their roots.

ANIMALS
The complexity and interaction of animal life in the soil performs two crucial functions: the breaking down and returning to the soil of organic matter; and the aeration and loosening of the soil, which enables roots to spread deep and wide, and oxygen, nitrogen, rainwater and other useful elements to penetrate deep down. It is the delicate balance between a myriad species that keeps soil healthy, productive and free from disease.

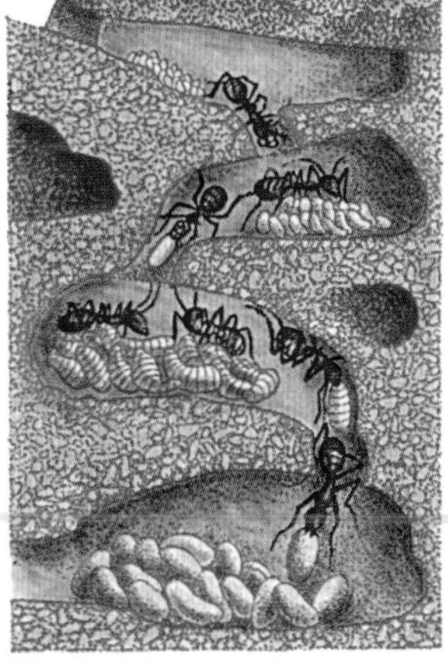

ANTS' NEST
The building of an ants' nest aerates the soil, but it may kill the plant above.

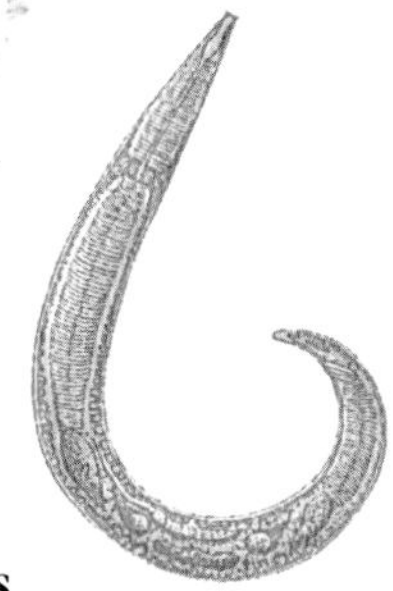

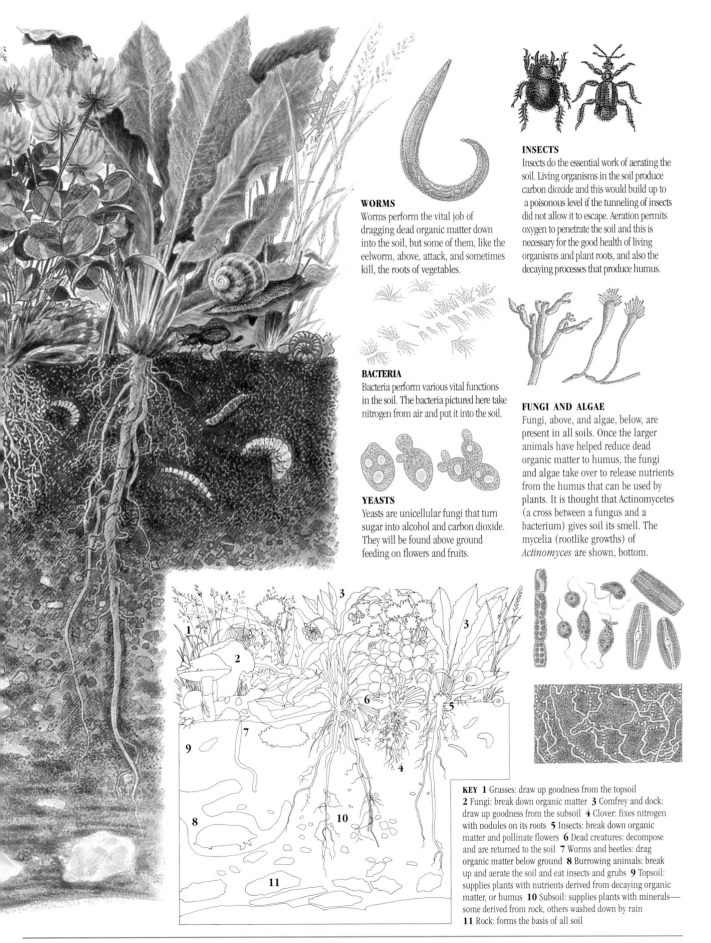

WORMS
Worms perform the vital job of dragging dead organic matter down into the soil, but some of them, like the eelworm, above, attack, and sometimes kill, the roots of vegetables.

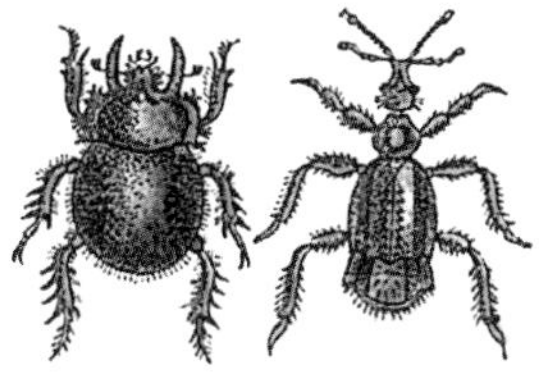

INSECTS
Insects do the essential work of aerating the soil. Living organisms in the soil produce carbon dioxide and this would build up to a poisonous level if the tunneling of insects did not allow it to escape. Aeration permits oxygen to penetrate the soil and this is necessary for the good health of living organisms and plant roots, and also the decaying processes that produce humus.

BACTERIA
Bacteria perform various vital functions in the soil. The bacteria pictured here take nitrogen from air and put it into the soil.

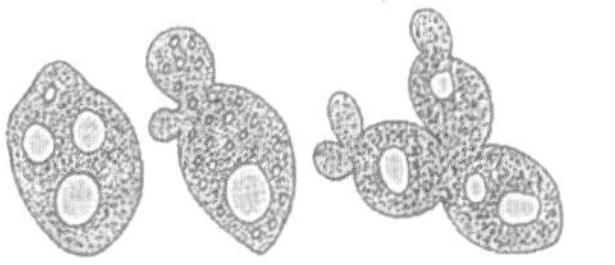

YEASTS
Yeasts are unicellular fungi that turn sugar into alcohol and carbon dioxide. They will be found above ground feeding on flowers and fruits.

FUNGI AND ALGAE
Fungi, above, and algae, below, are present in all soils. Once the larger animals have helped reduce dead organic matter to humus, the fungi and algae take over to release nutrients from the humus that can be used by plants. It is thought that Actinomycetes (a cross between a fungus and a bacterium) gives soil its smell. The mycelia (rootlike growths) of *Actinomyces* are shown, bottom.

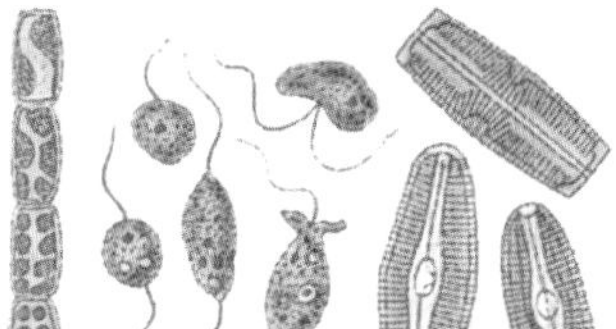

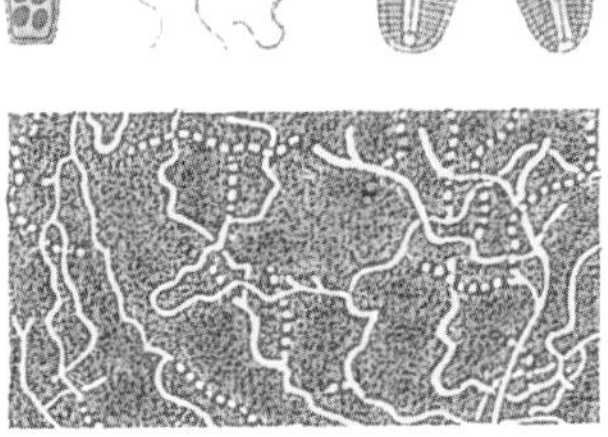

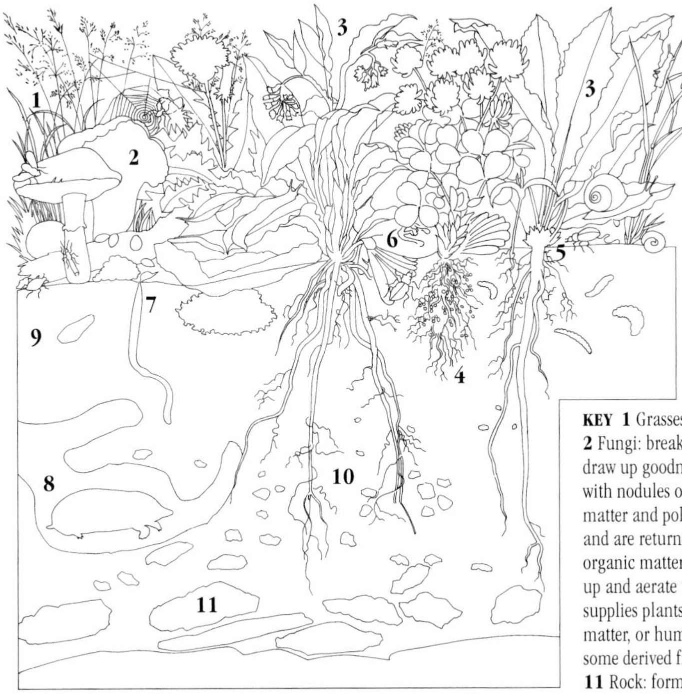

KEY **1** Grasses: draw up goodness from the topsoil **2** Fungi: break down organic matter **3** Comfrey and dock: draw up goodness from the subsoil **4** Clover: fixes nitrogen with nodules on its roots **5** Insects: break down organic matter and pollinate flowers **6** Dead creatures: decompose and are returned to the soil **7** Worms and beetles: drag organic matter below ground **8** Burrowing animals: break up and aerate the soil and eat insects and grubs **9** Topsoil: supplies plants with nutrients derived from decaying organic matter, or humus **10** Subsoil: supplies plants with minerals—some derived from rock, others washed down by rain **11** Rock: forms the basis of all soil the two is an intermediate layer known as subsoil, which is rock in the process of being broken down by natural forces. Some soils are the direct products of the rock underneath; others were brought to where they are by other forces. They may have blown, like the loess soils of North America and China; carried by glaciers, like much of the soil in North America and north of the Thames River in Britain; or washed by water, like many soils in river valleys.

Types of soil

To the practical gardener, the origin of his soil is of interest, but not vitally important. What is important to the gardener is the nature of his soil, wherever it came from: whether it is light, meaning composed of large particles like sand; heavy, meaning composed of very small particles like clay; or something in between. It is important to know: whether it is organic soil, which means it is composed of decaying vegetable matter; whether it is acidic or alkaline—sand is inclined to be acidic, clay alkaline; whether it is naturally well drained or not; what lies

underneath it—soil above limestone is very likely to be alkaline.

Fortunately, whatever your soil is like, you can improve it. There is scarcely a soil in the world that will not grow good crops of some sort if it is properly treated. Excess acidity is easily remedied by adding lime; excess alkalinity by adding compost or manure. Waterlogging can always be cured by drainage. Trace-element deficiencies can be cured simply by adding the missing trace elements.

Humus

Above all, everything about your soil can be improved by the addition of one thing: humus. Humus is vegetable or animal matter that has died and been changed by the action of soil organisms into a complex organic substance that becomes part of the soil. Any animal or vegetable material, when it dies, can become humus.

Humus has many beneficial effects on the soil. All of the following have been established experimentally by soil scientists—they are not just the optimistic conjectures of a humus enthusiast: humus protects soil from erosion by rain and allows water to percolate gently and deeply; it reduces erosion by wind; its slimes and gums stick soil particles together and thus turn a very fine soil, or clay, into a coarser one; it feeds earthworms and other useful soil organisms; it lowers soil temperature in summer and increases it in winter; it supplies nutrients to plants, because it contains all the elements that plants need and releases them slowly at a pace that the plant can cope with; it enables the soil to hold water like a sponge, and minimizes the loss of water by evaporation; it ensures that chemical changes are not too rapid when lime and inorganic fertilizers are added to the soil; it releases organic acids that help to neutralize alkaline soils, and help to release minerals from the soil, making them available to plants; it holds the ammonia and other forms of nitrogen in the soil in an exchangeable and available form—without it, nitrogen is lost quickly because of the action of denitrifying bacteria; it keeps down many fungal diseases and the notorious eelworm.

Clearly one of your main aims as a self-sufficient gardener should be to increase the humus content of your soil as much as possible. Soils ranging from the heaviest clay to the purest sand can be improved and rendered fertile by the introduction of sufficient humus. There is no soil that does not benefit from it, and there is no crop, that I know of, that is not improved by it.

Now, any organic material that you put into the soil will produce humus. Compost, green manure, farmyard manure, human excreta, peat, leafmold, seaweed, crop residues: anything that has lived before can live again. Bury it in the soil and it will rot and make humus. Leave it on top of the soil; it will rot, the worms will drag it down, and it will still make humus.

Humus is the firm basis of good gardening. It is possible to grow inferior crops on humus-deficient soil by supplying all your plants' chemical requirements, mainly in the form of nitrates, out of a fertilizer bag, but if you do this, your soil will progressively deteriorate and, ultimately, blow or wash away, as the topsoils of so much of the world's surface, abused by humankind, already have.

CHAPTER ONE

The Illustrated Index of Vegetables, Fruits, and Herbs

Containing the edible roots, stems, leaves, flowers, seeds, pods, and fruits that constitute the produce of the kitchen garden.

The Edible Parts of Plants

I was once extremely hungry in a jungle and almost overwhelmed by plant life. However, I could not find a single edible plant. I was made to realize that for human beings, few of the thousands of plants that grow on this planet are edible. Most of them are far too tough. Humans cannot digest cellulose, which is the basis of much plant tissue.

Of the comparatively few plants that humans can eat, most are only edible in part. The larger and more complex edible plants have, just like animals, evolved separate and specialized organs that are quite different from each other and serve quite different purposes. We, and other herbivorous and omnivorous animals, eat different parts of different plants, on the basis of what tastes best and does us the most good.

For the purposes of the gardener, the main parts of plants can be classified as: roots, stems, leaves, flowers, fruits, and seed. Most plants have all of these parts. There are some oddities that don't, like cacti, which don't have leaves—their stems serve instead. The tissues of which the various organs are composed are different in kind, and the botanist can tell, quite easily, whether an organ is, for example, a stem or a root. The non-botanist is in for a few surprises. This is because some plants have developed some of their organs for very particular purposes—storing nourishment through the winter, for example—and the resulting organ is often a unique and strange-seeming example of its type.

Many of the plants we eat, particularly root and stem vegetables, are naturally biennial. The plant uses a swollen root or stem to store in its first year of growth much of the energy that it will use in its second year to produce flowers and seeds. Gardeners harvest these biennial plants after their first year of growth, so as to get the full benefit of this stored-up nourishment before it is dissipated. This is why lettuces should not be allowed to "bolt," and why many vegetables should not be permitted to go to seed. If you leave a beet or carrot, for example, in the ground for more than a year, the edible roots will become tough, and shrink as the energy stored in them is used to make the plants flower.

Roots

When analyzing plants, it seems sensible to start at the bottom, with roots. Most roots have the specialized function of absorbing from the soil the nonorganic nutrients that plants need to grow and survive. These include: water, in which all the other nutrients have to be dissolved; nitrogen; potash; phosphates; and all the many trace elements that are essential to plants. Roots force themselves far down into the soil in their search for water and nutrients. Fortunately for humans, some plants also use their roots for storing food as well as gathering it. Gardeners are able to harvest this stored food to keep themselves going in the lean times of winter or drought.

ROOTS
Most edible roots are swollen taproots. Laterals, or side roots, grow from the tap root, and toward the ends of these are the microscopic root hairs that feed the plant by absorbing moisture and nutrients from the soil.

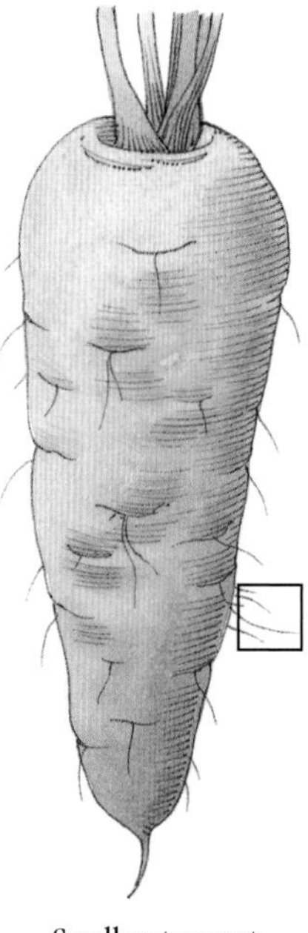

Swollen taproot

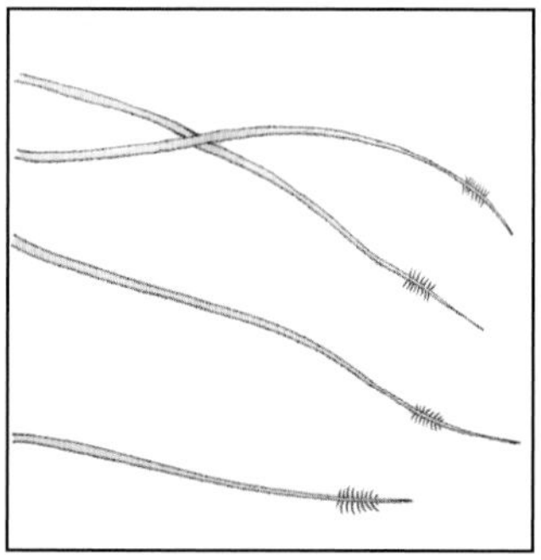

Edible roots are nearly all taproots. A taproot is the main support root of the plant, out of which grow the searching side roots and their absorbent root hairs. A number of common vegetables are swollen taproots. These include carrots, parsnips, radishes, rutabagas, turnips, and beets. Red beet is the beet that is eaten; sugar beet stores the plant's energy in the form of sugar and is grown commercially for that very reason.

Most plant energy is stored in the form of starch, but energy can only be transported around a plant in the form of sugar. This is because starch is not soluble. This fact is important to the gardener. If you want the sweetness of certain vegetables—new potatoes and sweet corn are cases in point—you must harvest the crop when the energy is still in the form of sugar and not wait until the plants grow older and the sugar has been stored away as starch. If you make wine, you will find that you need certain enzymes to turn the starch into sugar, for only sugar, and not starch, can be turned into alcohol by yeast.

Stems

There are some very unusual stems. Potatoes, for example, although they grow underground and are swollen to store food, are not roots but stems.

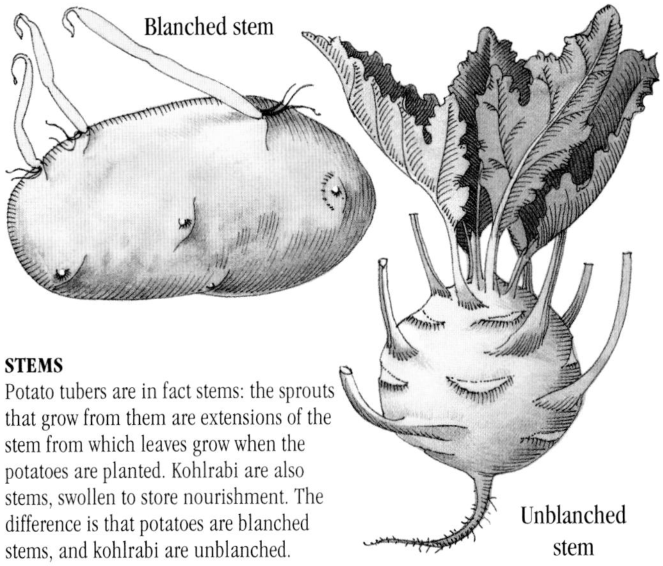

STEMS
Potato tubers are in fact stems: the sprouts that grow from them are extensions of the stem from which leaves grow when the potatoes are planted. Kohlrabi are also stems, swollen to store nourishment. The difference is that potatoes are blanched stems, and kohlrabi are unblanched.

Potatoes have all the morphological characteristics of stems. They don't put off lateral roots, and the "eyes" in them are in fact growth buds from which normal stems and leaves develop when the potatoes are planted below ground. Exposed to the light, potatoes immediately develop chlorophyll—turn green—as most stems do, so that they can engage in photosynthesis.

Photosynthesis is not only the very basis and mainspring of the gardener's art, but it is the one process that keeps every single living being, animal or vegetable, alive on this planet. It is the miraculous process—no scientist has ever been able to duplicate it—that uses the energy of the Sun to make carbohydrate or starch, which is the basis of all plant and animal energy.

Photosynthesis is carried out by molecules of chlorophyll—the green pigment of plants—and in the total absence of light, no green plant can live. Non-green plants (for example, Indian pipe) live, like fungi, as parasites or saprophytes on the living or dead tissue of other organisms. They lack chlorophyll and cannot derive energy from the Sun.

Many stems are very tough—consider the stem of an oak tree—because they have to support the upper parts of the plant in the air. Some stems are only pleasant for us to eat if they are blanched—that is, kept from the light so they do not develop chlorophyll and stay white instead of turning green. Potatoes are like this, and so are celery stems, sea kale, and chards, which are the stems of cardoons. The stem is the edible part of a rhubarb plant: the leaves are actually poisonous. Kohlrabi and celeriac are both stems, swollen so that they can store food. Notice the leaf buds and the scars left by leaves on the stems of kohlrabi.

Leaves

Leaves are often edible, and some of them have evolved to become storehouses of energy, just like the specialized roots and stems. For example, onions, leeks, garlic, and shallots are really layered clusters of leaf bases adapted to store food throughout the winter. The leaves of hearted cabbages behave in much the same way, and other plants, like lettuces, are halfway along this line of evolution toward tight clumps of leaves that store energy through the winter, so that, if you allow it to, the plant gets away the second year to produce a flowering head and early seed.

"MONOCOTS" AND "DICOTS" All plants are either monocotyledons (monocots) or dicotyledons (dicots). The difference lies in the seed, the leaves, and the mode of growth. The seed of a dicot—a bean, for example—is made up of two halves. You can see this for yourself: a fava bean will readily split in two if you just press into it with your fingernail. A monocot seed, on the other hand, is self-contained and cannot be split.

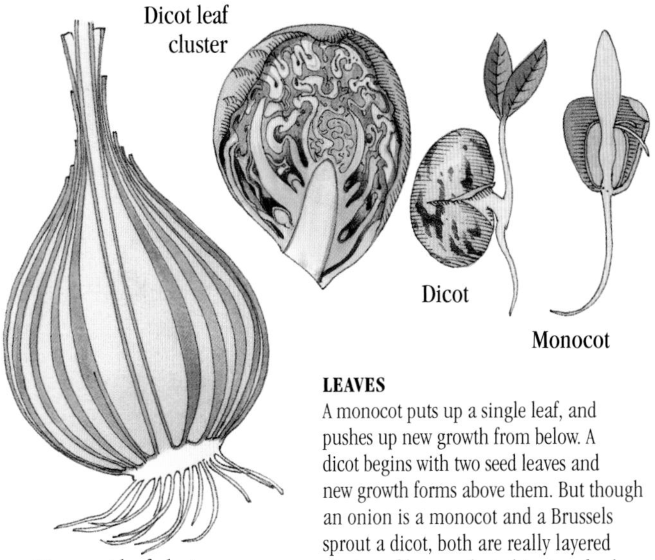

LEAVES
A monocot puts up a single leaf, and pushes up new growth from below. A dicot begins with two seed leaves and new growth forms above them. But though an onion is a monocot and a Brussels sprout a dicot, both are really layered clusters of leaves adapted to store food.

Dicots produce two seed leaves. Thereafter the stem grows upward and new leaves issue from a growing point at its tip, or from growing points at the tips of branches. Monocots grow in a completely different way. They begin by pushing up a single leaf—a blade of grass or an onion is a good example. Thereafter they continue to grow by pushing from the seed upward. The first leaf is pushed up and to one side and new ones push their way up beside it. Dicots add new growth above existing growth, while monocots push existing growth upward, adding new growth beneath it. Most vegetables are dicots, but some, notably onions, leeks, asparagus, and sweet corn, are monocots. An easy way to distinguish between a monocot and a dicot is to look at their leaves. If they have almost parallel veins, the chances are they are monocots, but if the veins of the leaves splay out away from each other, they are almost certainly dicots.

Flowers

Flowers are the next step upward in a plant, and they are not as a rule an important food for humans. But they are important to the plant, of course, for they ensure its posterity.

Sex reared its beautiful head in a new and elaborate form millions of years ago in the Jurassic Period, when insects and plants developed their amazing symbiosis. The plants provide nectar and other delights to attract the insects, and the insects unwittingly cross-fertilize the plants by going from one to the other, carrying pollen from the male organs of one flower to the female organs of another. The extraordinary elaboration of devices to attract the right insects, and ensure that they collect pollen on their bodies, and do not self-fertilize the flower they are visiting, but instead fertilize the next flower, has been the life study of many botanists, including, notably, Charles Darwin. It is a very good thing to keep bees, particularly if you have fruit trees, for not only do you get honey from them but they pollinate your trees and flowering plants.

Some flowers are pollinated by the wind. Corn is one of them, and because of this, you should plant your sweet corn in a wide block and not in one long row. You must try to make sure that when the wind blows from any direction, it blows pollen from one plant to another.

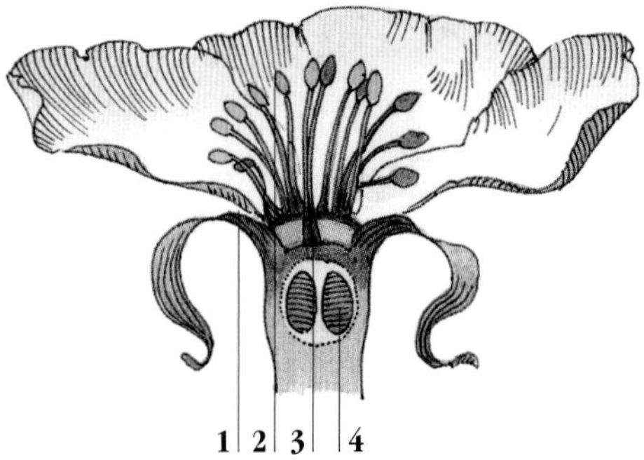

FLOWERS
Flowers are essential to the reproduction of plants. Their nectar **1** attracts insects that pick up pollen from the male part **2** of one plant and carry it to the female part **3** of the next. The ovary **4** of a flower after it has been fertilized is the fruit.

There are not many flowers that are edible. The main ones are cauliflower and broccoli, which are immature flowers. If you leave these to continue growing instead of eating them, they will produce inedible mature flowers, just as an abandoned cabbage will. Globe artichokes are flowers, although only a small part is actually edible. Nasturtium seeds make good substitutes for capers, and some herb flowers are good for flavoring and coloring. But, if you had to live exclusively on flowers, you would get very thin indeed. Seed and fruit, which come from the reproductive part of flowers, are more important to the self-sufficient gardener.

When the female element of a flower has been pollinated, the flower forms a fruit. The fruit grows and produces seed within it. The seed is spread far and wide by a number of amazingly ingenious methods that plants have evolved so as to propagate their species. A gardener who wishes to enjoy his craft will study all these things, for a knowledge of them will increase enormously the pleasure he gets from his labors. The more you learn about plants, the more you will wonder at the extraordinary cunning and elaboration that selection has evolved for their survival and perpetuation.

Fruits

To a botanist, fruit is the ovary of a flower after it has been fertilized. Fertilization causes the ovules, which are inside the ovary, to turn into seeds and the ovary itself to turn into a fruit. Some fruits don't bear much resemblance to what most people would call fruit. A whiff of dandelion fluff is a fruit, and so are all nuts. Tomatoes, eggplants, peppers, beans, and pods full of peas are all fruits; an individual pea, and a bean threshed from its pod for drying, are both seeds.

To a cook and a gastronome, and to most people, in fact, fruit means the sort of sweet succulent fruit that is eaten as a dessert. I have used this common classification in ordering this book. The only exception I have made is rhubarb, which though eaten as a dessert is not naturally sweet and is grown like a vegetable, which in fact it is.

Botanically, a blackberry or raspberry is not one fruit but a cluster of fruits. Each tiny globe that goes to make up such a "berry" is a complete fruit. The word "berry" means something different to a botanist. A tomato is a true berry, because its seeds are embedded in soft pulp. Grapes, gooseberries, and oranges are also berries. The fruits that contain single pits or stones—plums, cherries, and peaches—are called drupes. Fruits like apples and pears are called pomes. Only the core of a pome is a cluster of true fruits; the edible part surrounding the core is just a layer of stored food. Each tiny seed in a strawberry is a single fruit. They are all held together by a succulent mass. If you take the trouble to cut open pomes, berries, and drupes at various stages of their growth, you will see how all of this works.

Every part of the fruit is in embryo in the flower, and you can follow the development of each part of the flower as it becomes a fruit.

Generally speaking, fruits contain little nourishment, and most of the little they have is in the form of sugar. All the stored energy of the plant is going to go into the seed, not the fruit. If you tried to live on fruits alone, you would soon become seriously undernourished. Fruits tend to be rich in certain vitamins, however, particularly vitamin C, and this makes them valuable to humans. Many years ago a learned doctor said: "Apples have no nutritive value whatever and it does not matter whether you eat them or not." He was completely mistaken, but then he did not know about vitamins. Cobbett made the same mistake when advising his cottagers not to grow fruit trees. He thought they were good for nothing but giving children bellyaches.

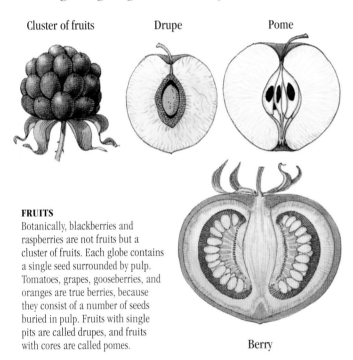

Cluster of fruits

Drupe

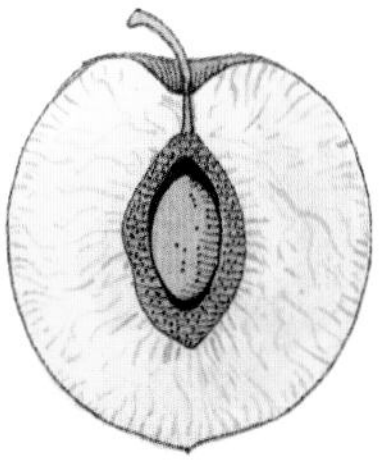

Pome

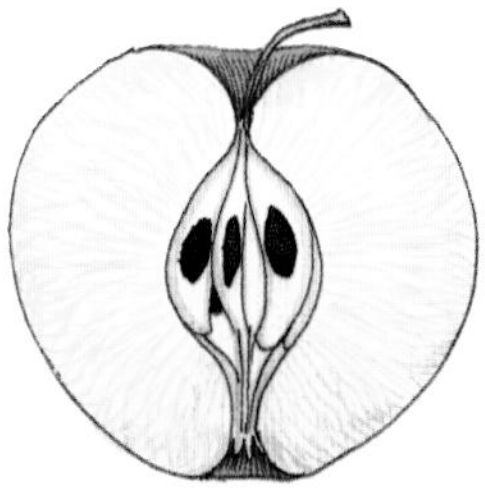

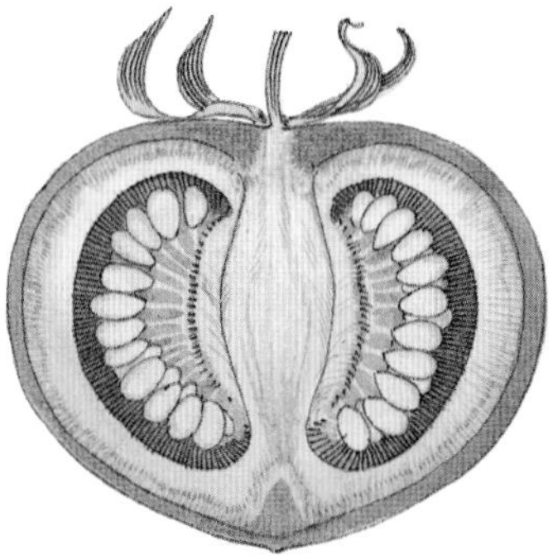

Berry

FRUITS
Botanically, blackberries and raspberries are not fruits but a cluster of fruits. Each globe contains a single seed surrounded by pulp. Tomatoes, grapes, gooseberries, and oranges are true berries, because they consist of a number of seeds buried in pulp. Fruits with single pits are called drupes, and fruits with cores are called pomes.

All the members of the squash tribe are fruits. Some, like melons, seem to have developed their large, water-filled fruit so that they can store water. They come from parts of the world with short rainy seasons. The stored water gives the seeds a head start—one good watering—to enable them to stay alive until the next rain. Other fruits have undoubtedly developed to attract animals. Apples, plums, peaches, and cherries have all evolved in this way. Man has taken many of these slightly succulent and somewhat sour wild fruits and, by artificial selection, evolved improved varieties, containing a lot of sugar, extra flavor, plenty of succulent flesh, and not too much acid. If you compare the wild crabapple with a Cox's Orange Pippin apple, you will understand what has happened. Most cultivated fruits have been improved beyond recognition from their wild form.

Seeds

If it were not for edible seeds, humans could scarcely survive. Some seeds have to pass through the guts of an animal in order to germinate. So it is important for plants that their fruit should be eaten. Many seeds are not strictly edible; humans eat them but they pass through them intact without doing them any good. It is the fruit that they benefit from.

Other seeds, especially the cereals, such as wheat, rice, and corn, are eaten for direct nutrition, and it is these seeds that keep humanity alive in many parts of the globe. Before an annual plant dies, it puts all the nourishment it has into its seeds, which are to carry its life on into future generations. Thus seeds are generally far more nutritious than any other parts of plants and, if you grind them or cook them and make them palatable, they will give you the energy to keep alive. Without seeds, a vegetarian would live on a very sparse diet indeed: it would be scarcely adequate to sustain life.

Seeds of certain plants, notably leguminous ones, are quite rich in protein. Pea and bean seeds of all kinds are very important. A vegan can enjoy a near-perfect diet by eating plenty of soybeans, some other vegetables, and a little comfrey, which is almost the only edible plant that contains vitamin B12. The great thing about seeds is that they are easy to dry and store.

Herbs

Many herb seeds are used to flavor food, for the essential oils and other virtues of herbs are often concentrated in their seeds. Many of the more aromatic herbs come from dry, warm regions. Their aromatic oils have evolved largely to protect them from dessication in dry, hot weather. The small leaves of many of them—often mere needles—are that size and shape to prevent moisture from being transpired by the plant too fast.

Some of the delicious flavors and aromas of culinary herbs are there for reasons that are not yet fully understood: attracting and repelling various insects may well be one of them. But plants are endlessly fascinating and mysterious, and happily, it is improbable that we will ever know all there is to know about them. It should be enough for us that thyme and rosemary smell and taste as they do, without worrying why.

Roots

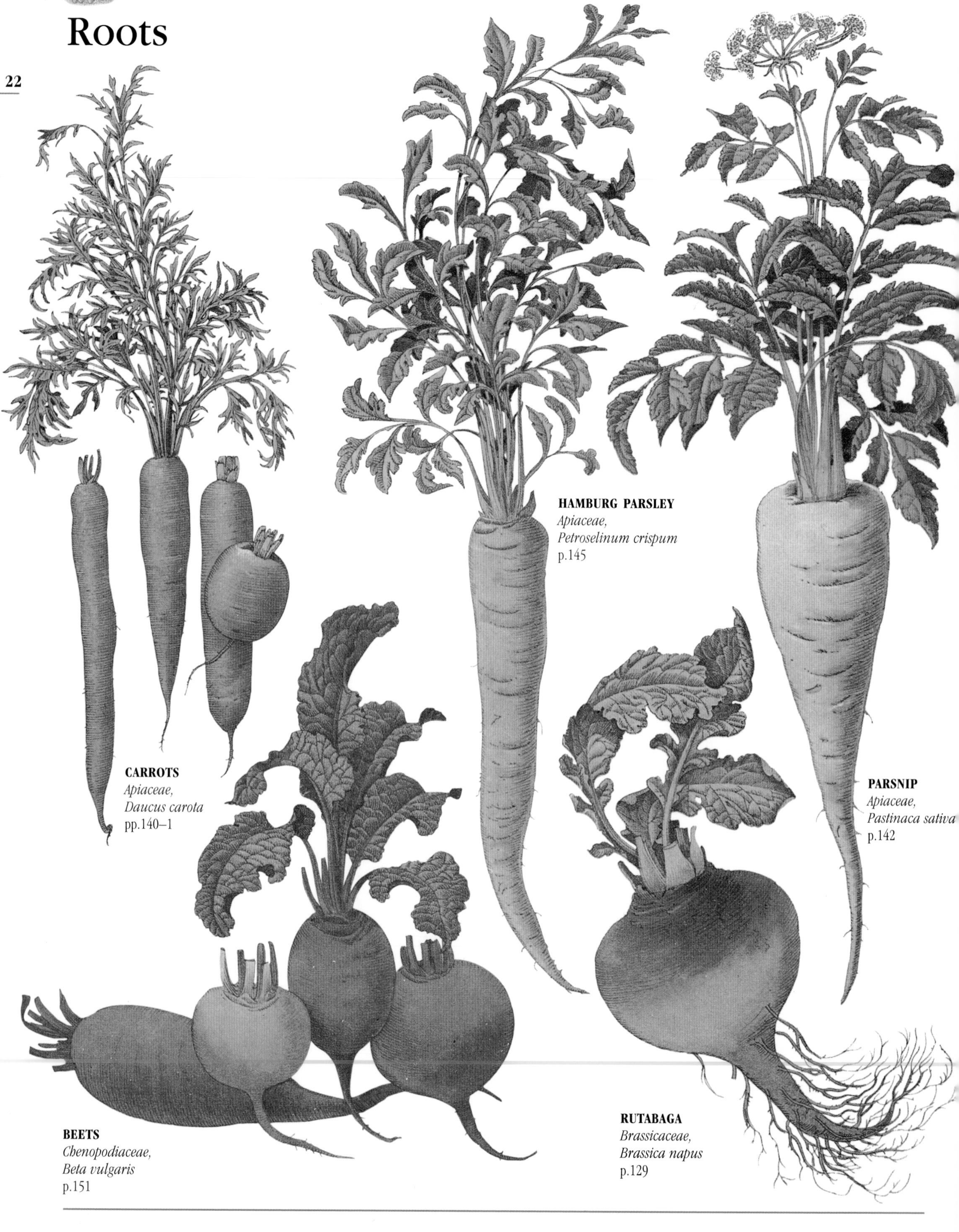

CARROTS
Apiaceae,
Daucus carota
pp.140–1

HAMBURG PARSLEY
Apiaceae,
Petroselinum crispum
p.145

PARSNIP
Apiaceae,
Pastinaca sativa
p.142

BEETS
Chenopodiaceae,
Beta vulgaris
p.151

RUTABAGA
Brassicaceae,
Brassica napus
p.129

TURNIPS
Brassicaceae,
Brassica rapa
p.129

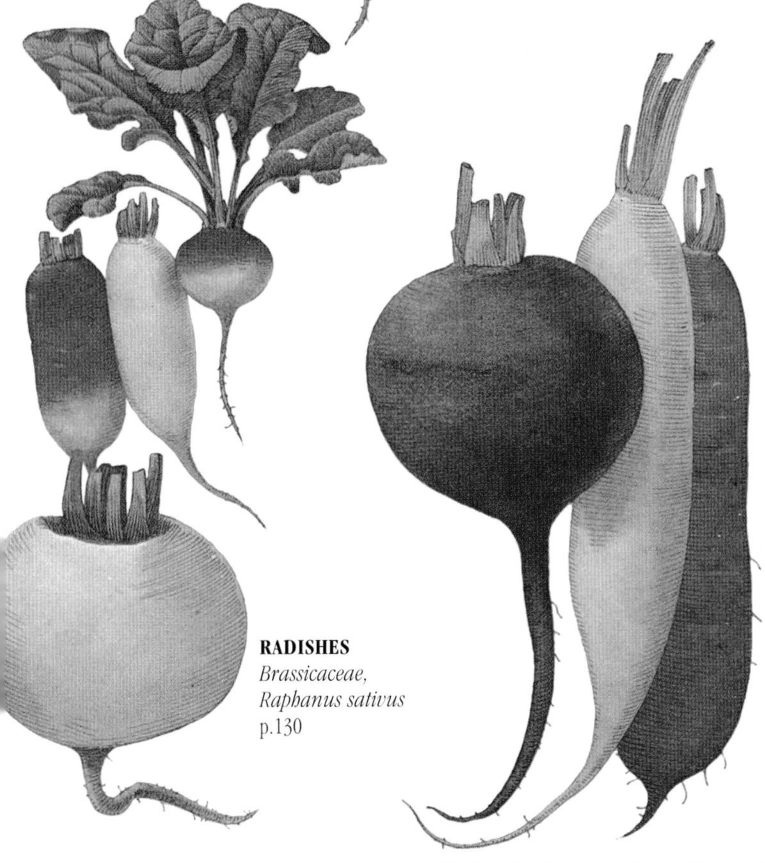

RADISHES
Brassicaceae,
Raphanus sativus
p.130

HORSERADISH
Brassicaceae,
Armoracia rusticana
p.197

SCORZONERA (BLACK SALSIFY)
Asteraceae,
Scorzonera hispanica
p.160

SALSIFY
Asteraceae,
Tragopogon porrifolius
p.160

Stems

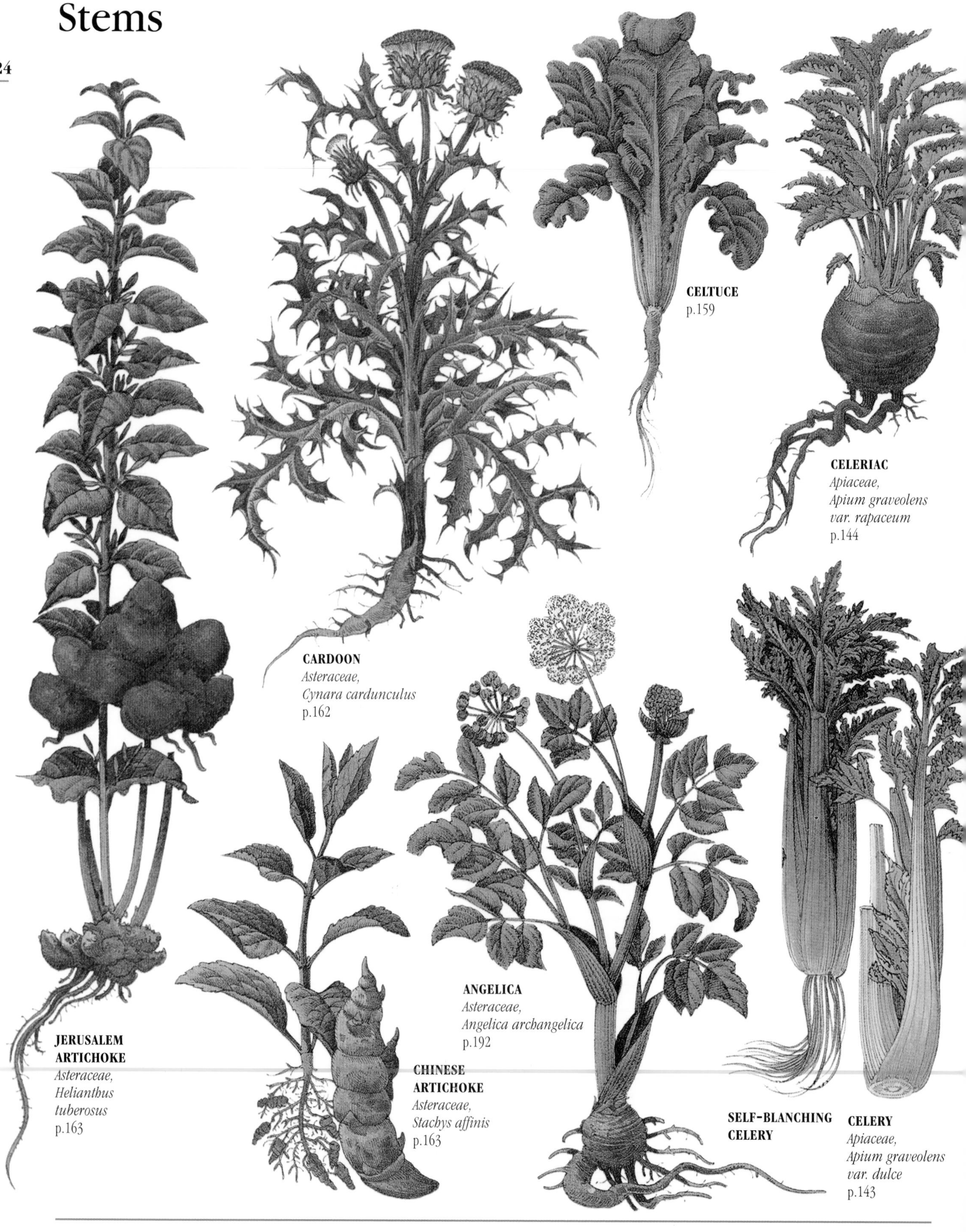

CELTUCE
p.159

CELERIAC
Apiaceae,
Apium graveolens
var. rapaceum
p.144

CARDOON
Asteraceae,
Cynara cardunculus
p.162

ANGELICA
Asteraceae,
Angelica archangelica
p.192

JERUSALEM ARTICHOKE
Asteraceae,
Helianthus
tuberosus
p.163

CHINESE ARTICHOKE
Asteraceae,
Stachys affinis
p.163

SELF-BLANCHING CELERY

CELERY
Apiaceae,
Apium graveolens
var. dulce
p.143

POTATOES
Solanaceae,
Solanum tuberosum, p.132

CHINESE CABBAGE
Brassicaceae, Brassica rapa
p.125

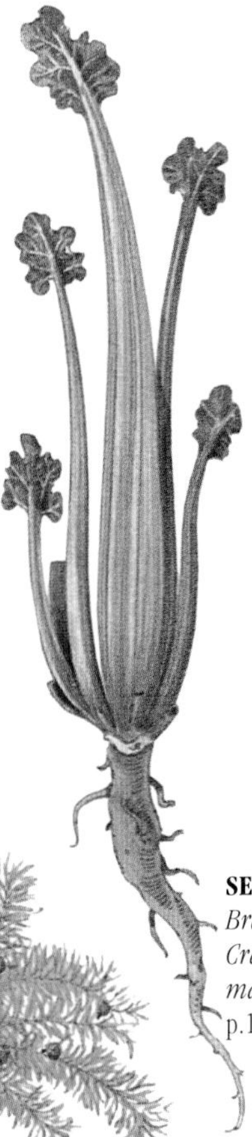

SEAKALE
Brassicaceae,
Crambe
maritima
p.131

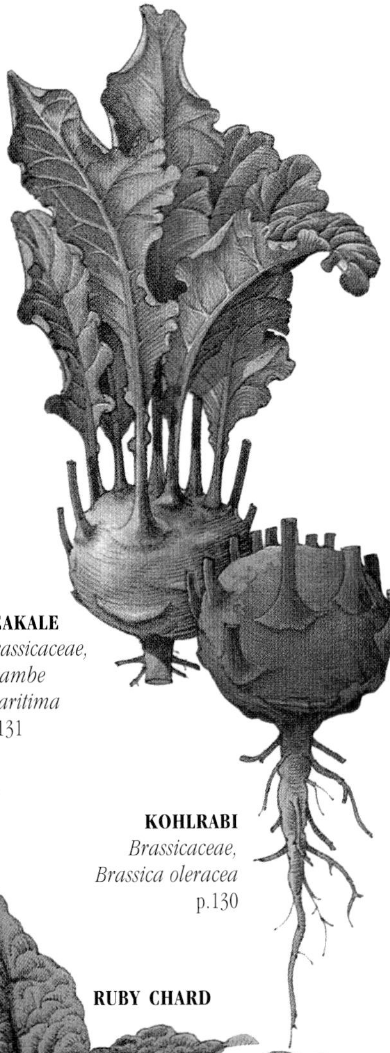

KOHLRABI
Brassicaceae,
Brassica oleracea
p.130

RHUBARB
Polygonaceae,
Rheum
rhabarbarum
p.165

ASPARAGUS
Liliaceae,
Asparagus officinalis
p.149

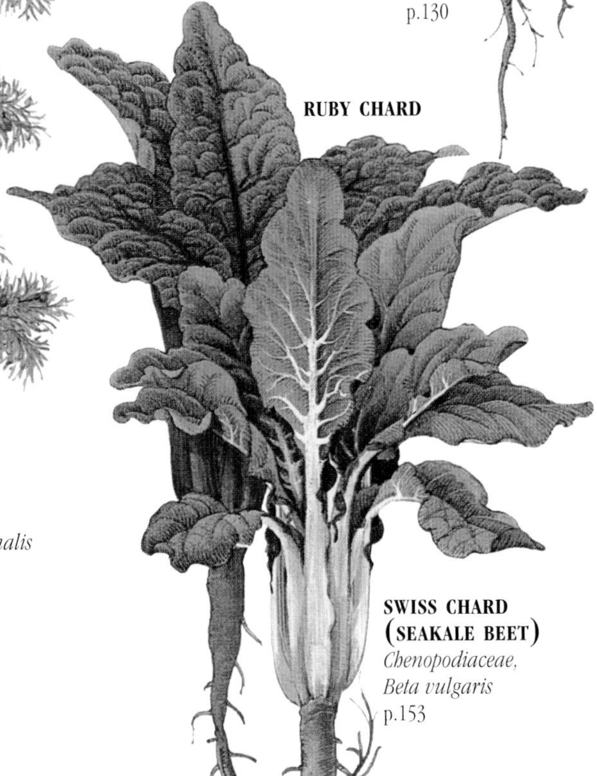

RUBY CHARD

SWISS CHARD (SEAKALE BEET)
Chenopodiaceae,
Beta vulgaris
p.153

Leaves

SPINACH BEET
Chenopodiaceae,
Beta vulgaris
p.153

FLORENCE FENNEL
Apiaceae,
Foeniculum vulgare
p.146

NEW ZEALAND SPINACH
Tetragoniaceae,
Tetragonia tetragonioides
p.152

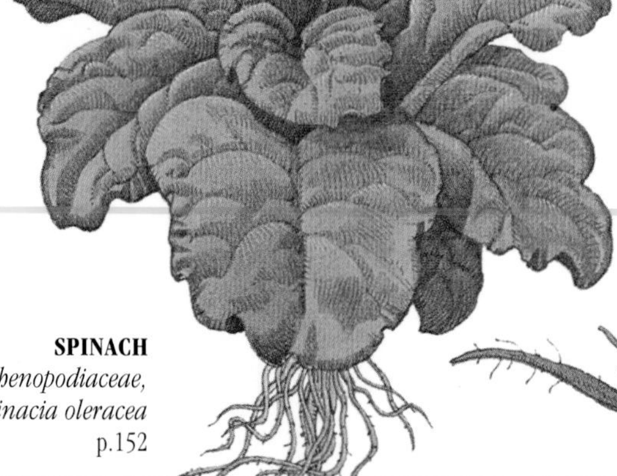

SPINACH
Chenopodiaceae,
Spinacia oleracea
p.152

CORN SALAD (LAMB'S LETTUCE)
Valerianaceae,
Valerianella locusta, p.159

CABBAGES
Brassicaceae,
Brassica oleracea, pp.122–8
RED CABBAGE
WHITE CABBAGE
SAVOY CABBAGE
CONICAL-HEARTED CABBAGE
ROUND-HEARTED CABBAGE
BRUSSELS SPROUTS
p.126
KALE
p.128
COLLARDS
p.128
CRESSES
Brassicaceae
p.131
WATERCRESS
Nasturtium officinale
CRESS
Lepidium sativum
LAND CRESS
Barbarea verna

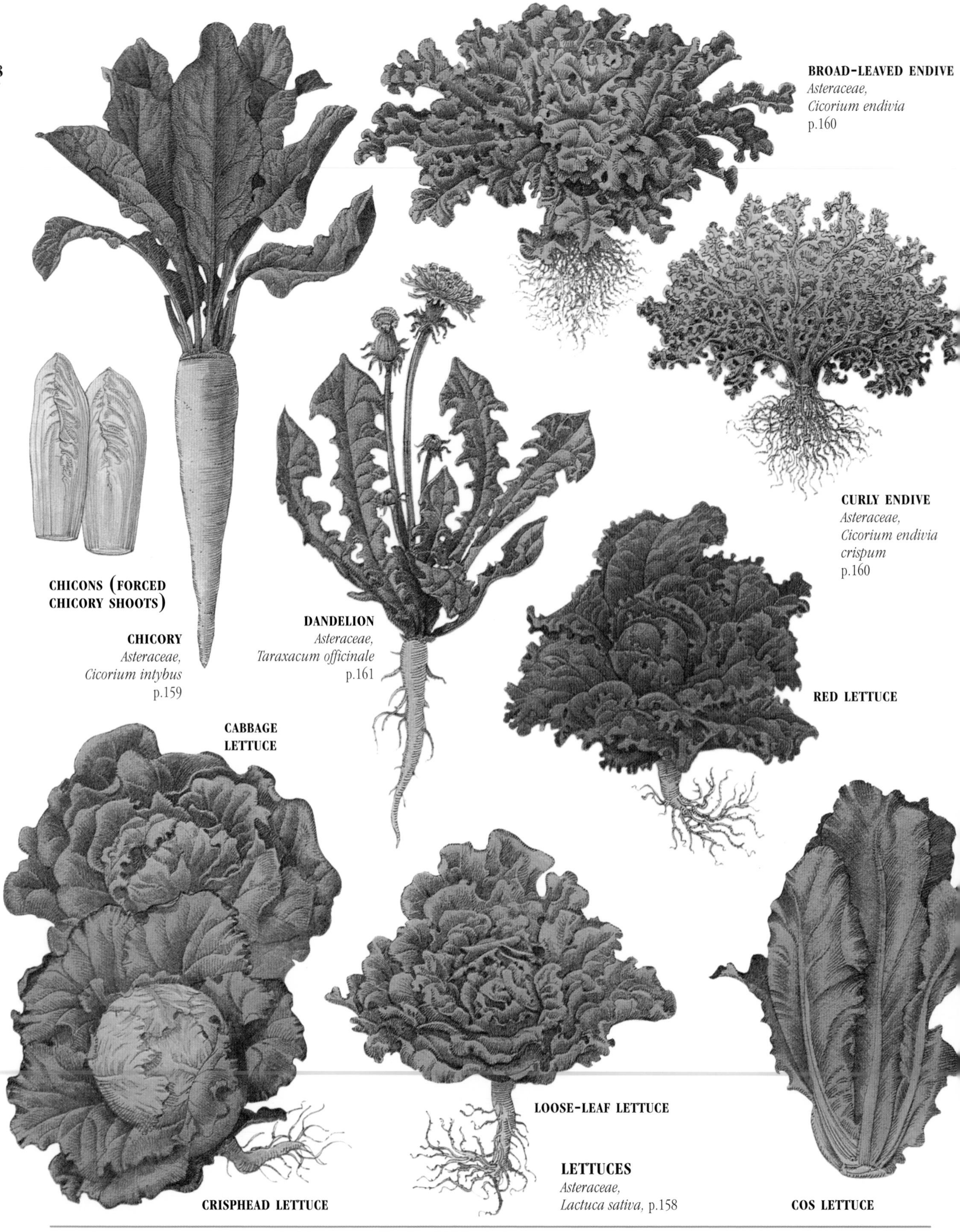
BROAD-LEAVED ENDIVE
Asteraceae,
Cicorium endivia
p.160
CURLY ENDIVE
Asteraceae,
Cicorium endivia crispum
p.160
CHICONS (FORCED CHICORY SHOOTS)
CHICORY
Asteraceae,
Cicorium intybus
p.159
DANDELION
Asteraceae,
Taraxacum officinale
p.161
RED LETTUCE
CABBAGE LETTUCE
LOOSE-LEAF LETTUCE
LETTUCES
Asteraceae,
Lactuca sativa, p.158
CRISPHEAD LETTUCE
COS LETTUCE

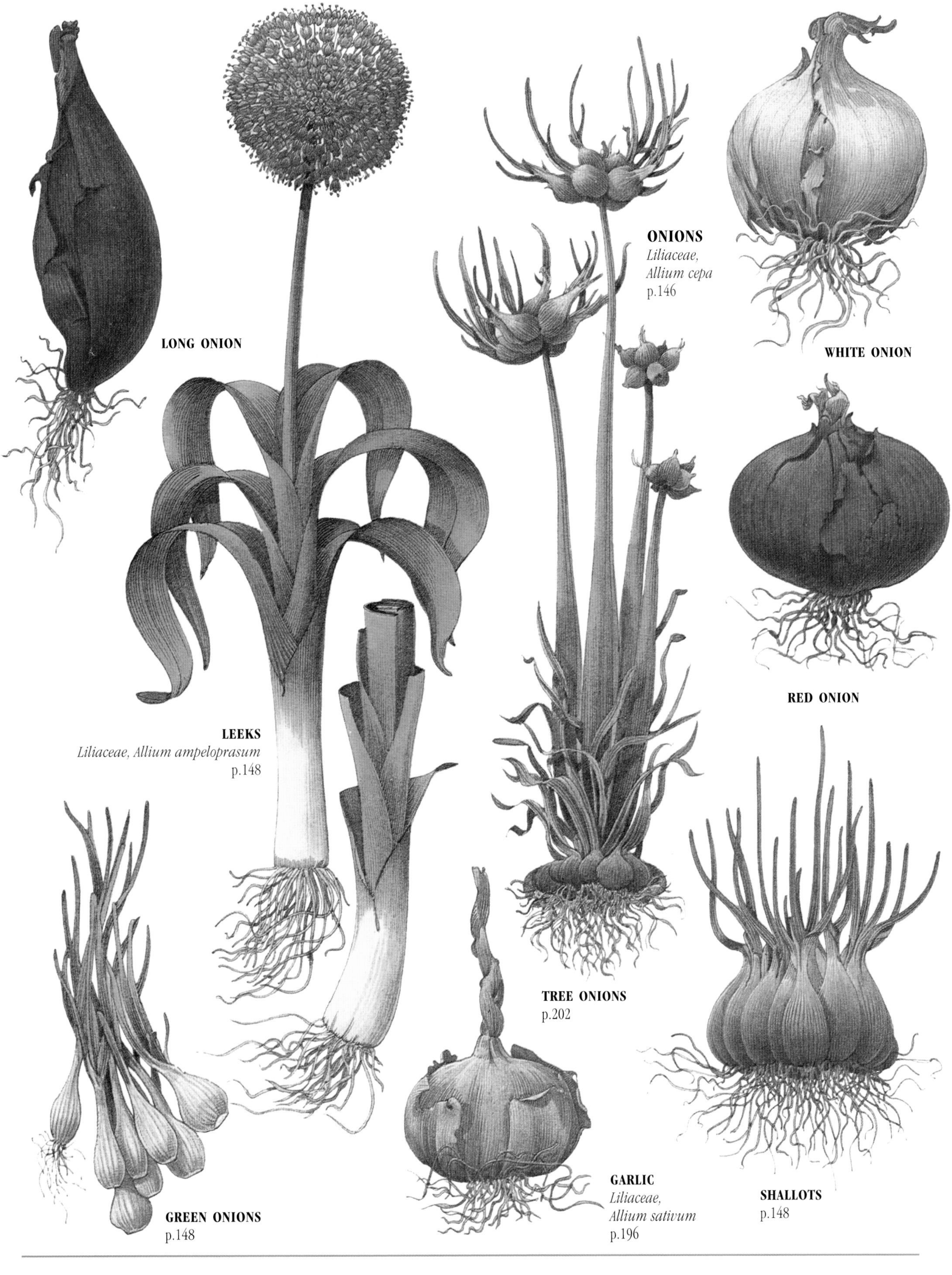
LONG ONION
ONIONS
Liliaceae,
Allium cepa
p.146
WHITE ONION
RED ONION
LEEKS
Liliaceae, Allium ampeloprasum
p.148
TREE ONIONS
p.202
GREEN ONIONS
p.148
GARLIC
Liliaceae,
Allium sativum
p.196
SHALLOTS
p.148

LOVAGE
Apiaceae,
Levisticum officinale
p.197

PARSLEY
Apiaceae,
Petroselinum crispum
p.200

COMMON PARSLEY

BROAD-LEAVED PARSLEY

CHERVIL
Apiaceae,
Anthriscus cerefolium
p.195

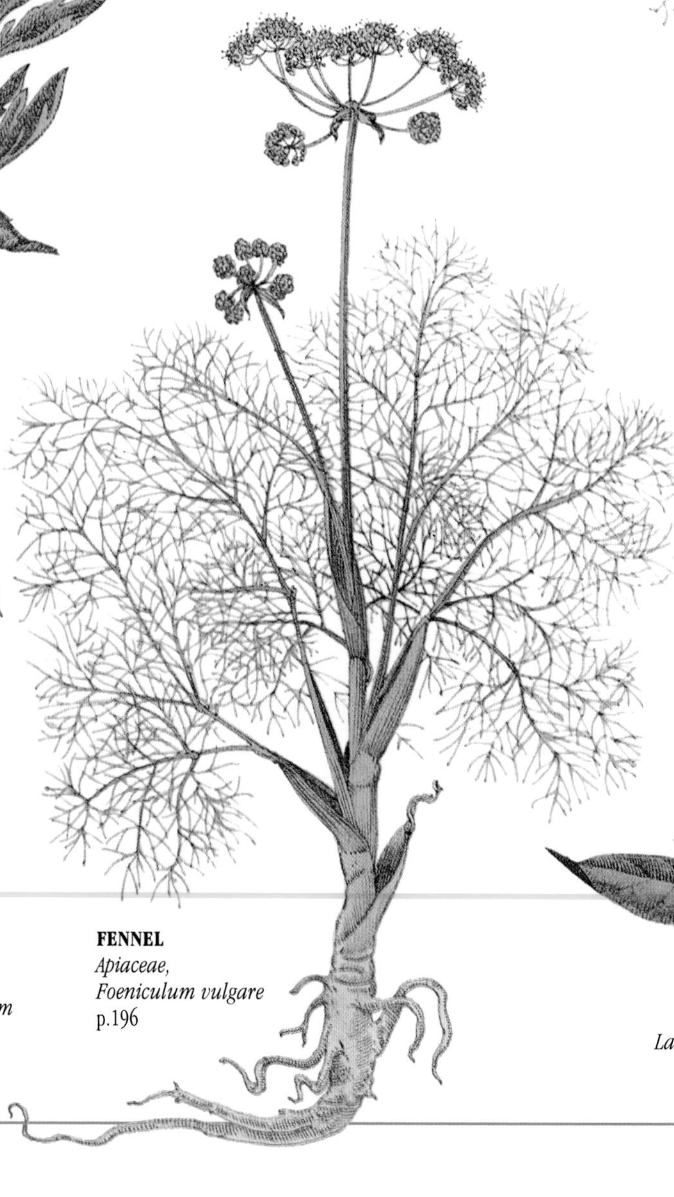

FENNEL
Apiaceae,
Foeniculum vulgare
p.196

BAY
Lauraceae,
Laurus nobilis
p.193

BURNET
Rosaceae,
Poterium sanguisorba
p.194

NASTURTIUM
Tropaeolaceae,
Tropaeolum majus
p.199

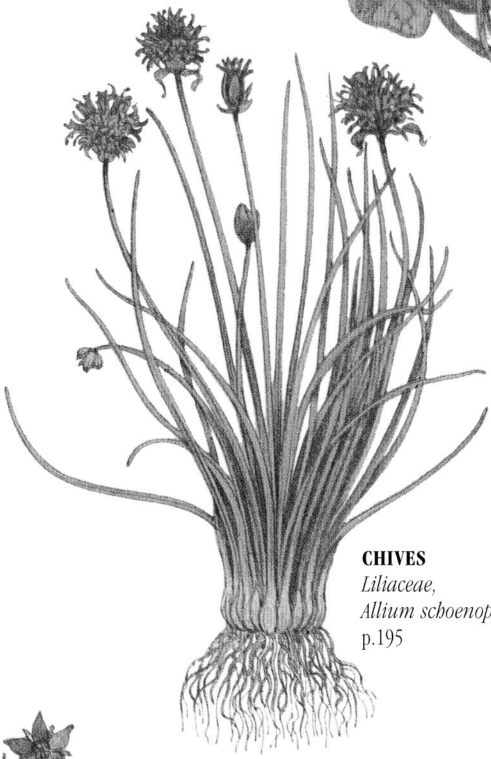

CHIVES
Liliaceae,
Allium schoenoprasum
p.195

SORREL
Polygonaceae,
Rumex acetosa
p.201

TARRAGON
Asteraceae,
Artemisia dracunculus
p.202

BORAGE
Boraginaceae,
Borago officinalis
p.193

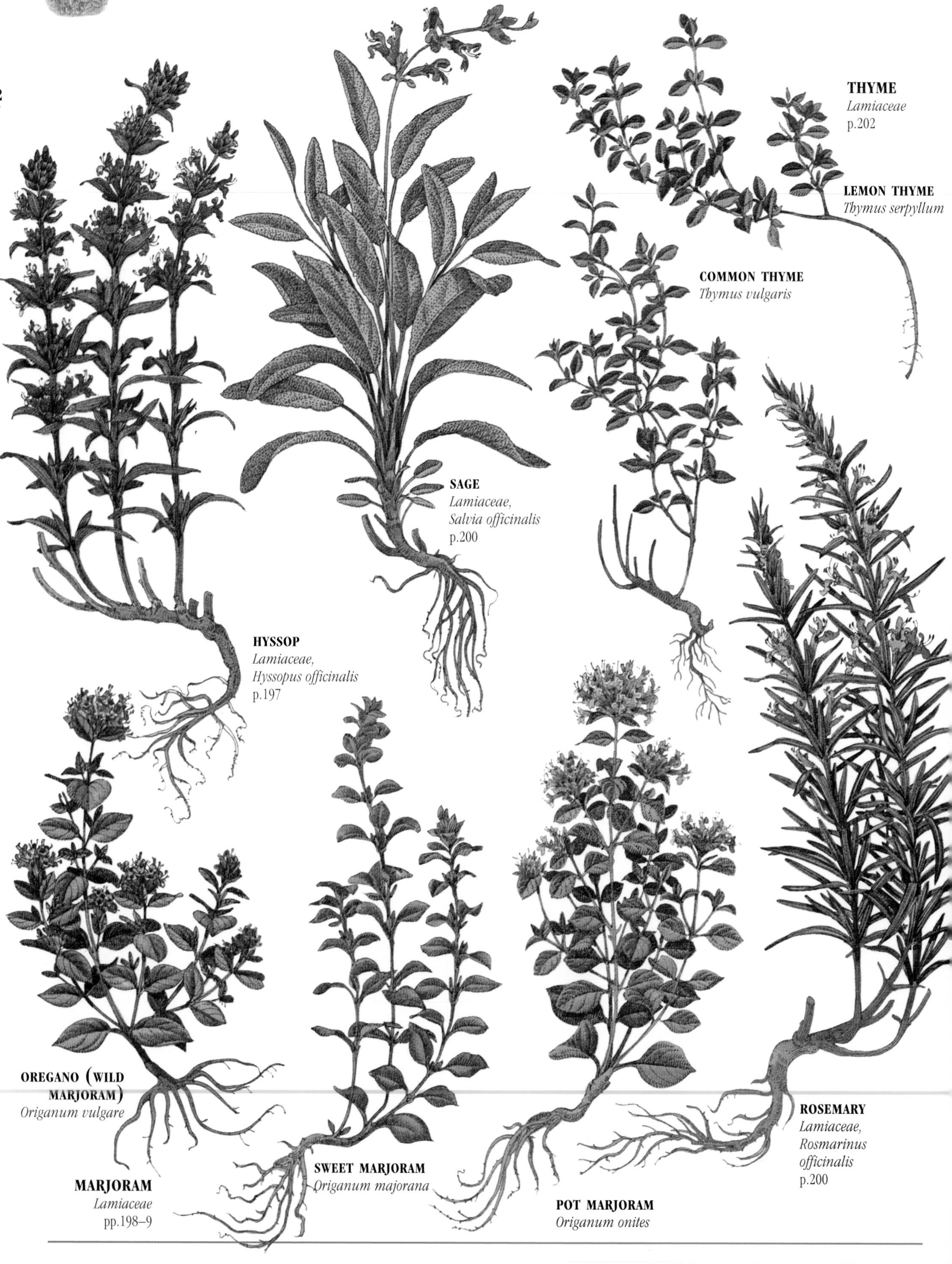
THYME
Lamiaceae
p.202
LEMON THYME
Thymus serpyllum
COMMON THYME
Thymus vulgaris
SAGE
Lamiaceae,
Salvia officinalis
p.200
HYSSOP
Lamiaceae,
Hyssopus officinalis
p.197
OREGANO (WILD MARJORAM)
Origanum vulgare
MARJORAM
Lamiaceae
pp.198–9
SWEET MARJORAM
Origanum majorana
POT MARJORAM
Origanum onites
ROSEMARY
Lamiaceae,
Rosmarinus
officinalis
p.200

MINT
Lamiaceae
p.198

COMMON MINT (SPEARMINT)
Mentha spicata

PEPPERMINT
Mentha piperita

APPLE MINT
Mentha rotundifolia

CATNIP
Nepeta cataria

BASIL
Lamiaceae,
Ocimum basilicum
p.193

LEMON BALM
Lamiaceae,
Melissa officinalis
p.192

SAVORY
Lamiaceae
p.201

SUMMER SAVORY
Satureja hortensis

WINTER SAVORY
Satureja montana

Flowers and Vegetable Fruits

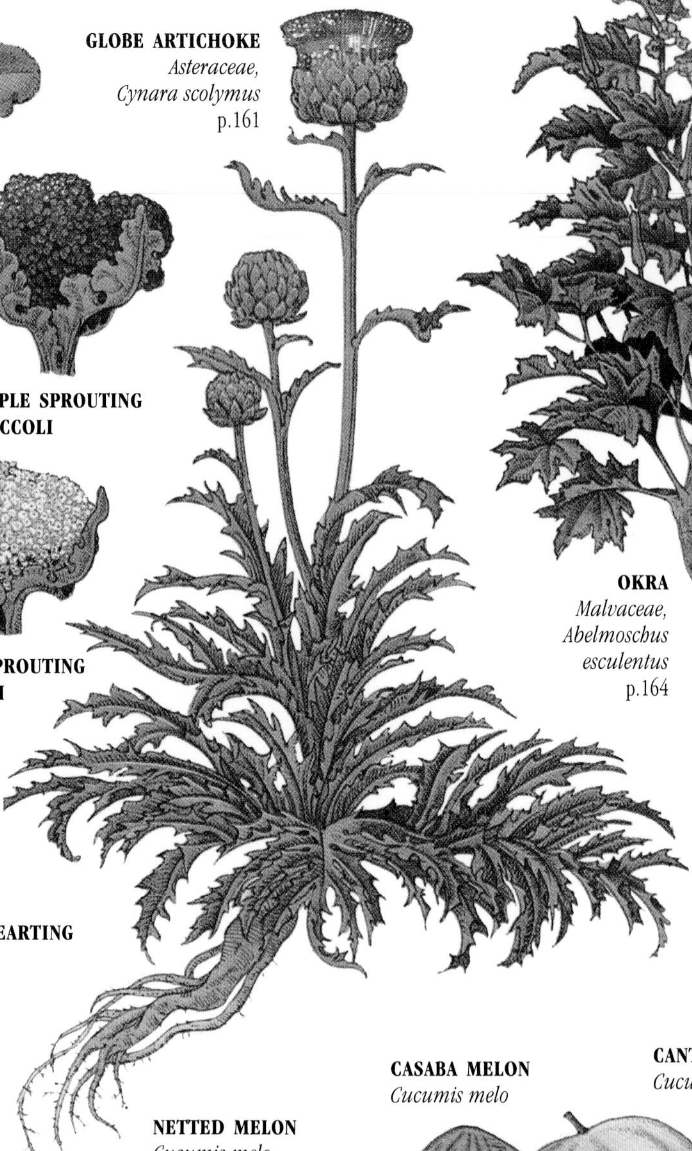

GLOBE ARTICHOKE
Asteraceae,
Cynara scolymus
p.161

OKRA
Malvaceae,
Abelmoschus esculentus
p.164

PURPLE SPROUTING BROCCOLI

WHITE SPROUTING BROCCOLI

CALABRESE

PURPLE HEARTING BROCCOLI

CAULIFLOWER

BROCCOLI & CAULIFLOWER
Brassicaceae,
Brassica oleracea, pp.127–8

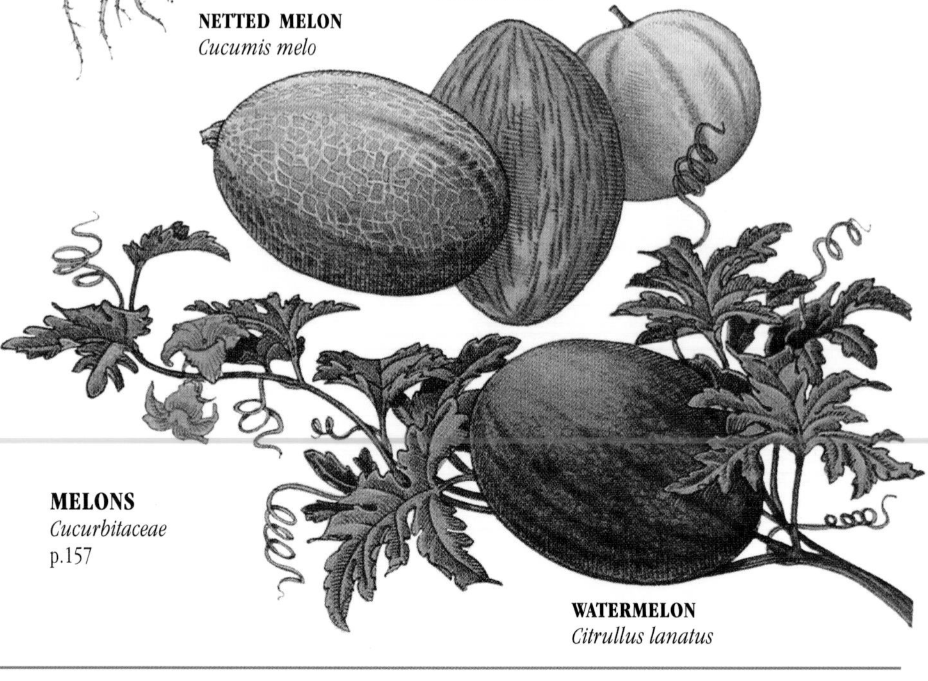

CASABA MELON
Cucumis melo

CANTALOUPE MELON
Cucumis melo

NETTED MELON
Cucumis melo

MELONS
Cucurbitaceae
p.157

WATERMELON
Citrullus lanatus

WHITE EGGPLANT

EGGPLANT
Solanaceae, Solanum melongena var. esculentum
p.139

PEPPER
Solanaceae, Capsicum annuum
p.139

CHILI PEPPER
Capsicum annuum

PEAR TOMATO & YELLOW TOMATO

TOMATOES
Solanaceae, Lycopersicon lycopersicum
p.137

CONTINENTAL TOMATO

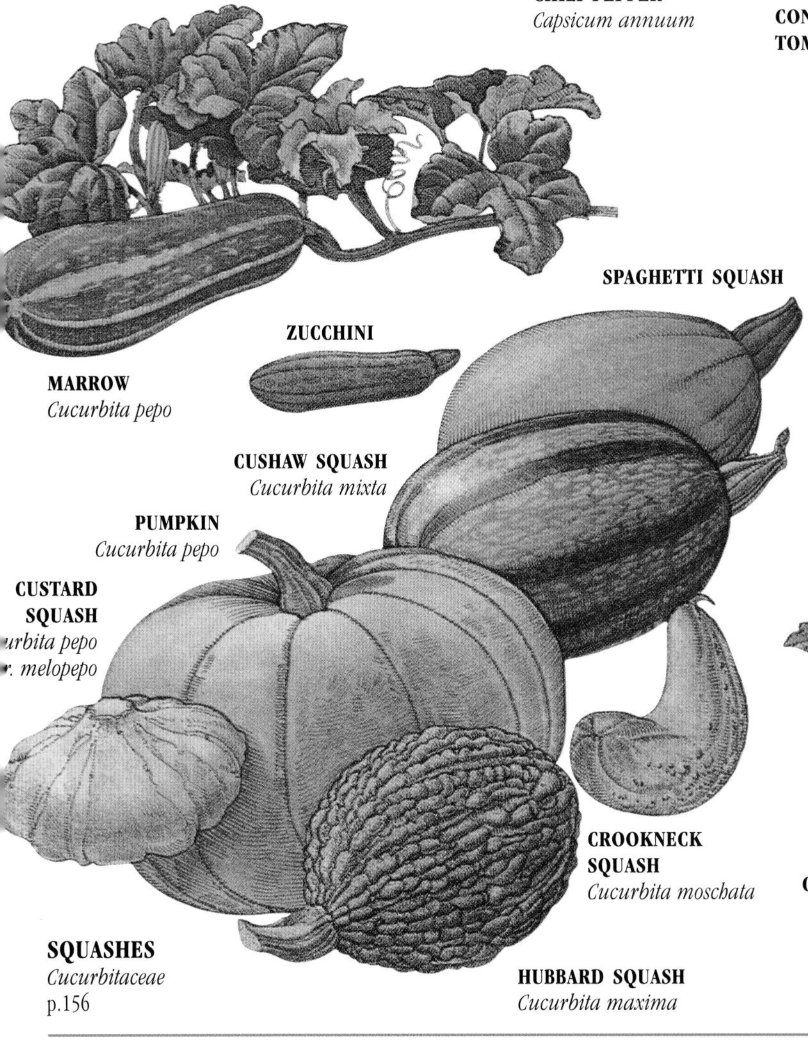

MARROW
Cucurbita pepo

ZUCCHINI

SPAGHETTI SQUASH

CUSHAW SQUASH
Cucurbita mixta

PUMPKIN
Cucurbita pepo

CUSTARD SQUASH
urbita pepo
r. melopepo

CROOKNECK SQUASH
Cucurbita moschata

HUBBARD SQUASH
Cucurbita maxima

SQUASHES
Cucurbitaceae
p.156

GHERKIN & RIDGE CUCUMBER

CUCUMBER

CUCUMBERS
Cucurbitaceae, Cucumis sativus
p.154

Seeds and Pods

BEANS
Fabaceae
pp.117–21

SOYBEANS
Glycine max
p.121

SWEET CORN
Poaceae,
Zea mays
p.163

LIMA BEANS
Phaseolus limensis
p.120

PEANUTS
Fabaceae,
Arachis hypogaea
p.121

MUSHROOMS
Agaricus campestris
p.166

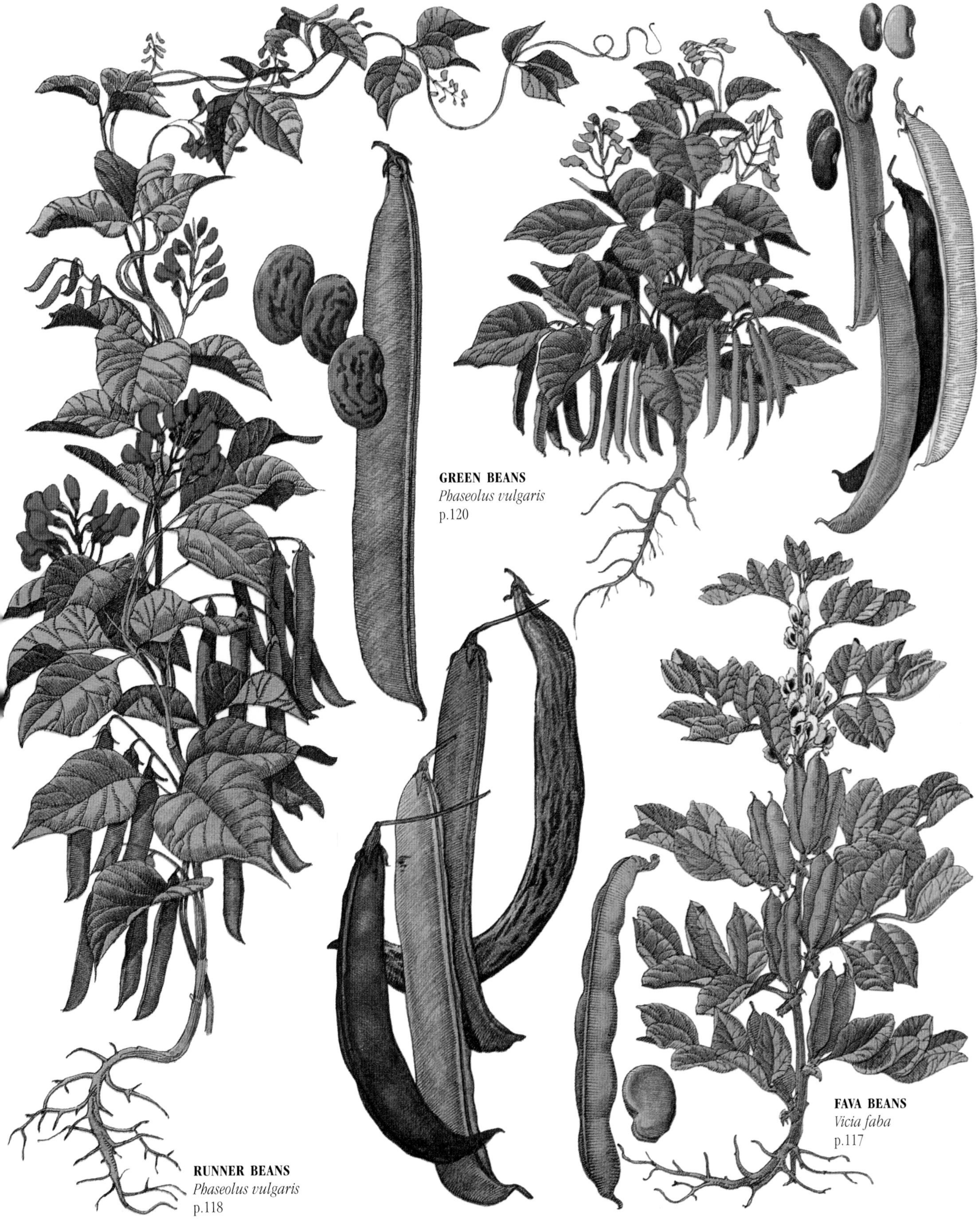
GREEN BEANS
Phaseolus vulgaris
p.120
RUNNER BEANS
Phaseolus vulgaris
p.118
FAVA BEANS
Vicia faba
p.117

DILL
Apiaceae,
Anethum graveolens
p.196

ANISE
Apiaceae,
Pimpinella anisum
p.192

CARAWAY
Apiaceae,
Carum carvi
p.194

MUSTARD
Brassicaceae,
Brassica nigra, p.199

CILANTRO
Apiaceae,
Coriandrum sativum
p.195

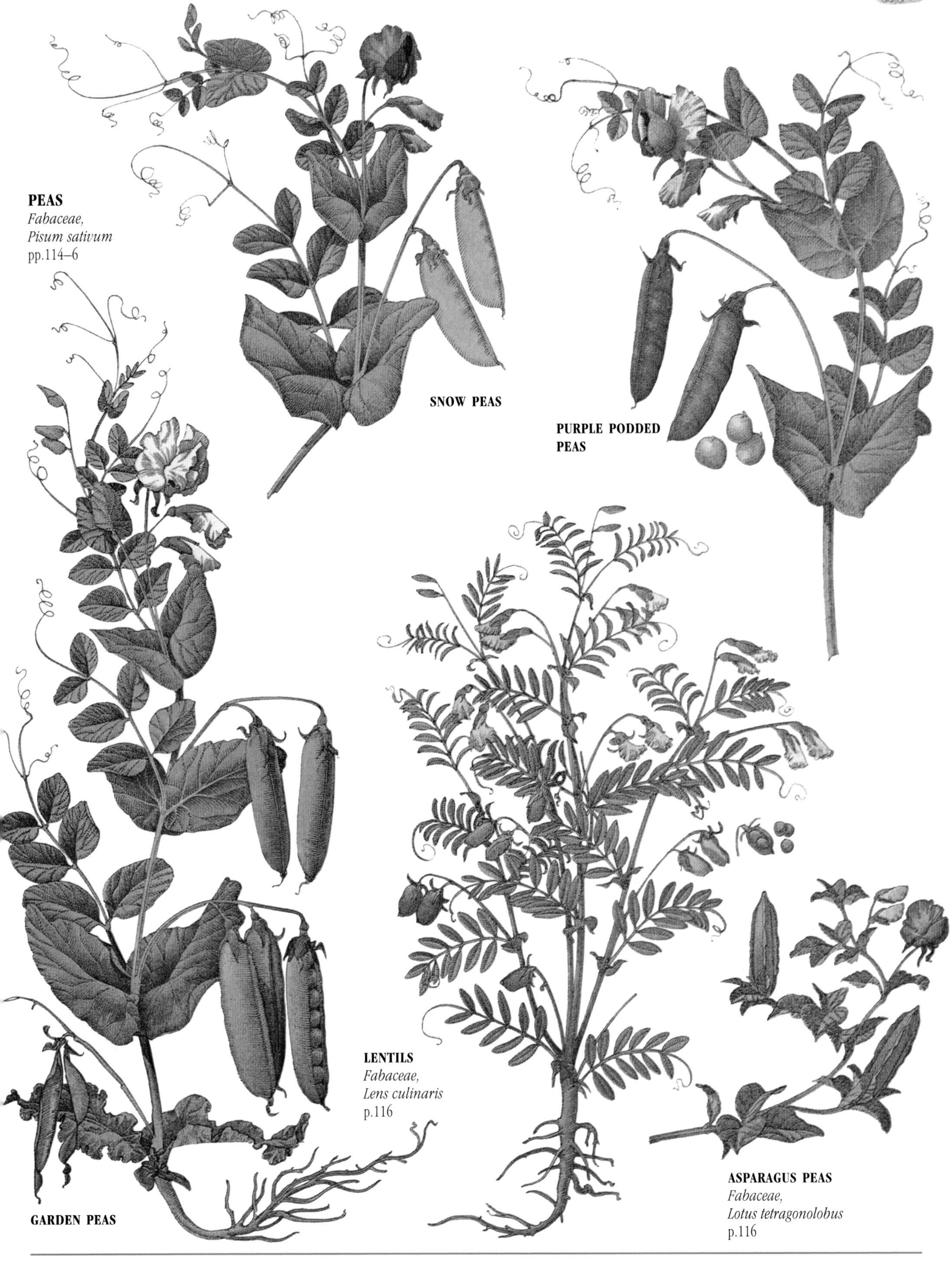

PEAS
Fabaceae,
Pisum sativum
pp.114–6

SNOW PEAS

PURPLE PODDED PEAS

GARDEN PEAS

LENTILS
Fabaceae,
Lens culinaris
p.116

ASPARAGUS PEAS
Fabaceae,
Lotus tetragonolobus
p.116

Fruits

PEARS
Rosaceae,
Pyrus communis, p.171

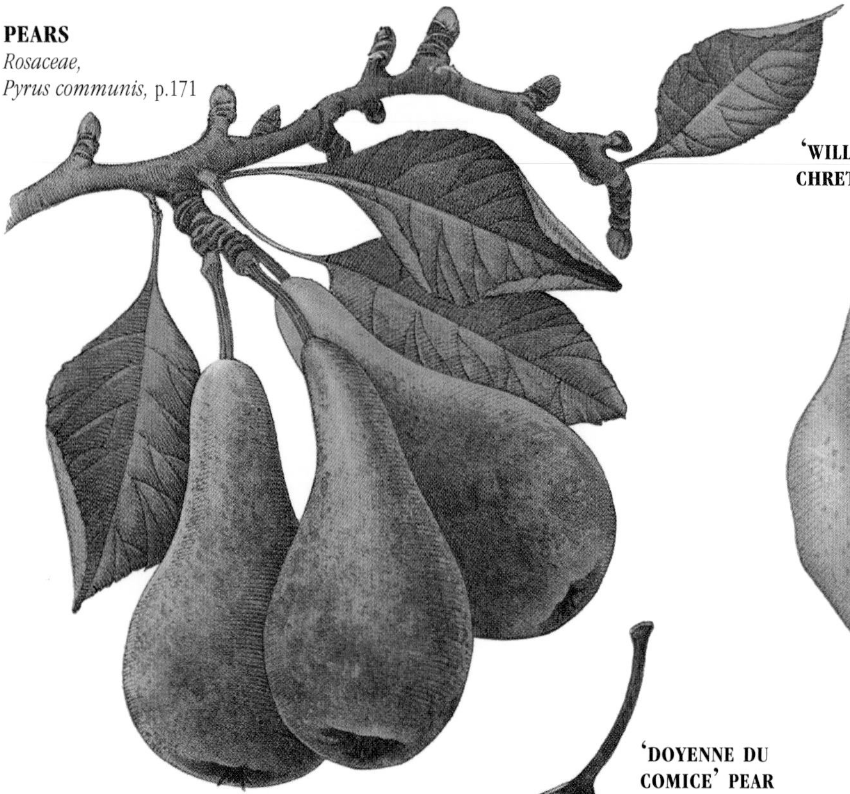

'CONFERENCE' PEAR

'RED BARTLETT' PEAR

'WILLIAM'S BON CHRETIEN' PEAR

'DOYENNE DU COMICE' PEAR

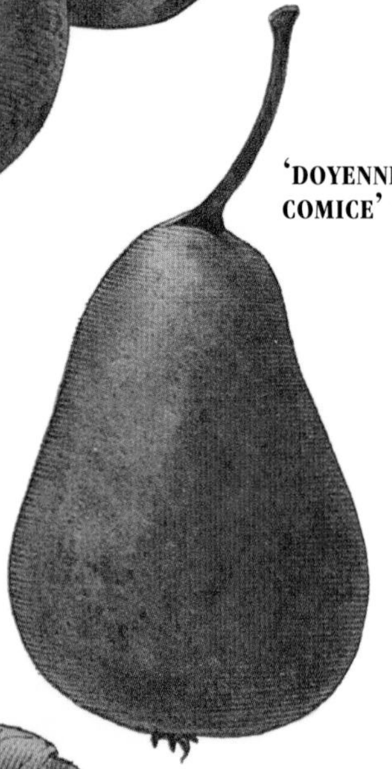

'ABATE' PEAR

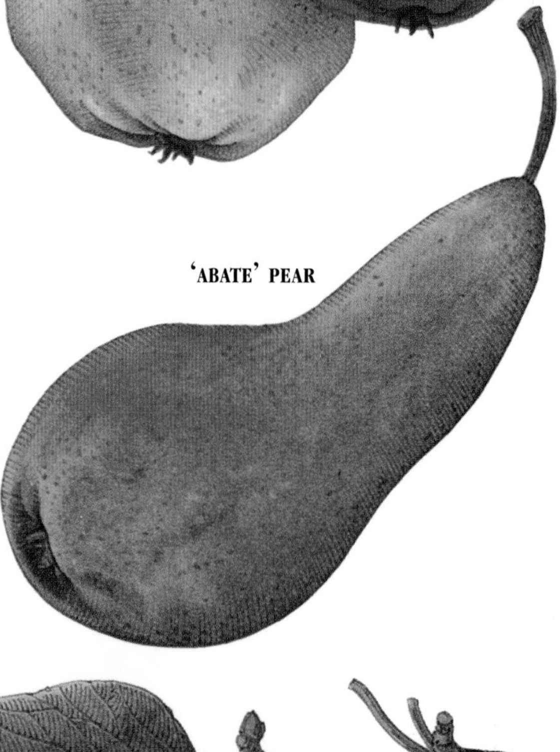

MEDLARS
Rosaceae,
Mespilus germanica
p.175

QUINCES
Rosaceae,
Cydonia oblonga
p.175

‘GRANNY SMITH’ APPLE

‘GOLDEN DELICIOUS’ APPLE

‘RED DELICIOUS’ APPLE

APPLES
Rosaceae,
Malus pumila, p.168

‘BRAMLEY’ APPLE

‘COX’S ORANGE PIPPIN’ APPLE

‘ALLINGTON PIPPIN’ APPLE

‘RUSSET’ APPLE

‘JONATHAN’ APPLE

GREENGAGES
Prunus insititia

PLUMS
Rosaceae,
Prunus domestica
p.174

APRICOTS
Rosaceae,
Prunus armeniaca
p.173

DAMSONS
Prunus insititia
p.174

PEACHES
Rosaceae,
Prunus persica, p.173
NECTARINES
MORELLO CHERRIES
Prunus cerasus
CHERRIES
Rosaceae
p.172
SWEET CHERRIES
Prunus avium
OLIVES
Oleaceae,
Olea europaea, p.187
AMERICAN RED GRAPES
GRAPES
Vitaceae,
Vitis vinifera
p.188
WHITE SEEDLESS GRAPES
EUROPEAN
RED GRAPES
BLACK GRAPES

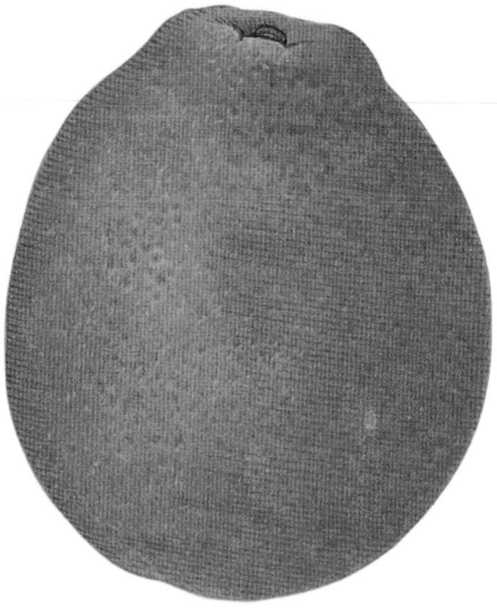

'JAFFA' ORANGE

NAVEL ORANGE

BLOOD ORANGE

SWEET ORANGES
Rutaceae,
Citrus sinensis, p.179

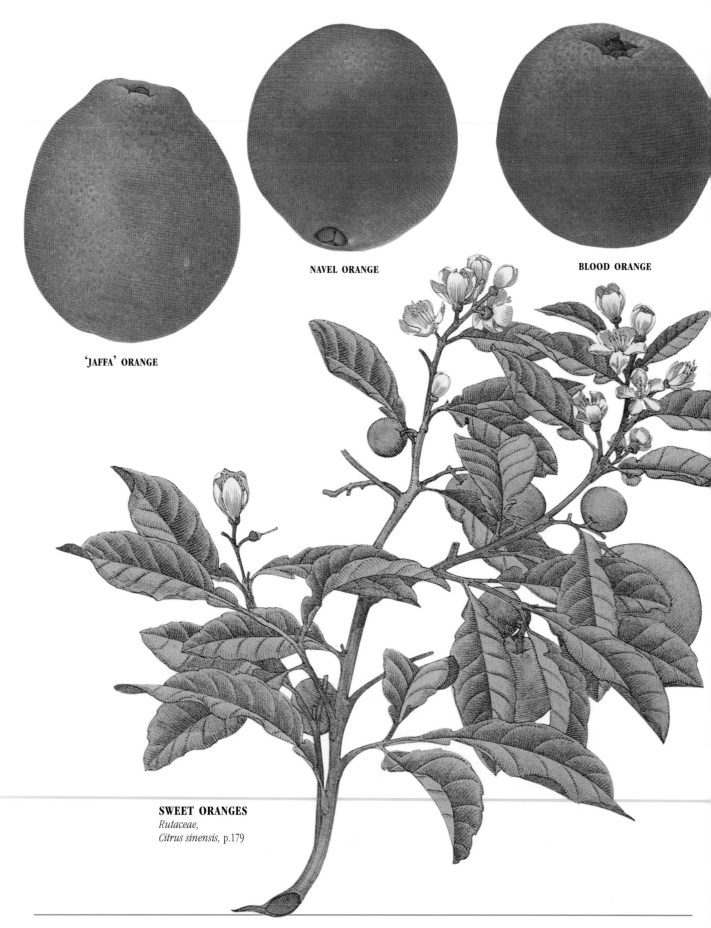

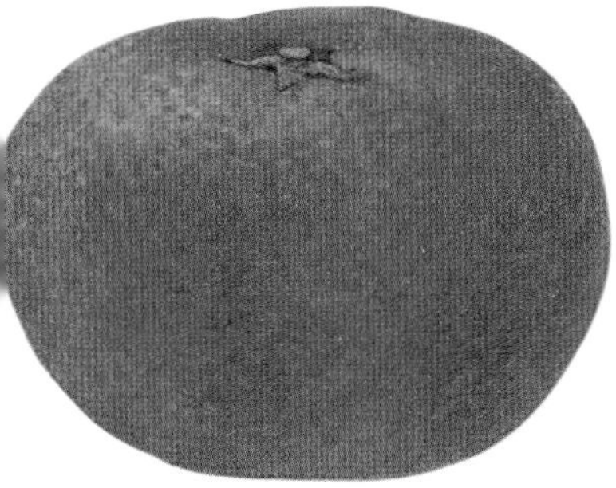

TANGERINE

MANDARIN ORANGES
Rutaceae,
Citrus reticulata
p.179

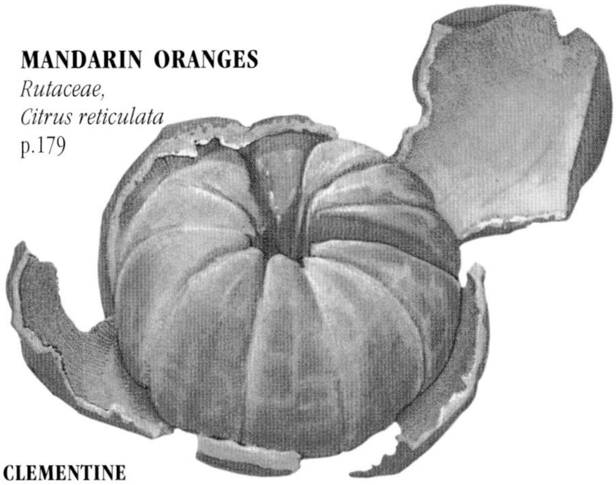

CLEMENTINE

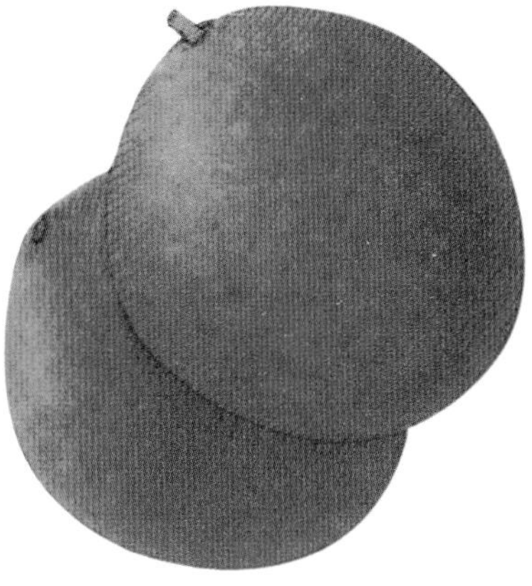

KUMQUAT
Rutaceae,
Fortunella margarita, p.180

GRAPEFRUIT
Rutaceae,
Citrus × paradisi, p.181

SEVILLE ORANGE
Rutaceae,
Citrus aurantium, p.179

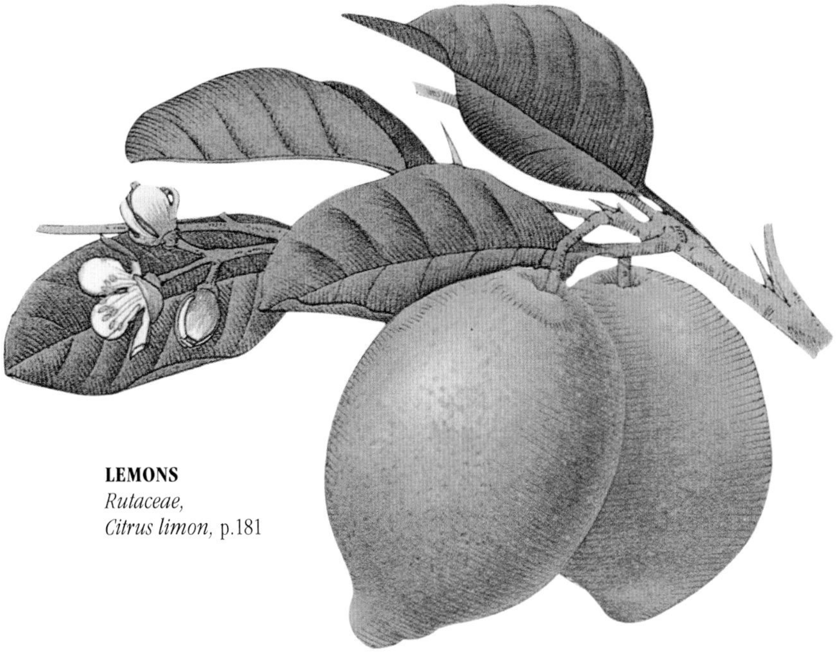

LEMONS
Rutaceae,
Citrus limon, p.181

LIME
Rutaceae,
Citrus aurantiifolia, p.181

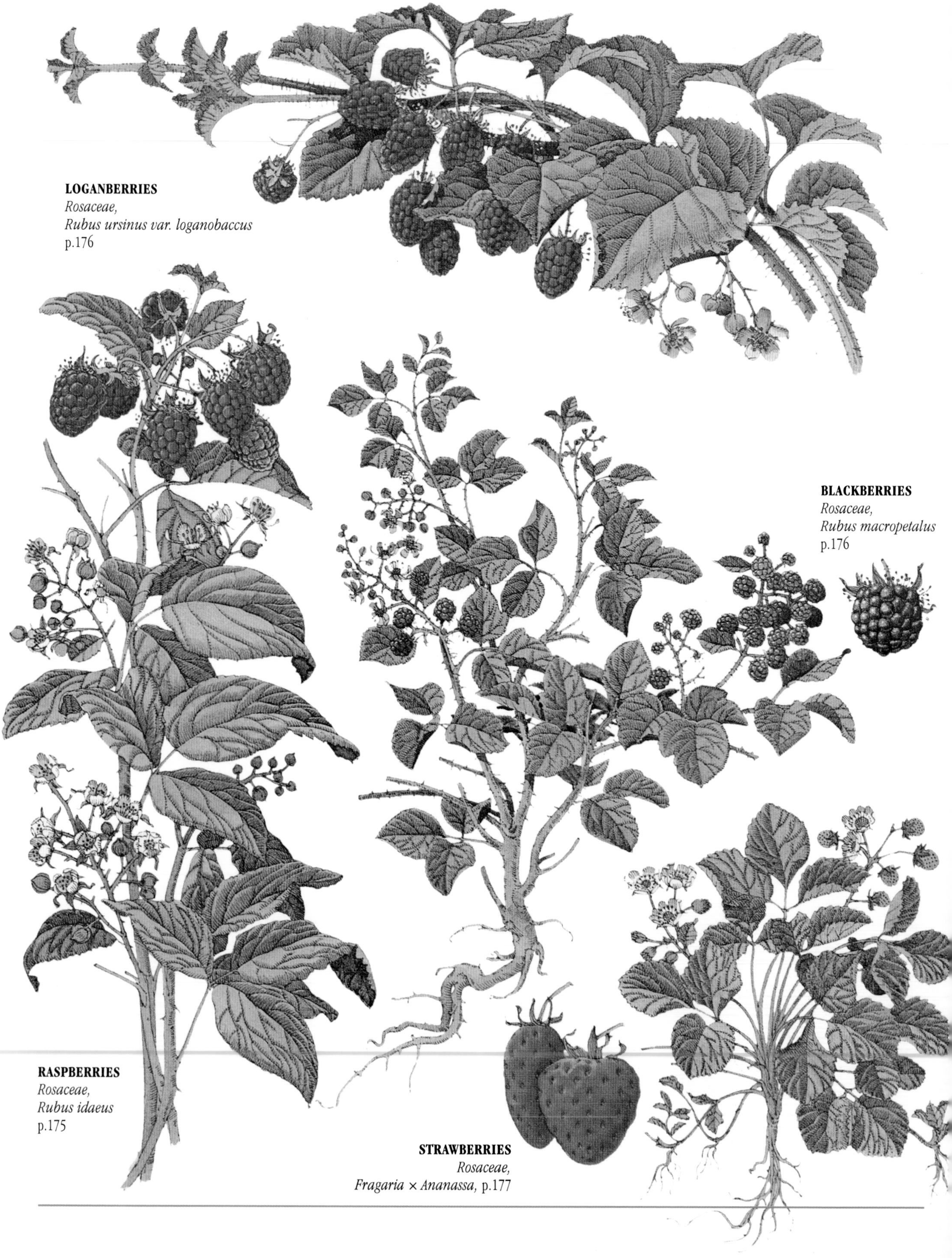

LOGANBERRIES
Rosaceae,
Rubus ursinus var. loganobaccus
p.176

BLACKBERRIES
Rosaceae,
Rubus macropetalus
p.176

RASPBERRIES
Rosaceae,
Rubus idaeus
p.175

STRAWBERRIES
Rosaceae,
Fragaria × Ananassa, p.177

FIGS
Moraceae,
Ficus carica, p.185

MULBERRIES
Moraceae,
Morus alba, p.186

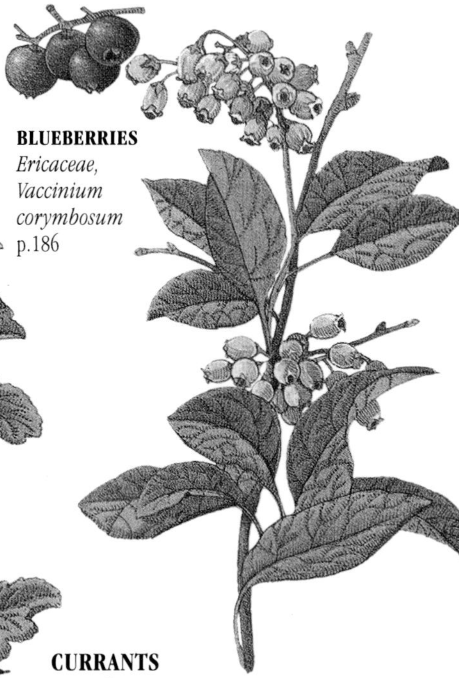

BLUEBERRIES
Ericaceae,
Vaccinium
corymbosum
p.186

RED CURRANTS
Ribes sativum

CURRANTS
Grossulariaceae
p.183

GOOSEBERRIES
Grossulariaceae,
Ribes uva-crispa, p.182

BLACK CURRANTS
Ribes nigrum

WHITE CURRANTS
Ribes sativum

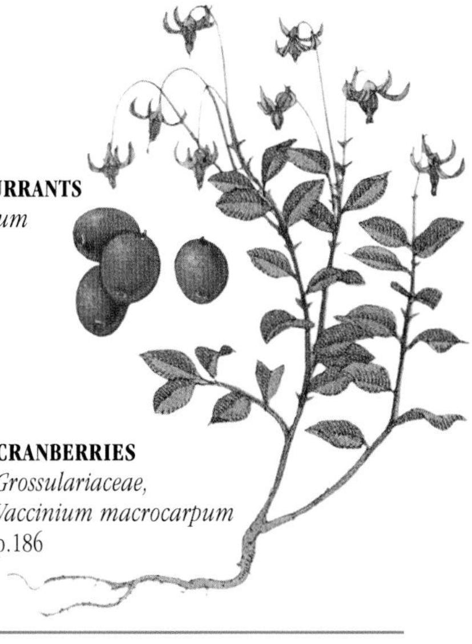

CRANBERRIES
Grossulariaceae,
Vaccinium macrocarpum
p.186

Green Manure Crops

GREEN MANURE
p.86

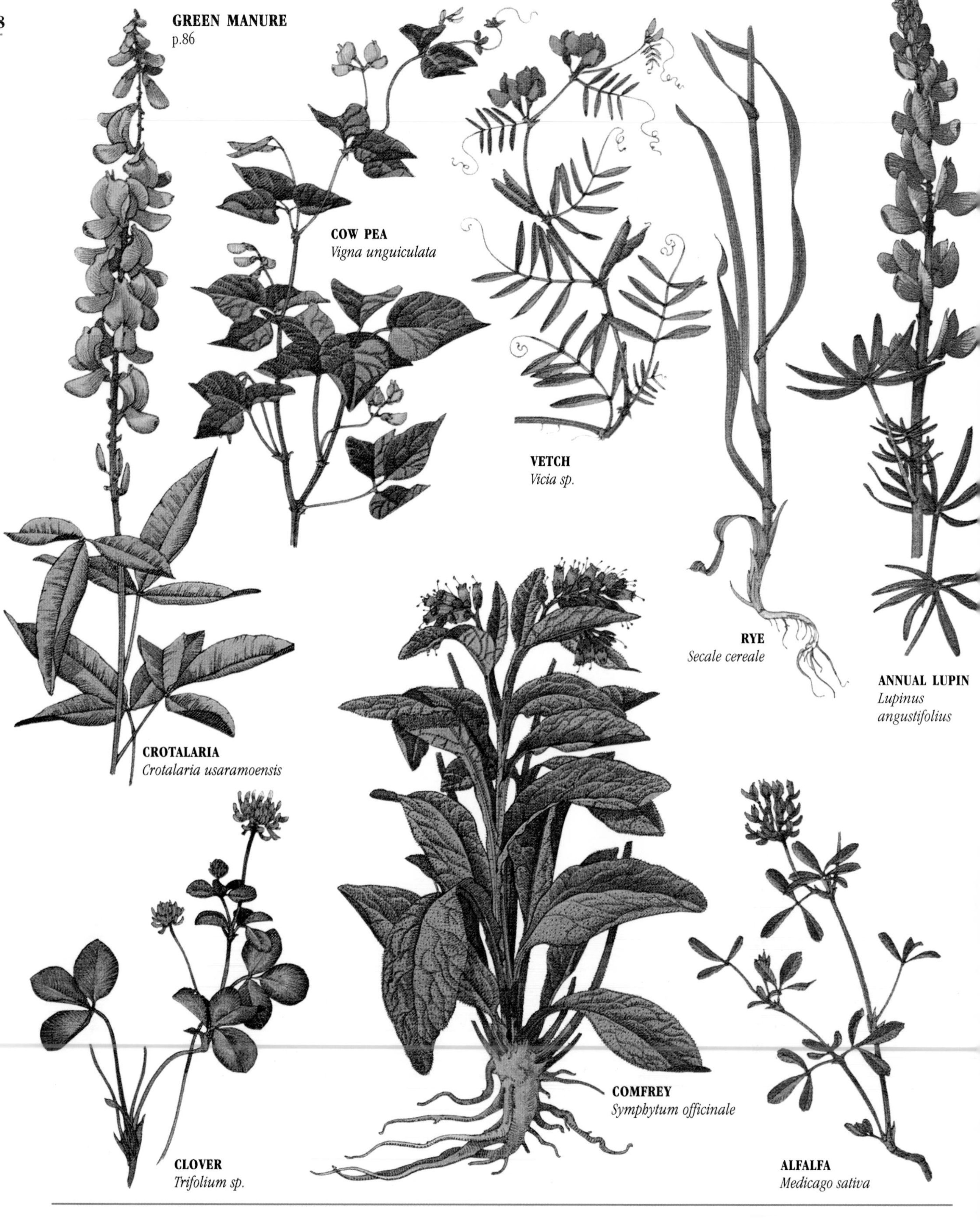

CHAPTER TWO

Gardening through the Year

Containing the cycle of the seasons, the effect of the seasonal changes on the garden, and a calendar of tasks for the diligent gardener.

Gardening through the Year

A man may live in a city all his life and scarcely be aware of the seasons: he knows it is winter when he comes out of his house in the morning and has to put on his gloves; he knows it is summer because he can have the window of his office open. But as soon as that man creates a vegetable garden, the seasons become all-important to him: they dictate the tasks he must perform each month, and they bring with them their own peculiar weather, which sometimes helps and sometimes hinders. If a gardener forgets some vital operation at any time of the year, he will find that he will suffer for it later on—perhaps twelve months later—when he has to go without some useful crop or buy it from the supermarket or farmer's market.

A philosophical gardener will say to himself: "There is no bad weather!" The rain that stops him from doing his spring digging is good for his early seed-beds; the drought that is shriveling his summer lettuces is giving him a chance to get out with the hoe and win the battle of the weeds.

There is no bad season: every season presents the gardener with a challenge and an interest of its own. All weather is good for somebody, or some plant, somewhere. The gardener cannot change these things. He must accept the challenge of learning to understand the seasons and of adapting himself to work within their never-ending cycle.

On the following pages, the seasonal activities in a typical food-producing garden are described in detail. The illustrations show how the same garden changes through the year. A key to these illustrations is provided below.

WINTER

During the spring, summer, and fall, you may feel you are under pressure. There is always an imperative: weeds to combat, seeds to sow, plants to plant, summer pruning to do, food to harvest, and so on. When you have harvested the last of the root crops in the fall, you may well heave a sigh of relief and feel you can relax.

Now you can stand back and take stock, congratulate yourself on what has gone right and not lose too much sleep over what has gone wrong.

KEY TO THE SEASONAL GARDEN
The garden that is illustrated on the following pages contains four plots for annual vegetables. These plots are cultivated according to a four-course rotation, so the vegetables grown in each vary from year to year. In the year illustrated, the plots are allocated as follows: **1** Plot A—Miscellaneous **2** Plot B—Roots **3** Plot C—Potatoes **4** Plot D—Peas and beans/brassicas. You will notice that in winter and spring, some vegetables from the previous year's rotation are still growing. The garden also includes: **5** Seed-bed **6** Holding-bed **7** Perennial vegetable bed **8** Cold frames **9** Soft fruit bed **10** Standard fruit tree **11** Bee hives **12** Greenhouse **13** Cordon fruit trees **14** Compost pile **15** Rhubarb **16** Herb bed.

Now is the time to tidy up, lay paths, build sheds, put up a new greenhouse, repair tools, and prepare for next year's food-growing campaign. But don't forget that now is also the time to feast. The winter is when beer jug and cider flask should circulate among those who have earned this merrymaking by their toil.

But even if life is less demanding now than at other times of the year, winter also has its tasks. In a mild climate, a quarter of your main garden will be growing *brassicas*, and as much of the rest of it as possible should be planted with green manure. I dig in grazing rye, which gives good bulk, and winter vetches, which provide nitrogen.

It is nearly always better for the ground to be covered with some crop in the winter: this prevents soil erosion, and locks up the soluble nutrients temporarily so that they are not washed out by winter rain. Mulching serves much the same purpose. Mulch the asparagus bed heavily: a friend of mine covers his thickly with seaweed every winter and gets a splendid crop, because asparagus likes salt as well as protection from hard frost; also mulch other perennial crops like globe artichokes and rhubarb.

When the frost grips the soil and makes it hard enough to bear a heavy weight, use the wheelbarrow to carry compost from the piles or bins out on to the land, where it will do more good than it would if it rotted away where it was. Out on the land, it will feed and sustain the earthworms, which are the gardener's best friend. Pull rotten leaves off Brussels sprouts and other big *brassica* plants, and put them into the very center of the compost pile.

In the early winter, dig up unproductive heavy clay so that the frost can get at it and break it down. But remember that in an organic garden you can improve the heaviest and most stubborn clay with yearly dressings of compost or manure. After a few years of this treatment, it will not be so necessary to dig it up every winter; the soil will break down naturally anyway in the spring.

In winter you should dig your deep beds (see p. 106) after first spreading a dressing of compost on top of them. Test the soil with a testing kit, if you think you need to, and apply lime to next year's legume break if the pH is below 6.5.

Winter is also a good time to cut sticks for peas and beans, because the leaves are fallen from trees and hedges. Look for slender branches with some twigs still on them and cut lengths four feet (1.2 m) long. Hazel sticks are the very best.

In January, sow onion and leek seeds in seed boxes and keep them in a warm place indoors. Onions grown from such a planting will ripen early and dry off well for winter storing; the leeks will grow as thick as your arm for early fall use. If you want to force early rhubarb, put some barrels or steel drums over the plants and bury these completely under manure with plenty of straw in it. The heat will force rhubarb for eating in early spring. In January, set seed potatoes to chit in boxes in a shed or greenhouse: they need light, but must never suffer frost.

In the greenhouse, be sure to provide extra ventilation on sunny or warmer days. There should be some ventilation all the time except in hard freezing weather. The air in a greenhouse should always be fresh.

Cauliflowers for harvesting in early summer will be growing in flats or seed boxes indoors, and you will have to prick them out into frames or under mini-greenhouses (see p. 111) as they get overcrowded. As you prick out, sow more seed in the seed boxes, for harvesting later in the summer. The cauliflower is always thought of as a difficult crop, but really it must just be sown indoors at the right time, and must be allowed to develop fast at all stages of its growth.

The great English horticulturist Alan Chadwick, who runs a training center in California, talks about the "breakfast, lunch, dinner" principle. What he means is that you first give your seedlings a good breakfast in the seed box—plenty of nourishing seed-starting mix. Then you give them lunch when you put them out in frame or holding-bed, into even richer soil. And when you finally plant them out in their permanent home—in Chadwick's case, always a deep bed—you give them dinner, the best meal of all. If you stick to this principle with cauliflowers, and never let them dry out, you should have absolutely no trouble and produce an excellent crop.

Sow summer cabbages now too, in flats or cold frames. Winter cabbages you will be eating, or giving to your neighbors, and the same applies to Brussels sprouts, broccoli, leeks, celery and kale. Rutabagas and parsnips you can leave out in the ground until after Christmas, unless you live in a very cold climate indeed; onions and garlic you will get from the strings hung up in your outhouse. Never store them in the warmth of the kitchen, or they will rot; hang one string at a time in the kitchen for immediate use. Don't leave bunches of herbs hung up near the stove to collect the dust, either; as soon as they are dry, shred them and put them in jars. If you are lucky you may still have tomatoes ripening away in a drawer. And, of course, you should have plenty of jars of tomatoes, sauerkraut, and chutney.

The Winter Garden

THE PERENNIAL PLOT
If you have left your asparagus ferns on the plants, now is the time to cut them off with a sharp knife and put them in the middle of the compost pile. Do the same to the globe artichokes, which by now will be dying down. Put a heavy dressing of compost, straw, seaweed, or other mulch on the perennial beds; this will protect the roots against the cold, as well as rotting and improving the soil.

TOOLSHED, TOOLS, AND FENCES
The winter is the time of year to attend to all garden hardware. Do all maintenance work now so as to spare yourself when the heavy work begins in spring.

SEED- AND HOLDING-BEDS
These should be covered with green manure by now. Leave them alone.

PLOT A
Brassicas
This is the bed you will depend on most all through the winter. At first there will be cabbages, self-protecting broccoli, and calabrese to pick and eat; later in the winter when your other cabbages have been eaten, savoys will come to the rescue, along with sprouting broccoli, which can stand a lot of cold weather; finally, curly and other kale will weather the storms and come through to feed you in the spring. Spring cabbages will have been planted out, and should now be protected if necessary by cloches or a plastic tunnel.

PLOT D
Potatoes/Peas and beans
Leeks, which will have gone in after the early potatoes last year as a catch crop, have grown well, are earthed up, and can be harvested through the winter. Fava beans and winter peas were sown in late fall, because this is next year's pea and bean plot.

COMPOST
Protect your piles of compost from winter rain by putting something on top of them: black plastic, old carpets, or even old sheets of corrugated iron. If you have time to turn your compost, so much the better; it will heat up again and mature all the quicker. Don't leave it in the bins too long once it has matured—get it onto the land.

THE COLD FRAME
Winter lettuces should be growing here. Take your chicory out and blanch it indoors in the dark. Open frames a little during warm days. Protect with mats at night.

FRUIT TREES
Plant new trees when the weather is clear and crisp. Do winter pruning and winter washing in the late winter months. Give your trees a heavy mulch or otherwise plenty of manure.

RHUBARB
Simply maintain the mulch over your rhubarb crowns. They will stay dormant, but need to be covered to protect them against very hard frost.

SOFT FRUIT

Toward the end of winter, you can prune your black currants and other soft fruit. Winter wash* if your bushes are attacked by aphids or borers in the summer.

THE GREENHOUSE

As the winter goes on, you will come to sow more and more seeds in seed boxes. As you clear the winter lettuce, dig and manure the soil in preparation for planting next year's crops in spring. It is a good idea to alternate your cucumber and tomato houses or, at least, the soil in which you grow these vegetables. Every winter, it may be necessary to remove all the soil underfoot and replace it with new topsoil to prevent disease. Grow bags may be cheating, but they will certainly save you all this trouble.

BEES

Leave your bees severely alone, but ensure that the hives are not blown over in winter gales.

PLOT B

Miscellaneous/Roots

This plot may still have a trench of celery. Pull this up before it starts to rot; the exact timing will depend on how cold it is in your area. The rest of the plot should be sleeping peacefully under a winter coat of green manure and you can leave it strictly alone until the spring. If you happen to get a series of fine dry days after Christmas, however, you may like to take a spade or a fork to it, and get some digging done before the big rush begins in spring.

PLOT C

Roots/Potatoes

There may well be some parsnips left in this bed, which last year held the root break. The parsnips you can leave in the ground until you want them, except in regions where there is snow or a very hard winter frost. So all of this bed will have been dug up, or will be on the point of being dug up: in other words, the bed lies fallow through the winter. Wheel manure out onto it whenever the weather allows, in preparation for it to become the new potato break in the spring.

HERB GARDEN

In the winter, if the weather is reasonable, you can transplant deciduous herbs (or indeed any others—but remember that you can transplant evergreens in the summer as well). Otherwise, now is the best time for the herb garden to rest.

The root cellar, or those old wooden crates in the garden shed, will be yielding up their store of potatoes, turnips, carrots, beets, kohlrabi, and so on. There is no point in surviving the winter to find you still have a ton of old roots that you don't really know what to do with. So be generous to yourself. But, at the same time, don't forget that the new potatoes very likely won't be ready as early as you think they will.

Now is the time you will be glad you salted down the runner beans. In the pouring rain, or when there is a heavy frost, or when a blizzard is blowing, think how nice it is not to have to go outside and pick Brussels sprouts or kale.

In the orchard, you can plant trees or bushes through the winter, as long as the ground is not too wet or frozen. Thin raspberry canes and tie them to the wires, and prune your black-currant bushes. Mulch or manure all soft fruit heavily, and tree fruit too, if you have the manure to spare. It is far better to suppress grass and weeds with mulch near soft fruit bushes or fruit trees than to hoe or dig. Digging is particularly bad for the spreading roots of raspberries and not good for currants either. I leave the pruning of fruit trees until February, and after pruning I spray with winter wash* (see p. 104). If your trees are heavily coated with lichen, it may be a good idea to spray with a lime-sulfur wash*. If you can't get this, one pound (400 g) of caustic soda dissolved in six gallons (23 l) of water will do the job.

By late February you will already feel the approach of spring and begin impatiently to start putting seeds in. If the ground is dry enough and unfrozen, parsnips and shallots can go in. And, under the protection of glass or plastic, you can sow all sorts of seeds: leeks, lettuces, onions, cabbages, cauliflowers, and Brussels sprouts. Sow celery seeds in seed boxes in a warm house. It is always best, particularly with plants you wish to start growing early, to plant little and often, rather than to risk starting them all at once and find later that nothing has grown because of a late frost.

DEEP BED IN WINTER
In winter, the deep bed that will grow next year's peas and beans should be limed if the pH is below 6.5. Sprinkle the lime on with a trowel and then rake it in, taking care not to walk on the bed.

SPRING

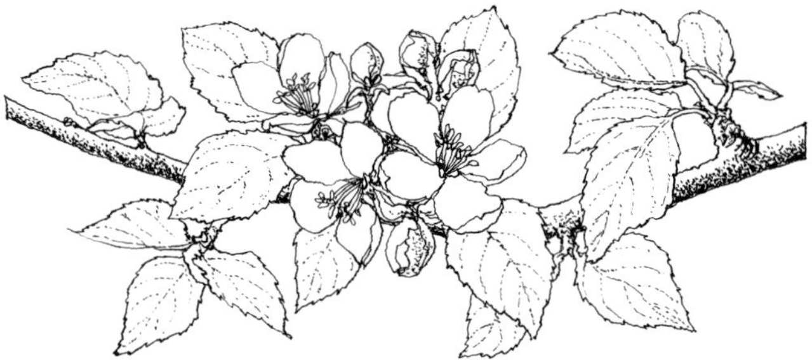

As winter fades, too slowly, into spring, the sap begins to rise not only in the gardener, but in the garden. And now is the time when you must resist a mounting feeling of panic. You cannot do everything that you need to do, but the spring is longer than you think: as long as you got things fairly straight last fall and kept them so all winter, you will find that by simply digging your garden quietly, all the spring jobs will get done in good order.

And don't be in too much of a hurry. If you put seed in very early, as early as many people tell you to, all too often cold weather, or wet weather, or dry weather, will set in and your seed will rot in the ground. Sow a little of each kind of seed early (very early if you have previously warmed up the soil with cloches or mini-greenhouses), and then wait until things warm up a little more before sowing the rest. Nature is not in too much of a hurry in the spring, and neither should you be.

Anyway, if you think about it, you do not want all your early potatoes or whatever very early. If you have just a few plants, or a short row, of very early earlies, these will be enough to give you a treat before the main crop of earlies (if such a thing can be) comes along a couple of weeks later.

As for the true main crop plants—winter *brassicas*, for example—these need to go into a dry, warm bed. If you put them into a freezing, wet one, they won't come up any quicker than if you waited two weeks longer before putting them in, and some of them won't come up at all.

Before spring actually arrives, you should be thinking about those two hardy plants, parsnips and Jerusalem artichokes. I always sow some parsnips and plant some Jerusalem artichokes in February if the ground is not frozen solid. The artichokes really don't mind what you do to them; if you put them in the ground, they will come up. However, parsnips, which come from seed, need seed-bed conditions. But don't sow seed in soaking wet ground just because the book tells you that February is when they ought to go in.

And if you are faced with a sullen, sour, thoroughly frozen surface, looking like concrete and nearly as hard, you will have a well-nigh impossible job turning it into a seed-bed. Wait for a spell of sunny, dry weather.

The ideal way to sow seed is in irrigation conditions. You sow your seed in ground as dry as dust (and dry ground is warm ground) and then you flood it. In temperate climates you can't do that, so you must just wait for those few warm days when the ground dries out to seed-bed conditions. If you have dug the land up during the winter, the loosened soil gets a better chance to dry out quickly in the first warm wind.

Here also the deep bed scores, as it scores in so many other ways. Because the beds are raised and the soil is very loose, the surface dries out very quickly and warms up at the same time. A quick fork-over should be perfectly possible without stepping on the soil, and is enough to let sunshine and air in to sweeten the soil.

There are lots of tasks in the garden in spring, and it is worth listing them chronologically. You can try putting in a few very early potatoes in the first week of March, particularly if you can protect the seedlings from hard frost later on. At the same time, sow leek, lettuce, onion, parsnip (if you didn't sow them in February), pea, radish, spinach, and turnip seed.

In the herb garden you must lift, divide, and replant many perennial herbs and sow the seeds of others. In the greenhouse or hot-bed, sow more cauliflower seed in pots or seed boxes, and also tomato and celery seed if you didn't sow them in February.

There are two ways of hastening growth in early spring. One is glass or plastic. You will achieve good results simply by spreading a large sheet of transparent plastic over a piece of dug ground and weighing it down with stones. The plastic keeps the rain off and allows what sun there is to warm up the ground. After two weeks or so, you remove the plastic, sow your seeds, water well, and replace the plastic again. As soon as the seedlings appear and make some growth, take the plastic off during the daytime and replace it at night. After hardening off the plants for a week or two, take the plastic away and put it over something else. The mini-greenhouse (see p. 111) is a modification of this and is a very good way of drying and warming the soil. Don't forget that all plants under such protective devices need some water.

The other way of speeding growth is to start things off indoors, in seed boxes, pots, or peat* pots. "Indoors" can mean a greenhouse, frames, or just the kitchen windowsill. Timing is the important factor. You have complete control over the environment until the seedlings are planted out in the garden. This must therefore occur at the right time—that is, not when a vicious frost is going to descend on you, not when the plants should have been happily growing out in their permanent positions for weeks anyway, and not when they have grown too weak and gangly in the pots indoors. If you can give the tender young plants the temporary protection of a mini-greenhouse or its equivalent the minute they are planted out, so much the better.

You can, of course, grow a lot of good food without mini-greenhouses, cloches, or any other protection, even in very cold climates. Staple vegetables, like brassicas, leeks, potatoes, and onions, will all thrive without protection. But in cold climates you cannot grow things like sweet corn, melons, pumpkins, zucchini, tomatoes, eggplants, or peppers without starting them off indoors so as to give them warmth for germination. You are, in fact, cheating the climate by extending it forward when you plant indoors and plant out later. You may well get a really hot summer even in the far north, but it will be a short summer and so you have to try to prolong it.

In April you may wonder what has hit you. This is probably the busiest month of the year. You should now sow the seed-bed with brassicas and leeks. This little seed-bed, even if it is no bigger than a table-top, is the most important thing in your garden. It holds what will eventually be your main winter vegetable supply.

Beets, carrots and, if you have room, second sowings of lettuces, peas, spinach, turnips, and

The Spring Garden

THE PERENNIAL PLOT
You should have removed the straw mulch from your globe artichokes and they will be shooting up fast. The asparagus is growing well and will soon be ready to cut.

SEED- AND HOLDING-BEDS
In your seed-bed you will have sown brassicas, leeks, onions, and lettuces. Part of the bed can be used as a holding-bed later in the year.

PLOT A
Brassicas/Miscellaneous
You will find that the cabbage tribe plants still gallantly surviving on this plot are a sorry sight. They have been supplying you with greenstuff all winter, they have suffered gales and frosts, but some of them are still there—the curly kale particularly and perhaps the sprouting broccoli. You will probably have picked all your Brussels sprouts, but the tops can still be used for spring greens. The spring is the true "hungry gap" : the Lenten fast when men and women begin to feel the dearth of fresh greenstuffs. So the remnants of your brassicas are useful to you now. Dig over the rest of the bed ready for planting tomatoes, lettuces, ridge cucumbers, spinach, sweet corn, squash, zucchini, and melons.

PLOT D
Peas and beans
The leeks that have grown through the winter are your great standby now. As your brassicas get fewer, and your stored onions run out, so the leeks come to the rescue. The fava beans and possibly winter peas that you sowed back in the fall are beginning to grow well, and give promise of early vegetables. As the leeks are cleared, you should sow early peas under plastic or cloches to protect them.

COMPOST
You will have built up a lot of compost over the winter, after almost emptying your bins in the fall for the great soil feeding that follows the harvest. Some can go into the soil now, especially into the potato plot.

THE COLD FRAME
In the cold frame you can sow early lettuces and early cabbages.

FRUIT TREES
Your fruit trees will be bursting into glorious flower. Try to keep birds away from them. Use nets (though this is laborious), scarecrows, pieces of mirror or shiny tin, loud noises, and anything else that works. Put a grease band around each tree to stop insects from crawling up.

RHUBARB
In early spring, cover your rhubarb plants with overturned pots or pails. This will force the rhubarb's growth, so that you can begin eating it earlier.

SOFT FRUIT
Prune your gooseberries early in spring. Set out strawberry plants in late spring.

THE GREENHOUSE
There is plenty to do in the greenhouse in spring. You must sow seeds of several kinds: celery, tomatoes, peppers, frame cucumbers, melons, and sweet corn are the most important. Put them in seed boxes or peat pots. Keep them watered and plant them out as the spring advances. If your greenhouse is heated, you will begin to get a crop from the tomatoes sown in early winter. As the weather gets warmer, begin to ventilate your greenhouse.

BEES
Your bees have been dormant throughout the winter. Now you can take the blanket and wire mouse guard out of the hive.

PLOT B
Roots
The green manure crop of vetch, or other winter legumes, is ready to be turned in with a spade or rototiller. Then the first root crops should be planted: parsnips first, then onions, and then a row of early carrots.

PLOT C
Potatoes
You will have planted early potatoes here, if possible in trenches full of manure or compost. Put one row under cloches or a transparent plastic sheet and you will have these to eat before the others. In late winter you should have brought out manure or compost and left it on the frosty ground ready for digging in with the main crop, which should be planted in the late spring.

HERB GARDEN
In early spring you can set out new herb plants grown from rooted cuttings indoors or in the greenhouse over the winter. This is also the time to lift, divide, and replant some of your older herb plants. Mint, thyme, chives, and sage benefit from this treatment.

radishes all go into their permanent quarters in April. These vegetables don't like being transplanted and don't need to be; you should sow them in their permanent beds.

April is also the time to plant out all those eager but pampered young creatures that you have been growing under glass: cabbages, cauliflowers, early leeks, and onions. Their places indoors may be taken by seeds of hot-climate plants such as melons, eggplants, peppers, and sweet corn, as well as by seeds of hardier plants such as summer squash. Sow all these seeds in good potting mix and find a warm place for them.

From now on you will have to keep a constant watch on plants outdoors so that they don't get overcrowded. You will need to thin seedlings carefully, unless you have planted them with a precision drill. The weeds too will by now be feeling the spring. Hoe them out or dig them in. Never let them get too rampant or you will have ten times the work dealing with them.

You may well be planting out some new strawberry plants in April. You won't get much fruit off them this year, but you will next year. You must keep a careful watch this month for insect pests in your tree fruit, and take the appropriate action if they start to get out of hand.

DEEP BED IN SPRING
In spring you can sow peas in the deep bed soil that was limed in winter. Sow in a triangular pattern so that the plants form clumps and not rows. The overall effect will be of very closely spaced diagonal rows. If you can't reach the middle of the bed, lay a board about five feet by three feet across the bed and sit on it.

When it comes to planting out plants that have been grown indoors, into frames or under mini-greenhouses, you should give them plenty of protection and not too much air for the first few days, until they have rooted themselves really well into the new soil. A transplanted plant is an invalid for a few days; be kind to it.

Small plants, put out into a deep bed that has been warmed for a couple of weeks with a mini-greenhouse, do far better than those put out in orthodox frames. Of course a true hot-bed is the very best of all; there is nothing like it for hastening things along.

In early May, be on the watch for signs of frost late at night. One of these frosts may nip your early potatoes right down and give Mr. Jones next door the opportunity he has been waiting for—his chance to gloat as he lifts some tubers the size of small marbles and you don't. If it looks like frost (a clear sky and no wind) cover those potato shoots with something; it hardly matters what. If you are caught out by frost, get out early in the morning and wash the frost off the potato tops and any other plants that may suffer, with cold water.

As ground becomes available, plant out main-crop brassicas from the seed-bed into permanent quarters; or, if you use the holding-bed method, plant them out into the holding-bed. The plants can live happily in the holding-bed until August, moving into their permanent quarters as land becomes free when you harvest other crops. Plant out a few leeks as well, so that you have early leeks in the fall. Main-crop ones will transplant much later.

Prepare your celery trenches in late April. This is one of the most useful plants of all because it can be harvested well into the winter. It makes a big difference in the flavor of winter soups and stews, and crisp white celery eaten raw with cheese is a rare delight in winter.

Keep on sowing successionally in May. All the things you sowed outside last month can be sown again now. Keep thinning young plants as they need it: choose rainy days to thin onions or carrots so as not to attract the wily onion and carrot flies. And keep hoeing. An hour of hoeing or weeding in May will save you days later on when the weeds have gotten the better of you. Hoe the weeds almost before they appear.

Watch tree fruit carefully for pests. Remove dead flowers and thin fruit during May. If you thin all apples, pears, and peaches this month, you will have a bigger weight of better fruit in fall. A fruit every five inches (13 cm) is what to aim for.

SUMMER

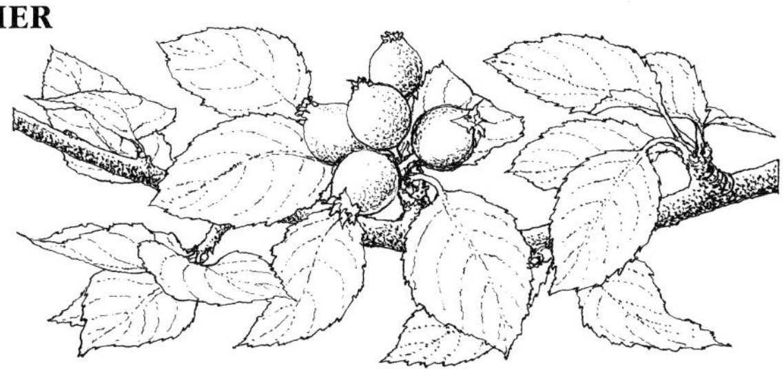

By now the strong rhythm of the growing season should have you in its grip and you will be carried along by the inexorable progress of nature. But the early summer does not immediately bring with it the end of the hungry gap.

It is now that you will be grateful, if you have been sensible enough to sow spring cabbages the previous fall, and nurtured them as they deserve to be nurtured. You will also be thankful if you grew lettuces in cold frames or under mini-greenhouses, along with radishes and some very early green onions. These vegetables will give you fresh vitamins at a time when you sorely need them. It is now that you will be watching those very early potatoes with great impatience, although their moment of glory will not quite have come. The first spears of asparagus will emerge now as well, and they can be eaten as soon as they are six inches (15 cm) high.

If you planted fava beans last fall, they will be growing well and tall by now. You might well nip the tops off some of these and cook them as leguminous spring greens. If you don't, the black-flies will only get them instead.

As far as planting and sowing goes, don't give up now. Successional sowing should be the slogan for this time of the year just as much as it is in the spring. Continue to sow lettuces, radishes, carrots, beets, rutabagas, and peas, little and often; in this way you will have fresh, tender young vegetables all through the summer and fall. It is from this angle that as a home gardener you are at such an advantage. You don't need vast quantities of anything, so you can put in half a row—even just a few seeds—and give yourself a constant supply of fresh young food.

As soon as all fear of frost is past and the ground is warm enough, in should go the tender vegetables that can't stand any frost at all, like seeds of runner and green beans, and soybeans too if you can grow them. At the same time, out go the seedlings that you have been nurturing in the greenhouse.

As June begins, the hungry gap eases up and your garden starts to hint at the abundance to come. It is a beautiful month in the garden. You may even be able to stand back and admire your own work for a moment or two every now and then, although this does not mean that you can relax—what gardening writer would ever suggest that? It is in June that the contest to see who lifts the earliest potatoes rages. The fall-sown beans will be edible this month, along with the peas that have grown through the winter. Spring cabbages will be a great standby and green onions should be available in plenty. If you haven't sown them specially, there will always be thinnings from your main-crop onion bed.

June is a bad month for insect pests, but then it is also the month when strawberries come in to cheer the heart and the taste buds. By the end of June you should have stopped cutting asparagus altogether; don't cut another shoot. Give the plants a chance to grow and store up food for next year.

Push the lids off your cold frames and prop up mini-greenhouses during the day. But keep closing them down at night. Tomatoes growing in greenhouses need plenty of ventilation. Cucumbers need far more humidity, but even they don't want to be suffocated.

As summer wears on, you will have to consider the vexed question of whether to water or not. In a maritime climate such as the Pacific Northwest, you can garden all your life, never put a drop of water on the land, and yet always have good crops.

In the freak summer of 1976—the worst drought ever recorded in the British Isles—the garden on my farm grew marvelously, and I never put a drop of water on anything. But the secret is that my land is full of humus and is constantly manured, so my land holds the water like a sponge. Even at the tail end of the drought, I could dig down half a spit and find moist earth; the roots of the vigorous, humus-fed crops all went down deep enough to reach it. If I had splashed water around from time to time, I would have caused roots to form on the surface where the water was, at the expense of deep penetrating roots, and I would not have had anything like such good plants.

But where I could water really thoroughly I did. For example, I diverted a stream through big plastic drainpipes and directed the water so that it flowed on to my main-crop potatoes, and thus I got a heavy yield of potatoes where gardeners who couldn't irrigate got very few.

So the rule is: water well if you water at all. Let the water sink right down to the lowest roots of the plants. This kind of watering does help; summer cauliflowers and lettuces particularly don't grow well during a drought.

The Summer Garden

THE PERENNIAL PLOT
Remember the rule that no more asparagus may be cut after the last day of June: the beautiful ferns must now be allowed to grow undisturbed. Globe artichokes are growing vigorously and will be ready to eat soon. Try to pick your globes very young, for you can then eat practically the whole flower, wasting nothing.

SEED- AND HOLDING-BEDS
The brassica plants you sowed in the seed-bed will all have left it by now. Some are in the holding-bed and others are in their permanent bed, Plot D. The seedling leeks, lettuces, and onions in the seed-bed are beginning to get overcrowded and should be planted out as soon as possible.

PLOT A
Miscellaneous
This was last winter's brassica bed, but from now on it will take most of the vegetables that don't fit naturally anywhere else. Tomatoes, sweet corn, spinach, lettuces, celery, and all the squashes fall into this category. If you suffer badly from eelworm, it will pay you to put outdoor tomatoes in with the potato break, so as to give a longer break between solanaceous plants.

PLOT D
Peas and beans/Brassicas
The fava beans and winter-sown peas will be almost finished by now, so you should soon be removing the plants, just leaving a few to produce seed for next year. As these crops are cleared, you will be moving brassica plants out of the seed-bed or holding-bed to take their place. Plant green beans in early summer. Along with peas and runner beans, they will begin to bear as the summer wears on.

COMPOST
At this time, when growth is at its most rampant and many plants must be pulled up or cut back, your compost piles will grow very fast. Through them you feed the soil, through that your crops, and through these yourself!

THE COLD FRAME
By now this will be full of cucumbers, peppers, eggplants, melons, and any other vegetables that need extra warmth. Make sure that they are well-ventilated on hot days and give them shade when there is a lot of strong sun.

FRUIT TREES
Thin out young fruit in early summer or you may find you suffer from the "June drop," which is when immature fruit falls to the ground. Later on you will have some pruning to do. Watch out for infestations of insects; otherwise leave fruit alone. Late summer will see you harvesting early varieties of tree fruit.

RHUBARB
Eat your rhubarb in early summer. Pull out the big stalks and leave the young ones. From late summer onward, give the rhubarb a rest.

SOFT FRUIT
If you are starting a new strawberry bed, high summer is the time to plant out seedlings, which will begin to fruit next year. Most of your strawberries will have been picked in early summer. Pick currants and gooseberries as they ripen.

THE GREENHOUSE
If you have two greenhouses, or one divided in half, one should be rampant with cucumbers and the other with tomatoes. Pinch out the growing points of both cucumbers and tomatoes, the side-shoots of tomatoes, and all the male flowers of cucumbers. Pick the fruit as it ripens; don't leave it to grow bitter and sap the vines. Tomatoes need more thorough ventilation than cucumbers, which like it very hot and humid.

BEES
The summer is the time when your bees need special attention. In the early summer, stop them from swarming; later on, take honey from them as they make it. Always see that they have enough spare combs to build on.

PLOT B
Roots
All you need to do here during the summer is hoe, watch onions for onion fly and carrots for carrot fly, and wait for things to grow. If you have space to spare, sow some more carrots, as well as rutabagas, turnips, and beets.

PLOT C
Potatoes
Most of this bed will be occupied with main-crop potatoes. You must weed them, and spray them with Bordeaux mixture* if you fear blight. The earlies will mostly have been pulled by now and their places taken by leeks.

HERB GARDEN
The summer is the time for harvesting herbs, for drying them quickly in the wind and shade, and for storing them in airtight jars when they are completely dry.

The deep-bed method is excellent from a water-conservation point of view: you need just half as much water for a deep bed as you need for a conventional garden. This is because the looseness of the soil prevents the water from rising too rapidly to the surface by capillary action and evaporating; it also permits the roots of plants to move downward very freely and reach water.

Rain falling on the deep bed sinks in immediately and does not run away, lie on the surface, or evaporate as it does in a conventional garden.

Deep-bed practitioners in California favor a good sprinkling every day or so, so as to keep a moist mini-climate under the leaves of their close-planted crops. But their climate is totally without rain in the summertime and they have to water. The same does not apply in areas with summer rain.

Don't leave any land idle in the summer. As soon as one crop comes out, another ought to go in. I even like to replace a single plant when it is harvested. Out comes a cabbage, in goes a lettuce. Out comes a lettuce, in goes another one or else some radish seed. If you think of nothing else you want to grow, grow a green manure crop, preferably a leguminous one like winter vetch. This is nothing but beneficial, for the land and for you.

If you can't even get around to that, rejoice to see nature do it for you by letting a fine healthy crop of chickweed and other annual weeds establish themselves. Let them grow: they take up nitrates from the soil, hold them safe, and prevent them from washing away. But don't let them seed. Dig them in when they are at the flowering stage, or before. They will then rot down quickly, form good humus, and release their stored nutrients into the soil again. Idle land is not only wasting its time, it is wasting its substance; land, like people, benefits from plenty of hard work.

DEEP BED IN SUMMER

By late summer you will have pulled your winter fava beans. The soil should then be forked over before brassicas are planted out. Fork half the deep bed from one side and the other half from the other.

Now, abundance takes hold in July. Start taking out every other carrot, onion, turnip, beet, and so on, so that you have fresh tender young vegetables to eat. This will give more room to the ones you leave in for winter storing. Keep picking peas and pick all the beans; do not let them get old and rough and stringy. Pick and pick again. To leave beans on the plant unless you want them for winter drying is wrong, because they only sap the strength of the plant and stop others from growing. The more you pick of peas and beans, the more you will get.

You need to tie your tomatoes to stakes now, pinch out their side-shoots, and feed them with manure or comfrey tea (see p. 103). Celery and leeks can do with liquid manure. Celery in particular must be watered if the weather is dry; it won't grow at all in drought conditions, or at best it will just bolt to seed. All brassicas can use plenty of nitrogen at this time of the year—now is the time for your top-dressing of chicken manure or other organic high-nitrogen manure if your land is not naturally sufficiently fertile. Don't put any on later than July: you don't want plants to grow too lanky and sappy before the cold weather sets in.

Now is the time also to lift shallots and dry them well before storing them. Spray main-crop potatoes with Bordeaux mixture* (see p. 104) if you are afraid of blight. Watch like a hawk for carrot and onion fly.

Don't forget to sow onion seed during the summer so that you will have strong plants to set out next year. If you have a cold frame or a mini-greenhouse, sow a few cauliflower seeds as well. If you keep these growing on quickly you will get real cauliflowers, as opposed to broccoli, which are pretend ones, in the autumn.

In frames, hot-beds and mini-greenhouses allow plenty of air, even for cucumbers. Pinch out cucumber side-shoots. Protect plants growing under glass against the sun by whitewashing the glass or setting up screens.

In August your work will be mostly weeding and harvesting. Cut cauliflower curds as they are ready; to leave them too long is only to waste them. Clear early potatoes and fill their space with leeks or green manure. Earth up celery, and earth up all brassicas as you hoe them; they all benefit from soil heaped around their stems. Plants that have been harmed by cabbage root fly, or even by clubroot, will often save themselves by putting out new roots from their stems.

Complete the summer pruning of your fruit trees in August. You can still undertake the budding of fruit trees in this month if you didn't do it in July. Root strawberry runners in small peat* pots for transplanting. August is a good month to establish more strawberry plants for fruiting next year, although early September is not too late.

Cut cucumbers and all the summer squash tribe as soon as they are ready, to keep the vines fruiting. Don't let anything go hard and bitter on the vine.

Lastly, hoe, hoe, and hoe. Your success or failure as a gardener depends more than anything else on how you use, or fail to use, your hoe. Hoe weeds when they are tiny, or even before they emerge. A good hoeing does more good than a good watering. Hoe early and often and you won't have to hoe so hard.

FALL

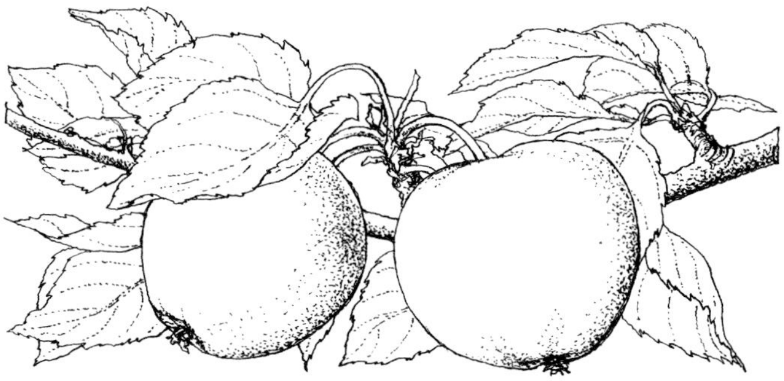

Fall is the time of fruitfulness and harvest. In comes crop after crop, either to be eaten right away or else preserved or stored; your store cupboards should get filled with jars, your root cellar with roots; and your crocks and carboys with wines fermenting away.

As you clear bed after bed of its crop, do not neglect to sow green manure (see p. 86) if you are not following one food crop with another, which is the best thing to do. Unfortunately, leguminous plants, which make far and away the best green manure crops because they fix nitrogen, are mostly summer things and tend to die down in the winter. I find that red clover, sown early in September, makes quite a good growth before the winter kills it; I am also getting good results from winter vetches, but the seed is expensive. An advantage of vetches is that, if you keep rabbits or poultry, the crop can feed these animals before it goes back into the land as manure.

You should consider the whole question of saving your own seed (see p. 91) very carefully, because seed is becoming more and more expensive.

In September you will be harvesting the last peas, and the fava, green, and runner beans. Salt down the green and runner beans and possibly dry some, and dry your fava beans. Summer squash, zucchini, and onions should be harvested when ripe and carefully stored.

Beets, carrots, rutabagas, and turnips may still have some growth left in them, so pull a few for eating fresh, but leave your main crop for storing in the ground. Celery and parsnips are not ready to lift yet, even for eating fresh, because both need a frost on them to give them flavor.

September in the fruit garden is the time to pick apples and pears. The soft fruit (apart perhaps from late strawberries) will already have been picked. But don't pick any fruit before it is ripe.

In your greenhouse, frames, or mini-greenhouses, water less and ventilate less, but don't let things get too tender. Like human beings, plants are healthier kept slightly on the cold side. Plants protected too much will not stand up well to any winter frosts.

October brings the major job of lifting and storing the potatoes. Beets, turnips, rutabagas, and carrots should also be lifted and stored now. Earth

DEEP BED IN FALL
In fall, you will have some young brassicas growing in place of the peas. You must hoe between these, but the leaves of the more mature brassicas in the bed now cover the soil; weeds cannot grow and moisture is conserved.

The Fall Garden

THE PERENNIAL PLOT
You will still be eating those delicious globe artichokes. Ignore all advice to cut down the asparagus ferns; leave them there to supply sap to the roots.

THE COLD FRAME
As you clear out the squash tribe plants, with the eggplants and the peppers, put winter lettuces in, along with chicory and other Asteraceae for winter forcing.

FRUIT TREES
Now is the time to harvest the later varieties of tree fruit. Attend to hygiene: burn all fallen branches and those fallen leaves that are diseased; put other dead leaves in the middle of the compost pile.

SEED- AND HOLDING-BEDS
These can be sown with a quick-growing crop, such as lettuces, radishes, turnips, or spinach. Otherwise sow green manure.

PLOT A
Miscellaneous
Your sweet corn should now be in full yield. Pick it before it gets too tough and rush it to the boiling pot. As you harvest crops from this bed, fork over the soil quickly and sow green manure seed: rye, or winter vetch, or both. Be patient with your outdoor tomatoes. They can ripen later than you think: I have often picked good ones in early fall.

PLOT D
Brassicas
As summer advanced you will have cleared the pea and bean crops and followed them immediately with the brassica plants from the holding-bed. These will be your winter standby and much depends on them. Do not give them nitrogen or rich manure at this stage or they will grow weak and sappy, and won't be able to stand the winter's frosts and gales.

COMPOST
Compost "makes" very quickly in the heat of the summer and by now you will have quite a lot. Get it out onto your beds and make room for more, because the time of harvest is the time when large quantities of plant wastes are available for the bins. Do not neglect to bring in what you can from outside too. If your neighbor does not want his grass mowings, collect them and compost them.

RHUBARB
Mulch your rhubarb well with compost, manure, leafmold, or seaweed and leave it alone. It needs a dormant period during the cold weather.

SOFT FRUIT
Don't be in too much of a hurry to prune your black currants. Let the sap go down into the roots first. Mulch the soil heavily with organic manure.

THE GREENHOUSE
On the ground you may plant lettuces as you harvest other crops. Clean and clear the shelves for propagating early seedlings later on.

BEES
You should have left the bees some honey, but the shortfall should by now be made up with sugar. Put in a blanket and wire mouse guard for the winter. In very cold climates you can wrap the hives in black building paper.

PLOT B
Roots
Parsnips and rutabagas will stand up to cold weather. All other roots must be harvested and stored in a basement, shed, or clamp. Get green manure seed into this bed if you possibly can.

PLOT C
Potatoes
In November, get the last of your main-crop potatoes out of the ground. The leeks that followed the earlies will be well earthed up by now. Follow the main crop with green manure: rye is probably the most practical.

HERB GARDEN
Now is the time to harvest carefully such herb seeds as you require. Let seeds get quite ripe; pull the plants out, and hang them up in an airy place to dry. Cut off any stems that have died.

up leeks and celery as required, and bend the leaves over the curds of your cauliflowers so as to protect them from the weather. Keep a vigilant eye out for slugs and snails, which love a warm, wet fall. October is also the month for sowing winter fava beans and peas.

Harvest all tree fruit before October is out, and rake up and compost all dropped leaves under the fruit trees. This piece of hygiene helps greatly toward preventing fungoid disease. If you put the leaves right into the middle of your compost pile, no harmful organisms will survive.

Seedling cauliflowers, sown in seed boxes in early October, should be pricked out into a frame. Some people sow straight into a frame, ignoring the seed-box stage; but I prefer to remember the "breakfast, lunch, dinner" principle—that at every transplanting, plants benefit by getting an even better meal from richer soil than they were growing in before. So I sow the seedlings in seed boxes first in ordinary seed-starting mix, then prick them out into frames full of very rich soil mix with plenty of manure in it, and, come spring, I transplant them outside into a deep bed, very well manured. Be careful not to let the frames get too damp and stuffy. Open them up on all fine days and close them only at night. Remember not to overwater in frames or greenhouses: growth and evaporation both slow down in the fall.

Once you have had a good frost, you can start pulling celery and parsnips for eating fresh. Both can be left in the ground and pulled when required until well into the winter.

November is a month you just have to put up with. The thing is not to feel defeated but to dig when you can dig, and when you can't, to get out regardless and tidy up. Pull dead leaves off the brassicas and put them on the compost pile; left where they are, they only harbor slugs. Cover winter crops such as celery with straw or other mulch to keep the worst of the frost out, and in the same way protect such tender perennials as asparagus and globe artichokes that have died down and are dormant in the ground. Seaweed makes a marvelous mulch for this purpose.

Ventilate your frames or mini-greenhouses when it is not actually freezing, but cover them with matting, old sacks, or straw when there is hard frost. Winter lettuces in the greenhouse or in frames should be kept warm, but not too humid and stuffy. Like so many plants, they must not be subjected to frost, but don't mollycoddle them.

THE VEGETABLE GARDENER'S CALENDAR

	January	February	March	April	May	June	July	August	September	October	November	December
Beans, fava		Sow spring seed			Harvest			Clear		Sow winter seed		
green				Sow under cover		Sow	Harvest					
runner					Sow		Harvest					
Beets				Sow		Hoe			Harvest		Clamp	
Broccoli		Sow as you harvest			Plant out							
Brussels sprouts	Harvest		Sow		Plant out					Harvest		
Cabbage, spring			Harvest				Sow		Plant out			
summer	Sow under cover		Sow				Harvest					
winter	Harvest			Sow		Hoe					Harvest	
Carrots			Sow			Hoe	Sow	Weed		Harvest		
Cauliflower	Sow in heat						Harvest		Sow under cover		Harvest	
Celery	Harvest		Sow under cover			Plant out		Earth up		Earth up	Harvest	
Kale	Harvest			Sow in seed-bed			Plant out			Hoe		Harvest
Leeks	Sow as you harvest					Plant out		Hoe			Harvest	
Lettuce	Sow and harvest most of the year								Sow under glass in winter			
Marrows (summer squash)				Sow under cover		Sow	Hoe	Harvest				
Onions		Sow	Thin		Weed			Sow	Harvest			
Parsnips		Sow			Hoe				Harvest			
Peas			Sow			Hoe		Harvest			Sow	
Potatoes		Plant out and hoe				Harvest		Spray	Harvest			
Radishes			Sow			Sow as you harvest						
Spinach	Sow and harvest all year											
Tomatoes			Sow under cover		Plant out		Hoe	Harvest				
Turnips/rutabagas				Sow			Hoe		Harvest			

CHAPTER THREE

Planning the Food-Producing Garden

Containing the organization and laying out of the vegetable plot, the herb garden, and the orchard, and the principles of rotating crops.

Planning the Food-Producing Garden

Lucky indeed is the man who can plan his garden starting with a piece of bare land. Unfortunately, most of us inherit a muddle of one sort or another, and there's no simple planning formula to help us. You have to take into account several factors, and of course each garden, and each gardener, has different requirements. But whatever you want of your garden, my advice is always to plan now, whenever now is. Don't put off planning. The sooner you do it, the sooner you will be able to harvest your first crops.

The individual's requirements

The first principle of planning is, obviously, to consider what produce you need and want. And don't neglect your personal tastes here: it's pointless to grow a huge crop of runner beans if you don't actually like them. So, before anything else, I suggest you make a list of the crops that you'd like in your ideal garden.

When I was a boy in England, farm laborers needed a great bulk of easily grown food to supplement the meager diet that their 28 shillings a week could buy them. So they grew potatoes on at least half their gardens in the summer, and always a long row of runner beans, than which nothing is more productive. Then they grew as many brassicas plants as they could in the winter. Many of them did not bother with much else.

But if poverty isn't an urgent consideration, I think it's better not to devote all your space to the bulk crops. Instead, grow vegetables like peas and sweet corn, which are best when eaten as fresh as possible. (There's a saying among sweet corn lovers that you can walk down the garden to pick the corn, but you have to run back with it once it's picked.) Apart from early potatoes, rutabagas, and beets, which are so delicious when young and tender, I recommend the gardener with limited space to buy his main root crops from the farmer's market, and use the space for more delicate vegetables that don't keep so well.

The true self-sufficient gardener, however, will want to be in a position where he has to buy no produce at all, and here we come to the tricky question of just how much you can expect to get from a given piece of land. Unfortunately, there's no easy way of estimating this, other than by long experience and an intimate knowledge of your own garden. The best general rule in my view is to grow as much as you possibly can. If you use intensive gardening methods such as the deep bed (see p. 106), you will be amazed how many people you will be able to feed from quite a small plot.

Garden geography

Once you have made a list of the produce you want from your garden, the next thing is to ask yourself whether the general geography of the garden will let you grow it.

EXPOSURE People worry too much about whether their gardens face north, south, east, or west. It's true that in the northern hemisphere, a south-facing slope warms up quicker and better than a north-facing one. But it's also true that a south-facing slope is hit as hard by a late frost as a north-facing one. In practice, I find that north-facing crops are really not so far behind south-facing ones as many people expect. And don't underestimate the value of a north-facing slope: for example, it is often better not to force early potatoes too quickly in the early season sun on a south-facing slope because they may not have the strength to survive a freak last frost.

On the whole, then, you need not be too concerned about the exposure of your garden. Remember that a good gardener will always do better in a north-facing garden than a bad gardener will in a south-facing one.

SHADE What is much more important than exposure is shade. When planning your garden, you must take into account how much of it will be in the shade, and for how long, during the course of the day, and the year. Certain plants just won't grow where it is too shady—just as others won't grow where there's too much direct sun. I suggest you make a rough sketch of the garden at different times of the year and color in the areas that get the sun all day, those that are sunny only half the day, and those that are in the shade all the time. When you have done this, you will have a clear idea of exactly how much space you will be able to devote to sun-loving and shade-loving crops, respectively.

On this question of sun and shade, I've heard it said that crop rows should always run north and south rather than east and west so that the plants in the rows do not shade each other. But this is bad logic. In the northern hemisphere in the summer the sun always rises and sets to the north of the east and west points, respectively, on the compass. In other words, the only time it shines directly from the south is at noon, when it is so high that shading between rows is minimal anyway. For most of the day, when it is lower and shading is important, the sun is shining from the east or the west. If anything, rows should be planted east and west to avoid too much shading. But personally I've never found it makes the slightest difference which way the rows go.

TREES There are certain features of the layout of your garden that may interfere with your planning intentions—for instance, that enormous tree in your neighbor's yard that overhangs your garden. Not only does it shade a large part of your available growing area, but the roots creep under your land and suck out nutrients.

I'm certainly not against trees—indeed, the nourishment that they take from your soil will eventually be returned when the leaves fall and begin to rot. But there are times when they really do inhibit what you can grow, and if your neighbor refuses to cut down a problem tree—as he probably will—you'll have to find some way of minimizing the inconvenience. It may be illegal to kill the tree by creosoting* the roots, even if they are on your side of the fence, but it is perfectly all right to trim branches that overhang your garden and roots that burrow under it.

SLOPING LAND If your garden slopes steeply, this can not only make gardening highly inconvenient, but can also lead to erosion of your valuable soil. The best solution here is to terrace the slope (see p. 241). Initially it's back-breaking work, but it's well worth it in the end: the garden will be far more productive, much easier to work in, the soil won't wash away, and a terraced garden can look both unusual and pleasing.

Climate

It is likely that your ideal list of the produce you want from your garden will already have been pared down by taking into account the geography of the garden. Now you'll have to pare it further by considering how many of the plants you want will grow well in your climate. While sometimes it can be fun to push your luck with plants that don't generally grow in your climatic zone, for the most part it's hopeless trying to compete against the climate—better to use it as an ally.

FROST The trouble is, of course, that it's not easy to predict with any certainty what the weather will be like from season to season—or even, as most of us know from long experience of weather reports, from day to day. The most experienced gardener can be caught out by a sudden hard frost in the middle of spring. Still, I suggest you try to work out roughly how long a growing season you can expect in your area: say, from the last frost of the spring to the first frost of the fall. (For climate maps, see p. 248.) It's always a good idea to make friends with a gardener who knows the area really well, and to ask him how long he expects the growing season to be.

Unless you decide to go in for extensive protection of your plants with greenhouses, mini-greenhouses, cloches, and so on, there's not much you can do to extend the growing season. But you may be able to take precautions against other climatic factors.

WIND Something you should include in your shade/sun chart of the garden is the direction of the prevailing wind, and an indication of which parts of the garden are sheltered or exposed. While some plants tolerate wind—though you might need to stake them—there are others such as most fruit crops, Brussels sprouts, and the taller brassicas that won't.

If the garden is particularly exposed to wind, the obvious thing is to build some sort of windbreak. A fence with gaps acts as a much more efficient windbreak than a solid wall since the wall creates swirling eddies in its lee that can do as much damage as unrestrained wind.

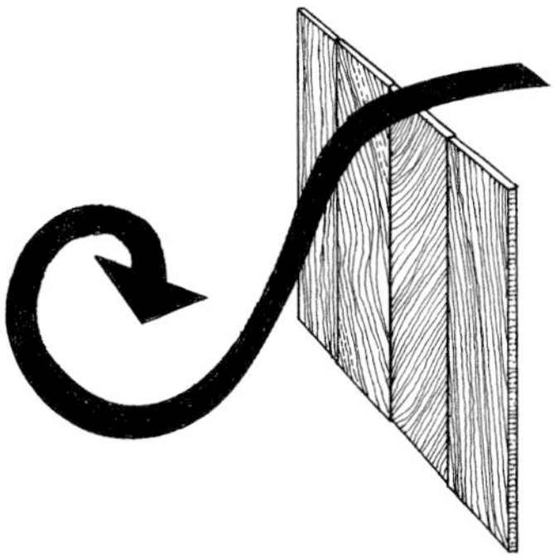

SOLID WOODEN FENCE
Wind swirls over a solid wooden fence, forming potentially harmful eddies.

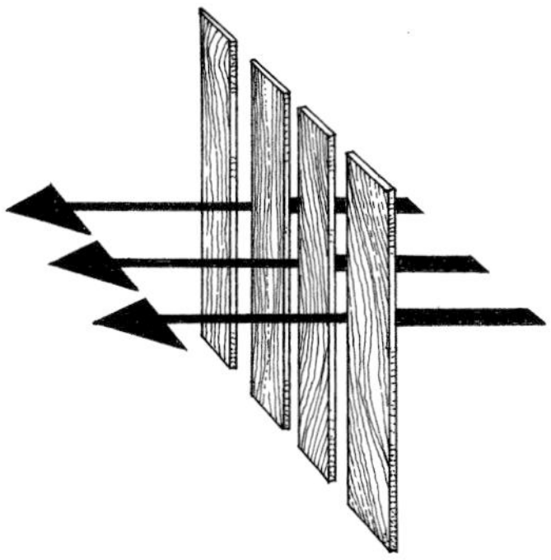

SLATTED FENCE
Wind passes through a wooden fence with gaps, but with its force drastically reduced.

If a fence is impractical, try planting a hedge (see p. 243), bushes or a line of trees—whatever will allow some wind through. Obviously, you'll have to give up some growing space for this, but it's quite possible that the increase in crops because of the shelter will make up for the loss.

WATER In most climates it's necessary to have a water source near your seed-bed to help the seeds establish roots. Once the roots are firm, regular watering is not so vital, except in very dry climates or during a drought. In these cases, I recommend laying water-lines. It's not a difficult job nowadays since plastic pipe is easily available. Just lead the pipe from wherever your water source happens to be. If your ground freezes in winter, bury the pipes underground—though normally plastic pipes don't burst as readily as metal ones. In fact, I think it's a good idea to bury the pipes anyway to keep them out of the way. And plastic can be affected by strong sunlight.

Soil conditions

One final, but very important matter you need to take into account when planning your garden is the nature of the soil. You may find that many of the plants you're thinking of growing need soil which is, for example, more or less acidic, than the soil you have in your plot. Fortunately, it's not difficult to correct the nature of your soil (soil tests and treatments are discussed on pp. 80-90). But I strongly advise you to get your soil balance right before you actually start planting. You can—and should—give your soil regular treatment even when it is occupied, but of course it's much easier to dig in, say, compost or manure when the land is bare than to wait for a time when it's covered by a crop.

Positioning the different elements

After considering the various factors that will influence your first list of the produce you would ideally like in your garden, you should now have a clear idea of what you can and what you can't grow. The next question, of course, is whether you have the space in your plot for the various crops. To estimate this, you'll need to take into account the basic elements that every garden should include. In my view, the best basic garden site should be worked around the following elements: a seed-bed; a holding-bed; an herb garden; a bed for perennials; a bed for perennial soft fruit; four beds for annual vegetables; and either an orchard or a smaller area to accommodate fruit trees. Garden constructions should include compost bins, a place to house animals, a garden shed, and, if you want one, a greenhouse. These form your basic working areas, but I think you should also allow for both a lawn and a small flower bed. You'll be grateful for a lawn where you can lounge in the sun when you've done the weeding, and it refreshes your senses—and your soul—to have a small plot containing some of those delightful old-fashioned cottage garden flowers.

The shape of your garden will obviously determine where you can put what. But there are a few siting rules that I think are worth noting. It saves both labor and frustration if you don't continually have to walk the length of the garden to perform one simple garden task.

SEED-BED The most important consideration in siting the seed-bed in my opinion is to have it close to a water source. Ideally, put it next to an outdoor faucet. If that is just not possible, then

GARDEN LAYOUT
The house is at the top end of this garden, closest to the herb garden, greenhouse, and garden shed, and farthest away from the livestock and compost. The lawn, screened from the compost by cordons, provides a "pleasaunce" at the end of the garden.

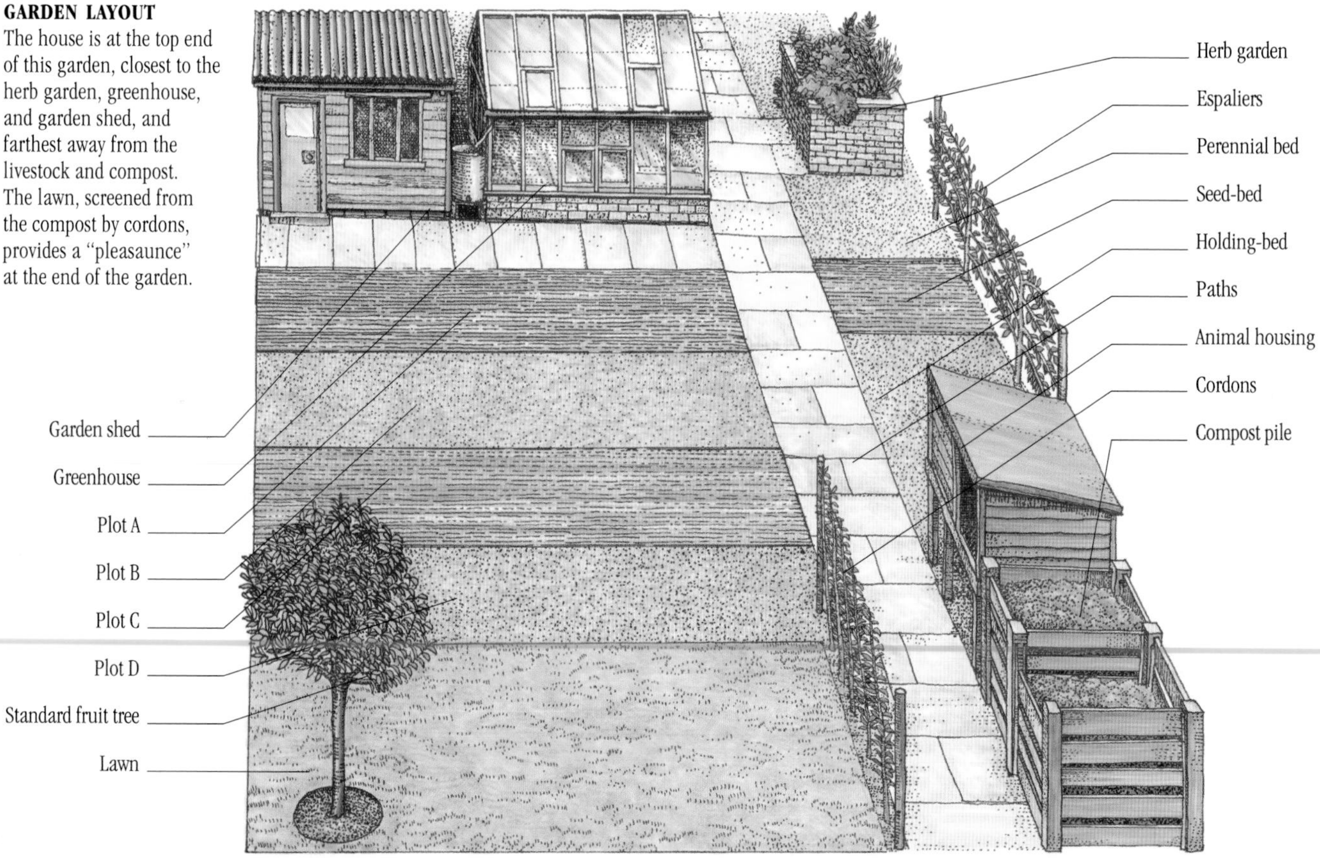

lay permanent water-lines (preferably buried out of the way) to the seed-bed.

HOLDING-BED Since you will be transferring your seedlings to the holding-bed from the seed-bed, try to have them adjacent—perhaps separated by a narrow path.

HERB GARDEN There's nothing worse when you're in the middle of cooking than to have to run to the end of the garden in a downpour in order to pick a handful of herbs. So put the herb garden as close as you can to the kitchen door.

COMPOST BINS Obviously it's best to have the compost bins near to the growing beds to save you from maneuvering wheelbarrows of compost up and down the garden. But there's a more important consideration here. If you keep livestock, then it's so much easier to be able to clean out the houses and dump the manure straight on to the compost pile. And since you won't want the animals too close to your own living quarters, I think the compost bins should be next to the livestock at the end of the garden.

BEEHIVES You'll no doubt want to avoid getting stung by bees if you keep them, so it's not a good idea to put your beehives with the other animal houses. Besides, bees like the sun, and they don't like being under dripping trees. So I suggest you site them somewhere high up—on a roof or a specially built platform. This will definitely save you from getting stung because the flight path of bees living high up is not going to be obstructed by a perspiring gardener.

GREENHOUSE AND GARDEN SHED I would advise keeping both the greenhouse and the garden shed near the house. In fact, the ideal in a small garden is to have a lean-to greenhouse against the house wall, since this saves space. If a greenhouse is too far away from the house, you may have a problem with your electricity supply for lighting and heating. If you want to use your garden shed for potting as well as keeping tools, consider putting it adjacent to the greenhouse—perhaps even interconnecting with the greenhouse.

THE SIZE OF THE GARDEN

Since there is no such thing as a standard-size garden, it's probably best to consider a basic garden plan and think of it as coming in three models: small, medium and large. Once you have worked out which vegetables you require, whether you have the conditions in which to grow the crops, and where you are going to site the various garden constructions, you must still accommodate everything in the available space.

The small garden

"Small is beautiful" is Dr. Schumacher's famous phrase, and a small garden can be as beautiful—and, if it's intensively cultivated, almost as productive—as a large one. In fact, it's often much easier to practice intensive gardening methods in a small garden: for example, a sackful of leafmold gathered in a nearby park or forest will make a significant difference to the fertility of a small garden, while you would need a lot more for it to be effective in a large one.

However, in a small garden you must learn to make use of every possible bit of growing space—and there are many more possibilities than you might think. In the first place, consider the third dimension of the garden: in a large garden there's no problem with spreading out horizontally, but in a small one it's an excellent alternative to garden vertically.

VERTICAL GARDENING Garden fences, fenceposts, walls—even the walls of your house—all provide vertical growing space. Peas, the climbing beans, tomatoes, cucumbers, and most of the squashes can all be trained upward with a system of ties and wires. Never let vertical space remain idle in the summer. And don't forget that you can grow down from a height by using hanging tubs.

USING SPACE IMAGINATIVELY
A hanging basket full of zucchini is making effective use of vertical space. Standing a bay tree in a tub on a plank on casters means it can be wheeled indoors.

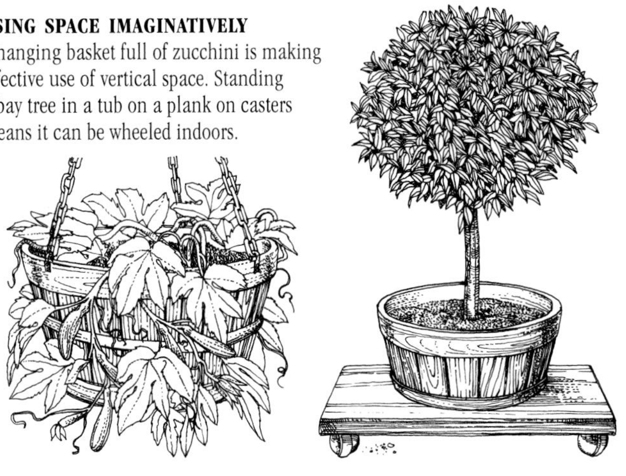

WINDOW-BOX GARDENING Remember that window ledges give you another horizontal plane for gardening, and it's very simple to construct suitable containers. It seems to me a pity always to put geraniums in window boxes: geraniums are nice, but tomatoes and lettuces are nicer to a hungry man, and they will both grow successfully in window boxes.

ROOF AND PATIO GARDENING A roof or a patio, if you have one, provides useful space for tubs and similar containers, or for grow bags filled with peat or other mixtures. (But remember to allow for the weight of the containers and the soil, if you contemplate using, say, an old garage roof that may not be very strong.) Fava beans, runner beans, broccoli, Brussels sprouts, cabbage, summer lettuce, and cucumbers will all grow well on a roof or patio; and if the climate is hot, so will peppers. Bay trees and lemon or orange trees planted in tubs that can be moved indoors in cold weather are also a good idea.

INDOOR GARDENING Nearly all vegetables and small fruits can be grown indoors, and, especially if you have a small garden, you should think of every windowsill, except those that face north, as an extension of your garden. Windowsills can be used to grow herbs, tomatoes, lettuces, carrots, radishes, and green onions, and almost any kind of container will do. Just punch holes in the bottom for drainage, and place it on a waterproof tray to catch seepage. Put three inches (8 cm) of gravel at the bottom of the container and fill it up with equal volumes of potting mix, garden compost, and good garden soil.

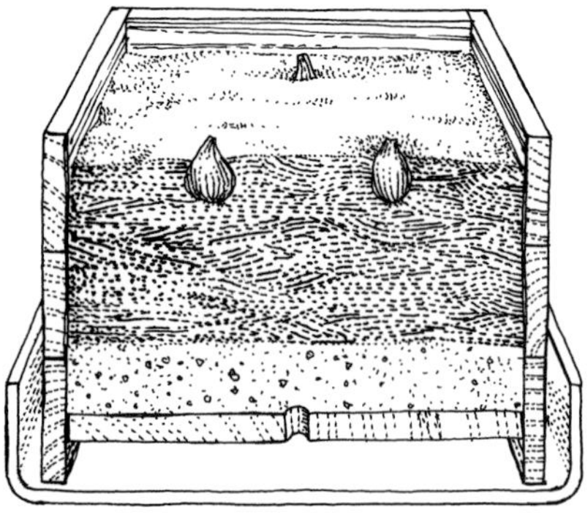

WINDOWSILL GARDENING
This cross-section through a window box shows shallots growing in equal parts of potting mix, garden compost, and garden soil. Three inches (8 cm) of gravel help drainage through small holes in the box, which stands on a waterproof tray.

When all your windowsills are occupied, consider other parts of the house too. Mushrooms, for example, grow very well under the stairs, in a basement or cellar—even in the bottom of an old armoire. For this, it's best to use special mushroom compost (see p.166), which you can either buy or make yourself.

Recently it's become quite common to garden under artificial lighting, and of course this can be arranged virtually anywhere in the house. Fluorescent tubes are best—a combination of two four-foot (1.2-m) tubes of forty watts each, one cool white and the other warm white. (Avoid using standard white or "daylight" tubes.) You can grow a number of plants successfully under lights: beets, carrots, lettuce, celery, cucumber, tomatoes, herbs, and mustard and cress. My only reservation about the system is that, given the current high price of energy, it isn't very cost-effective. But the investment is well worth it if you really haven't got any garden or outdoor growing space at all.

THE MEDIUM-SIZED GARDEN A medium-sized garden—I'm thinking of something like the larger type of suburban garden or a country cottage garden—allows the gardener greater freedom. But don't indulge the freedom too much. My advice is always to start small and intensively, and only gradually take in more and more land as you get the feel of the garden. Don't, for example, dig up the whole plot at the outset because you'll never cope without working it full time.

Even though there's more growing space, it shouldn't be wasted—for instance, by allowing an old, straggling and unproductive apple or pear tree to dominate. Cut such trees down (apple and pear trees make excellent firewood). It's a much better use of space to arrange that a fence or two are covered with cordon or espalier apple and pear trees (see p. 101), which in fact often bear a heavier crop than an old, neglected, standard tree.

The medium-sized garden will probably also accommodate a modest greenhouse, and that can be worth its weight in—well, tomatoes, anyway. You'll be able to start plants off early in the season, and may manage to avoid the dreaded hungry gap in the spring. Cold frames will also extend the growing season, and if you have the space you can look on them as fairly permanent. I think it's a good idea to have them adjacent to the greenhouse, so you can harden off the greenhouse seedlings conveniently.

A particular advantage of the medium-sized garden is that you can pay more attention to decorative and nonproductive areas—the "pleasaunce" where you can idle away a few hours in the sun. I must say I think the conventional ideas about siting it are mistaken: usually it's put next to the house because people think it's nicer to look out on flowers and a lawn rather than on the vegetable beds. But consider the view when you're actually out there—the back of the house, which is often pretty plain anyway, with all the gutters and vent pipes. I would have a lawn right at the end of the garden, screened from the vegetables by, say, an espalier fence, and with a few flower beds for fragrance, and perhaps some small fruit trees. Keep your tame rabbits here, and with all the birds twittering around you, you'll have a real haven as far as possible from buildings—your own, and those of prying neighbors, too. Remember, incidentally, that a lawn is not only a leisure area;

it is also a consistent source of good mulching material.

THE LARGE GARDEN No matter how much space you have in your garden, remember once again to start small. Master small areas first, and then you'll have the experience to work on a larger scale.

But even if you don't use the whole of your plot at the same time, you can still make it all work for you without expending much effort. Nothing improves the general heart of your land more than to lay it to pasture and graze stock on it for three or four years. At the end of this period, convert the pasture, bit by bit, into beds that can be taken up into your crop rotation. Grass and clover pastures are fine, but I suggest planting the deep-rooting crops like alfalfa or comfrey. These all mine up useful minerals from the depths of the subsoil, and whether you dig them in as green manure, put them on the compost pile, or merely feed them to the poultry, the minerals will eventually be spread around your garden. This method of resting the land will also ensure that any residual disease that remains in the soil from previous crops won't give you trouble when you come to plant your vegetables.

With a large garden, you can consider planting standard fruit trees instead of the dwarf varieties or cordon and espalier-trained trees that are space-savers in smaller gardens. The modern dwarfing root stocks probably will give you a higher yield of fruit more quickly than a standard tree will, but at the same time, few things look more beautiful than a fine orchard of large standard fruit trees. Standard plums and greengages are particularly delightful when they get really huge, and of course their fruits are so marvelous. It's always a temptation, in fact, to let big plum and greengage trees get out of control because they're so pleasing to look at, but remember that this will end up limiting your opportunities for cropping. And don't forget walnut trees: posterity will always bless you for planting a walnut.

PLANNING THE VEGETABLE BEDS

When I discussed the various elements of the basic garden, I recommended that you allow four beds for vegetables, and there is a very special reason for this. In my view, a crop rotation cycle is essential in growing vegetables, and since I would advise a minimum of a four-year rotation, you'll need four beds. More is better: for instance, if your garden is large enough, then you'll be able to rest some of the beds, as I suggested, by putting them to pasture and grazing livestock. And remember that if you're unfortunate enough to be hit by clubroot disease, the only way to get rid of it is to rest the land for nine years! The four-year rotation cycle, however, is the most practical, other things being equal.

Even if your garden is very small, don't look on the four-bed system as a luxury you can't afford. The need for a rotation cycle still applies, however small the garden, and it's not much more trouble to make four tiny beds than, say, two larger ones.

Siting the vegetable beds

Once you've decided in which area of the garden you're going to keep the vegetable beds, it is a matter of setting them out. It is not particularly important in my opinion which direction the beds run, unless you intend to use a rototiller to dig them. In this case, it's better if they don't run up to a fence or wall, because you will find it difficult to maneuver the machine. And remember to leave paths between the beds to make working easier. If your land is sandy and well drained, you probably won't need to pave the paths, but if it's muddy you will (see p. 241).

The four-year rotation cycle

The two major worries about growing vegetables are clubroot disease in *brassica* and eelworm infestation in potatoes, and these can easily build up in the soil if *brassica* and potatoes are planted year after year in the same bed. This is the primary reason for rotating your beds annually, though other crops will also benefit from being grown in different beds in successive years.

Planning a rotation cycle is a complex business, for you have to bear in mind whether the condition of the soil, after one crop has been lifted, is really suitable for the next crop to go into. There are four rules I recommend here. First, while potatoes need to grow in heavily manured soil, root crops tend to fork in such conditions: so keep potatoes as far as possible from the root crops in the rotation cycle. Second, peas and beans like well-limed soil, but potatoes don't: so avoid growing potatoes immediately after the legume break. Third, the brassicas do like lime, but only if it has been in the soil for some time: thus, it's best to plant your brassicas after the legumes when the lime has had a chance to establish itself in the ground. Fourth, what I call the miscellaneous crops (outdoor tomatoes, melons, squashes, lettuces, radishes and so on) will do better for a good mulching with well-rotted compost: since this will assist the root crops—and certainly it will avoid the problems of forking—it's a good idea to put in the root crops after the miscellaneous break.

To conform to these four rules, I suggest the following cycle for each of your four beds. But remember, of course, to start each bed off at a different point in the cycle.

FIRST YEAR Manure the bed heavily in the first year and sow potatoes. (If you like, devote a small area of the potato bed to spring cabbage, leeks, and turnips in order to save produce for the hungry gap in the following spring.) When you have harvested the potatoes, put down a crop of winter rye, which you can dig in early in the second year as a green manure (see p. 86).

SECOND YEAR After you have dug in the green manure crop, lime the soil fairly heavily and sow peas, beans, and the other legumes. Since these will be harvested from about midsummer onward, one possibility that should be considered seriously is to replace the legumes immediately with your brassicas, which have been growing at the same time, first in the seed-bed, then in the holding-bed. This may seem to be cramming plants in a little too optimistically, but it is a technique that I use successfully. Transplanting brassicas from seed-bed to holding-bed to final bed in such a short period of time actually seems to benefit the plants—and I always think it's wise to get plants accustomed to transplantation. The main advantage of this technique, of course, is that, in effect, you squeeze what would otherwise be one year of a rotation cycle between two others, and since you'll be harvesting the brassicas in late winter, this simply means that you'll be getting a lot more food from your garden. Many people devote an entire year to brassicas and have them in their final bed before the legumes are ready for harvesting. You can do this if you have space for five beds.

THIRD YEAR Assuming you have managed to get the brassicas in during the later part of the second year, in the third year you can go directly into the miscellaneous crops break. As the brassicas are ready for lifting, and when the weather begins to warm up, replace them with the miscellaneous crops, leaving the quicker-growing plants like lettuces until last. Remember that a good mulch of well-rotted compost will help the miscellaneous crops immensely. At the end of the third year, again I recommend putting in a winter rye crop for digging in as green manure.

FOURTH YEAR Root crops should be sown in the fourth year, when the manure that was spread for the potatoes is no longer exercising a direct influence in the soil. If your garden is large enough to extend the rotation cycle, then leave the plot lying fallow during the fifth year, or grow yet another green manure crop—preferably one of the deeper-rooting crops such as alfalfa or comfrey, because they will be of longer-lasting benefit to the soil. Otherwise, start the whole cycle all over again by manuring heavily and sowing potatoes once more.

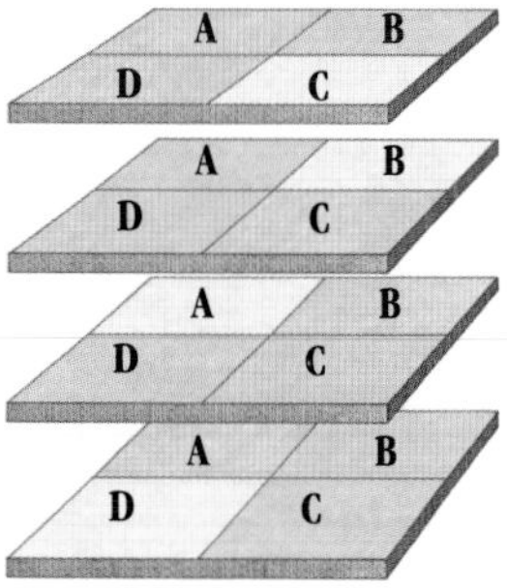

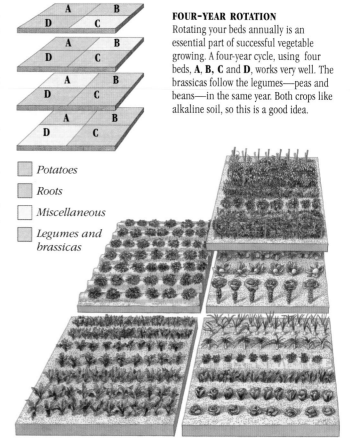

FOUR-YEAR ROTATION
Rotating your beds annually is an essential part of successful vegetable growing. A four-year cycle, using four beds, **A**, **B**, **C** and **D**, works very well. The brassicas follow the legumes—peas and beans—in the same year. Both crops like alkaline soil, so this is a good idea.

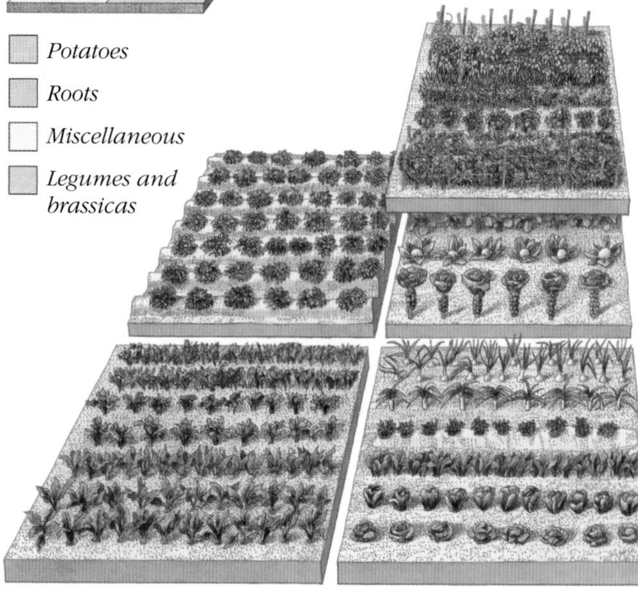

Alternative rotations

The four-year rotation cycle is, I should emphasize, just one of several possibilities. I find it suits me very well, despite a couple of objections that the purist might make. The first objection is that I put tomatoes in the miscellaneous crops break, not in the potato break, as is more common: thus, there are two solanaceous plants growing in the same bed in a space of less than four years. However, I don't grow very many outdoor tomatoes, and since I do like to have a lot of potatoes, I'm a bit reluctant to give over some of the potato bed to tomatoes. If you really are a purist, then cut back on the potatoes and plant the tomatoes with them. In this way you will be certain that you are not encouraging disease.

The second objection is that I plant radishes—which are cruciferous—along with the miscellaneous crops, instead of in the brassica bed, thus tempting fate to nurture or perpetuate clubroot disease in the soil. In fact I don't think clubroot is a serious possibility here, because the radishes are harvested quickly, before the disease really has a chance to establish itself. But never leave the radishes to get old in the ground, or you might have a problem with clubroot.

Try out my suggested rotation cycle, but of course if it doesn't suit you, consider some alternatives. For example, I know of very successful gardeners who always follow brassicas with the legumes, not, as I recommend, the other way around. Another possibility is to have a much less strict rotation cycle, where crops are jumbled up. The only rule here is to avoid planting the same crop in one bed in successive years. Personally, since I'm not good at remembering just what I've had in and where, I prefer a clearer system to work to. But if you keep good records, then the casual rotation might well suit you. I would advise making a map of your garden and noting what you have sown and planted, and when, together with details of how you have treated the soil that year.

PLANNING AN HERB GARDEN

At the Covelo Garden Project in California, you can find what is probably the most elaborate and sophisticated herb garden in the world. A large amphitheater has been excavated, and the inside slopes have been terraced: the terraces on the north side are facing south and those on the south side face north. At the highest point on the terraces, the soil is kept well drained and dry, while the lower parts surround a pond that provides moisture. So the optimum conditions have been created for growing all the culinary and medicinal herbs in the world—whether they like sunny, shady, wet, or dry conditions—and it presents the most pleasing and attractive scene.

No doubt when you plan your herb garden you will be content to settle for something less grand than the Covelo garden, although similar principles of siting the individual herbs will apply. Herbs have a great range of climatic and soil preferences—from the hot sun and dry sandy soils favored by the herbs that have a Mediterranean origin, like anise, basil, and oregano, to the partial shade and rich, damp soil favored by plants like lovage and mint. The individual preferences of the different herbs are described in the Cultivation of Herbs chapter (pp. 191–202). Obviously you will have to make do with less than ideal conditions for many of the herbs you grow, but when you plan your herb garden, try to select a site that offers at least a modest range between full sun and partial shade.

As I have already suggested, it is best to site your herb garden near the kitchen door. If sprigs of this or that fresh from the garden are within easy reach when you are cooking, you might resist the temptation to pick out a jar of dried herbs from your cupboard. Needless to say, fresh herbs are infinitely more fragrant and flavorsome than those that have been drying up over a number of years.

While herbs vary as to whether they prefer sun or shade, most do like a sheltered spot, so if you can manage to put the herb garden against your kitchen wall, this will certainly help.

Raised bed for herbs

An excellent idea for an herb garden is to build a raised bed with stone or bricks. There are a number of advantages here. First, the soil will be well drained and dry, and this is important for many herbs. Those that like damper soil and more shade can be planted at the base of the raised bed at the front or side. Second, the raised bed

RAISED HERB BED
You can build a raised bed made of brick or stone for your herbs. The soil will stay dry and well drained, which suits most herbs; the height of the bed means less bending; and the extra surface area encourages those herbs that like to straggle across walls.

entails less stooping for planting, tending and picking. And third, it provides more surface area for the straggly plants to trail over the walls.

If you build a raised bed against your kitchen wall, however, be sure to pitch it away from the house to stop water from pooling against the wall, or you could find yourself with a damp basement.

Rockery for herbs

Since many of the herbs are delicate and beautiful plants, another idea you can consider is to grow them in a small rockery (a miniature version of the Covelo herb garden, perhaps). If you're in a sandstone area, the pinks, reds, ochers, and whites of the stones you use in the rockery will complement the rich greens of the herbs. The rockery, too, will have the advantage of being well drained. It may be that you won't be able to keep the damp-loving plants like mint in a rockery. If so, grow your mint in that damp, shady patch that is found somewhere in almost every yard.

PLANNING THE FRUIT GARDEN

In *Cottage Economy*, William Cobbett strongly advised his cottagers not to grow fruit at all: in his view, it took up too much good land and served only to give children bellyaches. But then Cobbett wasn't as vitamin-conscious as we tend to be nowadays, and personally I think there's nothing to compete with fruit for providing you with gastronomic delight and with what your body needs at the same time. There are few pleasures equal to eating the first juicy strawberries of the year, or a fine, sharp fruit salad picked from your own garden. Besides, the sense of spiritual well-being that comes from walking through a glorious orchard, in full bloom or in full fruit, is one of the great luxuries of gardening.

However, planning a fruit garden requires a lot of hard thought. For one thing, you are dealing with three different kinds of plants: hard and stone fruit trees, soft fruit bushes, and ground plants. For another, you must take into account what else you are growing in the garden, since all fruit will draw a great deal of nourishment from the soil, and fruit trees create large areas of shade. Finally, and this is the important consideration, fruit-growing will take up a great deal of space in the garden.

How much fruit to grow

The size of your garden will have a fundamental influence on how much fruit you can expect to grow. Most gardens can—and should, in my view—accommodate some soft fruit. But if your garden is tiny, I doubt that the space-effectiveness of fruit trees—even the dwarf varieties or trained trees—makes them worthwhile. It is a pity, but you ought to think how many apples, say, you will get from a single cordon, and weigh this against the number of beans or potatoes you will get from the same piece of land.

Even if you have a garden with more space, you must think carefully about planting fruit trees. Remember that what seems to take up only a little space now will be a quite different proposition in ten years. Unless your garden is really enormous—big enough for a full-scale orchard—I would recommend that you avoid standard (full-size) fruit trees altogether. Several dwarf varieties, and, say, a few cordons and espaliers are a better use of space. This, of course, limits the kind of fruit you can have: hard fruit trees can be dwarfed or trained, but you can't do this with stone fruit trees such as plums, greengages, or cherries.

If you have a large area in your garden to devote to fruit, then standard trees can be considered. An area 175 feet (54 m) square, for example, will accommodate 16 huge standard apple or pear trees. When these are mature, they will produce up to eight bushels of fruit each. The initial drawback with standards is that you won't get fruit until at least six years after they have been planted, although you will get fruit for 40 or 50 years after that. Dwarf varieties fruit earlier but have a shorter life. In the same area, you could plant 64 half-standard trees and get about the same total yield (a half-standard gives about a quarter of the yield of a standard), and they will fruit four or five years after planting.

Soil for fruit trees

When choosing a site for a fruit garden, take into account the quality of the soil. Most fruit needs good rich soil, with plenty of manure or compost worked into it, because the trees and bushes quickly exhaust the nutrients in the soil. Figs will grow in poor soil, and peaches prefer light, sandy soil, but of course both these are warm-climate trees. All fruit requires well-drained soil, so if your land is wet, you will have to provide drainage (see p. 240). You will also need deep soil for fruit trees: standards and half-standards, particularly, send roots deep into the ground, although fan-trained fruit trees planted against a wall can put their roots out under unproductive land where a path or patio has been laid.

In a large orchard, you may think you will be able to grow at least some crops underneath the standards.

Fundamentally, you can't: a few daffodils might grow, but nothing edible. The best plan, if you want to use the space, is to graze sheep (just so long as they can't reach up to the leaves and branches of the trees). The manure makes a beneficial contribution to the soil in the orchard.

Generally, young fruit trees bear more fruit, more quickly, if the ground over the roots is left bare. Commercial fruit growers achieve this by spraying the area with herbicides, but I would suggest simply mulching the area heavily. However, if the ground is left bare, cut down on the amount of manure you put into the soil, since this encourages tree-growth at the expense of fruit formation. Don't clear the ground over the tree roots by mechanical means, because you may damage the roots near the surface.

Shade from fruit trees

Large fruit trees cast a considerable amount of shadow. If you can plant them at the north end of the garden, the problem is easily resolved, but if they have to go in at the south end remember that you will only be able to grow shade-loving plants such as rhubarb or mint immediately to the north of them.

Laying out a fruit garden

In view of the shading problem, I suggest you arrange the fruit garden in a stepped form. For example, to the north of the garden, put in a row of standard fruit trees, then, working progressively south, a row of half-standards, then bush trees, then some hard fruit espaliers. In front of these, put a few rows of raspberry canes, then a few currant bushes, and finally a strawberry bed. This is an ideal arrangement, of course, and obviously you would need an extensive garden to do it. But the principle is quite simple: have the taller plants to the north and the smaller ones to the south, and you won't have a shading problem.

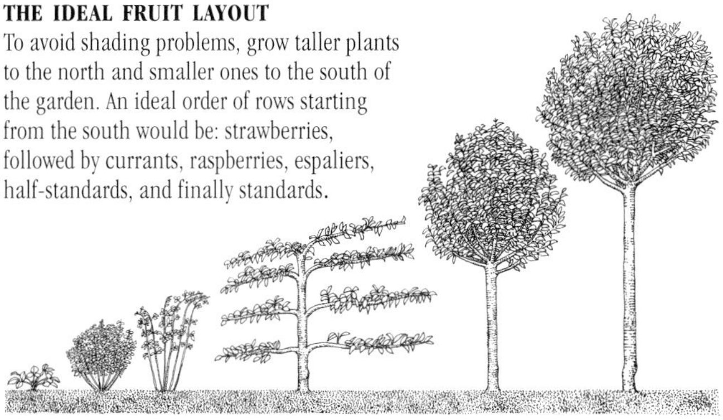

THE IDEAL FRUIT LAYOUT
To avoid shading problems, grow taller plants to the north and smaller ones to the south of the garden. An ideal order of rows starting from the south would be: strawberries, followed by currants, raspberries, espaliers, half-standards, and finally standards.

Choosing fruit trees

While deciding which fruit trees to plant is largely a matter of taste (and of the conditions in your garden), one important consideration is fertilization. It is no use planting, say, Cox's Orange Pippins alone, for they must have another variety of tree to act as a pollinator. One way to achieve this is to get what is called a "family" tree—that is, one rootstock that has had several varieties grafted on to it. Another solution is to grow several cordons or other small forms of different varieties. If you're lucky enough to have a friendly neighbor who also grows fruit trees, consult with him before selecting your trees. You'll both benefit if your trees pollinate each other. And of course, even if your neighbor isn't a friend, examine his trees anyway—he can't prevent the bees from carrying his pollen to your blossom, and he shouldn't want to, since he will profit from the arrangement as well.

In general, my advice is to select as wide a variety of fruit trees as possible, bearing in mind the need for fertilization. Grow some very early varieties of fruit, a main crop, and some late ones that store well. If you do this, it won't be difficult to achieve self-sufficiency in fruit. One final piece of advice, though, is that some varieties of fruit are particularly sensitive to locality. So be sure to consult a local expert fruit-grower about which varieties do best in your area.

Protecting the fruit

One of the primary factors to take into account in planning a fruit garden is whether you will be able to protect the fruit from birds. Birds are probably the biggest single hindrance to successful fruit growing. Plan to protect soft fruit completely—building a fruit cage (see p. 184) is probably the only really effective answer here. Cherry trees are particularly vulnerable to birds, and unless you net them completely, the birds will have stripped them bare before you have a chance to eat a single cherry.

You must also consider the damage that your trees may suffer from other animals. Hens will not attack fruit beds, but they do peck at the fruit, so keep them away from fruiting trees. Both geese and goats need to be kept away from all trees: they will bite the bark off a tree virtually on sight and kill it immediately. Solve the problem either by keeping them out of the orchard completely, or by circling the trunks with wire netting, which makes it impossible for the animals to reach the trees.

THE PLANNING REQUIREMENTS OF GARDEN VEGETABLES

	Prefers full sun	Prefers partial shade	Tolerates full shade	Prefers well-drained soil	Prefers damp soil	Tolerates dry soil	Prefers sandy soil	Prefers clay soil	Prefers rich soil	Tolerates poor soil	Prefers high-pH (alkaline) soil	Prefers neutral-pH soil	Prefers low-pH (acidic) soil	Requires long growing season	Requires short growing season	Benefits from frost	Tolerates frost	Tolerates wind
Artichokes, globe	•			•					•			•		•				
Jerusalem			•	•			•		•			•		•			•	•
Asparagus	•				•		•		•			•		•			•	
Beans, fava	•	•						•	•			•					•	•
green	•				•				•			•			•			
Lima	•				•				•			•		•				
runner	•			•	•				•			•			•			
soybeans	•				•				•					•				•
Beets			•	•	•				•			•						•
Broccoli			•		•					•		•		•			•	•
Brussels sprouts			•	•	•				•			•		•		•		
Cabbage			•	•	•				•		•			•			•	•
Chinese		•			•						•				•			•
Cardoons	•			•		•				•		•		•				
Carrots	•			•	•		•		•			•			•			•
Cauliflower			•	•				•	•			•		•			•	•
Celeriac			•	•	•				•				•			•		•
Celery			•	•	•				•			•		•		•		•
Chard, Swiss			•	•		•						•			•			•
Chicory			•	•	•					•		•						•
Corn, sweet	•			•					•			•		•				
Cresses			•		•										•			•
Cucumbers	•			•	•				•			•			•			
Dandelions			•		•					•							•	•
Eggplant				•					•				•		•			
Endive		•	•	•	•					•		•						•
Fennel, Florence	•			•	•				•			•						
Kale			•		•				•		•			•		•		•
Kohlrabi			•		•	•			•		•							•
Leeks		•	•	•	•							•		•			•	•
Lettuce			•	•	•		•		•			•			•			•
Melons	•			•	•		•		•			•			•			
Okra	•			•	•		•		•			•			•			
Onions	•						•		•				•	•				•
Parsley, Hamburg	•			•	•				•			•		•				•
Parsnips			•	•	•				•			•				•		•
Peanuts	•			•			•						•	•				•
Peas		•		•	•				•			•			•			
Peppers	•				•		•					•		•				
Potatoes		•		•	•				•				•	•				•
Radishes			•		•				•						•			•
Rhubarb			•	•	•				•				•	•			•	•
Rutabagas & turnips			•		•		•					•		•			•	•
Salsify		•		•	•				•			•		•				•
Seakale			•		•				•			•						•
Spinach			•		•				•			•			•			•
Spinach beet			•	•	•				•				•				•	
Squashes	•			•	•				•			•			•			
Tomatoes	•				•				•			•		•				

CHAPTER FOUR

The Essentials of Good Gardening

Containing the methods of digging, composting, fertilizing, soil testing, propagating, grafting, pruning, training, mulching, protecting against pests, storing, and gardening by the deep-bed technique.

Treating the Soil

Clearing overgrown land

Land left to itself in a temperate climate will turn first to grass, then to scrub, and then to forest. So if you start with badly neglected land, the first thing to do is clear it.

Long grass or tall weeds should be cut down with a scythe or bagging hook. Scything is faster, but for a small garden it probably is not worth buying a scythe. But don't try to cut bushes with either a scythe or a bagging hook or you will quickly ruin the blade. A long-handled slasher is best for bushes, or, if you haven't got one of these, a short-handled slasher, an ax, or even a hatchet. However, if you have to dig out the bushes later on anyway, it's better not to cut them down at all: if there's no top left on the bush, you will have nothing to heave against when you come to pull out the roots. After you pull bushes out, you should burn them because the ashes will give you potash for your soil.

You will need a wheelbarrow to move what you clear from the land. In my view, the most useful (and also beautiful) kind was the old professional gardener's barrow with a wooden wheel and wooden extension sides that could be installed for high, light loads and taken off for heavy ones. Nowadays, a good wheelbarrow to get is a builder's wheel barrow with a pneumatic rubber tire. These move a lot more easily, especially on uneven ground, than those that have solid tires.

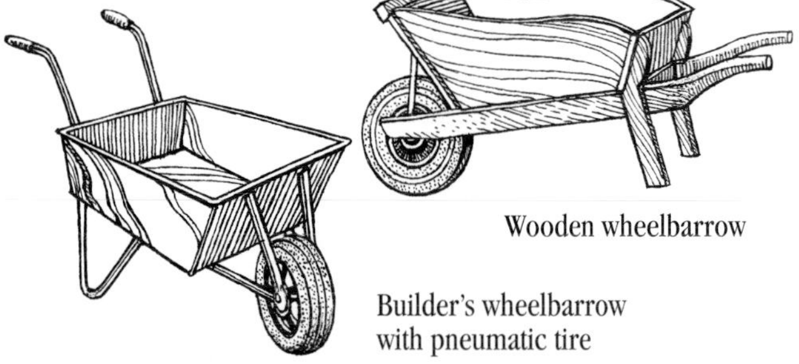

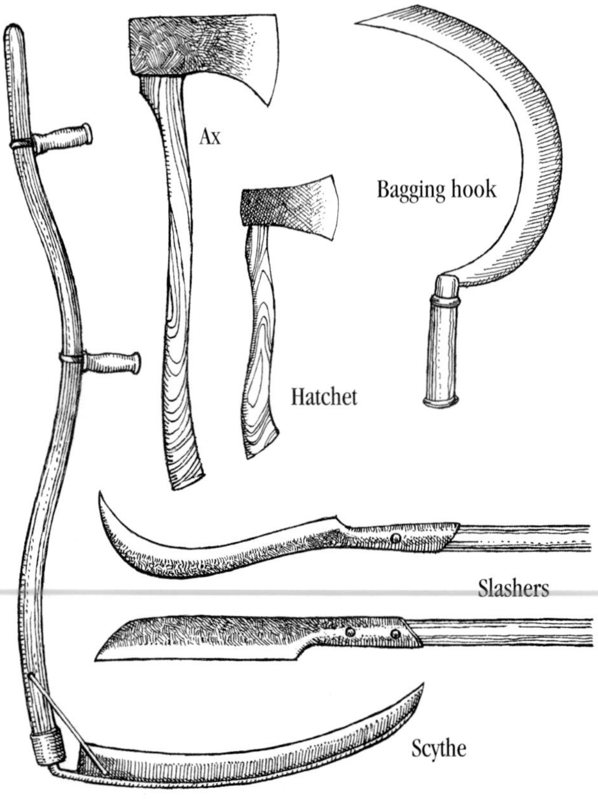

Establishing a vegetable garden

If you throw some seeds on top of the ground, the birds will eat them. If you dib a cabbage plant into a grass field, the grass will smother it and it will die. If you remove the grass, and all other wild plants, and dib a cabbage plant into the bare soil, weeds will grow up unless you stop them, and again the cabbage will be smothered. All our crop plants have been evolved over the centuries to be good to eat, to crop heavily, and to be nutritious, at the expense of other qualities, like hardiness and competitiveness among wild plants.

So, if you want to feed yourself from the produce of your soil, you must cultivate it. There is such a thing as the "no-digging technique" and I will discuss this, but if you inherit a normal garden, whether well-maintained or neglected, or if you want to establish a garden in a grass field, you will have to start by digging, or by turning the soil in some way.

Now, assuming you have a plot of old grassland that you want to turn into a garden, there are several ways you can go about it. One is with pigs. Run pigs over your plot (keep them in with an electric fence) and they will root up the turf and leave it in a condition that makes it easy to fork over and turn into a garden.

Double digging

The most traditional means of rescuing neglected ground is with a spade. And when you dig old sod for the first time with a spade, you must do a very thorough job of it. If you just turn the grass over, it will come up and grow again and you will have endless trouble. You cannot plant garden plants in half-buried sod. It is far better to do it by the time-honored method of "bastard trenching," now called double digging (see illustration right).

Once you have completed the double digging, your grass plants, roots and all, will be completely buried a spit deep. They will not grow up again to haunt you. Your soil will be loosened to a depth of two spits, which is good for aeration, drainage, and deep-rooting plants. And your new topsoil will be bare, broken up, and ready for planting with any crop immediately.

Some gardeners will tell you not to bury the scalped-off sod, but to lay all the pieces in a pile, upside down, on top of each other. They will rot down over time and make the basis of splendid potting mix. There is nothing wrong with this, as long as you remember that you are severely robbing the bed from which you take the sod.

You can double dig any garden that you take over; if it is not grassland, you can throw manure or compost into the trenches instead of sod. I would bury manure, compost, or sod once with a new garden, but I would never do it again, because I would rely on the earthworms to do it for me. Let them work their backs, and not me mine. Double digging is especially worthwhile if your land is heavy, because it will break up any hardpan (compacted, impervious layer) that there might be down below.

DOUBLE DIGGING
Divide the bed in half lengthwise. Mark out, with string if you like, a trench two feet (60 cm) wide across one half of the bed. Shave two inches (5 cm) of sod off it and pile this next to the top end of the other half of the bed. Dig soil out of the trench to a spit's depth, and put it next to the pile of sod. Mark out another trench next to the first. Scalp the sod off this, throw it into the bottom of the first trench, and break it up small with the spade. Dig a spit's depth of soil from the second trench, invert it, and throw it into the first trench. Continue on like this until you reach the other end of your plot. Cross over to the other side and work your way back (see illustration, right) and fill the last trench with the sod and soil from the first one.

Once you have dug your land over in this way, spread plenty of compost or manure on the surface. The earthworms will drag this deep down into the soil, and dig and aerate the soil for you themselves. Some old-fashioned gardeners will tell you to bury your manure or compost so as to incorporate it into the soil. This may well sound reasonable, but it is now known—and organic gardeners have proved it over and over again—that if you just lay compost or manure on top of the soil, the earthworms will dig it in for you. In a very short time it will just disappear. And the more humus you get into your soil in this way, the quicker it will disappear, because the more earthworms there will be and the more active, biologically, the soil will be. I bury manure when I plant potatoes, and I would bury manure if I took over an old, chemically worked and exhausted garden. I would never bury manure if I had dug in old sod, for the simple reason that sod is manure. I would spread manure or compost on the surface, though, for the worms to drag down.

It will pay you for the first few years—until you have put a lot of compost on your land—to dig once a year, and if you want to turn your newly created garden into a deep bed garden (see p. 106), the first year after double digging is a good time. For conventional gardening, use a fork for digging after the first year as long as your soil forms large enough clods. Just dig trenches one spit deep. Put the dug-out soil in a pile as you did before, turn the next spit over into the empty trench, and keep going until you have gone all the way around the plot; throw the first pile into the last trench (see below).

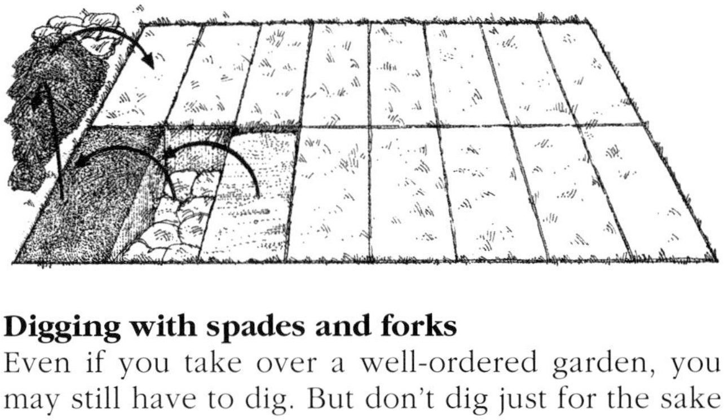

Digging with spades and forks

Even if you take over a well-ordered garden, you may still have to dig. But don't dig just for the sake of digging; dig shallowly and in moderation. Your soil will get deeply dug whenever you grow potatoes, or celery or the other vegetables that grow in trenches or on ridges. The soil needs to be loosened to some depth for root crops like parsnips or carrots, but you can do this without inverting the soil—just push a fork in and break it up. For shallow-rooted crops, a mere shaking up of the top four inches (10 cm) of the soil is enough. Time spent in what old-fashioned gardeners called "thorough digging" is wasted and in fact counterproductive.

The basic tool you need for digging is, obviously, a spade. Stainless steel spades are excellent but extremely expensive, and I doubt if the expense is justified, so long as you look after an ordinary spade properly (see p. 244).

There are two types of spades: the Celtic spade, with a heart-shaped blade and a long handle; and what I call the "Anglo-Saxon" spade, with a rectangular blade and a shorter handle that is shaped into a "T" or a "D" at the end (the "D" handle is by far the most comfortable for digging tools). A Celtic spade—so called because it is found in countries like Wales, Scotland, Ireland, and Brittany where the Celts settled—is the spade I prefer. Without too much backbending, you can work comfortably and quickly along the line of a

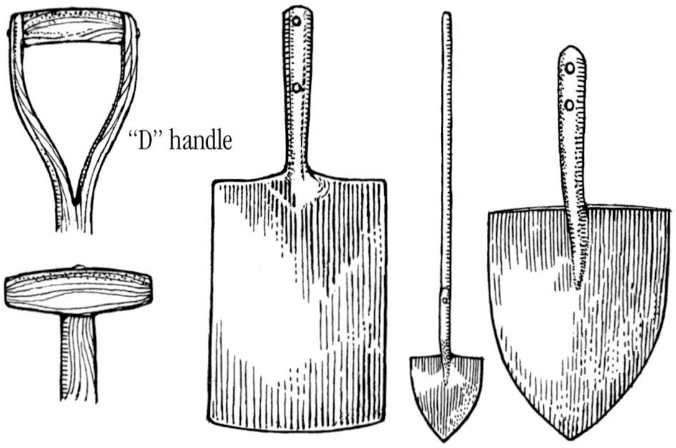

"T" handle　Common spade　Celtic spade

furrow for, say, potatoes, or you can dig a trench for, say, celery; and the Celtic spade is incomparably better for excavating holes. However, for precision digging, the Anglo-Saxon spade is best: it's difficult to dig down vertically with a Celtic spade since the blade naturally goes in at an angle. For double digging, or for digging a deep bed (see p. 106), I recommend using an Anglo-Saxon spade for turning the top spit, and then a Celtic spade for loosening the ground below.

For excavating hard soil, it's best to loosen it first with a pick, and then remove it with a shovel, which is like a spade, but larger and with curved edges. If you watch a professional gardener use a shovel, you'll see how the right knee shoves against the right hand to push the shovel into the soil.

The experienced gardener often uses a fork rather than a spade. The advantages are that it's

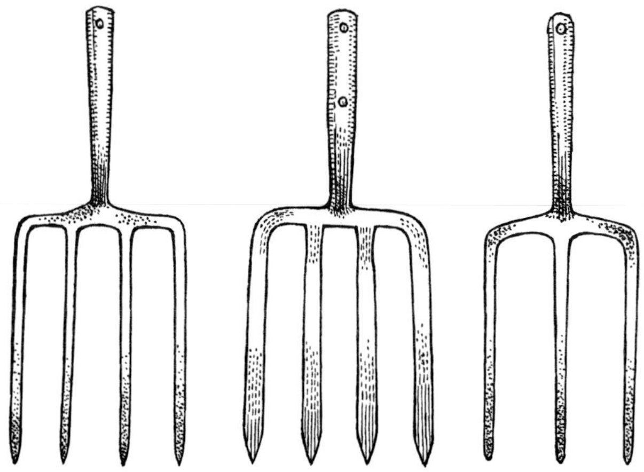

Common fork　Potato fork　Manure fork

AERATING THE SOIL
The worms will usually aerate your soil for you, but they tend to be less active in dry weather, and in land that has lain fallow for some time. In these instances, it is worth prodding your soil with a fork to aerate it.

likely to be quicker, breaks up clods better, and is easier to push into the ground than a spade. And if you are cursed with creeping root weeds such as couch grass, ground elder, or bindweed, then a fork is marvelous for raking the long roots out of the ground. It goes without saying, of course, that a fork is better than a spade for digging out root crops (a potato fork has flat tines to stop the spuds from being speared). A manure fork has three tines so that the manure will fall easily through on to the ground as soon as the fork is shaken. The only times a spade is really essential are for digging in grassland, turning heavy clay, or digging light sand.

Digging with a machine

There are two basic sorts of old-fashioned garden tractors: the kind that actually pulls a sort of plowshare along and inverts the soil as in a plowed field, and the rototiller type. The former has to be a very strong and heavy machine to be of any use, and I would only suggest it for people who have large gardens—say, half an acre (2000 sq m).

Rototillers, on the other hand, are smaller but nonetheless effective, and they accord more nearly with the ideal of the organic gardener, which is not to invert the soil too much but to leave the topsoil on top.

My own feeling about all garden tractors is that they are only really worth buying if you are going to grow and sell vegetables on quite a large scale. I have found, after most of a lifetime of doing it, that I can provide all the vegetables for a large family quite easily with hand tools alone.

Having said that, I freely admit that a rototiller will perform the equivalent of digging, the equivalent of forking, and go a long way toward breaking down clods, although raking with a hand rake afterward is generally necessary before you can plant seed. A rototiller will kill weeds, and is good at incorporating green manure into the soil. It will even reclaim old grassland for you as long as you go over the ground many times, say once a month for several months. This is because you must hit the tough grasses again and again, as they begin to recover, so that eventually you kill them and incorporate them into the soil. If you rototill old grassland and broadcast a good cover crop, like mustard, or rye in the winter, and rototill that when it is at the flowering stage, you will come close to producing good garden soil. If you take time and dig in two successive cover crops, your soil will be even nearer to perfection.

If you do decide to buy a rototiller—or rent one for a specific job, which to my mind is more sensible—you must bear in mind that there are two kinds: the kind that drives itself along by the turning of its rotor alone, and the kind that pulls itself along with driving wheels while the rotor just churns up the ground behind. The former probably does more work for less power but it works the gardener as well as the garden. Considerable strength is needed to control it. You have to hold it back to get it to dig deep.

The wheel-driven rototillers are generally bigger, more expensive machines, but they are far easier to manage; you just walk along behind them and they do all the work themselves. Some of them even have a reverse gear, so that you can back away from a hedge or path. Both kinds can be used with a range of attachments such as hedge clippers and even a circular saw.

"No-digging"

The basic principle of the "no-digging" technique is always to have at least two inches (5 cm) of well-rotted compost on the surface of your soil and to renew it from year to year, and you simply sow your seeds, or plant your plants, in that. That is all there is to it. Some no-diggers have not put a spade in their ground for thirty years and yet every year they get fine, disease-free crops—crops that are consistently much better than crops grown on chemically fed land.

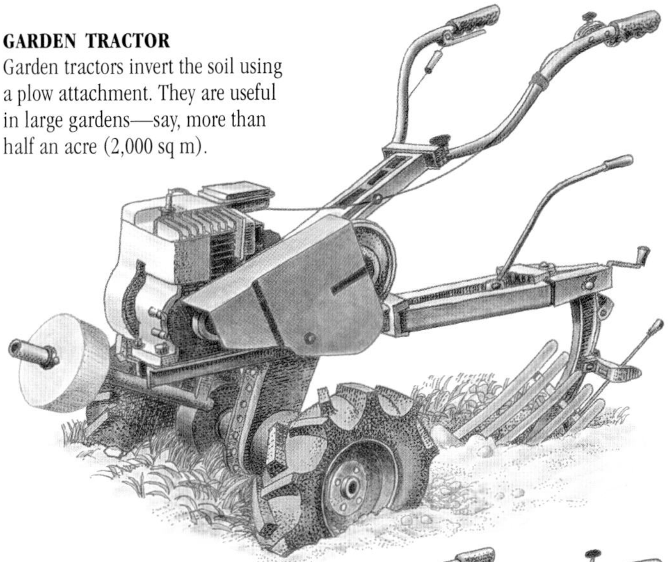

GARDEN TRACTOR
Garden tractors invert the soil using a plow attachment. They are useful in large gardens—say, more than half an acre (2,000 sq m).

ROTOTILLER
The rototiller, below, drives itself by the action of its tines. Wheel-driven rototillers are easier to use, but more expensive.

But no-diggers do use an awful lot of compost. In fact, they put so much on their soil that it is almost pure compost. All the no-diggers I know bring in organic material, in large quantities, from outside their property, because no matter what crops they grow, they cannot obtain enough organic material from their own lands to make the necessary compost. One no-digger I know gets tons of swept-up leaves delivered by the sanitation department. Another lives next door to a large flower nursery, whose proprietor is only too pleased to dump tons of what he calls "rubbish" over the fence.

I am not decrying these practices; I think they are marvelous. All gardeners should be constantly on the lookout for organic material and seize it whenever they can. But obviously every gardener in the world cannot do it; and you can be sure that if one plot of land is using this technique, it is robbing another plot of land that is either being doused with chemicals or left unproductive.

COMPOST

It is generally thought that compost is the invention of Sir Albert Howard, who first experimented with making it at Indore, in India, before World War I. But of course compost has existed ever since green plants invaded the land. Any vegetation that falls to the ground and rots "aerobically," meaning that it uses oxygen as part of the rotting process, turns into compost. (Vegetation that rots "anaerobically," or without oxygen, turns into peat and ultimately, under pressure, into coal.) But why, then, go to the trouble of making compost? Even when you pull weeds out and lay them on the ground, they rot and the earthworms pull them down into the soil, making compost. If you dig them in, the same thing happens but underground, and quicker. So why not just dig any vegetation you can find into the soil and let it rot and turn itself into compost?

The reason is that the bacteria that rot vegetable matter (by eating it) use a lot of nitrogen in the process. So when you dig vegetation into the soil, the bacteria seize all the nitrates and nitrites that are in the soil and use them to break down the vegetation. Thus they starve the soil of nitrogen to the detriment of the plants. The starvation is only temporary, for when they have finished their job, the bacteria die and release the nitrogen again, plus any nitrogen that was held in the vegetation. So you get it all back in the end. But you have to wait for it.

A far better method is to put all surplus vegetation through a compost pile. Here you supply the nitrogen required yourself (if you have it, or can get it) so that the putrefying bacteria get to work quickly to break down all the organic matter and—this is an important factor—generate a lot of heat. The compost pile should reach a temperature of at least 150°F (65°C). In fact, many a compost fanatic will take the temperature of his compost pile as a doctor might take the temperature of his patient. The heat in the compost pile is crucial: first, it kills most weed seeds and disease spores; second, it causes actual changes in organic matter and these are beneficial.

But what happened before Sir Albert Howard's invention? Well, farmers, of course, have made compost since time immemorial. They throw straw into the yard, and then let their cattle or pigs or poultry do the work. The animals produce vast amounts of dung and urine, which they trample into the straw. At that point, when the muck becomes consolidated, and therefore anaerobic, it may not rot down completely in the yard, so the farmer hauls it out and makes a "mixen" of it, which is British farmers' word for a dung-heap. This process aerates the muck thoroughly and it turns into compost, which is exactly what Sir Albert achieved with his heaps of vegetable matter at Indore. It was this muck that was the basis of good farming before the invention of artificial manures. In fact, it enabled farmers of the early nineteenth century to grow more wheat per acre than is grown as a national average even now, despite huge applications of power-derived fixed nitrogen.

But the average gardener doesn't have an animal yard, so if he wants to garden without large amounts of bought nitrogen, he must make his own compost. Plants grown in compost-rich soil grow tough and strong, and very resistant to most diseases and pests. Applied inorganic nitrogen, on the other hand, makes rapid and sappy growth that has no resistance to disease. Moreover, compost will keep the soil healthy.

Making compost

There are as many methods of making compost as there are rabid compost enthusiasts. Sir Albert Howard, for example, made a six-inch (15-cm) layer of green matter, then a two-inch (5-cm) layer of dung or manure, then a layer of soil, ground limestone, and phosphate rock, then another layer of green material, and so on to the top. He found that the optimum size of pile—which he didn't enclose in a bin of any sort—was ten feet (3 m) wide and five feet (1.5 m) high.

Another method, invented by Dr. Shewell-Cooper, who has also spent many years experimenting with compost, is to use a wooden bin, inside which the first layer of vegetation is laid directly on the soil so the worms can get into it easily. Alternate layers of vegetation and nitrogenous substances—dung, manure, and so on—are then added and eventually the whole pile is covered with a piece of old carpet. This method makes magnificent compost. If you use a black plastic sheet instead of carpet, you will get good compost very quickly. You must, however, keep the heap well watered under the plastic, because the bacteria and other organisms need plenty of moisture.

If you don't have a bin, and can afford the time and effort, it's a good idea to turn the compost. After the heap has gotten to its hottest and then started to cool down, turn it, putting the top and outsides in and the insides out. Sprinkle water on as you do this. The water and the aeration speed up decomposition and raise the temperature again. My own recommendation is that you use compost

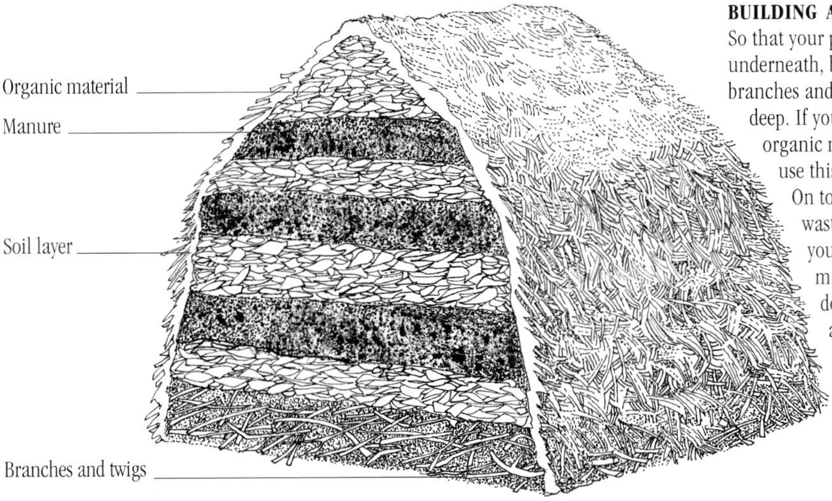

BUILDING A COMPOST PILE

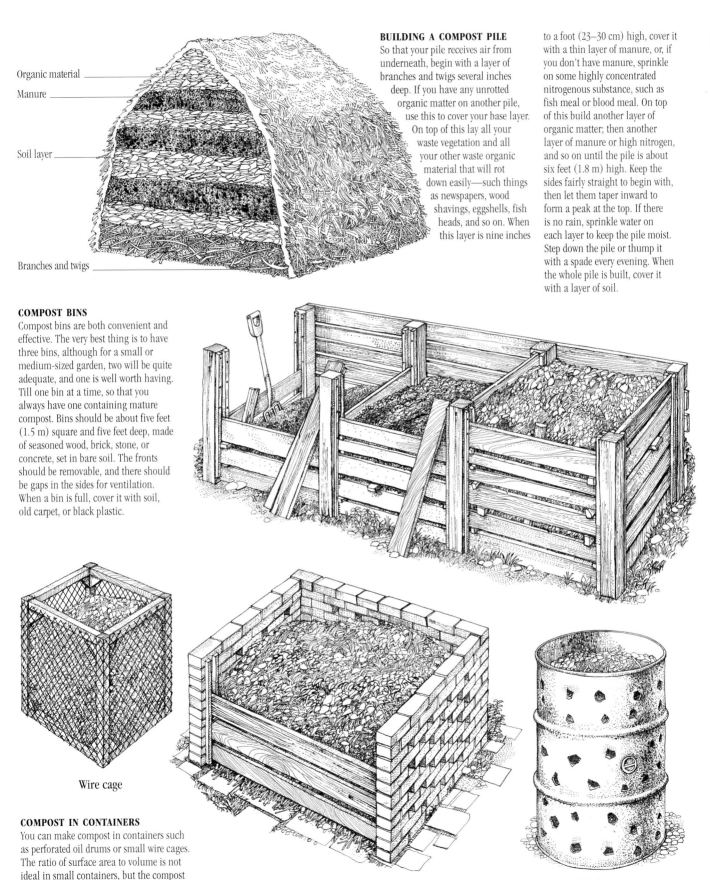

So that your pile receives air from underneath, begin with a layer of branches and twigs several inches deep. If you have any unrotted organic matter on another pile, use this to cover your base layer. On top of this lay all your waste vegetation and all your other waste organic material that will rot down easily—such things as newspapers, wood shavings, eggshells, fish heads, and so on. When this layer is nine inches to a foot (23–30 cm) high, cover it with a thin layer of manure, or, if you don't have manure, sprinkle on some highly concentrated nitrogenous substance, such as fish meal or blood meal. On top of this build another layer of organic matter; then another layer of manure or high nitrogen, and so on until the pile is about six feet (1.8 m) high. Keep the sides fairly straight to begin with, then let them taper inward to form a peak at the top. If there is no rain, sprinkle water on each layer to keep the pile moist. Step down the pile or thump it with a spade every evening. When the whole pile is built, cover it with a layer of soil.

COMPOST BINS

Compost bins are both convenient and effective. The very best thing is to have three bins, although for a small or medium-sized garden, two will be quite adequate, and one is well worth having. Till one bin at a time, so that you always have one containing mature compost. Bins should be about five feet (1.5 m) square and five feet deep, made of seasoned wood, brick, stone, or concrete, set in bare soil. The fronts should be removable, and there should be gaps in the sides for ventilation. When a bin is full, cover it with soil, old carpet, or black plastic.

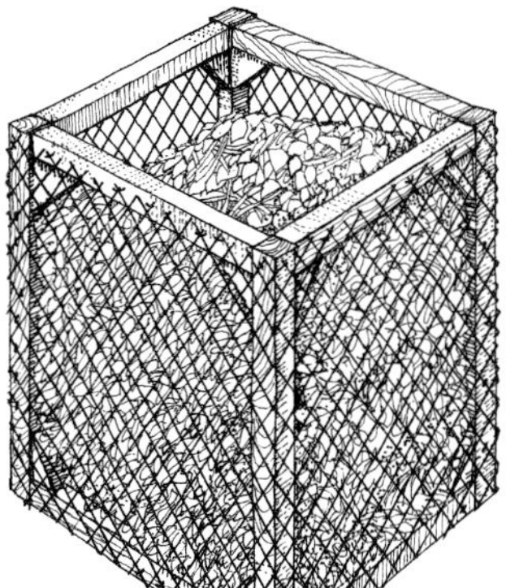

Wire cage

COMPOST IN CONTAINERS

You can make compost in containers such as perforated oil drums or small wire cages. The ratio of surface area to volume is not ideal in small containers, but the compost is certainly worth having nonetheless.

Brick compost bin

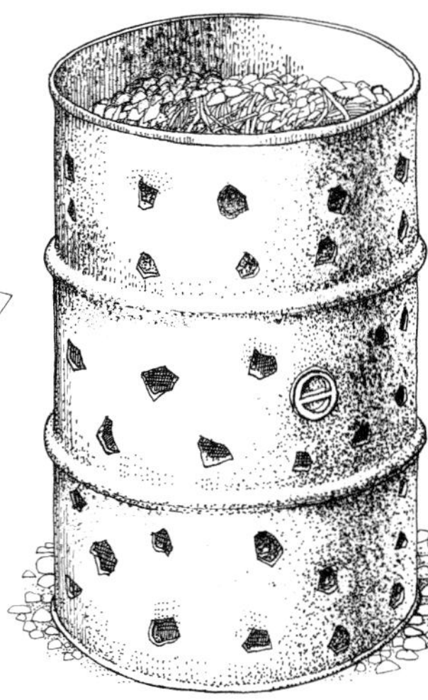

Oil drum

bins, three if possible, about five feet (1.5 m) square and five feet (1.5 m) deep. It doesn't really matter whether the bins are made of wood, brick, stone, or concrete, provided they are well-ventilated (though not too well-ventilated or they won't retain heat). If you have wooden bins, use well-seasoned wood; otherwise they will rot down along with the compost. And it's a good idea to make them so you can take out the front planks for easy access. If you fill one bin at a time, you should always have one that is ready for digging in, or spreading on your vegetable beds. In a very small garden you could use an oil drum punched with ventilation holes.

Fill the bins with a layer of any vegetable matter, say, six inches (15 cm) to a foot (30 cm) deep, then either a sprinkling of high nitrogen matter, or a two-inch (5-cm) layer of animal manure. Go on building in this way until the compost pile is complete. Water the material if it is very dry as you put it in. Don't bother to add lime, but throw in a few spadefuls of soil from time to time when you are building, since this will introduce worms and bacteria.

If you don't have manure and have to buy high-nitrogen material, then fish meal, blood meal, meat meal, or seaweed meal all do the job. As a last resort, use artificial nitrogenous fertilizer—though I've personally never yet come to that. Finally, when the bin is complete, cover with a layer of soil, a piece of old carpet, or a sheet of black plastic.

If you just can't get around to building a bin, or if your existing bins are full, then make a free-standing pile. Lay any branches and twigs you can find on the ground to a depth of, say, nine inches (23 cm). If you have it, dump any old but unrotted compost material from another pile on this, then lay organic matter on this, day after day, up to about a foot (30 cm) high. At this point, put on a sprinkling of highly concentrated nitrogenous material such as fish meal, or a two-inch (5-cm) layer of manure if you have it. Continue building like this as more organic matter becomes available. Keep the sides of the pile quite straight at first and step it down nightly. Water each layer if it is dry. When you've got as much as you want, encase the pile in soil or cover it with old carpet.

GREEN MANURE

The point of green manuring is to increase soil fertility by growing a crop for the express purpose of working or rototilling it into the ground. Alternatively, the crop can simply be cut or pulled and left as a mulch, where it will rot and be taken down by the worms. A less direct form of green manuring is to grow a crop that is then left to rot on the compost pile.

In any normal rotation, a good deal of land will be left bare in the winter and nutrients in the soil will constantly be carried away by rain, or nitrogen will be given off into the air in the form of ammonia gas. But if a green manure crop is growing on the soil—even just a crop of weeds—fixed nitrogen is immediately taken up and held in organic form so that it is lost neither through rain nor to the atmosphere. Subsequently, the nitrogen is released for use by the following crop after the green manure crop or weed cover has rotted down. And an added benefit of keeping the land covered with a crop is that the crop will provide resistance to erosion by rain or wind.

It is unfortunate that the best crops for green manuring, the legumes (which fix nitrogen in their root nodules), tend not to grow well through cold winters. It is best to use crops that produce lush green plants that will rot quickly. If possible, dig them into the ground when the plants are still young, preferably before they flower.

Of the common green manure crops, the following have specific values.

Mustard

This is much grown by gardeners in cool climates because it grows very quickly and thus does not use the ground too long. Mustard grows a good bulk. It seems likely that it suppresses eelworms, which are bad for potatoes.

Tagetes minuta

Although it is a bit too tough for direct green manuring, *Tagetes minuta* makes a fine bulk of composting material. It is particularly effective against eelworms, and it will also suppress couch grass, ground elder, and other perennial weeds. It is best to start the crop off indoors, and then plant out a foot (30 cm) apart in the spring.

Comfrey

Since it is perennial, comfrey is not a true green manure, but it is a compost plant. It sends its roots down deep into the subsoil, bringing up all sorts of nutrients. In good conditions, it will grow an enormous amount of leaf in one year, which will dry down to make one-tenth of the weight of good compost. It is worth growing a patch of it so that you can cut the leaves every year, and either put them on the compost pile or bury them green under your potatoes. Comfrey can also be used to make "comfrey tea" for feeding tomato plants (see p. 103).

Winter legumes

Winter vetches can be sown in late summer, after you have harvested the summer crops, and will grow until well into the winter and stand, if it is not too cold, until the spring, when they can be dug in. Like all legumes, winter vetches are valuable because they create fixed nitrogen.

Other legumes that will withstand mild winters include: rough pea, sour clover, Persian clover, fenugreek, crimson clover, bur clover, Austrian winter pea.

Summer legumes

Summer legumes are useful if you have a large garden, some of which you want to leave fallow through the summer. There is a wide variety of summer legumes to choose from, including sweet clover and lespedeza, both of which are much used in North America (the latter in the South); red clover, particularly used in Britain; *Crotalaria*, which is good for poor sandy soil in Mediterranean and southern climates; and cowpea, which grows almost anywhere in the summer. Annual lupines are best sown in a seed bed and planted out in the early summer a foot (30 cm) apart. Kudzu* is useful for reclaiming overworked land, but it is a perennial and should be left to grow for two or three years before being dug in, and it is only suitable in Mediterranean or southern climates. Alsike is a biennial and should be dug in during the first fall. Alfalfa, or lucerne, is a splendid crop that can put its roots down 40 feet (12 m). This means that it can really break up and aerate the soil, as well as bringing up nutrients from the subsoil to the surface. On a garden scale, you should dig it in long before it reaches maturity.

The choice of green manure crops will probably seem bewildering. What I would recommend, if you have enough land to let some lie under green manure during the summer, is red clover. For a winter green manure I would recommend a mixture of rye, which grows well in winter, with vetch. Indeed, a mixture of rye and vetch is good anywhere for land left fallow in the winter, since it increases soil fertility.

With the exception of *Tagetes minuta* and annual lupines, whose sowing directions I have already mentioned, green manure crops should be established simply by broadcasting seed sparsely on a prepared bed and then raking it in.

FERTILIZERS

In the nineteenth century, Justus von Liebig, a German chemist, made the discovery that plant growth depends on the presence of three main elements: nitrogen, phosphorus, and potassium. The discovery, however, had little immediate effect on crop husbandry at a time when farmers and gardeners had the convenience of easily available horse manure, which contains all three elements. But when motor transport replaced the horse, things changed dramatically—and, from the gardener's point of view, detrimentally. For today, the legacy of von Liebig's discovery is the utterly simplistic view that all you need do is to dose your plants with chemical fertilizers containing nitrogen, phosphorus, and potassium. The result of this has been that, although chemically fertilized crops do grow lush, their quality tends to deteriorate and they begin over the years to lack resistance to pests and diseases. So the chemists cope with this new problem by inventing all sorts of pesticides, fungicides, and bactericides. But of course they have to go on inventing them, because the pests and diseases quickly develop immunity to the poisons.

The good organic gardener doesn't have any need for chemical fertilizers. I never use them, and my gardens produce extremely high yields of good-quality crops and vegetables. As a growing number of gardeners are coming to realize, there are much better organic methods of ensuring that your land has the right amount of nitrogen, phosphorus, and potassium as well as the minor, or "trace" elements.

Nitrogen

Of the three major elements, nitrogen has a more dramatic effect on crop yields than the others. But before plants can use it, it has to be fixed, not free (as it is in the atmosphere). Nitrogen can be fixed chemically in a nitrate, but there are four things wrong with this. First, it is a very expensive process since it requires an enormous expenditure of power, and the price of nitrogen goes up every time there is a rise in the price of oil. Second, chemically fixed nitrogen causes too lush a growth and greatly weakens the resistance of plants to pests, diseases, and freezing temperatures. Third, it has been conclusively proved that excessive use of chemically fixed nitrogen lowers the quality of plants as it increases the yield. Finally, and this is the most important point, nitrogen is already being fixed constantly from the air in organic soils by nitrogen-fixing bacteria. If you add chemically fixed nitrogen, you cheat these bacteria out of a job and they fade away. So you destroy the soil's capacity to fix its own nitrogen and, instead of getting it free, have to pay through the nose for it.

Thus, I would advise the gardener to apply nitrogen only in bulky organic form: as manure; as compost; as leguminous green manure; or as the residue of leguminous plants (which fix their own nitrogen) from the compost pile (where it is legitimate to use high-nitrogen matter, such as fish meal or blood meal, to activate the pile and so release the plant foods in the composting material).

You can apply nitrogen in concentrated organic form such as fish meal, blood meal, seaweed meal, dried sewage sludge, or chicken manure, but do this only as an emergency measure as a tonic for a crop that desperately needs it. I am thinking here principally of brassica plants that have been hit by cabbage root fly and need something to get them over the bad period before they put out more roots and get a grip on life again.

The amount of fertilizer needed to correct nitrogen deficiency isn't critical. If you use cow manure, I would suggest about one pound (0.5 kg) per square foot (900 sq cm). Half this amount of poultry manure would be right; and with dried poultry manure, say, a fifth of the amount. These dressings will also improve the phosphorus and potassium content of your soil.

Phosphorus

If your soil is deficient in phosphorus, then, again, you could use a quick-acting chemical additive like superphosphate. But the organic gardener will use simple cow manure, which contains five pounds (2.3 kg) of phosphoric acid per ton. If you cannot get hold of enough cow manure, and a test shows that your garden is deficient in phosphorus, use ground rock phosphate—which may act slowly but goes on acting over many years—bone meal, dried blood, sewage sludge, or cottonseed meal. (The companies that supply these phosphate additives have to provide an analysis with them, so you'll be able to compare how much phosphorus you get for your money.) And remember that phosphate-rich organic additives have other organic side benefits. Incidentally, if you live near a steel industry center, basic slag is an excellent phosphatic mixture.

If you need to correct a phosphate deficiency in your soil, don't worry about putting too much phosphate-rich dressing on: it won't harm the plants, and it will go on doing good for many years. I would suggest about three pounds (1.4 kg) per hundred square feet (9.3 sq m).

Potassium

The third element of the trio is potassium or potash, which is specially necessary for root crops, though it improves the quality and stamina of all plants. Potassium is present in most soils, and clay usually has an adequate amount, but if you find you have a potassium deficiency then you can correct this without resorting to expensively-mined potash. Wood ash is specially rich in potassium, but farmyard manure or good compost has it as well as the other vital elements. It is common to apply greensand, greensand marl, or granite dust, each of which is an excellent source of potassium. As with phosphatic fertilizer, three pounds (1.4 kg) of dressing per hundred square feet (9.3 sq m) is about right.

Trace elements

In addition to nitrogen, phosphorus, and potassium, you may find your soil is deficient in what are called "trace elements": that is, elements that are essential in soil, though present only in minute quantities. These include magnesium, zinc, sulfur, manganese, molybdenum, and boron.

Well-composted soil is unlikely to be deficient in any of the trace elements, and, in general, a good dose of animal manure, sewage sludge, or seaweed compost should cure any symptoms of deficiency. But there are specific remedies for the lack of one or another of the trace elements. For magnesium deficiency, for instance, use Epsom salts: one ounce (28 g) dissolved in one gallon (5 l) of water per square yard (0.8 sq m). For sulfur deficiency, use sulfate of ammonia. For manganese deficiency, spray one ounce (28 g) of manganese sulfate dissolved in three quarts (3 l) of water over 30 square yards. For molybdenum deficiency, use a few ounces of sodium molybdate per acre. For boron deficiency, use one ounce (28 g) of borax dissolved in two gallons (9 l) of water over 20 square yards (17 sq m).

Soil testing

There are a number of soil-testing "kits" on the market that will adequately show up any deficiencies in nitrogen, phosphorus, potassium, and the trace elements. But many of them only give you the amount of inorganic fertilizer that the makers consider you should add to correct the deficiency. Otherwise, you can make use of the soil-testing services offered by your university extension. But probably the simplest way is to check your plants for symptoms of element deficiencies (see following table).

SYMPTOMS OF SOIL DEFICIENCY

SYMPTOM	ELEMENT DEFICIENCY
Leaves appear chlorotic (pale green or yellow color); older leaves turn yellow at the tips; leaf margins remain green but yellowing occurs down the midribs.	Nitrogen
Plants are stunted and dark green in color; older leaves develop a purple hue.	Phosphorus
Unnatural shortening of plant internodes (areas of stem between the nodes, or swellings in the case of grasses and sweet corn); leaf tips turn yellow and appear scorched.	Potassium
Older leaves turn yellow and then develop whitish stripes between the leaf veins.	Magnesium
New leaves develop whitish areas at the base on each side of the midrib; internodes appear shortened.	Zinc
Plants develop general chlorosis of the leaves.	Sulfur
Mottled effect on new leaves; in apples, a spotty chlorosis appears between the lateral leaf veins, and the chlorotic areas die, leaving holes.	Manganese
Brassica plants particularly show cupping, an inward curling of the leaves, and the leaf tips become wrinkled.	Molybdenum
Root crops, especially turnips and rutabagas, turn gray and mushy at their centers.	Boron

Lime

There is one more important element in soil—calcium, or, in gardening terms, lime. Lime is, in a way, the key element, because if the lime content of your soil is not right, then this is likely to affect the other elements.

Soil varies considerably in what chemists call its pH value, from extreme acidity to extreme alkalinity (see table, p. 90). And it is within this range that you may need to add—or withhold—lime in order to get the proper balance for plant growth.

The specific action of lime is to neutralize soil acidity. But as well as this, it has a number of beneficial effects. For instance, lime improves the structure of clay soils by causing the minute soil particles to "flocculate," or stick together in crumbs, so that the soil becomes softer and easier to work. (Clay is generally acidic, and this gives rise to deflocculation, where the soil becomes hard and impervious to water and air.)

Lime will also reduce the action of denitrifying bacteria, and thus save loss of nitrogen from the soil. It also releases phosphorus and potassium that get locked up in acidic soils. Where there is an excess of some of the trace elements, especially manganese, lime renders them insoluble, so they can't do the plants any harm. Finally, lime reduces the soil's takeup of strontium 90—which, with the proliferation of nuclear power stations, may well become a serious problem for gardeners.

But you need to be careful in applying lime, because too much of it is as bad as too little. If you overlime, you may cause deficiencies in some of the other soil elements, particularly phosphorus, manganese, zinc, and boron.

What you need to do first is to test your soil for its pH value. Simple pH testing kits can be bought at most gardening centers. And I would recommend fairly regular testing in order to check that the pH value is kept constant.

If you find you need to add lime to neutralize acidity (the optimum level is between pH 6.5 and 7), then you can get it in various forms. I prefer ground limestone (dolomitic limestone is best if you can get it), but slaked lime (limestone which has been burned and then slaked with water) is also commonly used. Chalk, too, is simply a soft limestone, and sea sand from certain coastal regions is rich in lime, which it gets from the shells

TESTING SOIL FOR pH
The simplest soil testing kits consist of two test tubes, a bottle of solution, and a color chart.
1 Fill one test tube a quarter full of soil. **2** Fill the other tube half full of solution. **3** Pour the solution into the tube with the soil in it. **4** Cork it up and shake it. **5** Allow the soil to settle, and compare the color in the tube with the colors on the chart.

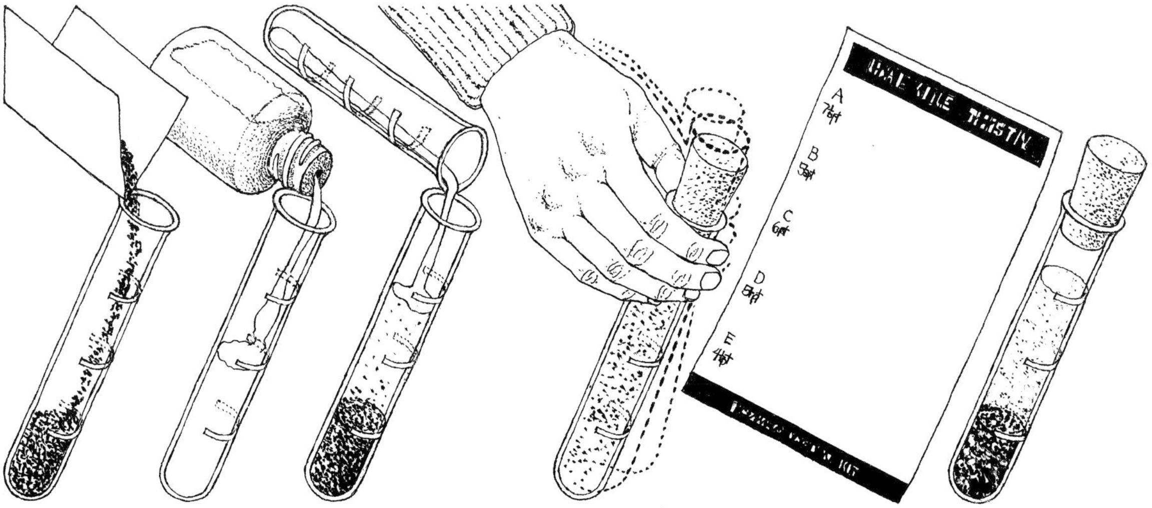

1 2 3 4 5

of mollusks. As a rule, it is just a matter of getting what is easily available, and lime isn't an expensive item. If you are really stuck, you could buy a bag of lime used for whitewash from a feed store or home improvement store.

How much lime you need to dress your soil with depends on the pH value of the soil, and on what plants you want to grow in it. Five pounds (2.3 kg) of lime spread over a hundred square feet (9.3 sq m) will increase the pH value by 75 percent. If your soil is very acidic, I would advise using this amount about every five years. Otherwise, if your soil has a pH value of, say, 5, and you want to neutralize it at 7, then dress it with two pounds (0.9 kg) of lime per hundred square feet (9.3 sq m). However much lime you need, make sure you avoid liming the soil at the same time as you add manure. It is best to leave a six-month gap. Generally, I apply lime before planting crops that particularly need it—the legumes, for instance.

RANGE OF SOIL pH VALUES

pH VALUE	LEVEL OF ACIDITY AND TYPICAL LOCATION	TYPICAL PLANTS
3.5–4	Extremely acidic. Rare, but it occasionally occurs in forest humus areas of the northeastern US	None
4–4.5	Still extremely acidic. In humid forest areas, or sometimes where there is wet, peaty soil	None
4.5–5	Acidic. Mainly in cold, damp areas	Blueberries, cranberries
5–5.5	Fairly acidic. Typical of unlimed soil in very wet climates	Potatoes, tomatoes, raspberries, strawberries, rye
5.5–6	Slightly acidic. Moderate climates with high rainfall	Grasses
6–6.5	Neutral. Moderate climates without too much rain	Most garden crops
6.5–7.5	Neutral. Hot, dry climates	Most garden crops, but not potatoes or tomatoes
7.5–8	Extremely alkaline. Semidesert areas of the western US	None

Routine fertilizing

After you have corrected specific deficiencies in your soil by direct dressing of the ground, I would advise you to add further regular but small amounts of phosphate and potassium fertilizers to your compost pile. Recent research, principally in Russia, has shown that phosphates certainly, and probably potash too, give better results when mixed with compost (or manure) since the acids in the compost help to release the elements. Also, by adding a phosphatic mixture to the compost pile, you help to fix nitrogen that would otherwise blow away as nitrogen gas. My suggestion is that you should estimate the area of your garden and add about a quarter of a pound (114 g) each of phosphatic and potassium fertilizers to your compost pile for every hundred square feet (9.3 sq m) of garden. Just buy the appropriate amount of the fertilizers every year, keep them in your garden shed, and sprinkle them into the compost pile as you build it, so that you get an even distribution throughout.

The final point I want to emphasize about fertilizers, though, is that virtually any organic manure will supply all the elements essential to soil (except, perhaps, lime). In fact, animal manures contain a great variety of chemical elements. If you keep livestock of any kind in your garden (animals or birds), you should never have any problem caused by lack of fertilizer. But remember that all organic manures should first go through the compost pile. Even human manure* is excellent, so long as it is put into the compost heap and buried under fresh greenstuff or other organic material. The heat of the compost, and the general bacterial activity, will destroy all pathogens, or harmful organisms in the manure long before you dig it into the soil. In my view, it's just a Victorian prejudice that you can't use good, honest human manure*—and there's no supply problem.

ELEMENTS OF ORGANIC FERTILIZERS (BY PERCENTAGE)

	NITROGEN	PHOSPHORIC ACID	POTASH
Bone meal	2–4	22–25	—
Fish meal	7–8	4–8	—
Dried blood	13.0	0.8	—
Hoof and horn	13–14	2.0	—
Fresh seaweed	0.6	0.2	2.0
Bracken (dried)	1.4	0.2	0.1
(green)	2.0	0.2	2.8
Tea leaves	4.2	0.6	0.4
Coffee grounds	2.1	0.3	0.3
Meadow hay	1.5	0.6	2.0
Straw	0.4	0.2	0.8
Compost from old mushroom bed	0.8	0.6	0.7
Wood ash (not rained on)	—	1.5	7.0
Soot	5–11	1.1	0.4
Farmyard manure	0.5	0.1	0.5
Horse manure	0.7	0.3	0.6
Poultry manure	1.5	1.2	0.7
Rabbit manure	2.4	1.4	0.6
Pigeon manure	5.8	2.1	1.8

Propagation

When a gardener is said to have a "green thumb," all that's meant is that he has the gift of sympathy for living plants. And where green thumbs really show up is in the various propagating processes. Put a tiny seed in concretelike clay and you'll suffocate it. Try to strike a stem in wet mud and it too will die. Let your seeds and cuttings dry out too much—or keep them too wet and short of air—and you'll undoubtedly kill them.

But if you think of your plants as living things, and treat them with the same sympathy you would any living thing, then there is no reason why you should not have a successful garden.

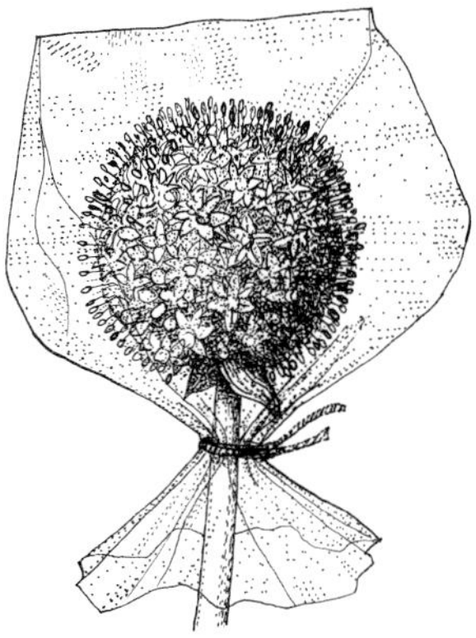

COLLECTING ONION SEED
Onion seed must be collected in the second year of the plant's growth. It is best to lift the onion plants in their first fall and keep them in pots indoors until spring. When the flowers appear, tie plastic bags around them. The seed will fall into the bags instead of blowing away in the wind.

Saving seed

Buying seed is a lot cheaper than buying vegetables, of course, but the true self-sufficient gardener will grow most of his own seed himself. If you grow a few more plants than you want for eating, and let the extra ones go to seed, then you will be able to collect the seed and go on sowing it year after year.

If you are saving your own seed, it's best to "rogue" your plants—in other words, pull out any that aren't true to type (unless, of course, they're better than type)—and use only the very best ones. It's only because for thousands of years gardeners have planted the best seed that we now have the plants we do instead of the weeds they were originally bred from.

Remember too, that since seed takes a long time to ripen, it's best to give the plants that are growing seed for you a head start by getting them off as early as possible in the year—even if you have to do it under glass. Warm-climate plants that are grown in cold climates (runner beans, for instance) have difficulty in ripening their seeds effectively in short summers unless they have help.

SEED OF BIENNIALS The problem with saving the seeds of biennials is that normally they make growth in their first year and seed in the next. Some of them "bolt," of course—that is, they shoot upward and seed in their first year—but resist the temptation to gather seed from these. Bolting beets—like bolting spouses—are no good.

Beet, carrots, parsnips, onions, turnips, and rutabagas are best lifted in their first fall, stored where it's cool, and replanted in late winter or early spring, when they will take root again, shoot up, and go to seed. Leeks I generally just leave in the ground: they shoot up taller than I am and make gorgeous round flowerheads that then go to seed. You can leave onions out all winter too, but it is safer to store them inside and plant them out in the spring. Salsify and scorzonera can be allowed to grow on for their second year and then they'll seed.

BRASSICA SEED Brassica seed, however, is far better bought; it costs very little anyway. For one thing, all brassica plants—together with that noisome weed charlock—can interbreed and so you don't know what the genetic gambler is going to pull out of the bag. For another thing, growing seed from these plants means leaving them in the ground for a very long time where they take up room and are liable to suffer from clubroot.

TOMATO SEED Tomatoes are easy to save for seed. Label a few fruits on your best (and earliest) greenhouse plants, and let them get fully ripe. Pick them, cut them open, and wash the seed out of the pulp. Then lay the washed seed on newspaper in a warm place to dry.

CUCUMBER SEED Cucumbers are more tricky, for you must help them to mate. Take a male flower (the one without a miniature cucumber behind it), cut off its petals so as to expose the stamen, and, with a small paintbrush, brush the pollen into the middle of some fully open female flowers. The female flowers stay fully open and receptive for two or three days, so, to make absolutely sure, brush pollen into them every day that they are like this. When the cucumbers are fully ripened, remove, wash, and dry the seed as with tomatoes.

SQUASH SEED Zucchini, pumpkins, melons and other gourds don't always have to be artificially pollinated. I have often just planted the seeds of squash-family vegetables that I have bought at the farmer's market and they have grown satisfactorily. It's also worth trying this with eggplants and green peppers. Leave the fruits in a warm place and let them get as ripe as possible—even to the point when they begin to rot—before extracting the seed.

LETTUCE SEED Lettuce is easy, but make sure you select the best specimens to breed from—and that does not mean the earliest ones to form seed-heads. Leave a row of good lettuces for seed, harvesting the smaller ones to eat, then rogue the ones that bolt first. Take your seed from a large and late-flowering plant; watch it pretty carefully, though, to catch the seed before the wind blows it all away. One lettuce plant will very probably keep you in seed for years.

PELLETED SEED If you decide to buy seed, it's worth considering pelleted seed. This is simply seed that has been coated with some nutrient substance, so that each individual seed is inside a little pellet of nourishment. This feeds the seed when it is wetted and starts to grow. The pellets make every seed the same size, which means you can sow such seed very easily in a seed drill. But even if you do not use a seed drill, pelleted seed has the advantage that very small seeds are easier to handle and can be sprinkled more thinly on the ground (or on seed-starting mix) than would otherwise be possible. But pelleted seed is very expensive, and sowing seed economically won't be important if you grow your own seed.

Forcing seed indoors

If you want to get out-of-season crops, or grow crops outside their normal climatic zones, you'll need to force your seeds indoors. There's nothing wrong with forcing seed, in my view: it's fun, and you get more to eat and a more interesting diet.

PROPAGATORS Seed must be forced in a propagator; the basic idea is to keep the temperature up and constant at about 70°F (21°C) and the humidity at the right level. The simplest form of propagator is just a shallow seed box covered with a sheet of glass with a folded newspaper laid on top. This is quite adequate, but it is better to have a hinged lid (preferably made of glass) with some device to keep the temperature up, and a thermometer to check it. You can make one of these yourself, or you can buy propagators of varying degrees of complexity.

To keep your propagator moist, either use a vaporizer for making mist, or, if you are using an ordinary seed box, lower it into water—about halfway up—for long enough to let the water soak up from below. Using a mist propagator is a good idea anyway, because if you douse tiny seeds, you may wash them away.

Tiny seeds must not be buried beneath the seed-starting mix. Sprinkle them sparsely on top and cover them with a layer of finely-sifted sand. Larger seeds should be covered with seed-starting mix to a depth of about three times their diameter.

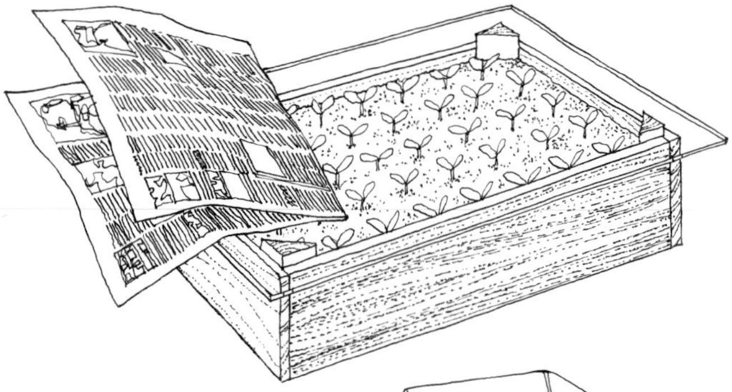

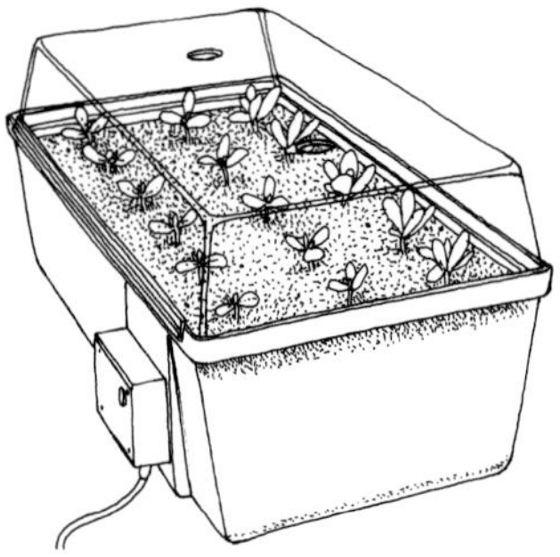

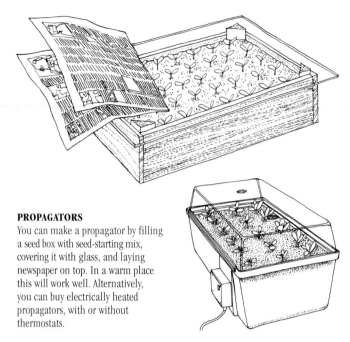

PROPAGATORS
You can make a propagator by filling a seed box with seed-starting mix, covering it with glass, and laying newspaper on top. In a warm place this will work well. Alternatively, you can buy electrically heated propagators, with or without thermostats.

As soon as the seedlings begin to show, uncover the propagator to give the seedlings light, and move it into progressively cooler and drier conditions until the seedlings are well grown (this hardens them off for planting outdoors). Don't water the seedlings until they are really well established, and then start sprinkling them from above. Since most seed-starting mixes are inert—that is, without nutrients—you should feed the seedlings with manure water (see p. 103).

Pricking out

When the seedlings are established, they should be pricked out—that is, replanted with more space between them (use a pointed stick to do this). For vegetables that don't like being transplanted, I would recommend sowing the seed in peat* pots rather than directly into the propagator. The peat pots can then be placed in the propagator. Later, when you plant them out, the peat rots down and the expanding roots push through.

Seed and potting mixes

Most gardeners buy their seed and potting mixes from the garden center or home improvement store, and this is the best thing to do unless you have the time to make your own.

If you wish, you can make your own mixes, starting with the three basic ingredients, loam, peat*, and sand. You get loam by first cutting turves of sod about four inches (10 cm) thick from a clean, well-drained pasture (try to find one with a medium-clay soil and a pH of 6.5 or less). Water the turves if they are dry, and then, in the early summer, start stacking them grass-side down. Make the stack about five feet (1.5 m) square and five feet high, and, as you build, put in alternate layers of strawy manure, mixed with ground chalk or limestone—you'll need about five pounds (2.5 kg) of one or other of these throughout the stack. When the stack is finished, cover it to keep the rain off, and in about six months the loam will have decomposed. Just cut it up and shred it with a spade. The loam now has to be sterilized, either by steam at a temperature of 212°F (100°C) or in an oven at 170°F (77°C).

The best peat* to use is either sphagnum moss peat or sedge peat. (Unless you're near a peat bog, you'll probably have to buy it.) The sand should be coarse and sharp, and it's best to use river sand.

To make seed-starting mix, combine two parts of loam to one part each of peat* and sand, and add about two pounds (0.9 kg) of superphosphate per cubic yard (0.7 cu m). To make potting mix, combine seven parts of loam to three parts of peat* and two parts of sand. To this, add either some very well-rotted compost from your compost pile, or alternatively about five pounds (2.3 kg) per cubic yard (0.7 cu m) of an additive that contains two parts by weight of hoof and horn meal to two parts of superphosphate and one part of sulfate of potash.

In North America, a potting soil known as the Cornell mix is widely used. To make this you will need two gallons (8 l) each of vermiculite and shredded peat* moss, to which you add two level tablespoons each of superphosphate and ground limestone, and eight heaping tablespoons of steamed bone meal or an equivalent amount of cow manure.

In general, of course, it's quite possible to grow many plants without seed or potting mixes at all, although they are definitely a good idea for celery and tomato seeds.

Finally, remember that when you've finished with the seed and potting mix, you can put it all into your ground to increase fertility.

Seed-beds and holding-beds

Often you will find that the land in which you want to sow seeds is already occupied by something else, so it may be necessary sometimes to sow first in a seed-bed, and transfer the seedlings later to a larger holding-bed. In fact, provided the plants are eventually planted out with care, most of them actually seem to benefit from transplanting. The idea that has grown up lately that plants should never be moved is, in my view, nonsense—as anyone who tries it will find out. In an average garden, the seed-bed can be quite small: say about a yard (1 m) square. A yard row of seedlings is actually a considerable number of plants. Ground for the seed-bed should preferably be light, dry and well drained, with plenty of peat, compost, or other organic material worked into it. I would advise raking in finely-rotted compost every year, and liming it lightly every two years to maintain a pH of between 6.5 and 7.

The holding-bed will obviously have to be bigger, since the seedlings you put in should be spaced about six inches (15 cm) apart. In general, treat the soil exactly as for the seed-bed, but here you must watch out for clubroot—because your holding-bed will almost certainly contain brassicas. Regular liming should prevent any build-up of the disease, and I'd also recommend rotating the holding-bed from time to time. There's no need to do this with the seed-bed if you put compost on every year.

ROLLERS If you're putting brassica plants and onions into the seed-bed or holding-bed, then the soil should be firm. A roller is the best solution here, and it saves time. But many gardeners get along perfectly well without one, and your boots are almost as good if you can learn that strange gardeners' dance that might be called the sideways tramping scuffle. Personally I would not put a roller very high on my list of essential gardening tools.

RAKES When you come to sow your seed in the bed, a good rake is indispensable, since small seeds need to be sown in fine tilth. The best type of rake

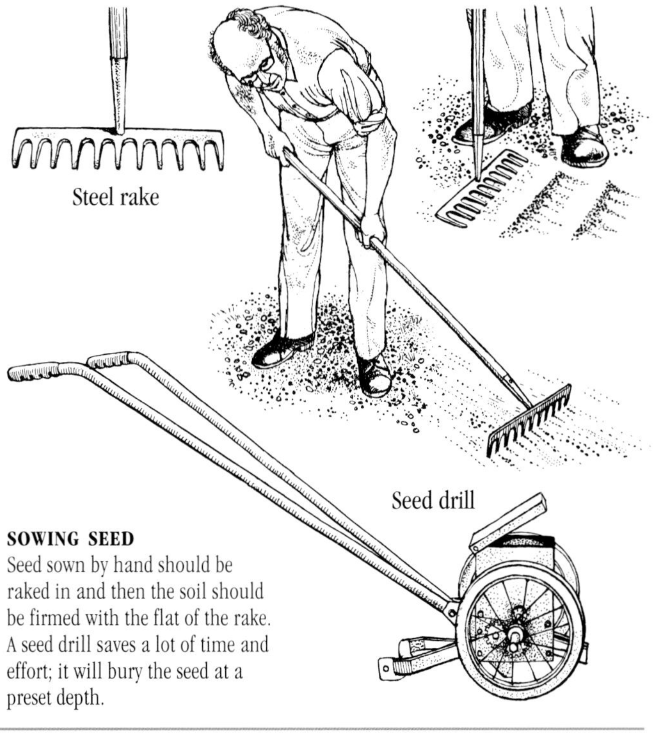

SOWING SEED
Seed sown by hand should be raked in and then the soil should be firmed with the flat of the rake. A seed drill saves a lot of time and effort; it will bury the seed at a preset depth.

is a strong steel one that is not too wide. There are cast aluminum rakes on the market that are cheaper, but after an hour of working fairly stony ground, they begin to resemble old men's smiles before the invention of false teeth. The rake is in very frequent use in the garden, so don't begrudge investing in a good one.

SOWING SEED To sow your seed, first fork over the surface lightly. Deep digging isn't necessary—in fact, keep the topsoil at the top. And only fork the bed over when the soil is dry. Next, rake the soil well, removing any stones, until it is reduced to a fine tilth. And, finally, score lines across the bed with the corner of a hoe before sprinkling seed on sparsely. Remember that small seeds should be sown shallowly: most brassica seeds, for example, need only about half an inch (1 cm) of soil or compost over them.

SEED DRILLS You can of course use a seed drill, and in big gardens they do save time—and save your back. They save you the trouble of making a seed furrow and sprinkling seed from the corner of a packet, or one by one from your fingers; instead, a seed drill lets you go along at walking pace, simply pushing the drill into the soil where it automatically drops the seed. Don't forget, though, that you can't actually see the seed going in, so if the drill happens to get clogged up, you may find that you have some embarrassing gaps when the plants come up.

A more sophisticated version of the seed drill is the precision drill, which works on the same principle but drops the seeds one by one at exactly the right distance apart. The advantage here is that you save an awful lot of seed, since otherwise, no matter how carefully you seem to be sowing it, you'll probably sow far too much and end up having to thin out the plants later. But precision drills are expensive, and many of them only work with pelleted seed anyway, so unless you're working on a market-garden scale, I wouldn't advise getting one.

COVERING SEED When you've got the seeds in the ground, rake the soil again lightly and tamp it down with the back of the rake. You should sow only when the soil is dry enough not to stick to your boots, but once the seed is in, it should be watered if no rain seems likely. Look out for flea beetles at this stage (see p. 124). A good shower of rain will disperse any that appear, but if they get bad and the weather is dry, dust the bed with derris or pyrethrum. And, above all, suffer no weeds to grow in the bed; any that do come up should be pulled out immediately.

Vegetative propagation

It is a good idea to propagate fruit trees and perennial plants, such as soft fruit bushes and many of the herbs, by taking cuttings. In this way you get a mature plant much more quickly than you do from seed.

The principle of striking cuttings is in fact quite simple: cut a piece from an existing plant, put it into the ground, and it will put new roots into the soil that eventually form a new plant. In fact, a cutting from almost any part of a plant will form another plant if it is nurtured in the right environment. The new plant is, of course, produced vegetatively. It will be exactly like its parent plant because there is no sexual crossing—genetically, it is the same plant.

HARDWOOD CUTTINGS Hardwood cuttings are taken from hard sections of the plant or tree's stem or branches. They are generally taken in the fall from the new season's growth, although with some species, such as the fig and the olive, two- or three-year-old wood can be used. Some hardwood cuttings should be buried in sand and stored indoors through the winter for planting out in spring. Others can be planted straight out in fall. Broadly, this depends on the hardiness of the species. (See the Cultivation of Vegetables, Fruit and Herbs, pp. 113–202.) The most delicate hardwood cuttings are best planted in a mist propagator (see p. 92) if they are to thrive.

SOFTWOOD CUTTINGS Softwood cuttings are cuttings taken from the tips of healthy young branches. Propagating from these is more risky, but it should work well with citrus trees and olives. Take cuttings in the spring and keep in a cold frame until they have rooted.

Both hardwood and softwood cuttings should be struck in moist, sandy ground, or in a special cutting mix made with three parts sand, one part leafmold, and one part loam.

PROPAGATING FROM CLUMPS To propagate clumping plants—these include rhubarb, globe artichokes, and shallots—you should dig up the clumps and split them into smaller ones, which you then plant separately.

PROPAGATING FROM RUNNERS Plants that put out runners—either overground, like strawberries, or underground, like raspberries—can be multiplied by first severing the runner and the new plant that forms at its end from its parent, then digging it up and transplanting it.

PROPAGATING FROM LAYERS Gooseberries are good examples of the type of plant that can be multiplied by layers. You simply bend a branch over and

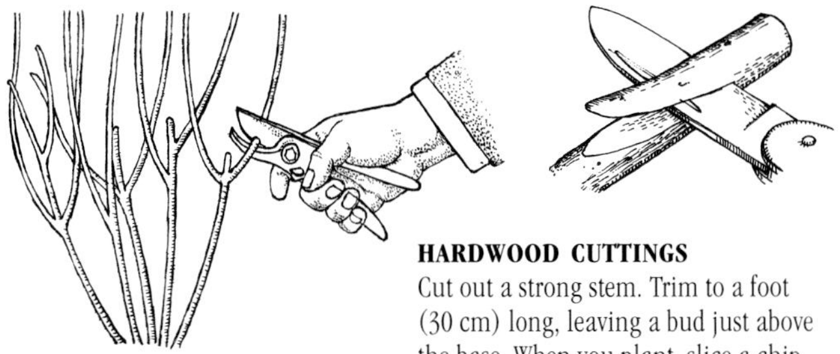

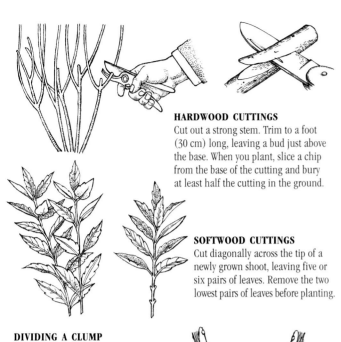

HARDWOOD CUTTINGS
Cut out a strong stem. Trim to a foot (30 cm) long, leaving a bud just above the base. When you plant, slice a chip from the base of the cutting and bury at least half the cutting in the ground.

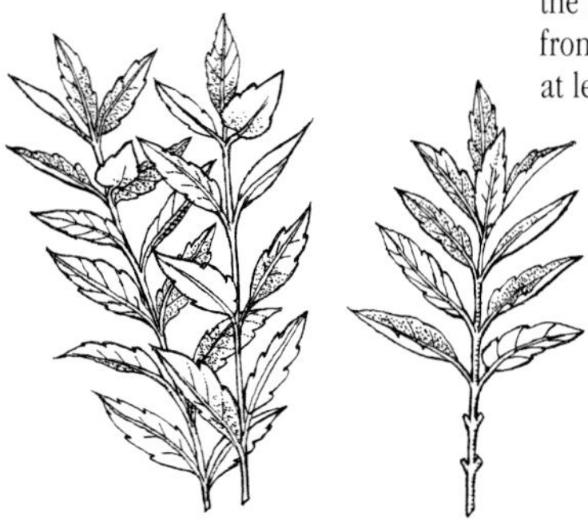

SOFTWOOD CUTTINGS
Cut diagonally across the tip of a newly grown shoot, leaving five or six pairs of leaves. Remove the two lowest pairs of leaves before planting.

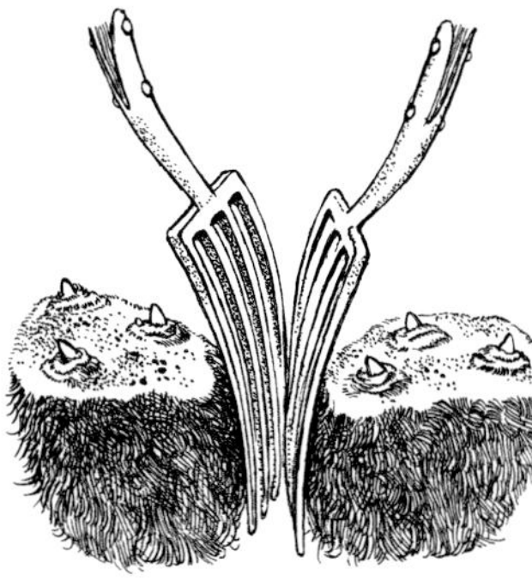

DIVIDING A CLUMP
Dig up the clump. Push in two forks back to back, and pry the clump apart between growing points before planting.

PEGGING LAYERS
Bend a branch and peg it to the ground. When it has rooted, cut through the original branch. Lift the new plant and move it to a new site.

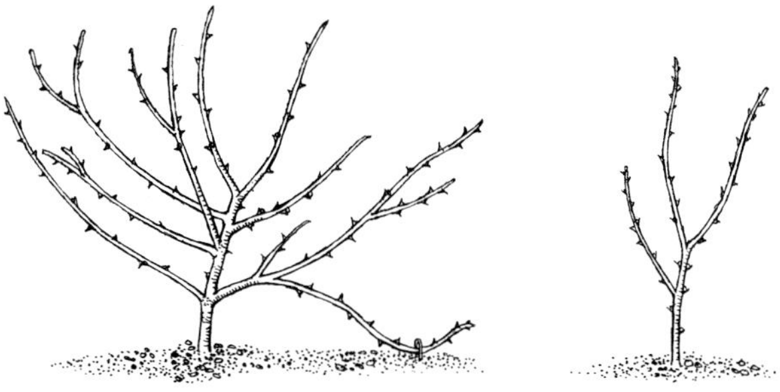

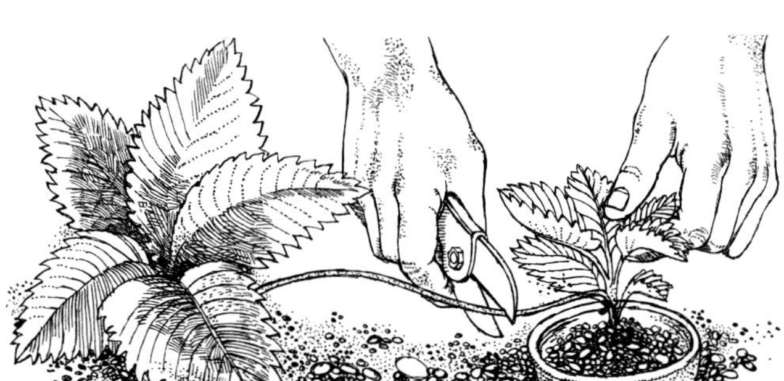

SEVERING RUNNERS
It is easiest to bury a flowerpot under a runner before it roots. Once it has rooted, cut through the runner, pull up the pot, and replant in a new site.

peg it down to the ground. The branch then sends down roots and, when they are well established, the new plant can be separated, lifted, and finally transplanted to its permanent position.

Protecting plants

Seedlings and young plants should be protected in the early spring until the weather is warm enough for planting out. There are a number of types of frames, cloches, and mini-greenhouses used for this purpose, all made from glass or transparent plastic. But don't forget the humble windowsill: a huge number of plants can be brought on in seed boxes placed on the windowsill of the average house. Or, of course, you get the same effect by putting the seed boxes on the shelves of a heated greenhouse. However, with both these methods of protecting the plants, you must remember to give the plants a chance to harden off before planting them out. I recommend a period of about two weeks in which you progressively give the plants more air and less heat. It can be fatal to take plants from a warm place and immediately transplant them outside into a cold one.

COLD FRAMES A cold frame is simply a wooden frame with a removable glass lid. It is particularly useful for hardening off your plants. Simply take the seed boxes from where they've been in the warmth and put them into the cold frame. Open the top of your cold frame on warm days and close it up at night.

HOT FRAMES A hot frame is commonly used for forcing garden crops out of season. It is similar to a cold frame except that it contains manure or compost (or both) to raise the temperature.

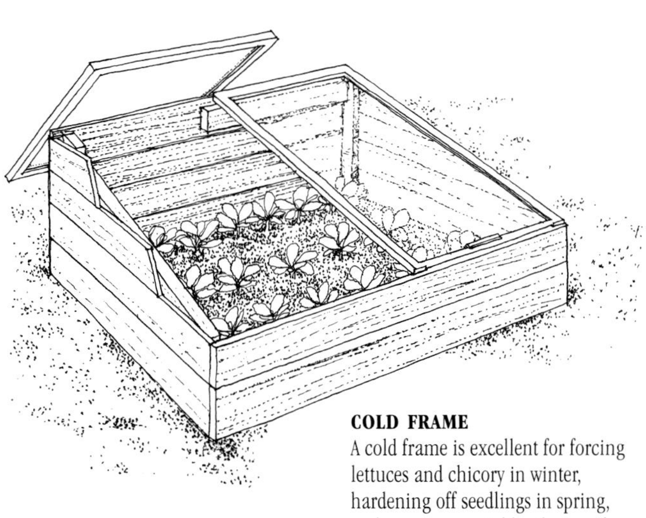

COLD FRAME
A cold frame is excellent for forcing lettuces and chicory in winter, hardening off seedlings in spring, and growing vegetables that need warmth—cucumbers, peppers, eggplants—in winter.

As a base, I think it's best to use the same compost as for mushrooms (see p. 166). Put this in the frame, and cover it with a casing of six inches (15 cm) of good loam. Make the compost up in the frame rather than outside on a heap so that you get the full benefit of the heat it generates. And after one crop has been taken out, put the spent compost on the garden and mix up another batch for the hot frame.

HOT-BEDS Nowadays there are plenty of under-soil electrical heating systems designed for forcing on early plants in frames or greenhouses. These of course all cost money and energy, both to install and to maintain. The true self-sufficient gardener will be more interested in the good old-fashioned hot-bed system of the old gardeners. It may seem like a lot of work—and a lot of manure—but it is pleasant work and it gives you an appetite to eat what you grow by it. It is basically a more elaborate and more effective version of a hot frame.

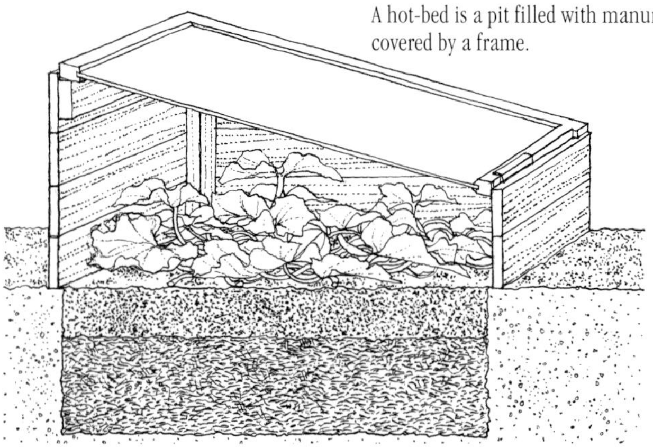

HOT-BED
A hot-bed is a pit filled with manure covered by a frame.

Dig a pit 18 inches (45 cm) deep, either in your greenhouse or where you intend to build a frame outside. Seven days before you want to plant in it, fill the pit with a mixture of one part by volume of loam to two parts of fresh horse manure. Moisten this and step it down. After three days, turn it and, if it's dry, moisten it again. After four days, "case" it: that is, add six inches (15 cm) of good rich soil. Meanwhile, if your pit is outdoors, erect a frame over it. If it is in your greenhouse, of course you don't need to bother.

When you plant or sow in the bed in spring, it will provide moderate and steady heat for as long as it takes to force your plants to maturity a month early. During the summer, you can get a good crop of tomatoes as a second crop, and after that winter lettuces, before you dig out the manure mixture and put it on the garden to increase fertility.

CLOCHES Cloches are used to bring on plants up to three weeks early in the spring, and I suggest you also use them year-round on delicate or out-of-season plants, because in this way you will derive the most benefit from them.

A good sequence in my view is to have lettuces under your cloches during the winter, early potatoes in the early spring, sweet corn, tomatoes, eggplants, and melons in the late spring, cucumbers and eggplants in the summer, and tomatoes again in the fall (take the tomatoes off their stakes and lay them flat on beds of straw under the cloches).

Cloches are made either from glass or transparent plastic. Personally I prefer plastic ones, since I constantly break the glass ones, and of course they are much cheaper. You can get them made from either hard or soft plastic. In my view the soft plastic tunnels are the best. They are readily available at garden suppliers, together with wire supports, though you can use willow wands

GLASS OR PLASTIC CLOCHES?
Glass cloches last a good deal longer than soft plastic ones and are a lot more stable. They also retain more heat at night. But plastic cloches are much cheaper and lighter than glass, and they are also unbreakable.

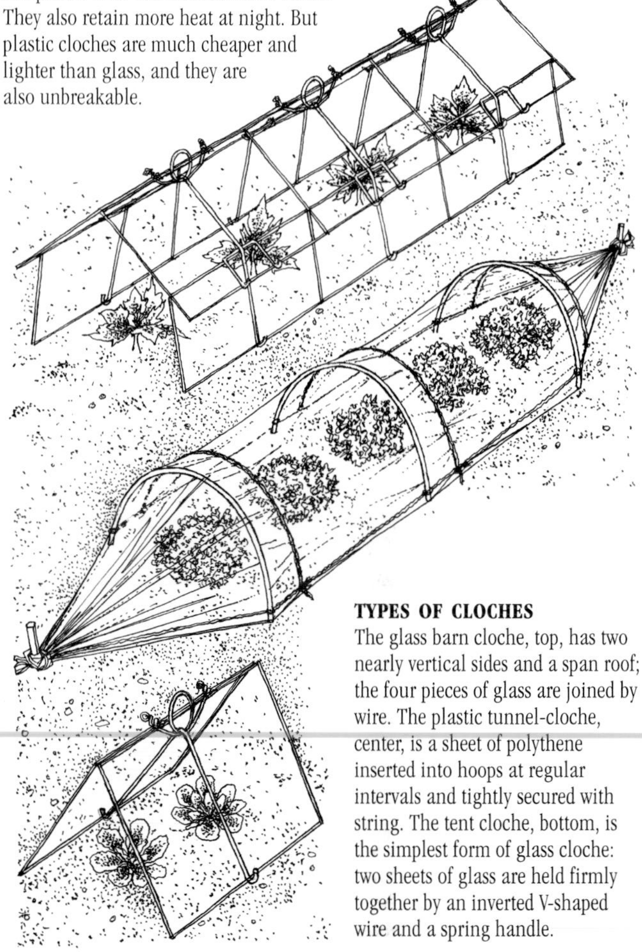

TYPES OF CLOCHES
The glass barn cloche, top, has two nearly vertical sides and a span roof; the four pieces of glass are joined by wire. The plastic tunnel-cloche, center, is a sheet of polythene inserted into hoops at regular intervals and tightly secured with string. The tent cloche, bottom, is the simplest form of glass cloche: two sheets of glass are held firmly together by an inverted V-shaped wire and a spring handle.

pushed into the ground at one end and then bent over. Plastic tunnels are simple to erect and fold away easily for storage.

MINI-GREENHOUSES The trouble with cloches is the difficulty of moving them for weeding, watering (although you can lay a soaker hose under them), keeping pests at bay, and harvesting, so I'm very much in favor of the mini-greenhouse. This can easily be tilted and propped up with sticks, or removed altogether. Instructions for building and using a mini-greenhouse are to be found on p. 111.

OTHER FORMS OF PROTECTION You can use inverted jelly jars as mini mini-greenhouses until your seedlings get too big (but remember to allow a period for the seedlings to harden off). Transparent plastic bags inverted over cut-up wire coathangers or twigs can also serve the purpose. I've managed to get new potatoes earlier than anyone else simply by laying a transparent plastic sheet over part of the potato bed. When the plants begin to grow, take off the sheet during the day and replace it at night. Keep putting the sheet on at night until all danger of frost is over. When the plants reach a height of six inches (15 cm), support the sheet on boxes or flowerpots.

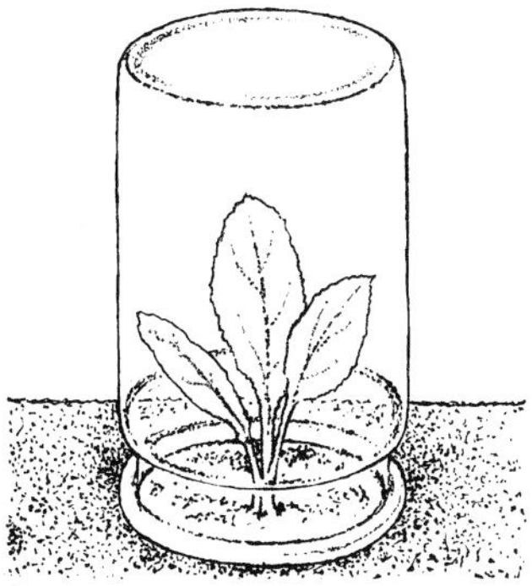

JELLY JARS
Use inverted jelly jars to protect young seedlings; prop the jars open when you want to start hardening off the plants.

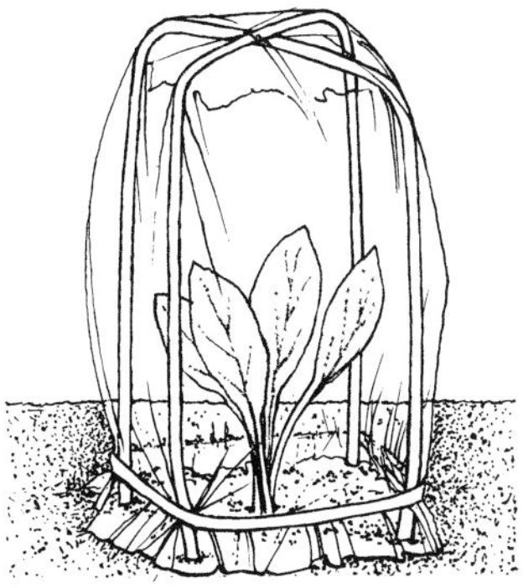

PLASTIC BAGS
A plastic bag inverted over a bent wire coathanger will do the same job as a jelly jar, even if it does not look as neat.

Starter solution

Plants may well suffer "shock" during transplanting. So it's a good idea to give them a tonic, called a "starter solution," each time you transplant.

A good starter solution can be made by quarter-filling a barrel or drum with farmyard manure (halve the quantity if you're using chicken manure), and filling it up with water. Stir it well and let it soak for two or three days. When you're ready to use it, dilute it with water until it's amber-colored and give each plant about a pint (0.5 l) of the solution immediately after transplanting; the result can only be beneficial.

Intercropping

To get the heaviest cropping from your garden, it is a good idea to practice intercropping when sowing and planting. As a rule, a small garden should be cropped as heavily as possible, though of course you must avoid crowding plants—a vegetable slum is very unhealthy. While there are several different possibilities for intercropping, I would particularly suggest the following.

Fava beans can be sown with cabbages. Space the cabbages a little farther apart than normal—say, 18 inches (45 cm) apart with about two feet (60 cm) between rows—and dib fava bean seeds into the row between the cabbages. The tall beans don't seem to interfere with the squat cabbages.

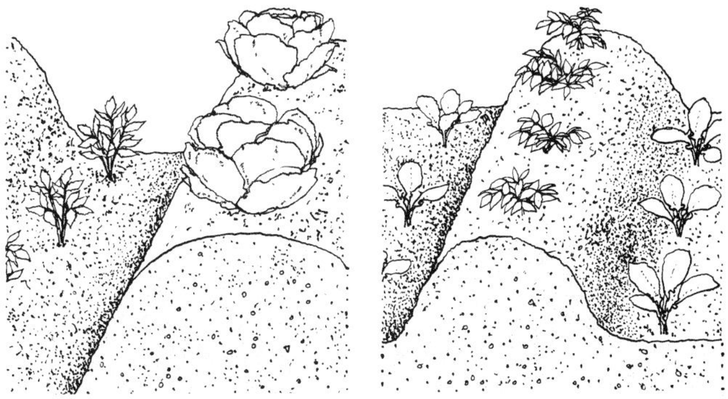

INTERCROPPING BETWEEN EARTHED-UP ROWS
After planting celery, you will have time to sow, grow, and harvest lettuces along the ridges on each side of the celery trench before it is time to earth up the celery, left. In the same way, plant baby cabbages between rows of earthed-up main-crop potatoes, right.

Winter brassicas such as Brussels sprouts, broccoli, or kale can be planted between the rows of main crop potatoes. Space the potato rows six inches (15 cm) farther apart than you usually would. When the potatoes are ready, be sure to lift them with special care so as not to loosen the soil under the brassica plants, and then ridge up the soil against the brassicas. Don't plant the brassicas until the potatoes have already been ridged up.

Lettuces can be sown or planted along the ridges thrown up on each side of the celery trench after the celery has been planted. The lettuces will have been removed and eaten before you have to earth up the celery.

Pumpkin and melon seeds can be sown in with sweet corn. The crops seem to suit each other very well, and I've always gotten heavy yields by intercropping them.

Planting a tree

As with all transplanting, putting a tree in demands your sympathy for the young sapling's needs. It is quite a complicated process, and awkward if you try to do it alone, so it is a good idea to get a friend to help.

First, dig a hole at least twice as big as the root ball of the tree you want to plant. It's often suggested that you do this a few days before the tree is to go in, but I think this is a mistake: the surface soil in the hole is liable to set into hardpan. And remember when you dig to put the topsoil in one pile and the subsoil in another. Put a layer of stones in the bottom of the hole to help with drainage. It is also a good idea to bury dead animals under newly planted trees. They provide the sapling with calcium and other useful elements released as they decompose. Shovel some manure into the middle of the hole and cover it with a mound of topsoil. Now spread the roots of the tree over the mound, cutting off any damaged ones. (In fact, a little root pruning is a good idea since it encourages the tree to put out new fibrous roots.) At this point you should drive a sturdy stake into the hole to support the tree. Get your friend to hold the tree upright as you do this.

Check that the tree is in at the proper depth: that is, the union between the scion and the root stock—the bulge or deformation at the bottom of the trunk—should be about six inches (15 cm) above the soil. In very dry climates, plant your tree six inches (15 cm) deeper. If the tree is too high, you'll have to make the hole deeper; if it's too low, increase the size of your mound of topsoil in the hole.

With your friend still holding the tree upright, replace the topsoil around the roots. (I recommend mixing the soil with compost in a ratio of three parts soil to one part compost.) Press the soil down firmly—but not so firmly that you tear

PLANTING A TREE

1 Dig a hole twice as big as the root ball of the tree to be planted. Put the topsoil in one pile, the subsoil in another.

2 Put a layer of stones in the bottom of the hole to help with drainage.

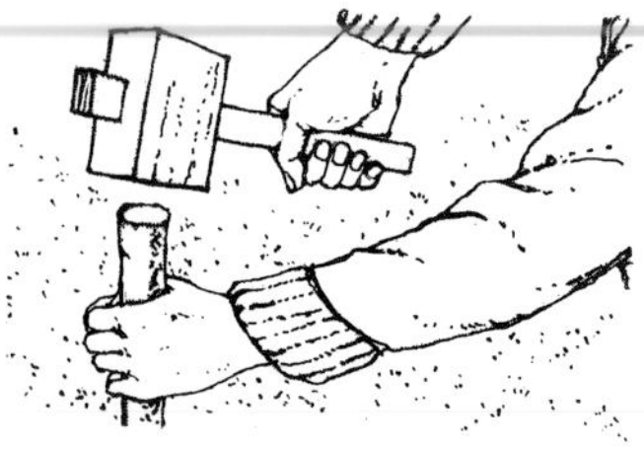

3 Spread the roots over a mound of manure and topsoil. Drive a stake in between the roots to support the tree, left. Check that the tree is at the proper depth by laying a plank across the hole and seeing where it touches the trunk.

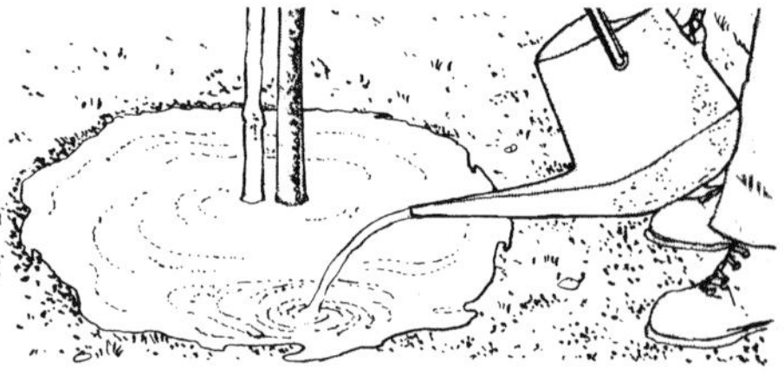

4 Replace the topsoil, pressing it down gently but firmly. Then give the whole area a really good soaking.

5 Mulch the tree with eight inches (20 cm) of organic matter, but don't let the mulch quite touch the trunk.

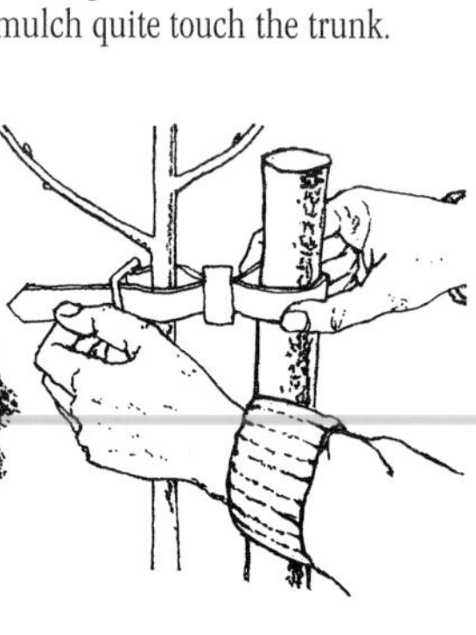

6 Tie the tree to the stake with old pantyhose, or else you can use an adjustable strap.

the delicate root hairs—and make sure you leave no gaps. Try to arrange it so that the roots lie as they naturally want to. As you go on filling the hole, gently step down the top of the soil to firm it around the roots.

Don't quite fill the hole; leave a depression in which water can stand, and then give it a really thorough soaking. Next, mulch the tree with about eight inches (20 cm) of organic matter such as old hay or straw, leafmold, manure, or compost, without letting the mulch touch the trunk.

Once you've planted the tree, tie it to the stake, but be careful about this. Don't ever use thin, hard string that will cut into the bark as the tree moves in the wind. The best thing I've found for tying is old pantyhose, but you can use webbing or leather collars, or wide braids of sisal bailer twine. And watch the tree as it grows to make sure that it is not strangled.

Grafting

The purpose of grafting is to get a fruiting spur, or scion, of one tree to grow onto the rootstock of another by bringing the cambium layers of each into contact with one another. (The cambium layer, which is the growing part of a tree, is the whitish area just under the bark.)

Most trees that are grown for their fruit are grafted, because the varieties that bear the best fruit are rarely the most hardy or vigorous varieties. Scions are therefore chosen principally for their fruit-bearing qualities, while rootstocks are chosen for their strength and their tendency to produce a tree of a given size—dwarf, semidwarf, half-standard, standard, and so on.

Since grafting is an asexual way of propagating plants, you must make sure that the scion is compatible with the root stock. Compatible pairings for different fruit trees are discussed in the Cultivation of Fruit chapter (pp. 167–190).

Grafting is an extremely old art, and several methods are now practiced. In my view, the three most important are whip grafting, chip budding, and cleft grafting. Whip grafting and chip budding are both methods of joining one-year-old scions to rootstocks that were planted out the previous year. In my experience, both methods work very well and there is nothing to choose between them.

Cleft grafting is a way of resuscitating an old or sick tree. All the main branches should be sawn off to within a foot (30 cm) of where they join the trunk. A cleft graft should then be made in the end of each sawn-off branch.

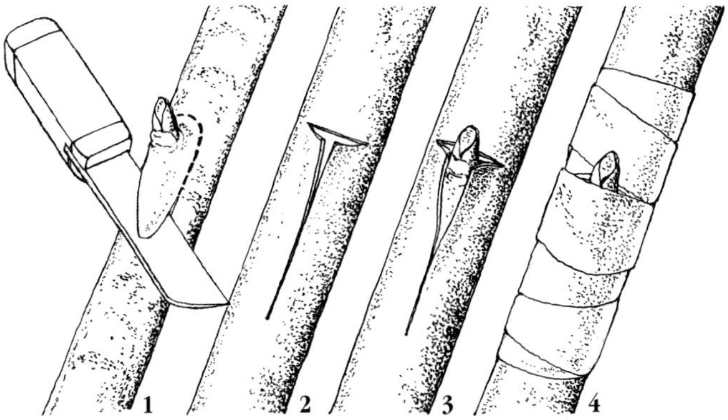

CHIP BUDDING

1 Take the scion from the current season's growth, and cut out a shield-shaped chip containing one bud.
2 Cut the root stock off a foot (30 cm) above where the bud is to go. Make a T-shaped slit an eighth of an inch (0.3 cm) deep.
3 Peel back the flaps of the slit and insert the chip.
4 Bind the joint tightly with plastic tape, leaving the bud itself exposed.

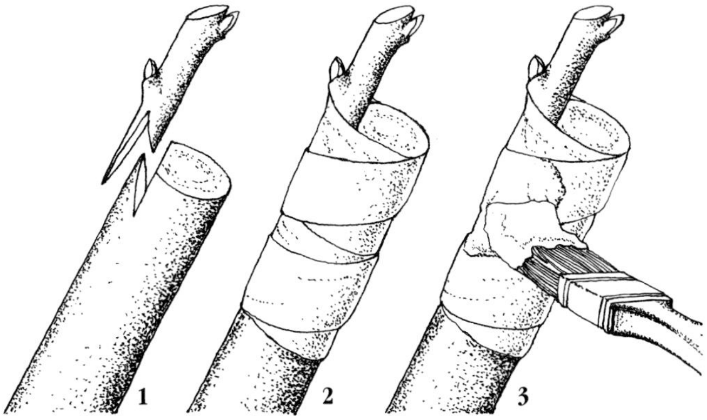

WHIP GRAFTING

Do this in early spring, using a length of dormant one-year-old scion with four buds on it. Cut the rootstock, planted the previous year, to within four inches (10 cm) of the ground.
1 Cut a notch in the top and a matching one at the base of the scion; fit the scion onto the root stock.
2 Bind the joint with either raffia or plastic tape.
3 Cover this and any other cut surfaces with grafting wax.

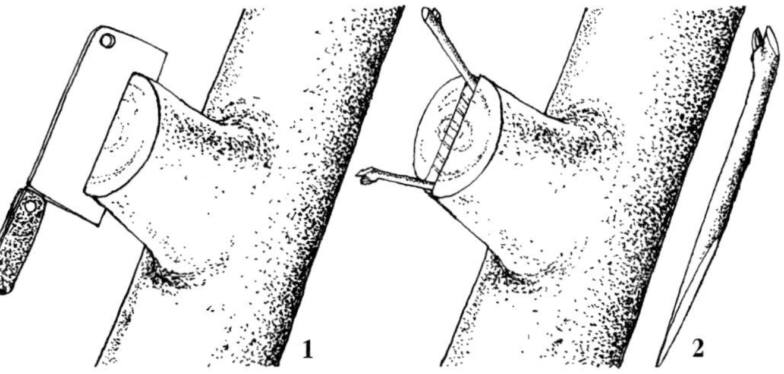

CLEFT GRAFTING

Saw through all the main branches a foot (30 cm) from where they join the trunk.
1 Take a sharp cleaver and use it on one of the sawn-off stumps. Cut two scions to a chisel shape at one end.
2 Force open the cleft and insert the two scions into this, lining up the cambium layer of scion and old wood. Let the cleft close up, thus clenching the scions into place. Pour hot wax all over to protect cut surfaces.

Care while growing

Pruning trees

There are three basic reasons for pruning trees: first, to remove damaged, diseased, and awkwardly placed branches; second, to shape the tree, for convenience of cultivation; and third, to increase the crop and improve its quality.

PRUNING UNWANTED WOOD In the first case, the general principles are to cut out any dead or unhealthy branches, and any that are overcrowded. You should also cut out any branches that point in toward the middle of the tree, and all suckers—the long, straight, vigorous shoots that will never bear any fruit. This heavy pruning should be done in late winter—though never when the temperature is lower than 20°F (–7°C). Make sure you cut out branches flush at the joints, and paint the wounds with one of the commercial tree paints, or with any oil-based paint.

PRUNING FOR SHAPE Pruning and training for shape should for the most part also be done in the late winter, and it is important to establish the general shape, or "scaffold," when the tree is still young. The scaffold is formed by the leaders, or main branches that spring from the trunk (sub-branches that grow from the leaders are called the laterals). It's best to keep the number of leaders that form the scaffold to a minimum.

How you prune will depend on the general shape you want. For example, if you want a branch to spread from the middle of the tree, cut it down to an outward-pointing bud. If you want to prevent the tree from spreading, cut it down to an upward-pointing bud and try to gather the tree together. In each case, it's important to cut to a quarter of an inch (1 cm) above the bud.

Until recently, experts have advocated the open-centered cup or goblet scaffold, where the spreading branches allow light to get to the middle of the tree. But, along with an increasing number of growers, I favor the basic Christmas-tree-shaped scaffold, called a pyramid or spindle. The reason for this is that the short branches of the pyramid shape are less likely to break if they have to carry a great weight of fruit, or of snow and ice, than the spreading branches of the goblet shape.

There are, however, several specific shapes that have advantages in certain circumstances. In small gardens where space is at a premium, I recommend shapes such as the dwarf pyramid, half-standard, and bush, or, if you can grow against a wall or fence, the espalier, fan, and cordon. Dwarf trees fruit earlier, but don't live as long as full-size ones, and they should be pruned very carefully. A dwarf pyramid must be cut back each spring so that it is no higher than you can reach. In late summer each year, shorten all branch leaders to six inches (15 cm), cutting to outward-pointing buds, and shorten laterals arising from the leaders to three inches (8 cm). Any shoots emerging from what is left of the laterals should be shortened to one inch ($2\frac{1}{2}$ cm). Do this pruning in the first week of August. In large gardens and orchards, full-size stock is better and less complicated to cultivate.

PRUNING TOOLS
A pruning saw cuts through thick branches and tapers to a narrow end that will reach awkward places. For small branches, use pruners instead.

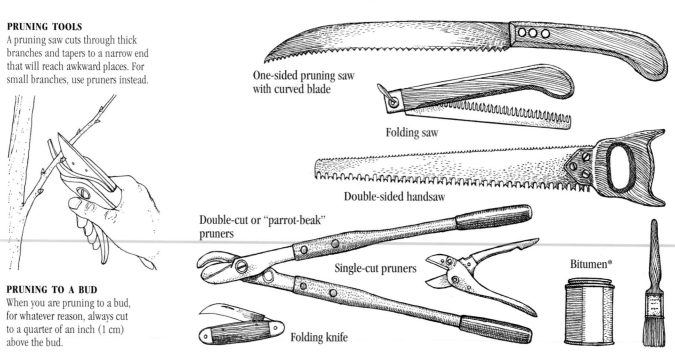

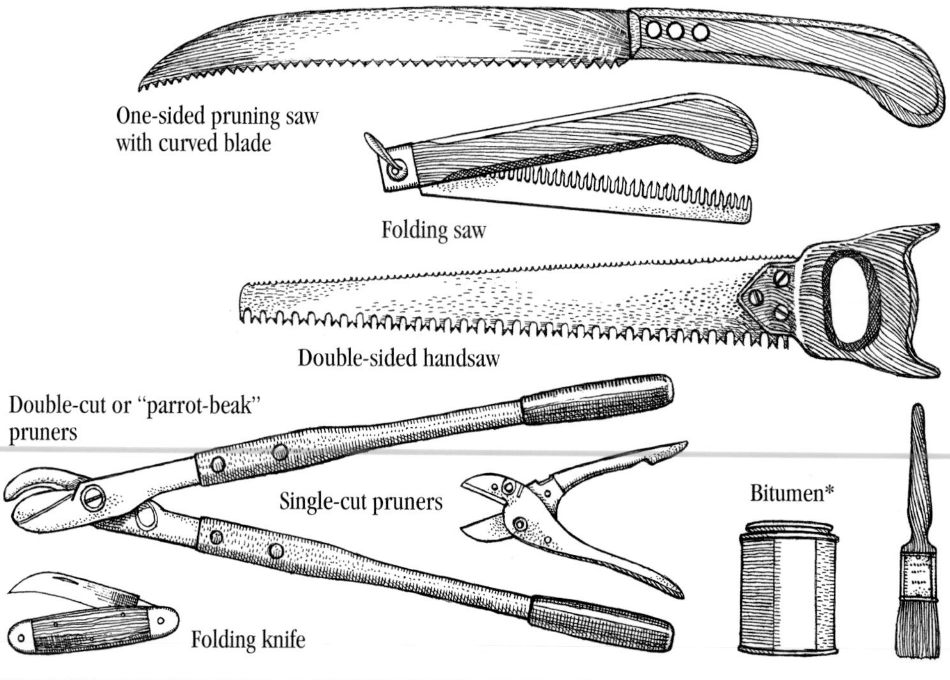

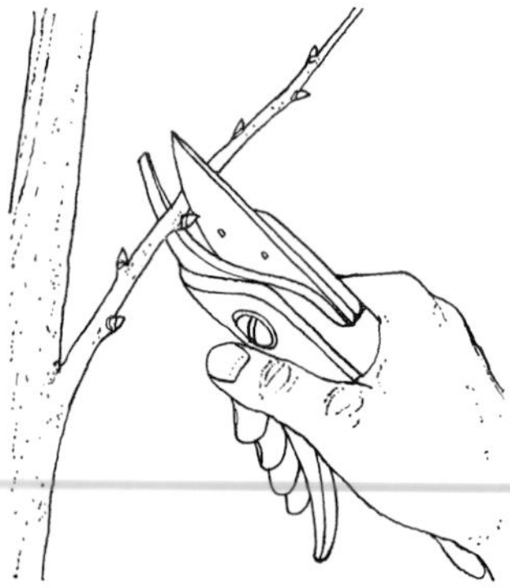

PRUNING TO A BUD
When you are pruning to a bud, for whatever reason, always cut to a quarter of an inch (1 cm) above the bud.

In general, winter pruning has the effect of encouraging tree growth. If a tree is growing weakly, then heavy pruning is advisable—as much as half the tree may be pruned away. But cut very little off if a tree is growing vigorously, because you are liable to make it get straggly.

PRUNING TO ENCOURAGE FRUITING Summer pruning has the opposite effect from winter pruning: that is, it inhibits tree growth by encouraging the fruit spurs to develop rather than the lateral branches. Trimming back the new season's growth helps the tree to fruit more heavily and earlier. If the fruit spurs are overcrowded, you should thin them out to allow the fruits to flourish. With summer pruning, never cut into old wood, and remember if you're cutting back to a bud to cut to a quarter of an inch (1 cm) above the bud.

PRUNING TIP-BEARERS Certain varieties of apple and pear trees bear fruit at the tips of their branches. Prune these very little; just cut out surplus branches. If you tip the leaders and the laterals, the tree will cease to bear fruit altogether.

TREE SHAPES

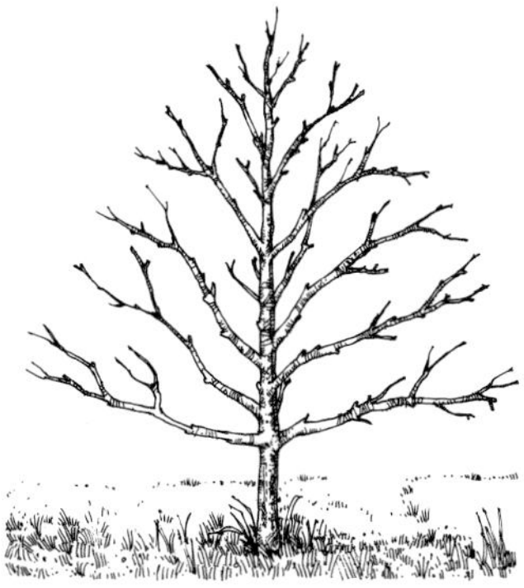

PYRAMID
This is the basic "Christmas-tree" shape. It is increasingly favored over the goblet, because the side-shoots are kept very short and are less likely to break under a heavy weight of fruit or snow.

GOBLET
The shape of the tree resembles an open-centered cup or goblet; prune so that the arms are outward-spreading and the light is let into the middle of the tree.

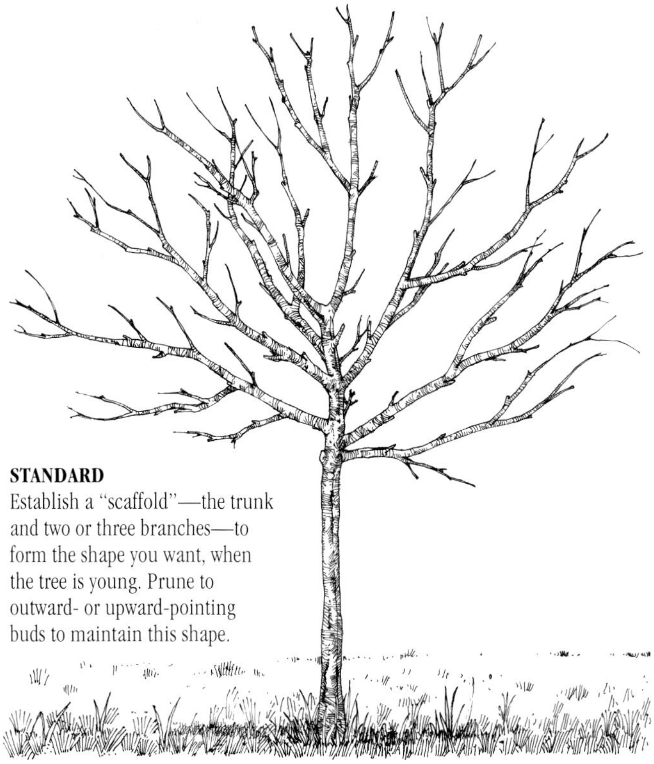

STANDARD
Establish a "scaffold"—the trunk and two or three branches—to form the shape you want, when the tree is young. Prune to outward- or upward-pointing buds to maintain this shape.

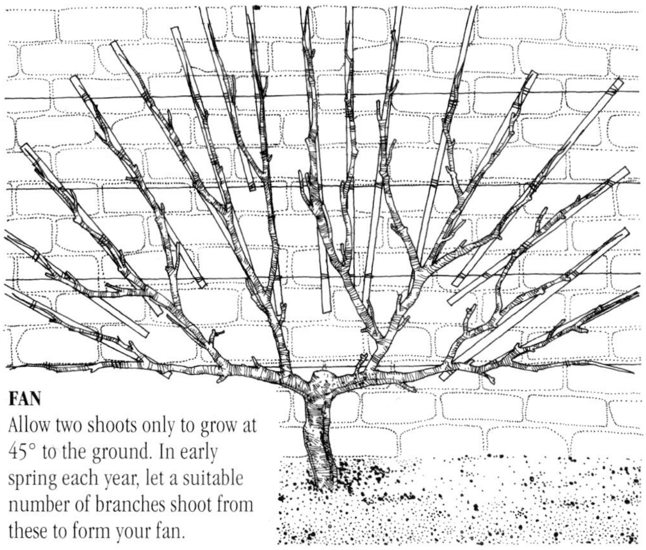

FAN
Allow two shoots only to grow at 45° to the ground. In early spring each year, let a suitable number of branches shoot from these to form your fan.

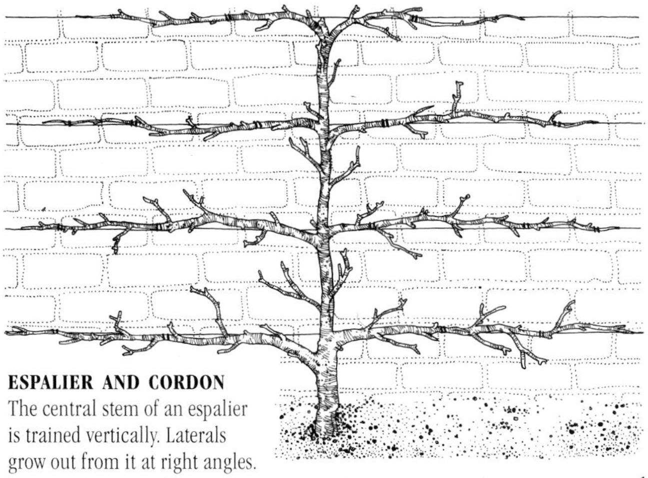

ESPALIER AND CORDON
The central stem of an espalier is trained vertically. Laterals grow out from it at right angles. Train cordons at a sharp angle to the ground and keep the laterals very short.

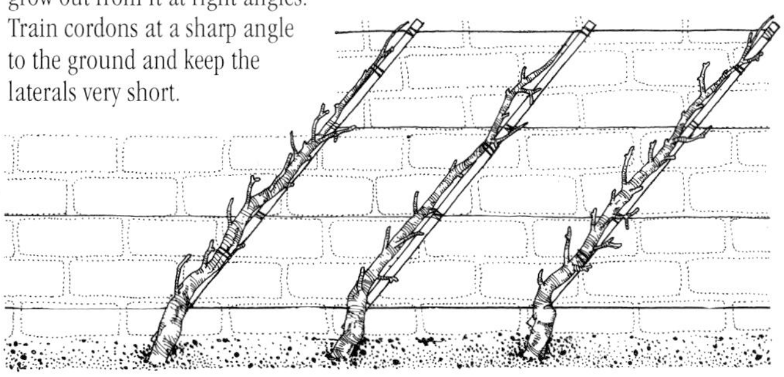

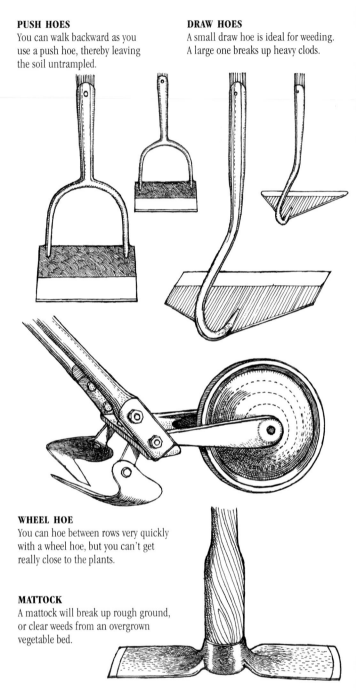

PUSH HOES
You can walk backward as you use a push hoe, thereby leaving the soil untrampled.

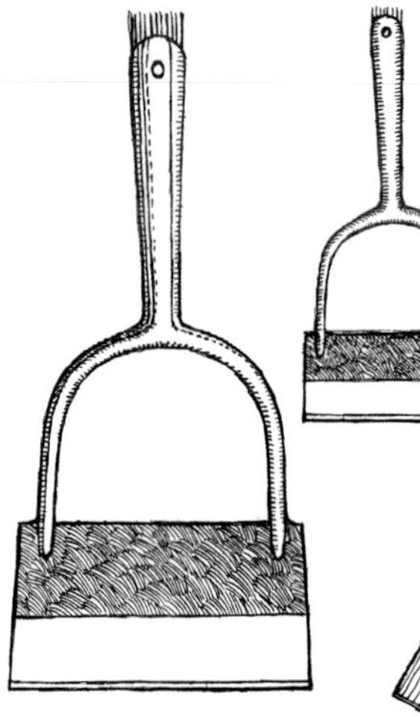

DRAW HOES
A small draw hoe is ideal for weeding. A large one breaks up heavy clods.

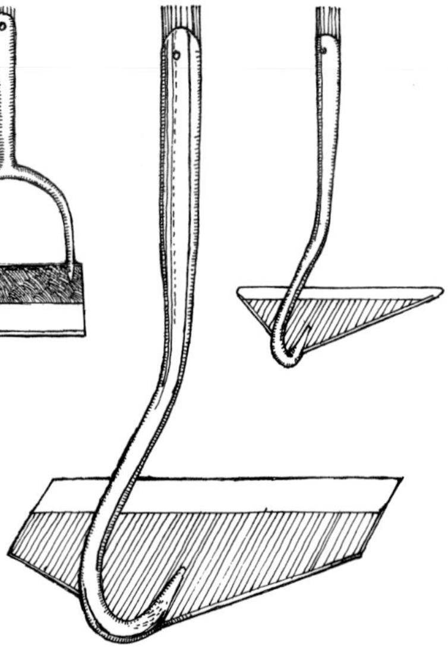

WHEEL HOE
You can hoe between rows very quickly with a wheel hoe, but you can't get really close to the plants.

MATTOCK
A mattock will break up rough ground, or clear weeds from an overgrown vegetable bed.

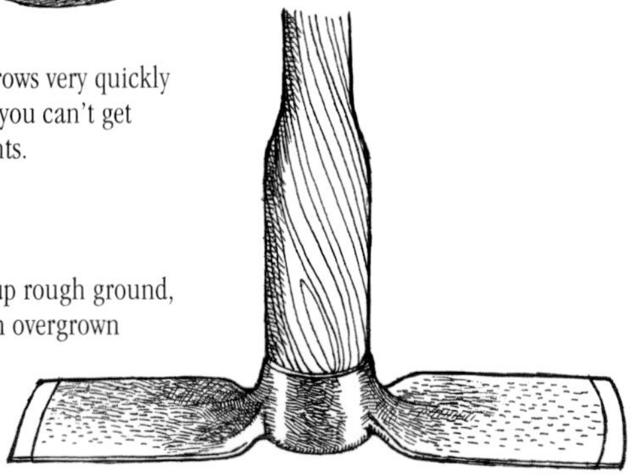

Watering

Watering is necessary where rainfall is not sufficient to maintain good plant growth. In regions with higher rainfall, you need hardly water at all. In desert climates, however, you can grow practically nothing without frequent watering. But even in wet climates, it is often desirable to water young plants that are struggling to survive in dry periods. And there are few crops that do not show an increased yield when they are watered even in rainy climates. The metabolic process stops without water, for it is water that carries every useful element, all the way through the plant's body.

Seedlings and tiny plants should be watered little and often. Larger plants should be heavily watered but not so frequently. If there is enough water to sink deeply into the ground, it draws the roots of the plants deep too, to where there is more nourishment. An advantage of the deep bed method (see p. 106) is that it enables water to percolate deeper and quicker, thus encouraging strong root development.

Weeding

Weeds are in a strong ecological position: they were evolved by nature to occupy just the places where they are found. Crops, on the other hand, were evolved partly by humankind to provide good food, and as a result their capacity for survival has been somewhat neglected. Don't believe people who tell you to tolerate weeds among your crops. No crop plants ever do their very best with competition from weeds.

On the other hand, if you leave your ground fallow and don't plant a green manure crop to dig in and improve fertility, the weeds will actually do this for you. Let them grow under such circumstances—but never let them seed. "One year's seeding is seven years' weeding." Always dig weeds in before they seed.

Hoeing is the usual method of destroying weeds. The secret is to hoe the weeds before they declare themselves, or very soon afterward. Keep using your hoe regularly. It takes little time to whip over a pretty bare piece of ground with a hoe, but a long time to hack through ground heavily infested with weeds.

HOES There are several kinds of hoes; the two most common are the draw hoe and the push, or Dutch, hoe. Personally, I've always preferred the draw hoe since I learned how to hoe from professional farmworkers—and you never see them tackling a field of sugar beets with push hoes. The advantage claimed for the push hoe is that you can walk backward as you use it and so do not trample on ground you've just hoed, but I've never found this very significant. Recently, however, I've discovered that a push hoe is quite useful for working deep beds (see p. 106).

I recommend having two draw hoes: a good heavy one for chopping up unoccupied land and loosening the surface; and a light, sharp-cornered one for whipping out weeds that grow up between plants, or for singling out individual plants from a bunch.

There are also wheel hoes—some of them beautifully crafted tools with wooden shafts. The tool bars in fact take several kinds of attachment, such as plow blades (which I think are useless, since they only scratch the surface), ridgers, raking spikes, and harrows, as well as hoe blades. It's true that wheel hoes can save time and effort, and they work quite well. I don't use a wheel hoe, however: for one thing, I find I can't get as close to plants with a wheel hoe as with a hand hoe.

MATTOCKS It's worth mentioning the mattock along with the hoe. As a sort of cross between a spade and a heavy hoe, it's good for breaking up the surface of rough ground. You can use it to make a seed-bed in a hurry, by going over the ground first with a mattock and then raking it. And if you return from a vacation to find your garden in a horrible mess, a light mattock is often a better tool to use than a hoe.

If weeds get too tall to hoe easily, you will just have to pull them. Either put them on the compost pile or leave them to rot on the ground as a mulch. If the weather is wet, the compost pile is better, because weeds left as a mulch may start rooting again.

Mulching

Mulching is the technique, much favored by organic gardeners and farmers, of covering the soil with some organic material. The benefits of mulching are that it prevents the evaporation of soil moisture by stopping the wind from blowing directly on the soil and the sun from shining on it; it prevents weed growth; and ultimately the mulch rots down and adds to the humus content of the soil. Mulches are particularly useful under fruit trees, canes, and bushes, since they suppress weeds without damaging surface roots.

Many different organic materials can be used as a mulch. Weeds that have been hoed or pulled out of the ground and left between the crop plants (or just on bare ground) make a good mulch, and so do nettles and bracken that have been dumped from elsewhere.

Wood chips, shavings, or sawdust form a mulch that is excellent for suppressing weeds, but there is one danger: when wood products rot down, they draw nitrogen from the soil in order to feed the bacteria that break down the cellulose of the material. This results in a temporary but serious loss, and you have to add extra nitrogen to do the job. Wood products also take a long time to break down and add their nutrients to the soil.

Recently, I've come across the technique of using black polythene or other inorganic substances as a mulch. I am very much against this: it seems to me quite wrong to deprive the soil life of all sun and air. Ultimately, all soil life is killed by such practices and the soil turns to sterile dust.

Mulches are, unfortunately, fine homes for slugs and snails. So, if you are mulching, you must remember to take special precautions by trapping and killing them.

Top-dressing

The practice of putting fertilizer on land once the crop has already started is known as top-dressing. It is favored mainly by inorganic gardeners, who use highly soluble substances such as sulfate of ammonia. However, these substances are quickly leached out of the soil, so, if you do use them, it is best to apply them to the crop little and often.

In my view, really fertile soil doesn't need top-dressing. The bacteria in the soil constantly break down organic substances and convert them into nitrogen that is made available to the plants. But certain very nitrogen-hungry crops, such as the brassicas, do benefit from a top-dressing of organic nitrogen (poultry manure, blood meal, fish meal), particularly if they have been checked by drought or root maggots, for example. If phosphates, potash, or trace elements are needed by the soil, they are better applied before the land is occupied by a crop than added to it at a later stage.

MANURE WATER When you apply organic manure, compost, or green manure to the land, it gets moved around by worms and rots down, so it eventually becomes available to the crop. But if you want to give a quick boost to a particular crop, you might wish to consider the technique of soaking organic material in water, thus dissolving out some of its useful constituents, and then pouring the water on the crop. Do this by putting some manure or compost in a container such as an old oil drum. Fill this up with two to three times the quantity of water, and leave for a week or so. Manure watering is excellent when the fruit is forming on crops, such as tomatoes, cucumbers, and squashes. But you can grow very good crops, consistently, outdoors, without resorting to such means, if you have built up a good organic fertility in your soil.

COMFREY TEA When made into a tea in the same way as you make manure water (see above), comfrey leaves will provide a rich, crash dose of potash for those crops that particularly need this element.

Controlling Pests and Diseases

INSECT PESTS

I would never recommend using chemical insecticides. All life is one, and as a rule anything that destroys one form of life will damage another—perhaps even your own. There are, however, certain vegetable substances that will kill or deter harmful insects without damaging your crops or, most important, your insect predators. Remember: to kill a pest and its predator is very stupid. The pest will return and rage in an even more uncontrolled way.

NICOTINE MIXTURE* If used with great discretion, a form of insecticide that is worthwhile is a solution made by boiling four ounces (114 g) of cigarette butts in a gallon (4.5 l) of water. Strain and bottle this, and, when diluted in two parts of water, it's a good standby for getting rid of leaf-miners, weevils, and caterpillars. But it is a strong poison and may destroy some useful insects too. Don't leave the solution lying around, and always wash after using it.

APHID DETERRENTS Aphids do not need the same drastic treatment as most other insect pests. A squirt of water—especially soapy water—is usually enough to knock them to the ground and they won't be able to get back on to the plant again. If aphids are a particular nuisance, then I recommend you spray with either derris or pyrethrum, both of which are obtainable at garden shops. Quassia*, if you can get it, will kill aphids but not the helpful ladybug. As an aphid repellent rather than poison, a good idea is to boil up the leaves and stems of certain plants that aphids don't attack, such as marigolds, asters, chrysanthemums, coriander, anise, and rhubarb. Make fairly strong solutions of these and experiment to see which is the most effective. Rotenone is another plant extract that can be dusted on to get rid of aphids and some other insects.

FRUIT TREE SPRAYS For fruit trees, winter wash sprays* of various kinds can be bought from garden suppliers, but it's easy to make your own. They are effective against aphids, red spider, mealybugs, scale insects and other pests that lay eggs on the tree bark. Mix a gallon (4.5 l) of light oil with two pounds (0.9 kg) of soft soap. Boil this up and pour it back and forth until it is very well mixed. Dilute this with 20 gallons (90 l) of water and spray the trees immediately.

GREASE BANDING Grease bands prevent ants and other pests from crawling up tree trunks. How to make one is explained on p. 170. In general, ants are useful in the garden, but they can be harmful when they "herd" aphids onto bushes and trees.

WASP JARS Wasps are harmful to certain ripe fruits, such as plums, and, if left unchecked, can ruin a crop of grapes. There are several wasp destroyers on the market that you put into the nest holes, but I find that hanging jars, filled with sweet stuff like syrup, in the trees the wasps are attacking is just as good. I admit, though, that you will need a lot of jars.

MILLIPEDE TRAPS An effective device against millipedes—not centipedes, which are goodies—is a number of old tin cans punched full of holes and filled with potato peels. Bury them in the soil; every few days, lift them out and put them under the faucet to drown the millipedes.

SLUG AND SNAIL CONTROL You can trap slugs by sinking some old plates or bowls in the ground and filling them with a little sugar or beer. But you'll need a lot of traps to keep the slugs down. Otherwise, try going out at night and sprinkling salt on every slug you find. The most effective way of killing snails is simply to step on them, but that always seems to me a pity—if you eat them, what was a curse becomes yet another harvest to add to all the rest.

ANIMAL PESTS

MOLE DETERRENTS I used to trap scores of moles, but I gave it up long ago. They lift up a row of seedlings now and again (which you can simply press back into the ground), but otherwise they don't do any real harm. So for most of the time I just let them burrow away.

During droughts, however, moles can be a pest when they tunnel along potato ridges where the soil is softer and full of worms. Row after row of potatoes can get lifted. A plant called caperspurge is supposed to repel moles, though the only time I ever tried it the moles just heaved it up. Alternatively, bury some empty bottles in the ground so that the wind can blow across the tops of the necks. The vibrations of the sound spread through the ground and scare away the moles, who have very sensitive hearing.

MOUSE CONTROL Cats are best for keeping mice down; otherwise mice can be deterred from eating pea and bean seed by soaking the seed in kerosene*. If you suffer from them very badly, warfarin poison is safe for all animals except rodents.

GOPHER CONTROL Gophers can be a real garden enemy. Tiny windmills that make a clicking noise are said to deter them, although I have seen gopher damage quite close to such windmills. A more effective remedy is simply to put a gopher snake down the gopher's burrow.

Harvesting and Storing

BIRD CONTROL For the most part, birds are useful since they eat insects, but bullfinches* are an exception. They will debud bushes and trees, and personally I do not have any scruples to prevent me from shooting them.

Birds can be prevented from attacking most crops with fine black thread stretched over the garden: the birds fly into the thread and frighten themselves. The traditional fruit cage (see p. 184) is effective but expensive. A good idea, I think, is to use a mini-greenhouse (see p. 111) with plastic or wire netting instead of the transparent plastic sheeting. Put this over the bed that is being attacked. It's easy to remove, or you can just prop it up while you're working. And of course you won't be duplicating equipment because you'll use the mini-greenhouse at other times of the year for its main purpose.

FUNGUS DISEASES

It's possible to garden all your life without resorting to a single fungicide spray and still get good crops. But there are times when even the best organic approach won't be enough to protect your potatoes, tomatoes or even your fruit trees.

BORDEAUX MIXTURE* In blight years, potato blight can reduce your crop by as much as half. To confine it, use Bordeaux mixture made by dissolving about half a pound (225 g) of copper sulfate in about five gallons (23 l) of water. Next, make a "cream" of five ounces (150 g) of quicklime mixed with a little water and pour the cream into the copper sulfate solution through a fine sieve. Test the mixture with a clean knife: if the blade comes out coated with a thin film of copper, add more cream to get the copper fully dissolved. If you make Bordeaux mixture yourself, use it within a day or two. Alternatively, you can buy a can already made up at a garden center.

BURGUNDY MIXTURE* For potatoes that are already suffering from potato blight, use a Burgundy mixture. This is made like Bordeaux mixture, except you should use about two pounds (0.9 kg) of washing soda instead of the slaked lime.

HARVESTING

To get the fullest flavor from your vegetables and fruits, harvest them just before they reach maturity. As a rule, with most plants, the sugars that provide so much of the flavor begin to turn to starch at full maturity. New potatoes, for example, taste much sweeter than old ones—and sweet corn is so dull when it gets old that I generally give it to the chickens. However, you can't put everything into the cooking pot just when it's right for eating, and vegetables and fruits for storing should be harvested when they are well and truly ripe.

STORING

SEEDS AND PODS It's essential that all seeds and pods are bone dry before they are stored, whether they are eventually to be eaten or used for seed. Hang them upside down by their stalks in a well-ventilated place under a roof. When seeds are dry, thresh them out by hand—knocking the plants over the rim of a barrel is a good way—and hang them up in bags made of calico or other loose-woven fabric.

STEMS AND LEAVES In all but the coldest climates, those vegetables of which you eat the stems and leaves, such as the *brassicas*, celery, leeks, spinach, and lettuces, can be left in the ground until required. In the coldest climates, the brassicas, celery, and leeks can be stored in a cool cellar.

VEGETABLE FRUITS Members of the squash tribe are best hung in nets indoors at a temperature around 45°F (7°C). Green tomatoes should be stored at 60°F (16°C) in a drawer or some other dark place. Peppers can be dried indoors or out. Dry them on their vines and hang them up until you want them.

STORING ROOTS AND TUBERS
Bury a large-diameter section of drainpipe—say, 18 inches (45 cm) long—in well-drained soil. Leave a few inches exposed at the top and put small stones in the bottom for drainage. Fill it up with any root crop and cover it with a wooden lid. In frosty weather, lay straw on the lid and weigh it down with a stone.

ROOTS AND TUBERS For potatoes and other roots, and tubers a root clamp (see p. 136) is best, but on a smaller scale you can store roots in a pipe buried in the ground. Carrots and beets can be stored indoors in containers filled with sand. Potatoes can be stored indoors at 45°F (7°C). Keep them in the dark, otherwise they will turn green. Other roots and tubers can be stored indoors at about 37°F (3°C); they must be well ventilated.

The Deep-Bed Method

It would be unthinkable to publish a book on gardening for self-sufficiency at this time without fully describing the new method of gardening—or rather the very old method now revived—called variously the Chinese Method, the French Intensive Method, the Biodynamic/French Intensive Method or, by some of its practitioners in North America, just the Method. The word "biodynamic" applied to gardening is, of course, tautologous, because all growing things operate biodynamically: that is, they live and they move. I call this method, quite simply, the Deep-Bed Method, because this describes it exactly.

In the nineteenth century the French "maraichers," or market gardeners, were working as close to Paris as they could get on small patches of expensive and scarce land. However, they had unlimited supplies of horse manure—for at that time Paris moved on horses—and they developed a system of gardening of a productiveness that has never been surpassed. It is not surprising that Chinese gardeners, also working near cities and therefore compelled to produce as much as they could off a limited amount of land, arrived at the same solutions as the French did.

Alan Chadwick, an English actor who studied gardening first under Rudolph Steiner and then at Kew, started experimenting with deep bed cultivation in South Africa. He moved to California in the 1960s and established a four-acre (1.6-hectare) organic garden using this method at Santa Cruz University. Having established this he moved to the Round Valley in northern California, where he now runs a seven-acre (2.8-hectare) garden and has 60 students working with him. It was Chadwick who coined the name Biodynamic/French Intensive Method, using the word biodynamic because it relates to Rudolph Steiner's approach.

Meanwhile, several Chinese immigrants to the US had also been practicing the Deep-Bed Method and one of them, Peter Chan, wrote a book about it: *Better Vegetables the Chinese Way*. Give or take a few inessentials, the two methods are the same.

Digging a deep bed

The method is this. Drive four posts in at the four corners of your proposed bed and put a string around them. The bed should be five feet (1.5 m) wide and as long as convenient, but remember that to make it too long is to give yourself a long walk to get around it because you never step on it. Twenty feet (6.1 m) long is about right; this gives you a hundred-square-foot bed,

SPREADING MANURE
Before you start to dig, lay a good covering of manure all over the top of the bed.

DIGGING THE FIRST TRENCH
Starting at one end of the bed, dig a trench a spit deep. Put the soil in a wheelbarrow.

LOOSENING THE SUBSOIL
Dig your fork deep into the trench and wiggle it around as you loosen the subsoil.

which is convenient for making calculations about yields and so on. (The people who have been researching the method so far have been using the hundred-square-foot bed as a standard for calculations and comparisons.)

Lay a covering of manure on top of the proposed bed. The digging is basically double-digging, but you must be sure to loosen the subsoil. Take out a trench a spit wide and a spit deep at the top of the bed. Dig your spade or fork into the bottom of the trench and work it around so as to loosen the subsoil as deep as you can. Dig out a second trench next to the first one and throw the topsoil, and the manure that lies on it, into the first trench. Work the subsoil in the bottom of that too. Move on to the third trench and throw the topsoil into the second trench. Continue in this way until you reach the bottom of the bed. Then throw the soil you took out at the top of the bed into the empty trench that will be left at the bottom of the bed. The bed is then well and truly dug.

You can of course split the bed in two down the middle (see p. 81) and then you don't have to barrow the soil from the top to the bottom.

Thereafter—and I must repeat even to the point of tedium because it is the key to the whole matter—never step on the bed nor let anybody else step on it until you come to fork it over the next year again.

John Jeavons, another Californian deep-bed practitioner, has written a very good little book about the method (*How to Grow More Vegetables than you ever thought possible on less land than you can imagine*) and has carried out very careful controlled experiments for four years at Palo Alto. He estimates that it takes from six to ten hours to dig a hundred-square-foot bed for the first time. He believes in double-digging his deep beds every year, and finds that after the first year this does not take more than six hours, because the texture of the soil has been so improved since it has not been stepped on.

Peter Chan does not recommend digging again after you have done it once, and my own experience tends to make me agree with him: provided you put on plenty of manure or compost every year, and fork the land over once, the roots and earthworms will ensure that the subsoil does not get compacted again, and it is the compaction of the soil that inhibits plant growth. I find I can fork over a hundred-square-foot well-established deep bed, one spit deep, in ten minutes and it is light work.

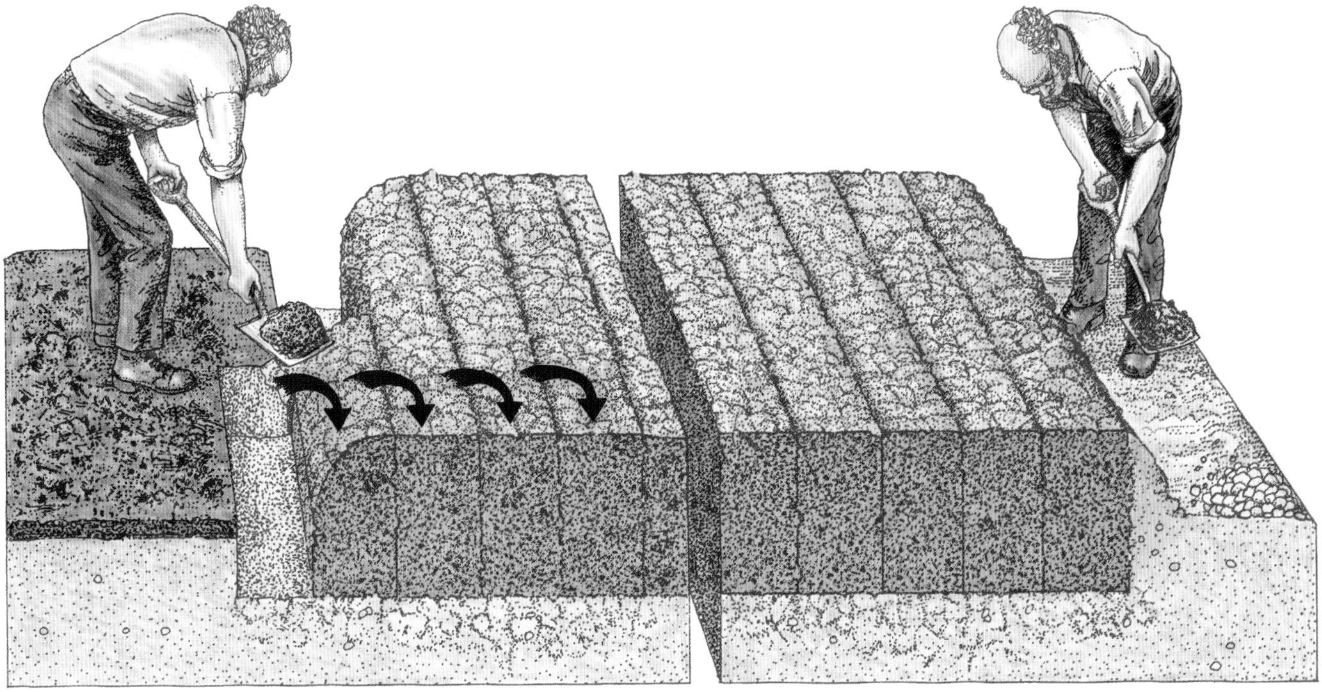

DOUBLE-DIGGING THE DEEP BED
Dig a second trench next to the first one, throwing the topsoil and manure into the first trench. Work the subsoil in the bottom of the second trench. Dig a third trench and repeat the process.

MAKING THE PATH
As you dig, throw all pebbles and stones to the side of the bed. When you finish, throw the topsoil from the path-to-be back on top of the bed. Spread the stones evenly over the surface of the path.

If you have several beds side by side, leave narrow paths between them. These paths are never dug, simply walked on. Some people leave their paths very narrow, but I find this makes it difficult to maneuver a barrow between the beds. So I make them 18 inches (45 cm) wide. You may feel you lose land by having all these paths, but the much closer spacing of the deep bed makes up for this loss of land and in fact you lose a strip of land almost as wide as this between every row in conventional gardening. Also, as the years go by, your deep beds will get more and more convex until they stand perhaps 18 inches (45 cm) above the paths.

As you dig your deep beds, throw any stones you find in a pile on one side. Then, when you make the paths, throw the topsoil from the path on the bed and scatter the stones in their place.

Some deep-bed gardeners build small walls of brick, stone, or timber around their beds to hold the sides up. In my view such arrangements only harbor slugs and are not necessary, because the beds do not, in practice, erode. Ordinary good organic practice dictates that the soil should be covered for most of the time by a crop, even if it is just a green manure crop, and this will hold it together. In any case, well-manured soil will not erode even if it is left bare in high beds.

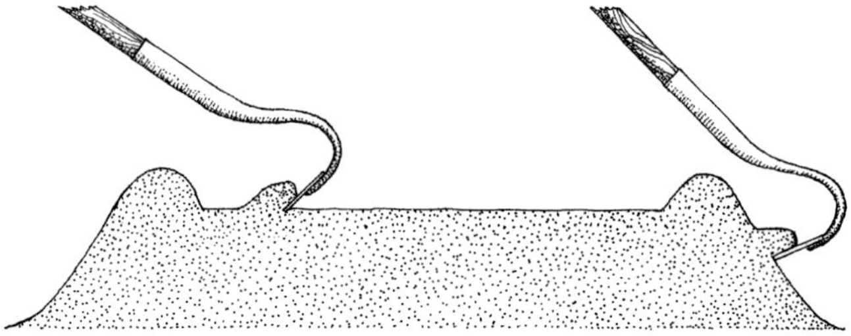

DEEP BED IN DRY AREAS
In very dry areas it is worth shaping the beds with a draw hoe, so that they rise steeply at the sides and form a basin shape at the top. The moisture is then conserved when the beds are watered. In wetter regions the looseness of the soil in the deep bed is enough to ensure that rain or spray percolate into the soil at once.

Sowing and planting in a deep bed

Now, in your newly dug bed, either plant out plants from a seed-bed or seed boxes, or simply sow the seed direct into the ground, just as you would with a normal bed. The difference is that you sow or plant four times as densely, or thereabouts, as you do when gardening in the traditional way. The reasons for this are explained in detail later, but have to do with the fact that you never walk on the soil between the rows so that the soil remains loose and uncompacted.

You do not sow or plant in rows with wide spaces in between. Instead, you work to a triangular pattern so that your crops grow in clumps. The overall effect is of very closely spaced diagonal rows. And in almost all cases you should allow much less space between the plants in all directions than you would between plants in traditional rows. The basic objective is to space the plants so that their leaves are just touching when they are mature.

You do not need the normal spaces between rows, because you never walk between the rows. The soil is loose and untrodden, so the roots of the plants can go down straight and deep—when you pull a plant grown in this way, you will be amazed by the size and length of its roots—and therefore the plants do not need nearly so much space at the surface.

The fact that the plants' leaves just touch when they are mature means that they create a mini-climate that conserves moisture in dry weather. You should find that you use from a quarter to a half of the water you use in conventional gardening. Weeds are of course suppressed by this close planting; you can hoe gently from the sides of the bed before the plants meet each other or, even better, just hand-weed. Out of this soft deep soil weeds come out so easily, roots and all, that there is no trouble getting rid of them. Weeds are really no problem with deep-bed gardening.

An obvious question is—what about crops like brassicas and onions, which gardeners have always believed need firm soil? What about all those exhortations to stamp and dance around on beds before planting them? Well, in husbandry only one argument is of the slightest weight and that is experience: what actually happens. I have grown crops of brassicas by this method; the vegetables have been magnificent and the yields have been extremely impressive.

All that you do when you plant out brassicas or onions is plant considerably deeper than you would normally and then press the ground fairly firmly down around the plant with your hands. Only the top few inches of the soil are thus compressed but in practice it seems to be enough. This does not, in this very loose soil, make onions "bull-neck"—in any case you can gently remove the soil from around the onions later on.

The deep-bed practitioners favor frequent transplanting of plants before they are put out in the beds, but they always plant into, if possible, better and looser soil than the plants were in before. Thus, if you prick plants out from a seed

CHICKEN WIRE FRAME
A frame with a one-inch (2.5-cm) mesh of chicken wire will help you space seeds and seedlings accurately. Plant through the centers of evenly spaced hexagons.

CREATING A MINI-CLIMATE
If the seeds have been correctly spaced, the leaves of the plants should just touch each other when they are mature. This creates a mini-climate that conserves moisture in dry weather.

SPACING OUT SEEDS
Sow seeds in the deep bed in a triangular spacing pattern, with each seed the same distance from those surrounding it.

USING A BOARD
If you dislike stretching to the center of the bed, squat on a piece of board five feet by three feet (1.5 x 0.9 m). Your weight, evenly distributed, will not compact the soil.

box into another seed box to give them more space, make sure that the soil in the second box is at least as good, preferably better, than that in the first box.

The spacings for sowing and planting different crops are given in the chapters on the Cultivation of Vegetables and Fruit (see pp. 113–190). These spacings should be used as a guide only; every person should experiment for himself and use his common sense. After all, how big is an onion? How big is a carrot? Everybody knows, and provided the tops of the plants are given sufficient room, their roots will be all right because there is plenty of room below a deep bed.

With conventional gardening the roots cannot penetrate the compacted soil below, and must spread out laterally, where they compete mercilessly with each other, and get cut and damaged every time you walk near them. With deep-bed gardening, the roots go way down into the loose soil, with nothing to stop them, and there is nothing to damage them when they are there.

Deep-bed yields

Careful records have been kept, at Santa Cruz University in California and at other places, and the deep beds have been found to yield, quite consistently, four times the crop produced by conventional gardening.

I, like many other gardeners in Britain, had read these figures but did not fully believe them. So I took the trouble to go to California to see for myself, and spent five weeks there searching out every example of deep-bed gardening that I could find. Seeing is believing and in this case I am completely convinced of the superiority of this method. Four times the crop is about right and I never saw a case of this gardening being practiced correctly that did not, more or less, bear out this figure. On the strength of this I returned to Wales and tried it for myself, and I have now proved, through personal experience, that it certainly does work.

Deep beds for perennial vegetables and herbs

Obviously you cannot dig perennial beds every year, but it is well worth creating a deep bed by digging the land very deeply once—perhaps three spits deep—but leaving the subsoil underneath. After that, never step on the bed.

Deep beds for fruit

All soft fruit bushes, and also fruit trees on dwarfing rootstocks trained as goblets or dwarf pyramids, can be grown successfully in deep beds. Alan Chadwick is experimenting with dwarfed fruit trees with other crops growing under and around them. This is a new technique (the French deep-bed gardeners never bothered to grow tree fruit on their deep beds), but it seems successful.

If you like big fruit trees, you can plant them in circular deep beds—one tree to a bed. Simply mark out a circle around the likely drip-line—the area to be overhung by the tree. Double-dig this circle around the edges and dig very deeply—four spits would be ideal—at the spot where you are actually going to plant the tree. Plant the tree in the normal way (see p. 98).

It is known that the roots of trees advance much more quickly in unconsolidated soil. It is also easy to observe that roots tend to come upward toward the surface. If you can keep the soil within range of a tree's roots soft and open, you can give the roots the conditions they need for rapid growth without constantly digging into them with a spade or cultivator. The only way you can achieve this is by not stepping on the soil, ever, at all, after you have done the initial deep digging.

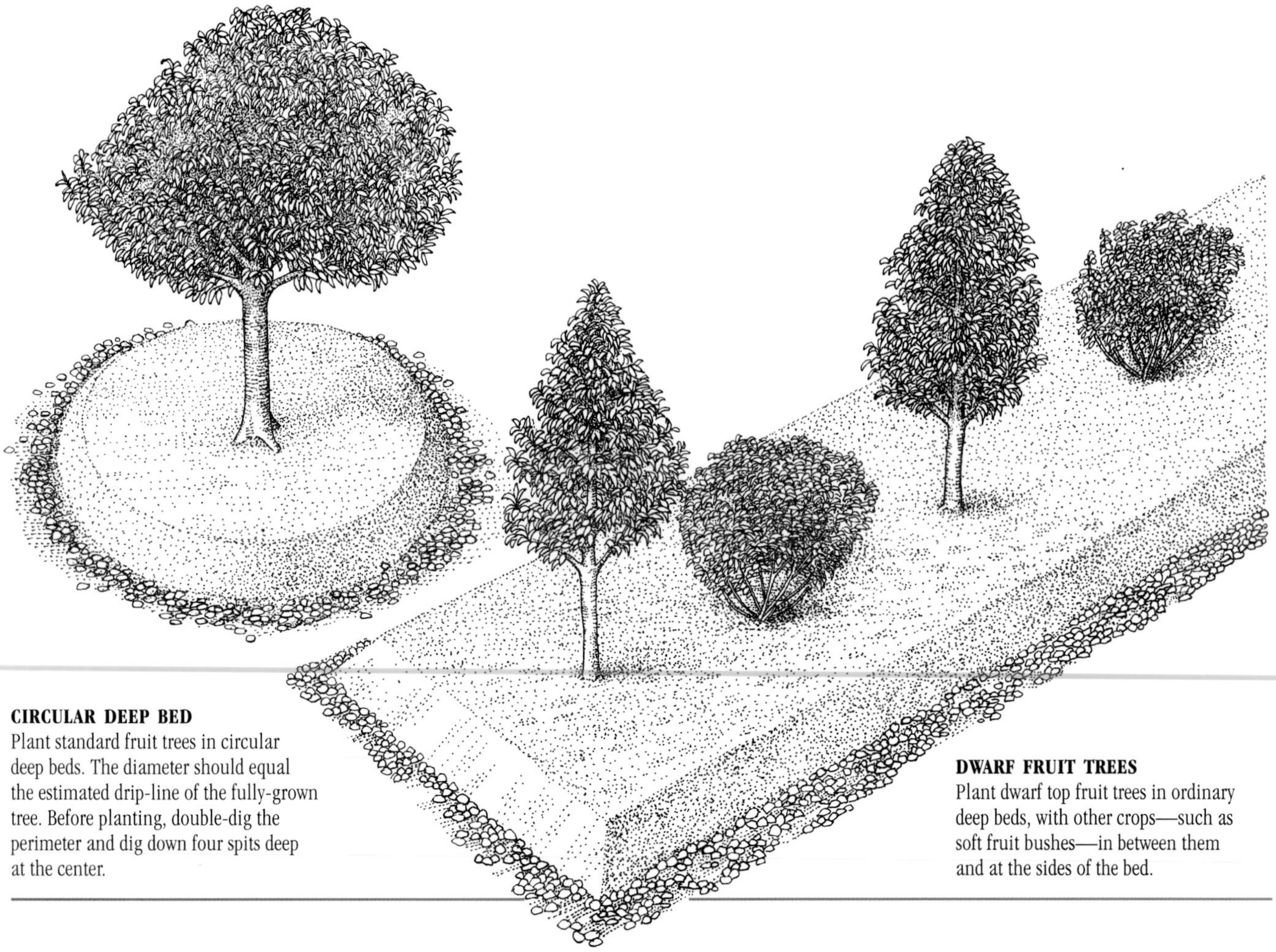

CIRCULAR DEEP BED
Plant standard fruit trees in circular deep beds. The diameter should equal the estimated drip-line of the fully-grown tree. Before planting, double-dig the perimeter and dig down four spits deep at the center.

DWARF FRUIT TREES
Plant dwarf top fruit trees in ordinary deep beds, with other crops—such as soft fruit bushes—in between them and at the sides of the bed.

Mini-greenhouses

The idea of the mini-greenhouse is to bring lightness, mobility, flexibility, and economy into the business of protecting plants from the weather. The trend in the last twenty years has been toward cheap plastic structures, principally plastic-covered tunnels. These work, but are troublesome. You cannot hoe under them without removing them and this is always awkward and time-consuming.

The mini-greenhouse, on the other hand, is very light and can be moved easily by two people from one bed to another. A proper rotation of crops that are usually grown indoors can be practiced without the laborious soil changes, or expensive soil sterilizations, that have to be carried out in fixed greenhouses. In other words you never have to grow either tomatoes or cucumbers two years running on the same bed.

Building a mini-greenhouse

A convenient size for a mini-greenhouse is 20 feet by five feet (6 x 1.5 m), because this will fit over a standard-sized deep bed, and besides it is fairly easy to handle. Use two inch by one inch (5 x 2.5 cm) wood, and for permanence make proper mortise and tenon joints. Brace the structure with ordinary thin binding-wire; this hardly adds to the weight and does not cost much, but it increases rigidity enormously.

The wires should be pulled quite tight. Tie one end of a wire to a top corner of the mini-greenhouse, drill a hole in the strut at the opposite corner and thread the free end of the wire through the hole. Haul the end of the wire as tight as you can with a pair of pliers. I always grip the wire tight with the pliers just where it emerges from the hole and twist the pliers around the strut so that they act as a lever to pull the

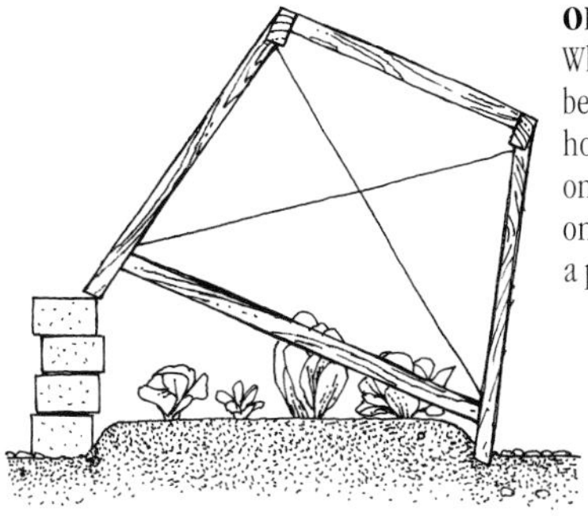

OPENING MINI-GREENHOUSES
When you want to work on a deep bed under a mini-greenhouse—for hoeing, thinning, harvesting, and so on—all you need to do is to prop up one side of the mini-greenhouse on a pile of bricks or on a piece of wood.

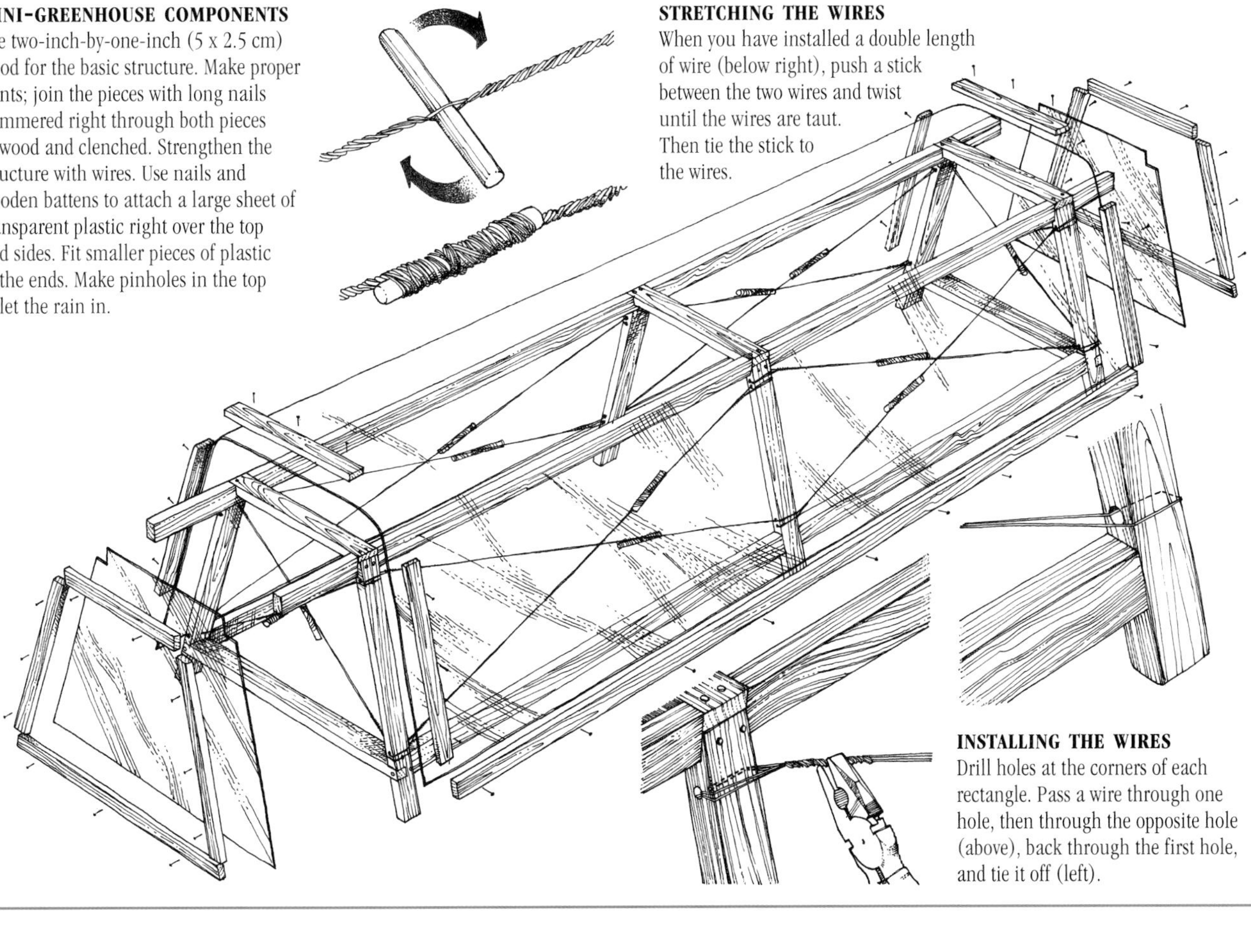

MINI-GREENHOUSE COMPONENTS
Use two-inch-by-one-inch (5 x 2.5 cm) wood for the basic structure. Make proper joints; join the pieces with long nails hammered right through both pieces of wood and clenched. Strengthen the structure with wires. Use nails and wooden battens to attach a large sheet of transparent plastic right over the top and sides. Fit smaller pieces of plastic to the ends. Make pinholes in the top to let the rain in.

STRETCHING THE WIRES
When you have installed a double length of wire (below right), push a stick between the two wires and twist until the wires are taut. Then tie the stick to the wires.

INSTALLING THE WIRES
Drill holes at the corners of each rectangle. Pass a wire through one hole, then through the opposite hole (above), back through the first hole, and tie it off (left).

wire even tighter. You can then just wind the end of the wire around its own standing part.

Alternatively, you can use a much longer wire, double it back, and poke its end through a hole near the top corner where the wire came from; stretch it there and tie it off. In this way you have tensioned wires side by side and you can stretch them further by pushing a stick through them and winding it around and around.

When the two stretches of wire have been twisted around each other, fix one end of the stick to the wires with another short piece of wire so that the bracing wire does not unravel. Beware any loose ends of wire that will puncture your plastic sheeting. Use a single large sheet of plastic to cover the top and sides of the structure, and small piece at either end. Attach all the edges of the plastic to the structure with wooden battens. Don't leave any ends unattached: if you do, the plastic will rip in the wind. Make a row of tiny holes in the top sheet of the plastic to allow rain water to sprinkle down onto the plants below, instead of forming a lake.

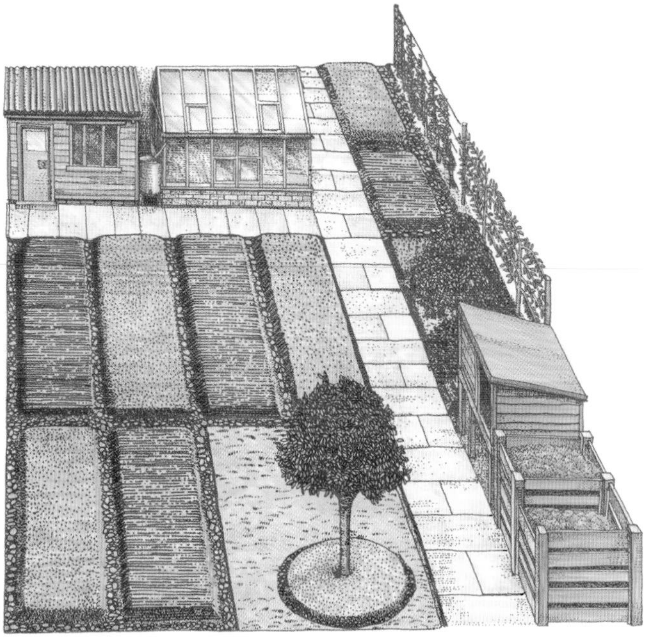

CONVERTING YOUR GARDEN TO DEEP BEDS
This is the garden shown on p. 70, replanned so that the same crops can be grown in deep beds. There is room for six beds 20 by 5 feet (6 x 1.5 m) deep: four for the vegetable rotation, one for perennial vegetables, and one for soft fruit. The old perennial bed, in front of the espaliers that now stand in a deep bed, now contains alternate dwarf fruit trees and soft fruit bushes. The herbs are now also in a deep bed, and a standard fruit tree has been planted in a circular deep bed. The seed- and holding-beds remain the same.

Using a mini-greenhouse

Use the mini-greenhouse as much as possible. Use it in winter to cover winter lettuce. Move it on to newly planted cauliflowers after that. When they have gotten a hold, move it to protect early potatoes and then on to green beans, perhaps. In the warmer weather use it on tender vegetables like eggplants, melons, and sweet corn. And you can cover your mini-greenhouses with bird netting instead of plastic to protect your seed-beds.

PLANNING A DEEP BED GARDEN

If I were to take over a new garden, I would have no hesitation about turning it into a deep bed garden as soon as possible. I am, in fact, turning my existing garden into a deep bed garden at the moment. There is some capital labor—work that only has to be done once—involved in doing this, and maybe this should be done little by little, one bed at a time when the time and energy are there. However, the aim should be a whole garden of deep beds.

Once your garden is completely given over to deep beds, you will need to do much less work on it than you used to do in your conventional garden, but the results should be considerably more impressive.

Except that you will be more or less tied to beds five feet (1.5 m) wide or less, and that it is a good idea not to have beds so long that it becomes a bore to walk around them, there is no difference between the general planning of a deep-bed garden and that of an orthodox one. The rotations will be just the same. The general layout will be similar, although in larger gardens each division of the rotation may take up two or more separate deep beds, because of the width limitation.

Your perennials and herbs will be planted in deep beds not more than five feet (1.5 m) wide. If your herb bed—or any other bed—is up against a wall it will need to be about three feet (90 cm) wide, because you will only be able to reach into it from one side. Your soft fruit bushes can share a deep bed with some dwarf fruit trees, and you can plant standard fruit trees in circular beds.

You will very likely find yourself doing far more in the way of interplanting. You might even try the sort of mixed hard fruit, soft fruit, vegetables, and flower beds that Alan Chadwick is trying in California. In your deep beds you are exploiting another direction in gardening: downward. You are making it possible for the roots to go deep, and saving them the necessity of spreading out laterally and thereby competing with one another for space. You can cram plants closer together.

Remember that this form of husbandry is in its infancy in the West (no matter how long it may have been practiced in China and Japan) and there is plenty of scope for learning and experimenting.

CHAPTER FIVE

The Cultivation of Vegetables

Containing the sowing, growing, and harvesting instructions for members of the families Fabaceae, Brassicaceae, Solanaceae, Apiaceae, Liliaceae, Chenopodiaceae, Cucurbitaceae, Asteraceae, Poaceae, Malvaceae, and Polygonaceae.

Fabaceae

Peas, fava beans, runner beans, green beans, Lima beans, soybeans, and peanuts are all members of the *Fabaceae* (formerly *Leguminosae*). For those who wish to grow as much of their own food as they can in a garden, this family is surely the most useful of them all. It provides more protein than any other. It is hard to see how a vegan, or for that matter any person who aims to be completely self-sufficient without much meat, can subsist in a healthy state without the *Fabaceae*.

The other useful thing about the *Fabaceae* is their nitrogen-fixing ability. Organic gardeners who don't like spending their money on expensive nitrogenous fertilizers (which make the soil lazy about fixing its own nitrogen) find that peas, beans, and the clovers are the answer. For the *Fabaceae* are the plants which, *par excellence*, fix nitrogen in the nodules on their roots. Pull out any healthy leguminous plant and examine its roots. You should find small pimples or nodules. If you were to cut these open and examine them with a powerful microscope, you would see bacteria. These live symbiotically with the plant. The plant feeds them with everything they need except nitrogen: they fix nitrogen from the air (combining it with oxygen to form nitrates), and this they use themselves and also feed to the host plant.

If you grow any leguminous plant, and dig it into the soil when it is lush and green (at the flowering stage), it will rot down very quickly, providing its own nitrogen to feed the putrefactive bacteria, and this nitrogen will then be released into the soil. It is worth growing clover for this very purpose. If you put leguminous plants on the compost pile, they will have the same beneficial effect. If you have a lawn, remember that if you put nitrates on it, you will encourage the grasses but suppress the clover. If you put on phosphate you will encourage the clovers at the expense of the grasses.

Leguminous plants should account for at least a quarter of your acreage each year and there is nothing wrong with having far more than that. They are not acid-loving plants, so if your soil is acidic, give it lime. They also like phosphate and potash. But in good garden soil that has been well manured or composted over the years, and in which any serious lack of lime, phosphate, or potash has been corrected, you can grow peas and beans without putting anything on at all.

Peas

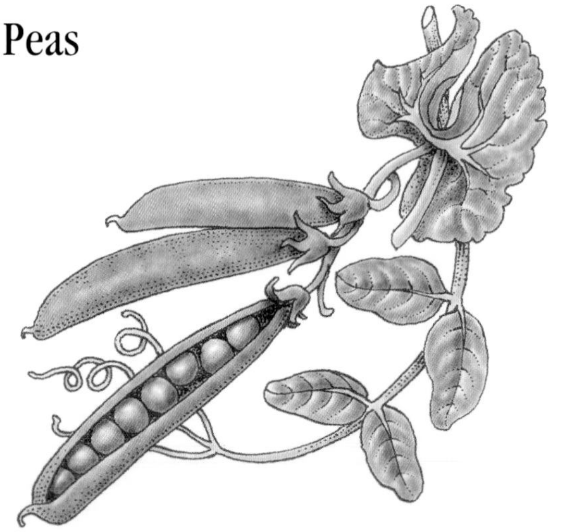

GARDEN PEAS

The first fresh peas of the summer eaten raw are one of the great rewards of growing your own vegetables. And later in the season, of course you can cook them and dry them. Whatever you do, they are a great source of nourishment. Dwarf peas are a good idea for a small garden.

Soil and climate

Peas are not too fussy as regards soil; light soil will give you an early crop, heavy a late one. Rich loam is best, and any soil can be turned into this by constant composting. As for climate, peas are not a tropical crop and will grow well in cool climates, with plenty of moisture, but too much rain when they are ripening will give them mildew. In hot latitudes they generally have to be grown in the spring or fall, to avoid the very hot part of the summer. As small plants they are frost-hardy, therefore in climates where frosts are not too intense they can be sown in the fall for a quick start in the spring. They will not grow fast and produce flowers and pods, however, until the arrival of spring and warmer weather.

Soil treatment

Peas need deeply cultivated ground. If you are trying to grow them in land that has previously been gardened inorganically you should try to spread seven to ten cwt (350 to 500 kg) of manure or compost on every 100 square yards (84 sq m). Put

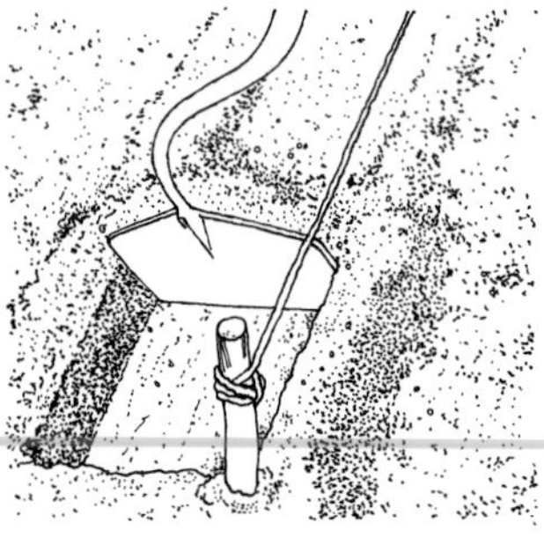

DRILLS FOR PEA SEED
Use the flat blade of a draw hoe to make broad drills about two inches (5 cm) deep and four inches (10 cm) wide in the soil.

SOWING PEA SEED
Sow evenly, leaving an inch or two between seeds. If necessary you can keep mice away by dipping the seed in kerosene* before planting.

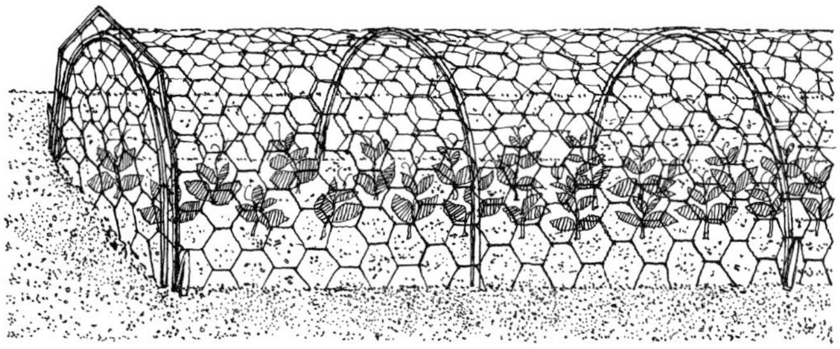

PROTECTING PEAS FROM BIRDS
Wire-netting pea-guards, which you can easily make yourself, will protect seeds and seedlings from attack by birds. So will mini-greenhouses (see p. 111) such as you use for deep beds; but you should cover them with wire rather than with plastic.

this on the land the fall before, and possibly 25 lb (11 kg) of slag or ground rock phosphate per 100 square yards (84 sq m) and 10 or 12 lb (4.5 to 5.4 kg) of wood ash. Peas don't like acidic soil; if the pH is about 6.5, that is all right. If it is below this, lime it; a quarter of a pound (100 g) per square yard is about right.

If your soil is not yet sufficiently fertile and you cannot bring in enough compost or manure from outside, you can still grow excellent peas over trenches taken out the previous year and filled during the winter with kitchen garbage and other material that will readily decompose, such as old newspaper. The organic gardener's aim, though, should be to raise the whole of his garden to a high level of fertility, so that such piecemeal treatments as this are unnecessary.

Propagation

I make broad drills with the flat of the hoe, about two inches (5 cm) deep, and four inches (10 cm) wide. I then sprinkle the seed in evenly, so there is an inch or two between each seed. I then rake the soil back into the trench from each side and bang down firmly with the back of the rake—or, if the land is puffy and dry, I tread it with my boots. A good soaking of the drill, if the soil is dry, will then start them growing. If you use the deep-bed method (see p. 106), allow three inches (8 cm) between plants all ways. You leave this distance in the deep bed because you are not sowing in rows, of course, but in clumps.

Many people speed up germination by soaking the seed, for as long as forty-eight hours, before they plant it. It should be remembered that all these seeds that are large and edible, like peas and beans, are a standing invitation to rodents and birds, and so the sooner they start to grow, the less time there is for them to be eaten by something. Birds may have to be kept away by thin black threads, or, better still, inverted wire-netting pea-guards. And if you are troubled by mice, dip the seed into kerosene* just before planting. The mice don't like the smell.

Now peas take about four months to grow to maturity: perhaps three-and-a-half if you plant early varieties or if you like your peas very young like I do. Sow them successionally, every two weeks from March to July (sow earlies in July), and you will get fresh peas all summer.

Care while growing

All but the smallest dwarf peas are better if they have sticks to grow up. Any fine branches with some twigs left on them will do for this. Hazel trimmings make ideal pea sticks. If you need a hedge between your garden and the next one, use hazel. It will give you nuts as well as pea sticks. If you just can't find pea sticks then use wire netting. Get the coarsest mesh you can (it is cheaper)—say, three feet (90 cm) wide—and make an inverted "V" of it so that a row of peas climbs up each side. This method has the advantage that many of the peas hang down inside the wire where the birds can't get at them. If the

BUILDING WIRE PEA FRAMES
You can build a "fence" of wire netting for peas to climb up, or you can build an inverted "V" shape and train a row of peas up each side. The peas will dangle into the middle where the birds can't get them. Use wide-gauge netting—it is cheaper and you can get your hands through the mesh quite easily.

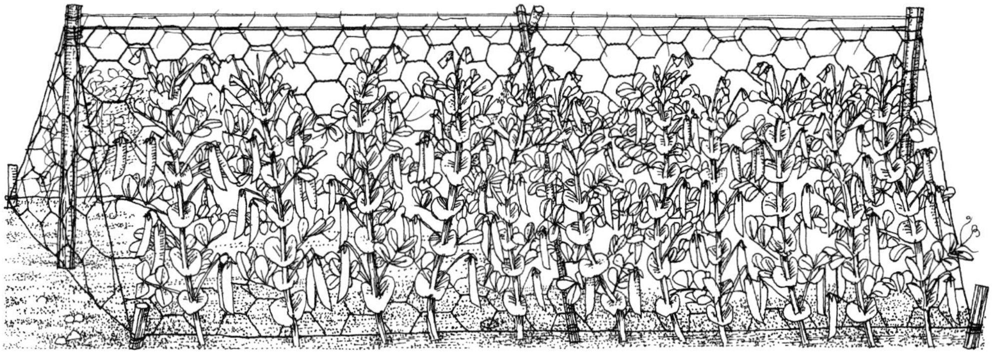

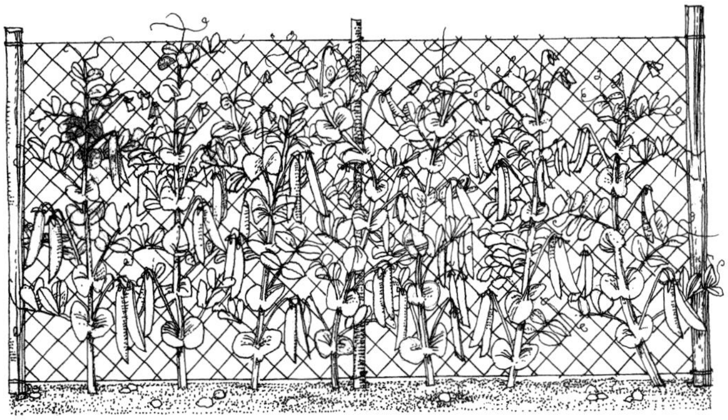

TRAINING WITH PEA STICKS
Any branch with a few twigs left on it will make a pea stick. Hazel branches are especially good, as they will provide you with nuts, too. Cut them to a length of about four feet (1.2 m), sharpen the thick end, and drive well into the ground beside each plant.

wire is wide-gauge, you will be able to get your hand in to pick the peas; otherwise you can put your hand down through the gap in the top. There are plenty of dwarf pea varieties nowadays, which are supposed to need no support at all. They are worth growing in a small garden, but the yield is low and unless you take precautions (see p. 104) slugs will attack the peas near the ground.

Peas don't like drought, and watering in dry weather always pays in more peas, but remember that soil rich in humus retains water more efficiently.

Pests and diseases

PEA AND BEAN WEEVIL This creature is the color of soil, falls off the plant and feigns death when you disturb it, and is nocturnal; it hides under clods of soil during the day. No-digging gardeners suffer from it a lot because the compost with which they have to cover their ground gives it splendid cover. It nibbles around the edges of young pea leaves and often eats out the growing centers. Dusting the plants with lime*, while the dew is on them, is a deterrent, or you can do the same with soot*. If you have neither, spray the young plants and the surrounding ground with quassia* spray, or nicotine*.

PEA MOTH This is a small brown moth that lays eggs on young pea pods. The larvae bore in and eat the peas. If you dig, or cultivate, the soil frequently, but very shallowly, during the winter you can get rid of these pests, for the birds (chiefly robins and starlings) will come along and eat the pupae—thereby breaking the moth's life cycle.

PEA THRIPS These tiny browny-black insects make minute holes in the leaves. The plants become yellow and shrivel up. A thorough drenching with soapy water will get rid of them.

MILDEW In very damp weather, pea leaves and pods may go white with mildew, and then rot. Sticking the peas well, so that they can climb high, helps prevent this. Don't water the foliage of peas in hot muggy weather. Spraying with Bordeaux mixture* sometimes works. Otherwise there's not much you can do about it, but it is not disastrous.

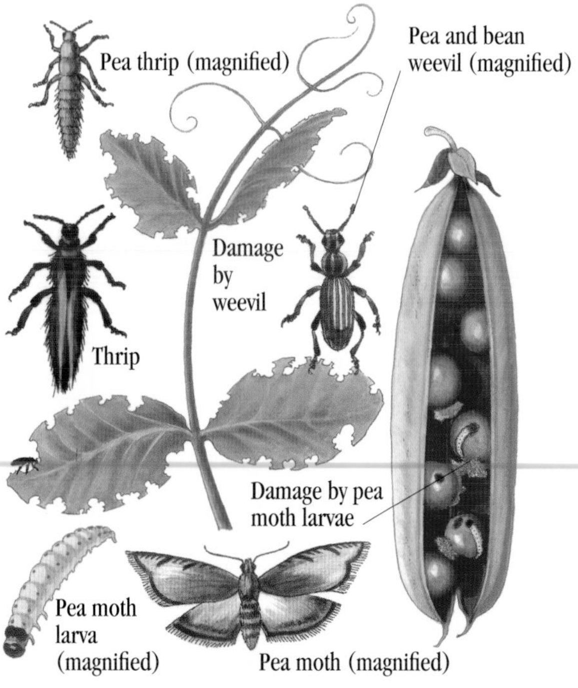

Harvesting

Always use both your hands to pick peas! Put the basket on the ground, and hold the vine with one hand and the pod with the other.

Very young peas taste quite exquisite raw and contain high doses of vitamins A, B, and C. They are very sweet because they contain sugar. A few hours after the peas are picked this sugar turns to starch, which is why store-bought peas taste dull and dried peas completely different. If you pick peas and freeze them immediately, you can preserve this sugar, which is why frozen peas don't taste too bad.

I like to eat fresh peas all summer and then enjoy dried peas, rather than frozen peas, during the winter, so that I come to the first fresh peas in May or June with a fresh palate and enjoy what is then an exquisite gastronomic experience. The palate jaded by "fresh" peas all year round never has this great sensation.

As fresh peas grow older and tougher on the vine, you have to boil them. When your peas get too tough to be perfect boiled, leave them on the vines and just let them go on getting tougher. Wait until they are completely ripe, as hard as bullets; then pull the vines out and hang them up in the wind but out of the rain.

When the vines are thoroughly dry, thresh the peas out of them; either rub the vines hard between your hands, or bash them over the back of a chair. Put the peas away, thoroughly dry, inside covered containers. When you want some in the winter, soak the dried peas in water for a day or two. Then boil them with salt until they are soft, and eat with boiled bacon. A plate full of peas and bacon, in December, and you are fit to go out and dig for a few hours.

Pea pods make the basis of a good soup. Boil them well and press them through a sieve.

SNOW PEAS

These are also called sugar peas. Cultivate them just like ordinary peas. The difference is that you harvest and eat the pods with the young peas inside them, because they lack the hard membrane that lines the ordinary pea pod. Start picking and eating the pods as soon as they are about two inches (5 cm) long and when the peas inside are tiny flat bumps. You then have a long picking season because you can go on picking until the peas inside are quite big. Personally I prefer ordinary peas, but snow peas are nonetheless worth growing and eating.

ASPARAGUS PEAS

These are not true peas, but you can treat them the same way, although you must plant them later—say, in April—and then, if you live in a cold place, protect them with cloches. Support them just like peas. Cook the complete pods when they are about an inch long. They are quite delicious.

LENTILS

Lentils are closely related to peas and are excellent for drying. However, they are low-yielding and are really only worth growing if you have space to spare after allowing for your staples. They like the same climatic conditions as peas, and do best on sandy loam. Propagate them and care for them exactly as though they were garden peas. When the plants are well ripened, pull them out and hang them up in a shed. Thresh when required.

Fava beans

For the self-supporting gardener, fava beans are one of the most important crops. They really will feed you and your family right through the year, and if you have dried fava beans and potatoes, you will not starve. The old English bean was the "longpod," and this is still the best bean to grow if you really want food that will keep you in high-protein vegetable nourishment right through the winter as well as the summer (unless your climate and conditions enable you to get a good yield with soybeans).

All the other beans, like runner beans, green beans, and dwarf beans, came from the Americas, especially the hot parts of South America. They are all very frost-tender. The good old longpod fava bean however, can stand up to a fierce winter and get away early in the spring. It stands up straight and tall, needs no support, and produces a heavy crop of fine, big kidney-shaped seeds, which can be either cooked and eaten fresh, or dried and kept for the winter.

Tic beans, horse beans, and cattle beans, which are all grown by farmers to provide high protein grain for feeding animals in the winter, are varieties of the same plant. Their seeds are smaller and white, but they are heavy-cropping. They are more liable to get the dreaded chocolate spot than fava beans and they are said not to be edible before Christmas. But then they are said not to be fit for human consumption yet I eat them freely every winter.

The ordinary longpod fava bean, however, is the best thing for the gardener; but if you can get a handful of tic beans from a farmer, why not experiment with them and see what happens?

Soil and climate

Fava beans like strong soil, even heavy clay, but compost-rich soil suits them no matter what the original soil was—clay or sand. Their behavior in different climates is very similar to that of peas (see Peas). If you can plant out your peas in the fall, then you can do the same with your fava beans, for they are rather more hardy than peas.

Soil treatment

Beans don't like too much acid; like peas, they find a pH of 6.5 ideal. Potash does them a lot of good, so if you have a limited supply of wood ashes, put it on your fava bean patch at a rate of about four oz (100 g) to two yards of row. Comfrey too, dug in as a green manure, is good, both for the potash that it contains and for its capacity to hold water.

Dig the soil deep and well, digging in ashes and comfrey leaves if you have them, and as much manure or compost as you can spare. I like to plant fava beans after main-crop potatoes, so the soil is already full of manure that was put in for the spuds, and it has been well worked. Lime, then, is only necessary if the pH is much below 6.5.

Propagation

It is far and away better, if you live south of the heavy snow line, to sow your fava beans in the fall—in, say, October or November. If the birds get them, or it is too cold where you are, sow them as early as you can in the spring—in February if the ground is not frozen then. It is best to soak the seeds in cold water for twenty-four hours before planting them; this softens them up and gives them a head start over the birds.

Take out drills about three inches (8 cm) deep with the hoe, drills about two feet (60 cm) apart, and put the seeds in six inches (15 cm) apart. Or—and here I think is a very good tip—sow them in rows six feet (1.8 m) apart and, later, sow green or dwarf beans in between the rows. Fava beans make a fine nurse-crop for these more tender plants, keeping the wind off them. Instead of digging drills you can put out a garden line and make a hole with a trowel for each seed. If you use the deep-bed method (see p. 106), leave four inches (10 cm) between plants in all directions.

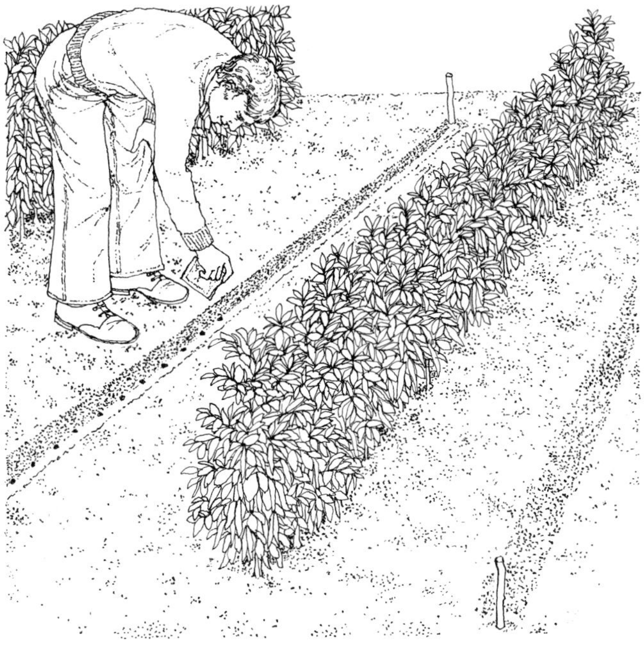

INTERCROPPING
Tough, tall fava beans make an excellent nurse-crop for smaller plants like green beans. Plant the fava beans in rows six feet (1.8 m) apart in fall or early spring. When the weather warms up in early summer, fill the spaces between the rows with low-growing green beans.

Care while growing

As the young beans grow, it is a good idea to earth them up a little with the hoe. Keep them clear of weeds, of course. In windy and exposed positions, it is worth driving in a stake at the corners of the rows, and running a string from stake to stake around each row, to prevent the beans from being blown over. In most gardens, though, this is not necessary. Any sort of mulching between the plants is valuable.

Pests and diseases

CHOCOLATE SPOT This looks exactly like its name, and if you get it there is nothing you can do. If you get it early, it will lower your yield considerably, but a late attack is not so bad. To guard against it you need plenty of potash in your soil. If you keep getting it in fall-sown beans, you must give up fall sowing and sow your beans in the spring instead, because these are less liable to attack.

BEAN RUST You are not likely to get this. The symptoms are small white spots on leaves and stems in the spring. Spray with diluted Bordeaux mixture* (see p. 105), and burn all your bean straw after harvesting to destroy the spores.

BLACK FLY OR APHID If you sow spring beans you are very likely to get this, and in fact unlikely not to get it. It does not trouble fall-sown beans very often. The aphids can only pierce the skin of tender young growing points and the winter beans are generally grown enough by the time the aphids are around. If aphids do attack you will find them clustered on the tips. If the beans have already grown high enough, pick these tips off; this will deny the fly its food. You can cook and eat the tips; they are tender and juicy. If you get a really bad attack, spray the plants hard with soft soap and kerosene* solution. Bees love bean flowers, so it is important not to spray with anything that will harm bees. Pyrethrum sprayed at night will kill the aphids and not harm the bees in the morning.

PEA AND BEAN WEEVIL (See Peas).

BEAN MILDEW Spray with Bordeaux mixture*.

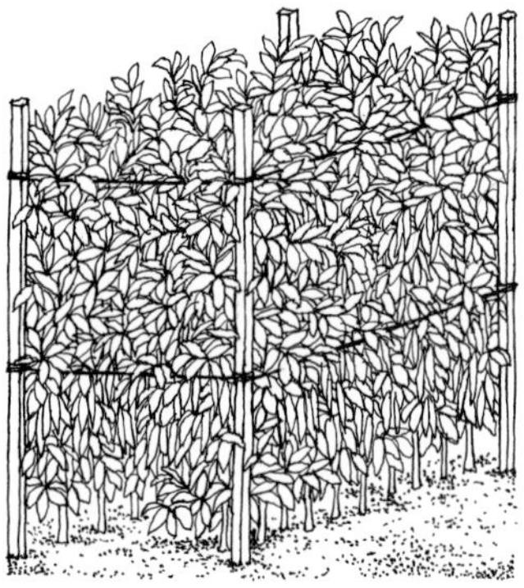

SUPPORTING FAVA BEANS
If your fava bean patch is exposed to wind, support the plants with strings run between stakes, which are driven in at the ends of the rows.

BANISHING BLACK FLY
Black fly or aphids on your fava beans will almost certainly cluster on the tips. If the beans have grown high enough, simply pick the tips off.

Harvesting and storing

The tips of growing fava beans, nipped off in the spring to thwart black fly, are one of the first and most tasty fresh greens of the spring. Soon after eating these, you can start pulling the very small pods and cook them as they are. Later, the pods get too tough for this, so you split them open, remove the beans from their silky beds, and cook the seeds. When the seeds get too tough for this, let them dry on the plant, and harvest by pulling out the whole plant. Then hang it in an airy but dry place. Shuck the seeds out when they are dry and store them for the winter. Soak them for at least twenty-four hours (twice that long is not too much) before cooking and then boil them well. Eaten either with butter or with bacon, they will give you strength to face the winter.

Runner Beans

The runner bean was brought to Europe from North America in the seventeenth century by people who thought it was beautiful rather than nutritious. The fact that it is both makes it ideal for the small-scale vegetable garden.

Soil and climate

The runner bean is not frost-hardy and it prefers a warm, sunny climate, although this is not essential. In warmish climates, it will survive the winter, underground, and grow up again in the spring as a perennial. In cold climates you can get a crop by sowing in peat* pots indoors and planting out after the last frost. The runner bean needs plenty of moisture at its roots, and its flowers will not set without an occasional shower of rain, or spray from a hose. It will grow in most soils but likes rich ones; it benefits from plenty of humus and plenty of moisture. And it does not like acidic soils. 6.5 pH is ideal, so lime if necessary.

Soil treatment

The classic method is to take out a deep trench in the fall or winter and fill it with manure or compost, or else spend the winter filling it with kitchen refuse and anything else organic you can find. In the spring, cover the trench with soil and plant on that. The compost, or whatever it was, will have subsided as it rotted and so there will be a shallow depression for the water to collect and sink into, and runners like plenty of water. If you don't dig a trench, you must still dig deeply and put in plenty of compost.

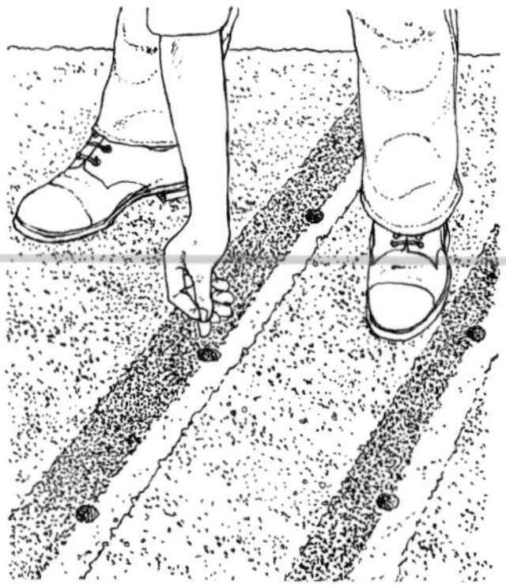

SOWING RUNNER BEANS
To get as many plants as possible into the space available, sow seeds three inches (8 cm) deep, ten inches (25 cm) apart in two rows one foot (30 cm) apart. Stagger the seeds in the rows. You can speed germination by soaking the seed in water before planting, although you should not do this if you suspect halo blight.

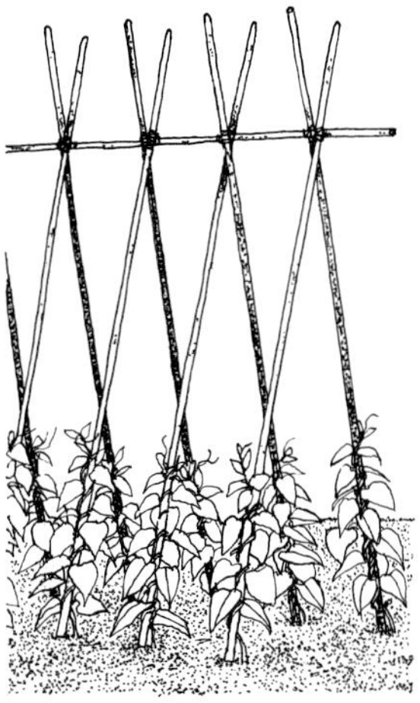

STAKING RUNNER BEANS
Each runner bean plant must have its own stake, anywhere between seven feet (2 m) and 12 feet (3.5 m) high. With two rows; you can use the crossed pole method, where pairs of poles are tied together to form forks near their tops and the whole row is made secure with a pole laid through the forks and tied down. If you want your beans against a wall, fix wires close to the wall by attaching them to long nails top and bottom. The wigwam method is excellent, even in a tub or large barrel. Sow eight to twelve seeds in a circle; when the plants come up, give them each a pole and tie them all together near their tops.

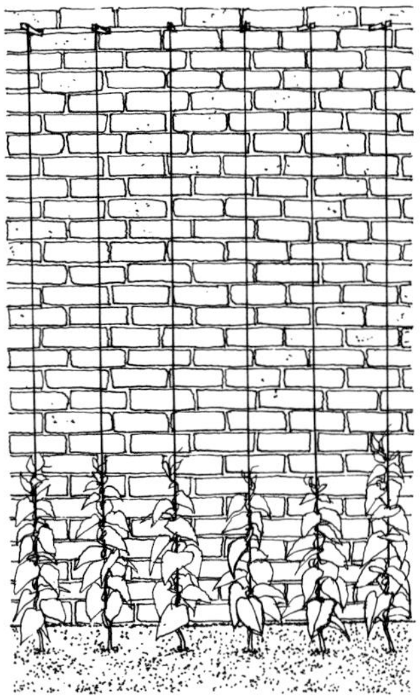

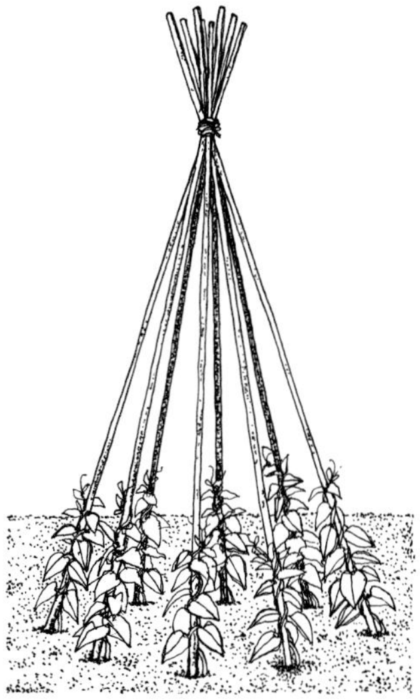

Propagation
If you want early runner beans, sow in peat* pots in the greenhouse, or in a sunny window, during early April. A temperature of about 55°F (13°C) is fine. Most people just sow them outdoors *in situ* during May after the last frost. I draw two drills a foot (30 cm) apart and sow the seeds three inches (8 cm) deep, ten inches (25 cm) apart in the rows and staggered. If you use the deep-bed method (see p. 106), you must still plant in a row so that all your plants receive plenty of light. Therefore, plant no closer.

Care while growing
Weed them, of course, water them in dry weather, and before they are many inches high, stake them with stakes at least seven feet (2 m) high. They will climb as high as 12 feet (3.5 m) if you give them long enough stakes, and personally I like them to do it. The more beans, the better, and you can always stand on a box to pick them. Put in stakes when two true leaves are well opened.

There are several methods of training runner beans. As long as you have plenty of space—enough for two rows of plants—you can use bamboo or bean sticks (like pea sticks, only longer) to build a row of crossed poles, with their apexes in between the rows. Tie them together where they meet and strengthen by tying canes along the top. If you are short of space you can plant your beans in a circle and build a wigwam of poles around the circle. (This is better in my opinion than using dwarf varieties in a small space.)

Another method is to plant the seeds along the bottom of a wall and train the plants up wires against the wall. I love to see a really high screen of beautiful runner beans, sited so as not to shade anything that must have sun (plenty of plants, like lettuces, grow well in shade) and screening some "unpleasaunce" in the garden.

If the flowers are out in a dry period, spray them lightly with water—preferably water that is not too cold. This helps the flowers to set.

Pests and diseases
With good organic soil you are unlikely to get trouble, but you might just have one of the four varieties below.

BEAN DISEASE This is a fungal disease that causes black spots to appear on the pods of runner and green beans; later on, the spots develop reddish outlines. At the first sign of this disease, spray with Bordeaux mixture*, but if the disease gets a strong hold, root out the affected plants and burn them. Never save seed if you have had an attack of bean disease.

HALO BLIGHT Caused by bacteria, this results in semitransparent spots surrounded by a yellow halo on leaves. Spray with Bordeaux mixture*. Do not save seed for next year, and if you have any doubt about your seed do not soak it before planting.

MOSAIC DISEASE This shows itself as yellow blotches on the leaves. It lives in clover, so do not plant runner or green beans in soil where old clover has been dug in. If you do get it, pull up the plant and burn it.

MEXICAN BEAN BEETLE This is a brown spotted beetle a little bigger than a ladybug. If your beans are prone to attack, you will find they will attack your main crop more than your early crop. So put in your main crop as early as you can. If you are attacked, pick off and kill any beetles you see.

Harvesting and storing
Pick runners, like peas, with both hands. Hold the vine in one hand and pull the pod with the other. Harvest on the "pick and pick again" principle. Keep picking them when they are young and tender, and suffer them not to get old and haggard on the vine. If you keep picking, they will keep coming—they are the most generous of crops.

If you can't eat them all, salt them (see p. 215). But remember, only salt them when they are young and tender.

PICKING RUNNER BEANS
Pick beans and peas with two hands. If you pick with one hand, you can do lasting damage to the vine. Pick them while they are young, and you will be encouraging more to grow. Don't ever leave them to get old on the vine: if there are too many for you to eat all at once, salt them. Frozen runner beans don't taste as good as salted ones.

Green Beans

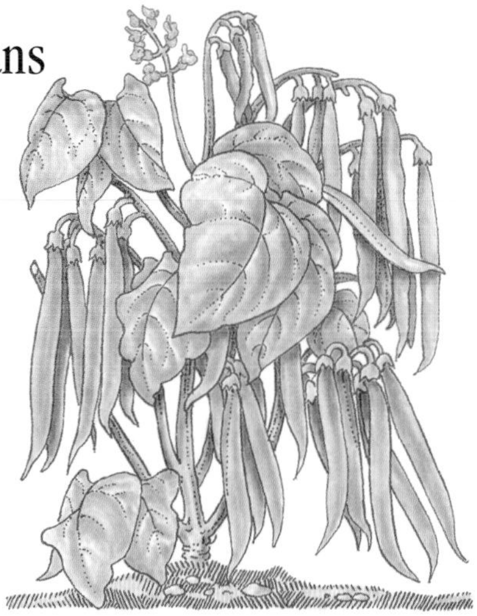

Green beans are nothing like as hardy as fava beans. When fully ripened and dried they are haricots and form a rich source of winter protein.

Soil and climate
Plant when the soil has warmed up in the summer. They prefer lightish soil, or soil that has been well-improved by compost, and a pH of about 6.5.

Soil treatment
Don't lime for them if your soil is not too acidic. The more humus you can incorporate when digging the soil, the better.

Propagation
When sowing in drills, have the rows two feet (60 cm) apart and sow a foot (30 cm) apart in the drills—deep-bed method four to six inches (10–15 cm) apart (see p. 106). It is worth sowing two seeds in each station and removing the weaker of the two plants after they germinate. If you want early beans, start them off in seed boxes in your greenhouse.

Care while growing
Keep the bed well weeded and the soil loose.

Pests and diseases
CUTWORM Cutworms can be kept away by placing a three-inch (8-cm) cardboard collar around the stem of the plant (see p. 124). Bend the cardboard so that it is half an inch (1 cm) from the stem all the way around. Allow one inch (2.5 cm) below ground and two inches (5 cm) above. Alternatively, place a ring of wood ash around each plant.
WIREWORM If you are troubled by wireworms, try to trap them during the winter by burying cut pieces of potato six inches (15 cm) deep at intervals of about a yard. Mark them with sticks, carefully dig them up each evening, and destroy the wireworms that you will find there.

Harvesting and storing
Like all beans, green beans are grown for two purposes: for the green pods with immature beans inside and for the ripened beans that can be dried into haricots. To dry beans, you just let them ripen, hang the vines upside-down in a shed, and thresh them when you want them. To harvest, pick the beans by hand. They can be stored green in salt.

Lima Beans

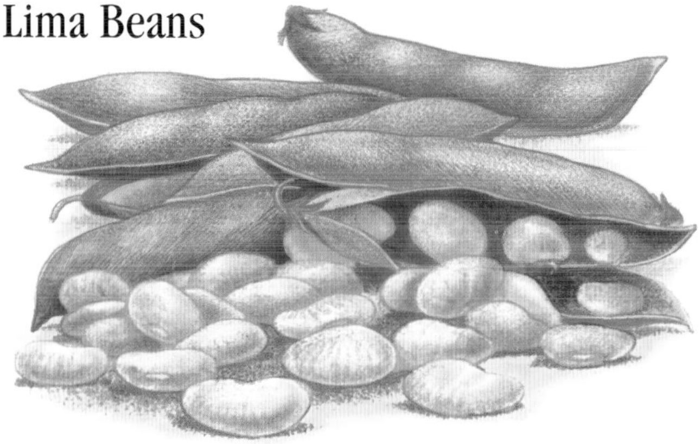

These are tropical American beans which have been adapted for growth in warmish temperate climates. They can be cooked and eaten when green, or dried for winter storage. There are bush and climbing varieties.

Soil and climate
Unless they are started off under glass, they need three months of pretty warm days and nights. The seed needs warm soil to germinate, so don't plant until two or three weeks after the last frost. And bear in mind that the first fall frost will cut them down. If you have this sort of climate, they are worth growing because they are very heavy-cropping. Limas like lightish soil but will grow in any soil except heavy clay. Atypically for beans, they prefer a slightly acidic soil; a pH of 6 is about right.

Soil treatment
Limas should follow a well-manured crop such as potatoes or celery. Simply dig the soil fairly deeply and, if you can spare it, mulch with compost.

Propagation
Sow the seed about three feet (90 cm) apart for bush varieties and eight inches (20 cm) apart for climbers. The former should be in rows 30 inches (75 cm) apart; the latter should be in one row. For the deep-bed method (see p. 106) allow one foot (30 cm) between bushes, or six inches (15 cm) between climbers, which should still be planted in one row. In colder climates, plant indoors in peat or paper pots and plant out in warm weather.

Care while growing
Mulching is very valuable, and the beans must also be kept well watered.

Pests and diseases
Lima beans are pretty tough. If they do suffer, it will be from something that attacks runner or green beans (see Runner Beans and Green Beans).

Harvesting and storing
For eating green, harvest on the "pick and pick again" principle, once the beans are swelling in the pod. Don't pick them too late because, like runner beans, they get tough. If you want to dry them for what are called "butter beans," leave the pods on the plant until the plants are dry. Pick the beans by hand, or thresh by walking on the plants.

Soybeans

You can eat soybeans green in the pods, or dry them. Either way, they are very high in protein. The beans can be crushed for their oil, and the flour that is left can be added to the flour of cereals to make a high-protein bread.

Soil and climate
Soybeans grow well only where it is warm. They don't mind slightly acidic soils, like high organic matter, and will grow in quite moist conditions.

Soil treatment
Soil with plenty of humus in it just needs a light forking. Otherwise, dig thoroughly and lime for a pH of 6.5.

Propagation
Sow them outdoors in early summer; a good rule is to sow when the apple trees are in full bloom. Sow an inch (2.5 cm) deep and three inches (8 cm) apart in the rows—deep-bed method (see p. 106) four inches (10 cm) apart. Where the beans have not been grown before, the seed should be inoculated with nitrogen-fixing bacteria, because it is likely that the right bacteria do not exist in the soil.

INOCULATING SOYBEAN SEED
Where soybeans have not been grown before, the soil may not contain the right nitrogen-fixing bacteria. Prepare seeds by stirring them up with water in a bowl. Add nitrogen-fixing bacteria to the water-coated seeds, making sure each seed is thoroughly covered with the bacteria. Careful inoculation will increase your yield by up to a third—and improve your soil as a bonus.

Care while growing
Hand-weed rigorously, and mulch with compost if you can.

Pests and diseases
Soybeans are very hardy, but they can suffer from various fungus diseases (see Runner beans). These can be prevented by proper crop rotation.

Harvesting and storing
Pick soybeans green and eat them whole, or wait for them to ripen, in which case steam or boil the pods for a few minutes before shelling them. Otherwise, pull the plants and hang them up to dry.

Peanuts

Peanuts, also called groundnuts or monkey nuts, are very rich in the vitamins A, B, and E. They grow extensively in the southern states of the US but can only be grown in colder regions with glass protection at each end of their season. As they are quite cheap to buy and as there are so many other things we really need our glass for, they are hardly worth growing in cool climates.

Soil and climate
Peanuts need a warm growing season of over four months; five is ideal. They like sandy soil and, unlike most legumes, they like acidic soil—a pH of 5 is about right.

Soil treatment
Dig the soil deeply and incorporate plenty of compost. Never lime for peanuts.

Propagation
You can plant peanuts, shells and all, or shell them and plant the nuts. Plant shells eight inches (20 cm) apart, nuts four inches (10 cm) apart. For the deep-bed method (see p. 106) allow four inches (10 cm) and three inches (8 cm), respectively. In warm climates, plant four inches (10 cm) deep, but in cool climates make it only one-and-a-half inches (4 cm). To give them the longest possible growing season in cool climates, they should be planted at about the time of the last probable frost. You may need to start them off under glass, if you live in a very cold place. Sow them in rows 30 inches (75 cm) apart.

Care while growing
The yellow flowers are the staminate ones; the productive pistillate flowers are inconspicuous, and after being fertilized they bury themselves in the ground and develop into peanuts. Raise the soil in a circle around the plant so that the fruits forming at the ends of their stems can easily bury themselves. Peanuts will only ripen below ground.

Pests and diseases
Peanuts don't seem to suffer from much at all.

Harvesting and storing
In a warm climate, pull the vines when the leaves turn yellow and hang them in a dry airy place. In cooler climates, leave them until after the first frosts—the nuts will continue to ripen underground even after the leaves have frosted away. Before eating them, roast your peanuts in their shells for 20 minutes in a 300°F (150°C) oven and leave them to cool—a vital part of the peanut roasting process.

Brassicaceae

Cabbages, Brussels sprouts, cauliflowers, broccoli, kale, kohlrabi, rutabagas, turnips, seakale, cresses, and radishes all belong to the *Brassicaceae* (formerly *Cruciferae*), which is one of the most important families, for it includes the genus *Brassica*, the cabbage tribe. This contains a great variety of plants that have been bred by humankind to a profusion of different forms, most of which are very good to eat. The reason for the peculiar succulence of the brassicas is that nearly all the cultivated members of it are descended from the sea cabbage, and this gives them certain important characteristics. One is that they share with desert plants the ability to make do on very little fresh water; another is that they are adapted to store what water they can get. It is this last fact that makes them so succulent. They guard the water they get under a waxy, waterproof cuticle.

Another characteristic of the brassicas is that they are biennials: that means they store food in themselves during their first year of life and then flower and go to seed in their second. The stored food and energy of the first year's growth is available to us and our animals all winter.

Seed-bed for brassicas

There are spring cabbages, summer cabbages, and summer cauliflowers, but you can take it that most of your brassicas will be for winter use. So you will find yourself, around early spring in temperate areas, establishing a seed-bed (see p. 92). This might be an area—depending on the size of your garden—as big as a table top. Work it to a very fine tilth, score parallel lines six inches (15 cm) apart with the corner of a hoe, lightly sprinkle seed along the rows, then cover the seeds with fine compost or soil, firming with the side of the rake. Plant a row each of cabbage, red cabbage, Brussels sprouts, winter cauliflower, sprouting broccoli (including calabrese), and, for good measure, leeks. I know the latter are not brassicas, but they go in there just the same. You must keep this seed-bed well watered, and when the plants are about five inches (13 cm) high, plant them out in their permanent beds, or if you are trying to get two crops, their holding-beds. In cold climates you must sow these seeds in seed boxes (flats) indoors, and plant them out later.

Cabbages

COMMON CABBAGES

You can grow cabbages all year in mild climates, and in climates with freezing winters you can easily store them the winter through in a shed or basement. And if they are organically grown, and not overcooked, they are a delicious vegetable and you need never get tired of them.

They come with round or conical hearts, but this makes no difference to the way you grow them. Winter main-crop cabbages are very high-yielding: it is not unusual to get, on a field scale, forty tons per acre (100 metric tons per hectare). On a garden scale, you can reckon on getting from a pound to a pound and a half (500–700 g) of cabbages per foot (30 cm) of row. So they are a good crop to grow, even if you only have a small garden.

Soil and climate

Cabbages will grow almost anywhere, but in hot, dry areas they can only be grown in the fall and winter. They will stand winter frosts down to 20°F (–7°C): below this temperature, it is better to store them. They are greedy plants and like good soil, with plenty of organic matter in it and plenty of nitrogen and lime.

Soil treatment

Unlike the other *brassicas*, cabbages like deeply dug ground with plenty of humus worked into it. If they follow the *Fabaceae* (pea and bean family) they will not need lime. If they don't follow the *Fabaceae,* they may well fare better if you do add a generous helping of lime.

Propagation

If you want cabbages year-round, you have to divide them into three groups: winter, spring, and summer.

SPRING CABBAGE Sow seed in the seed-bed (see above) in midsummer in cooler climates and late summer in warmer ones. Plant out into a permanent bed in early fall 18 inches (45 cm) apart or 12 inches (30 cm) in the deep bed (see p. 106). Don't fertilize too heavily with nitrogen before the winter. Instead, give them a boost when the spring arrives. These plantings will keep you going all through the late spring and early summer if you live in a cool climate.

SUMMER CABBAGE Cabbages are apt to be forgotten in summer, for there are so many other things to eat at this time, but they are excellent for eating raw in salads. You can grow them quite easily as long as your summers are not too hot and dry. Choose a fast-growing summer variety and sow in your seed-bed (see left) in early spring, or, in cold areas, sow indoors in late winter. Plant out when they are tiny, about two inches (5 cm) high, in very good soil, and keep them well-watered. Plant in staggered rows 18 inches (45 cm) apart with 18 inches (45 cm) between rows.

WINTER CABBAGE These will form by far the greater bulk of your cabbage plantings and will be sown in the brassica seed-bed (see left) in early spring, or indoors in flats or peat* pots in cold climates. When they are about five inches (13 cm) tall, you can either plant them straight out in rows two feet (60 cm) apart—15 inches (38 cm) in the deep bed—with 20 inches (50 cm) between cabbages, or you can double-crop—that is, take two crops of different vegetables from the same bed in quick succession. To do this you must plant them out firmly in a holding-bed (see p. 93), a piece of the garden set aside for them, with each plant about six inches (15 cm) from its neighbors. Then, when ground becomes available as you harvest your early potatoes or peas and beans during the summer, you can plant out your cabbages and any other *brassicas* that are in the holding-bed. This double transplanting does not seem to do them much harm.

Planting out

Plant out cabbages and all *brassicas* firmly. Make a hole with a dibber, put the plant in at the same depth as it was in the holding-bed, and firm the soil around it with either the dibber, your hand, or your boot. Swilling *brassica* roots in a bucket of thin mud with a handful of lime in it before planting out helps them a lot. This brings the roots into instant contact with the waiting soil. I knew an old gardener who dipped his plants in a paste of half soil and half cow dung with a handful of soot in it; he grew magnificent cabbages.

Care while growing

Cabbages must suffer no check. They must have ample water, ample nitrogen, and no weed competition. If you are going to the extravagance of using some organic high-nitrogen manure like blood meal, meat meal, cottonseed meal, or chicken or rabbit manure, then the *brassica* crops are good ones to put it on. Use it as a top-dressing when they begin to grow. If the plants are checked—say, by cabbage root fly—urge them on with a dressing of this kind. You may save them. Don't put nitrogen on just before the winter; it drives them on too fast, making them sappy and susceptible to frost damage. Earth up the stems as the plants grow.

Pests and diseases

Alas, these are plenty and virulent. But you may, if you are a good organic gardener and lucky, avoid them all.

CLUBROOT One of the most troublesome things in the garden, but many people live all their lives and never see it. Your garden either has it or it hasn't. Beware, though, if it hasn't got it, because it can get it: by your buying in infected plants; your bringing in manure from a contaminated source; you can even bring it in on your boots, after visiting a neighbor's garden that has it. Don't put the stems of bought cabbages on your compost pile unless you have inspected them first and made absolutely sure they have no clubroot.

When you have got clubroot you will find lumps or malformations on the roots of your wilting cabbages. You can get this with all the *brassicas*. Cut a few of your root swellings open. If there is a maggot inside one, what you have probably got is cabbage gall weevil, not clubroot. Rejoice. At least that is preventable. But you can, of course, be plagued by both.

Clubroot is caused by microscopic spores of a fungus that can lie dormant in the soil for up to seven years. The disease can be eradicated if the land is rested completely from cruciferous plants for seven years, and that means no cruciferous weeds, either—so no shepherd's purse or charlock. The disease thrives in acidic soil, so lime helps to reduce it. If you can get the pH up to 7 you may get rid of it. But many gardeners have to live with clubroot (I have done so for several decades) and just grow *brassica* crops in spite of it. Nonorganic gardeners dip the roots of their plants in calomine* at the planting-out stage. Calomine is a highly poisonous mercuric compound. The mercury is persistent in the soil and over the years inevitably builds up to serious proportions. Furthermore, the treatment is only occasionally effective. I fear that the plants are often infected invisibly at the seed-bed stage in which case nothing will cure them.

Preventive measures are: strict rotation so that cruciferous plants don't recur more often than once in four years; liming; burning of all affected roots; putting half a mothball (camphor) down each hole before planting; putting a half-inch (1-cm) length of rhubarb stem down each hole before planting; putting an equal mixture of wood ashes and crushed eggshells down each hole. I have not had complete success with any of these, but they may help depending on your particular circumstances.

Another line of attack, which is being researched by the Henry Doubleday Association in the UK, is to douse the land that is not being planted with brassicas with water in which brassica plants have been boiled. The effect of this is to wake

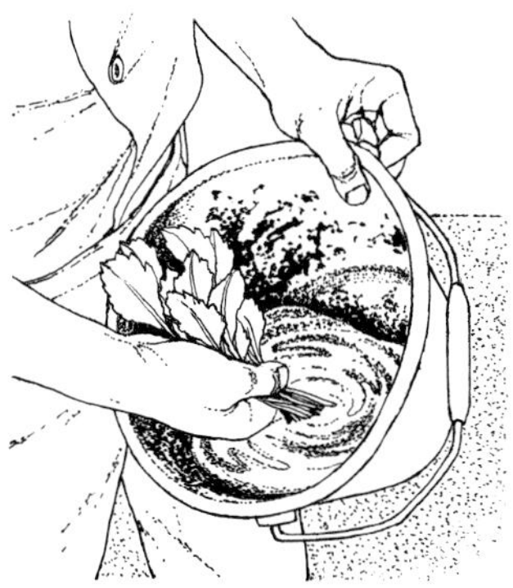

SWILLING, DIBBING AND FIRMING
Prepare cabbages for planting by swilling the roots in a bucket of thin mud mixed with a handful of lime. Remember that cabbages should be planted firmly. Use a dibber to make a hole to the same depth as the plant was growing in the holding-bed. Then pack the soil around the plant and heel in firmly with your boot.

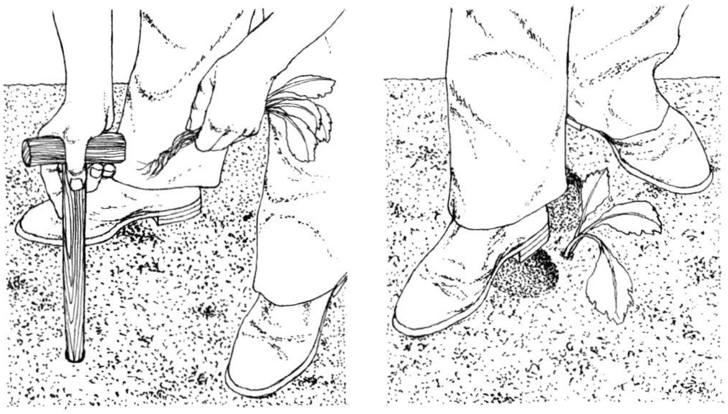

the sleeping spores by fooling them into thinking that *brassicas* have been planted. But there are no *brassicas* and the awakened spores, being unable to go dormant again, die.

Yet another approach, which is worth trying, is to sprinkle affected ground with 65 lb (30 kg) of quicklime per 100 sq yd (84 sq m) and then leave the ground brassica-free for at least five years.

CABBAGE ROOT FLY This attacks cabbages and cauliflowers but is less likely to go for Brussels sprouts or broccoli. If your plants wilt and you pull them out and cut into the roots and stems and find maggots, those are cabbage root fly maggots. When plants are badly affected their leaves appear bluish, with yellow edges. The fly, which looks like a house fly, lays its eggs on the top of the ground near the plants. The larvae hatch out, dig down through the soil, and then burrow up into the stem. Once there, nothing will shift them completely and they can kill the plant. Poisons don't help because they kill predators but on the whole tend to miss the maggots.

Small squares of tarred felt put like collars around each plant can obstruct the maggots. Either slit a five-inch (13-cm) square piece of tarred felt from one side to the middle and slip each plant into the slit, or else fold the felt in half, snip a "V" out of the middle, and thread the plant through the resulting hole. The flies lay their eggs on the felt and the maggots can't get down into the earth. A smear of kerosene on the felt is a good idea.

If plants do become infected, I hate to say it but a teaspoonful of nitrate of soda, or some other high-inorganic-nitrogen substance, works wonders. It not only helps the plant to start growing quickly and make new roots, but it also seems to disperse the maggots. Banking the soil up around the stems of affected plants also seems to help them; the plants can put out new and healthier roots. Kerosene* just sprinkled on the ground around each plant, once a week until they are large and healthy, also acts as a deterrent. Burn all infected roots after the plants are lifted and fork the soil over frequently in winter to allow the birds to help themselves to the pupae that are lying dormant in the soil.

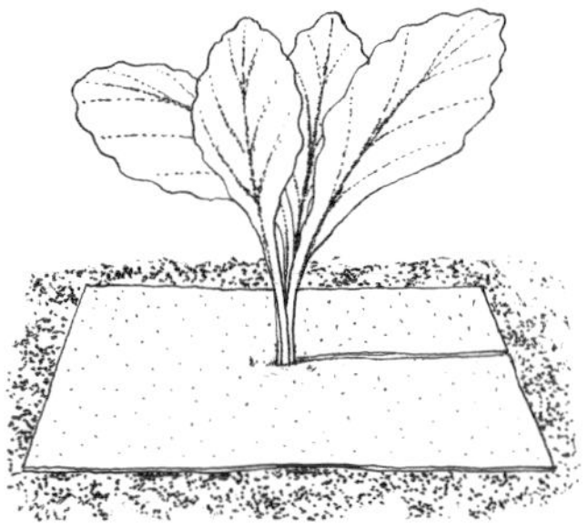

A TARRED COLLAR
Slit a five-inch (13-cm) square of tarred felt from one side to the middle and slip it around the plant. A smear of kerosene or grease on the felt is also a good idea. The cabbage root fly will lay its eggs on the felt, but the maggots will not be able to get down into the soil to burrow up inside the plant stem.

A SEED-PACKET COLLAR
An old seed packet, torn at both ends and placed over your cabbage plant, is effective protection against the destructive cutworm. Alternatively, surround each plant with a ring of wood ash.

CABBAGE GALL WEEVIL These sometimes attack plants in the seed-bed, and you will see small galls on the roots when you come to transplant. If there's only one gall, cut it open and kill the maggot. If there are more, burn the plant.

CUTWORM These minute worms frequently nip small plants off at ground level. Keep them away with a ring of wood ash or by placing a cardboard collar around each plant. A simple collar can be made by tearing both ends from an old seed packet.

CABBAGE WHITE BUTTERFLY The caterpillars of these can completely ruin a stand of cabbages if allowed to go unchecked. The best way to get rid of them on a small scale is to pick

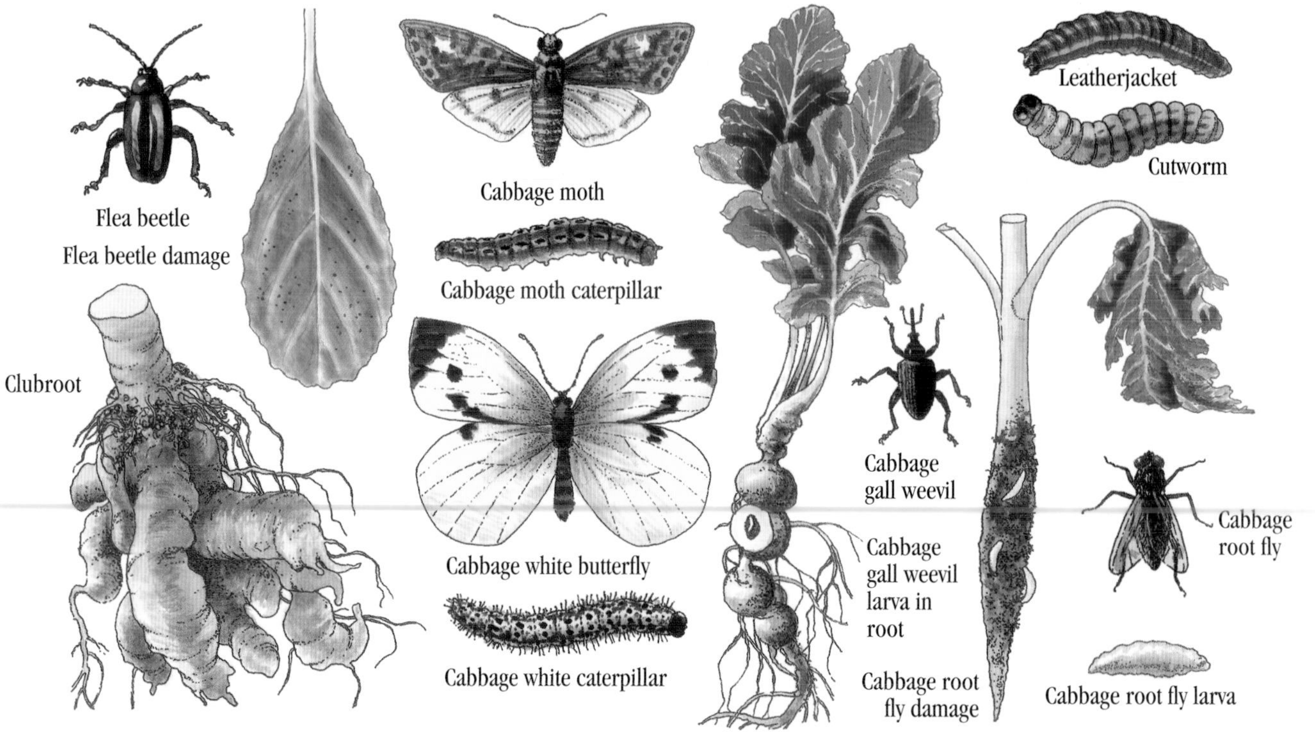

them off. Or soak with soapy water.

CABBAGE MOTH Pyrethrum or derris spraying will kill these larvae, which eat the hearts of cabbages and whose droppings can cause mold.

LEATHERJACKETS These are the gray-brown legless larvae of the cranefly. They sometimes eat the roots of brassica seedlings in late spring, causing the plants to wilt and die. All you can do is dig the ground very thoroughly and frequently in early spring so that the birds can eat the larvae.

Harvesting and storing

When you have cut a cabbage, pull the root out immediately; if you leave it in, it will encourage disease. If the plants are healthy, you can bash the stems with a heavy mallet or sledgehammer, or else run the garden roller over them. The object is to crush them, so that they can be put on the compost pile or buried in a trench over which you will plant next year's runner beans.

Cabbages can be stored by putting them on straw in a frostproof shed or basement and covering them with more straw. In mild climates, just leave them growing until you want them: 20°F (–7°C) won't hurt them.

SAVOYS

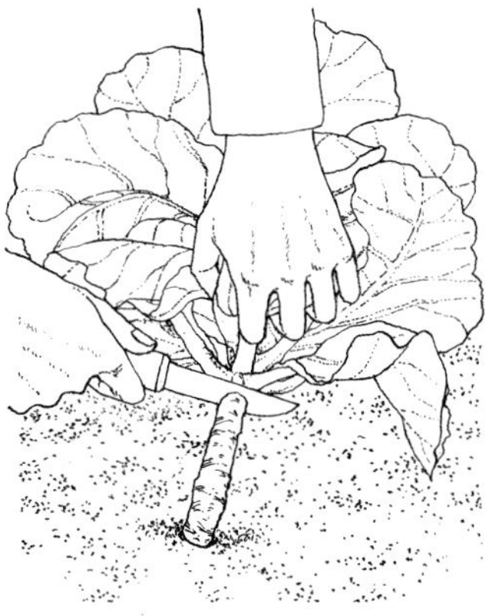

HARVESTING CABBAGES
Cut your cabbages at the top of the stem with a sharp knife. Remember to pull both the stem and the root from the ground.

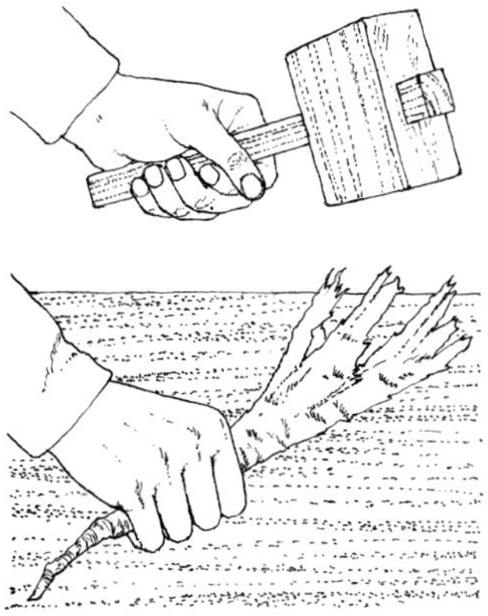

BASHING CABBAGE STEMS
Bash the uprooted cabbage stems with a mallet. The crushed stems can be left to rot for compost or buried in a trench.

These are the hardiest cabbages, and for late winter and early spring use, they are the most valuable. Treat them like winter cabbages, but don't eat them until the winter is well advanced and most other vegetables are gone. They fill a gap in the early spring.

RED CABBAGES

These need an extra-long growing season. Sow seed in a seed-bed in early fall in drills six inches (15 cm) apart. As the seedlings appear, thin them in the seed-bed. Plant out in spring 24 inches (60 cm) apart, and make sure the soil is firm, otherwise the cabbages will not develop large and compact heads. Cut red cabbages in the fall, and remember that although you can eat them raw, if you cook them they need far longer than ordinary cabbages—up to two hours.

Chinese Cabbages

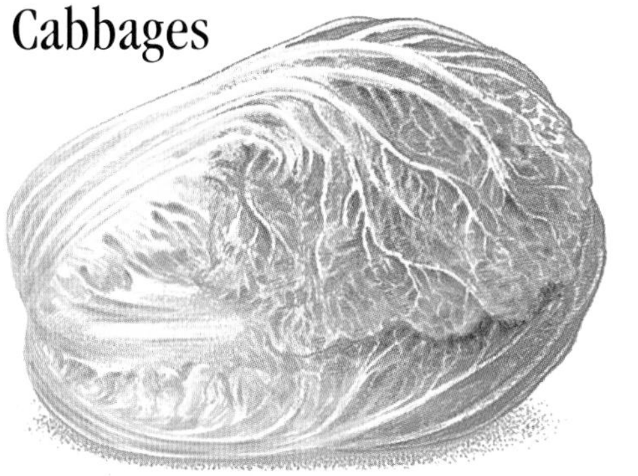

There are two kinds of Chinese cabbage—pe-tsai and pak-choi. Pe-tsai comes to a head; pak-choi doesn't and you just eat the leaves.

Soil and climate

Chinese cabbage is not so winter-hardy as cabbage, but nevertheless it does not do well in too much heat. It does not like acidic soil.

Soil treatment

They like plenty of humus. A four-inch (10-cm) covering of compost on the soil forked in before planting is ideal.

Propagation

Sow Chinese cabbage seed in situ in late summer or even later if you have no winter frost. Either broadcast and single later or sow thinly in rows.

Care while growing

Chinese cabbages need plenty of water. Mulching when the plants have grown to six inches (15 cm) helps to conserve moisture. Tying with raffia at top and bottom also helps to conserve water. Thin to about nine inches (25 cm) apart. If the plants grow big and crowded, uproot half and eat them. They suffer very little from pests or diseases; what they do have will be something that afflicts cabbages (see Cabbages).

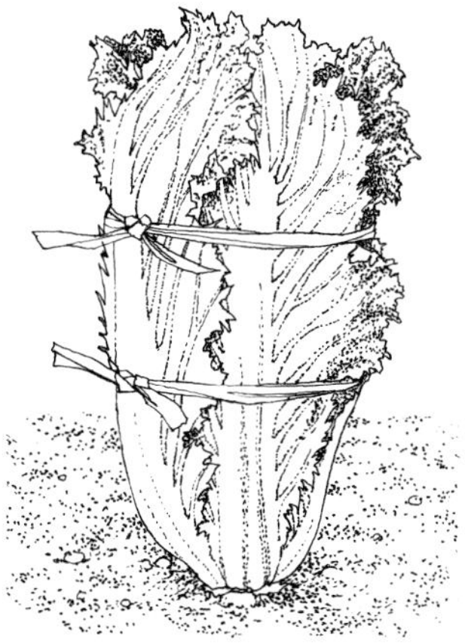

TYING A CHINESE CABBAGE
You must never allow a Chinese cabbage to go without water. When it has grown to about six inches (15 cm), a good mulching will help to conserve moisture. But as soon as the heart begins to form, it is an excellent idea to tie the leaves together top and bottom with strips of raffia. This will keep the plant sufficiently moist and at the same time blanch the inner leaves. Later on, thin the plants to about nine inches (25 cm) apart; if they still grow crowded, uproot half and eat them.

Harvesting

Uproot them and eat them, either in salads or as cooked greens, as soon as they have a good heart. This can be as soon as ten weeks after planting.

Brussels Sprouts

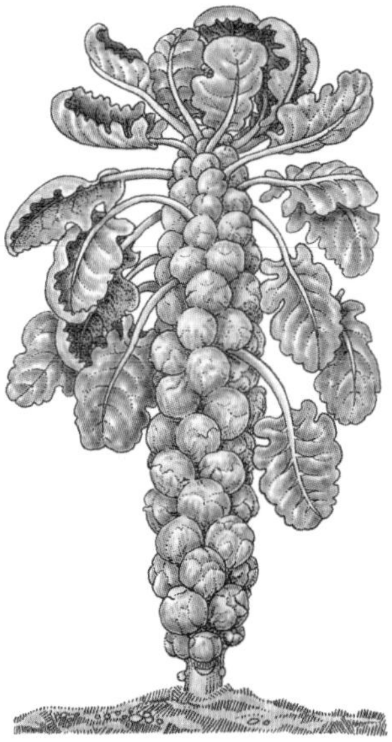

In my view, Brussels sprouts are the tastiest of the brassicas, and they will stand through the winter in temperate climates, allowing you fresh sprouts right through to the spring.

Soil and climate
Sprouts grown in countries that have no frost are tasteless, but where there is frost they make a noble standby for late winter and spring. Sprouts grow well in any good soil.

Soil treatment
Deep cultivation and plenty of manure or compost are the rule. They need lime if the soil is acidic. The traditional way to grow them is in very firm soil. Deep-bed gardeners can transplant them into soft deep beds (see p. 106) by planting them deeper than usual and pressing down the soil around them with one hand before and during planting.

Propagation
Traditionally, seeds are sown in an outdoor seed-bed in early spring. But you can get better results by sowing indoors in seed boxes (flats) from midwinter onward. Prick them out into frames if you sowed them early, then when they are about five inches (13 cm) tall plant them in a holding-bed (see p. 93) and then again into their permanent quarters, where they should be three feet (90 cm) apart in rows three feet (90 cm) apart. Like other brassicas, they seem to benefit from transplanting. Deep-bed gardeners should allow 20 inches (50 cm) between plants.

Care while growing
Brussels sprouts gain a lot from being earthed up while they are growing, and mulching is also good for them. Like all other brassicas, they don't like weed competition. Brussels sprouts grow very tall, so in very windy situations in exposed places, it may be necessary to stake them, but in most places a good earthing up will provide sufficient support. It is important to strip the lower leaves off when they begin to turn yellow.

Pests and diseases
Brussels sprouts are prone to all the diseases that afflict cabbages (see Cabbages).

Harvesting and storing
Seed sown in midwinter, indoors, will give sprouts as early as September, but these will not have the flavor of the later ones that have the benefit of a touch of frost. Like all vegetables, other than roots, they can be picked just as soon as they are ready. Pick the bottom ones first, then pick upward along the stems as more ripen, and finally eat the tops of the plants. If you have chickens, hang the denuded plants upside down in the chicken run. You can go on eating fresh sprouts in mild climates right into the spring, hence the classic idea of a British garden in February is of a bare and untidy mud patch, with a sprinkling of snow on it and a few dozen tattered Brussels sprouts, naked as to their lower parts, bravely leaning against the freezing wind, sleet, or hail.

In countries of intense cold or deep snow, you can harvest the sprouts by pulling the plants out of the ground before the worst weather comes, and store them by heeling them into soil or sand in a basement or storeroom. If it is cold enough they will keep for months like this.

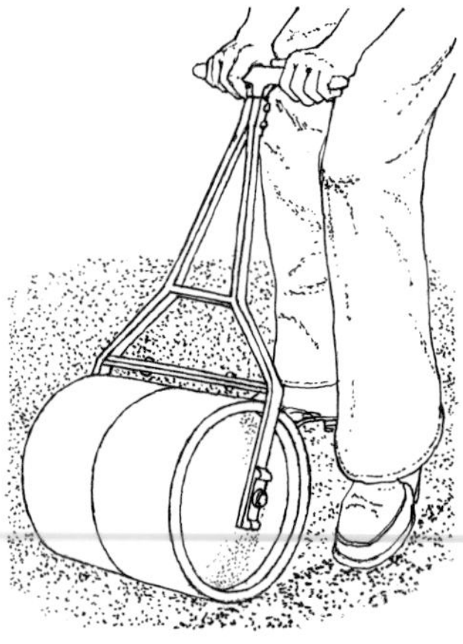

ROLLING THE BED
Sprouts need firm soil or they will grow loose-leaved. Prepare the bed for planting by stamping on it or rolling with a heavy garden roller.

EARTHING UP
Deep cultivation is essential for sprouts. At regular intervals, earth up around the base of each plant with a hoe.

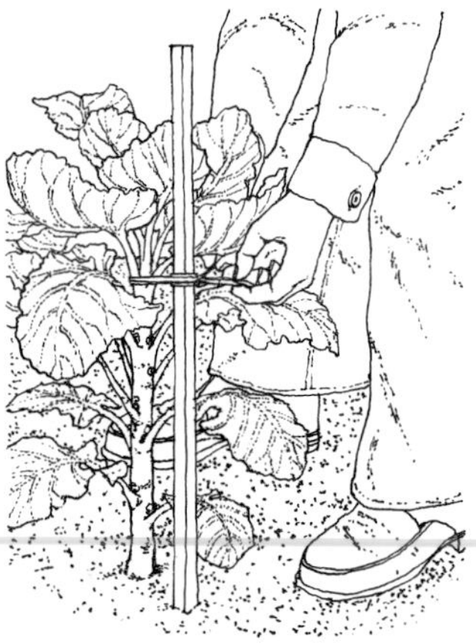

STAKE SUPPORTS
Sprouts can grow very tall, so it's a good idea to stake the stem of each plant—especially if it is growing in a windy position.

PICKING SPROUTS
When the sprouts have grown to the right size, pick first from the bottom of the stem, then work upward to the top of the plant.

Cauliflowers

EARLY AND MAIN-CROP CAULIFLOWERS

To develop cauliflowers, and also broccoli, man has taken the basic biennial cabbage plant, which naturally flowers in its second year of growth, and bred it to flower in its first. Cauliflowers have heads made up of tightly bunched white or purple flowers; the purple ones, sometimes called purple hearting broccoli, turn green when cooked. When you grow cauliflowers, remember that you are asking biennial plants to get through the whole of their life-cycle—growth, storage of nourishment, and flowering—all in one short season.

Soil and climate

Cauliflowers do best in temperate climates. They prefer heavy, moist soil with plenty of humus. They simply won't grow in poor soil, or in bad conditions.

Soil treatment

They need very firm soil, so don't pull out the pea or bean plants that may precede the cauliflowers in the bed, because this will loosen the soil. Hoe them off instead, leaving the roots in the ground. Then roll or tramp. Like all brassicas, cauliflowers don't like acid, so you must lime as necessary. Fork on a good dressing of fish manure (see p. 90) two weeks before planting. They also need some potash.

Propagation

You can get very early cauliflowers by sowing seed in midwinter, indoors, and planting out as soon as the ground has warmed up after the winter—probably in the middle of spring. With luck you will be eating them by midsummer.

For your main crop, sow seed in your brassica seed-bed (see p. 122) in early spring. Sow a quick-maturing variety so that you can begin to harvest in late summer, alongside a slower variety that can be harvested in fall and early winter. Plant out as soon as the plants have three true leaves as well as the two original seed leaves. If you are double-cropping, plant out in your holding-bed (see p. 93) and move them to their final position after peas or beans in midsummer. Examine the plants when you take them out of the seed-bed. If they are "blind," meaning they have no tiny bud in the middle, throw them away. They won't make curds, as the flowers are called. For the final planting-out, allow two feet (60 cm) between rows with 20 inches (50 cm) between plants. Allow 15 inches (40 cm) between plants if you use the deep-bed method (see p. 106).

PLANTING OUT CAULIFLOWER
Young cauliflower plants can be removed from the seed box as soon as they have grown three true leaves in addition to the original two seed leaves. But check that the plant isn't "blind"—meaning that it has failed to develop a bud in the center. If it is blind, then it won't be of any use to you—without the bud, your cauli just won't flower.

Care while growing

Do plenty of hoeing. Top-dress the growing plants with high nitrogen if you have it, but if your ground is really organic, with plenty of humus in it, a mulch of compost will do very well. Don't let your cauliflowers dry out because they must keep moving.

PROTECTING THE CURDS
Sunlight striking cauliflower curds can cause not only discoloration, but sometimes even a bad taste, so you need to protect them from the sun. Cover them by bending or breaking some of the outer leaves of the plant, and tying these in position over the plant with string.

If you don't want the trouble of blanching your cauliflowers, try growing purple ones. They last longer in the ground than the white-headed variety and need no blanching. The heads are deep purple on top and turn green when cooked; the flavor is like a mild broccoli.

Pests and diseases

Caulis share pests and diseases with cabbages (see Cabbages).

Harvesting

Harvest your cauliflowers as soon as they develop solid curds. The first should appear in late summer. If you wait, the curds will loosen and deteriorate. Cut well below the head if you want to eat the cauliflower immediately. If you pull them up by the roots you can store them in a cold basement for up to a month.

WINTER CAULIFLOWERS

Winter cauliflower is also known as self-protecting broccoli, but it looks and behaves like cauliflower, except that it is hardy and comes to harvest from midwinter onward. Sow in late spring and treat like main-crop cauliflower.

Broccoli

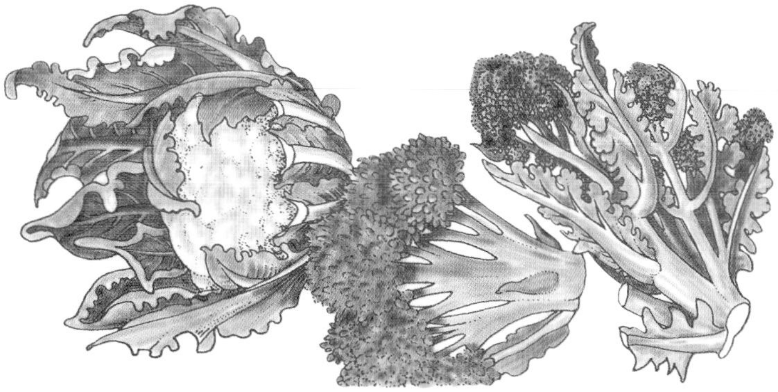

PURPLE AND WHITE SPROUTING BROCCOLI
The broccolis are easier to grow than cauliflowers. The great virtue of the purple and white sprouting varieties is that they are winter-hardy and can be harvested from late winter on, when kale and sprouts will be your only other greens.

Soil and climate
Good soil is not essential and they will grow in any but the very coldest climates.

Soil treatment
Give them very firm ground; even grassland from which you have just stripped the sod will do. Sprinkle with lime.

Propagation
Sow broccoli in your *brassica* seed-bed (see p. 122) in spring and plant it out when you have room, preferably after peas and beans. Let it wait in the holding-bed (see p. 93) a month or two if necessary. Allow 18 inches (45 cm) between plants and 30 inches (75 cm) between rows. For the deep-bed method (see p. 106), allow 18 inches (45 cm) all ways.

Care while growing
Mulch between the rows in summer. Mulch with straw in winter and stake the plants if they grow tall.

Pests and diseases
Broccoli can develop a mild case of clubroot. It does not suffer badly from pests.

Harvesting
Pick, and pick again, from the little flowering shoots as these become available in late winter. Keep picking right through spring and even into summer.

CALABRESE
Calabrese is a broccoli with green sprouts. It has more flavor than the other varieties, but the plants are less hardy. Sow it and plant it out along with the broccolis. Start picking it when the green flowerheads appear, until the first frost.

Kale

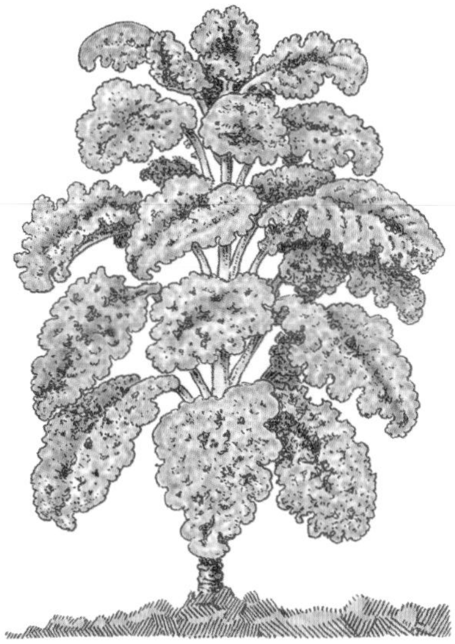

Kale is very winter-hardy and is often the last of the winter *brassicas* left standing. There are many varieties, both crinkly- and smooth-leaved, including collards. Kale is a non-hearting cabbage plant. All you get is the green leaves, and these are much better after a frost. Although they don't have the same delicacy of flavor as other *brassicas*, they are very nutritious and rich in vitamins. Cooked kale, provided it is only lightly cooked, has twice as much vitamin C as the equivalent weight of orange juice.

Soil and climate
Grow it wherever there are frosts. Any soil will do, but it crops best on rich soil.

Soil treatment
Kale likes fertile soil, which need not be especially firm.

Propagation
Grow kale for eating in your *brassica* seed-bed (see p. 122) and plant it out, when you have room, 20 inches (50 cm) apart in rows 30 inches (75 cm) apart. If you use the deep-bed method (see p. 106), allow 15 inches (40 cm) between plants in all directions.

Pests and diseases
Kale has a strong built-in resistance to clubroot. You can protect it from cabbage root fly and cabbage white butterfly by spraying with nicotine* in the fall. Kale is very tough, but all the same, it can suffer from any of the cabbage pests and diseases (see Cabbages).

Harvesting
Harvest from January onward. Pick leaves and side-shoots, but always leave some on the plant. Your crop will last you till May if you are very lucky. If you have a surfeit, kale makes a fine feed for hens, rabbits, goats, pigs, or any other domestic animals that are not carnivores.

Rutabagas & Turnips

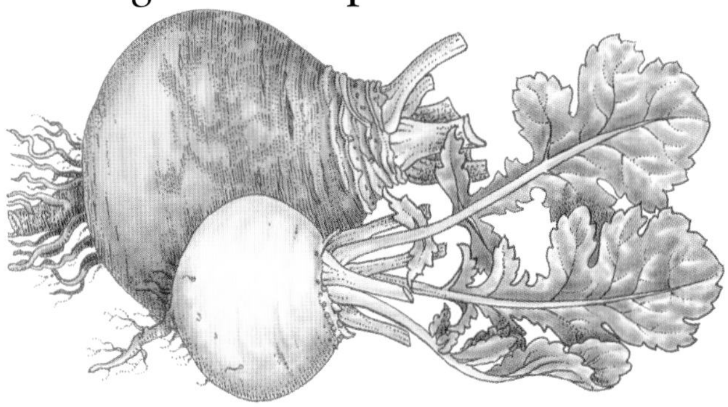

Rutabagas and turnips are both brassicas in which the first year's nourishment is stored in the root instead of, like the other brassicas, in the stem or the leaves. Rutabagas are orange whereas turnips are white, and rutabagas have their leaves coming out of a neck on top of the root, whereas turnip leaves grow directly from the root. Turnips will not stand severe frost, but rutabagas are a lot more hardy.

Turnips will come to harvest between 60 and 80 days after sowing. Rutabagas take about a month longer. Both can be sown late in the summer. They therefore make a good "catch crop"—a crop put in late after you have cleared the ground of something else. This is another way of getting two crops from the same plot in one year. When you are choosing what variety of turnip to plant, don't overlook the fact that some varieties can do double duty by producing both edible leaves and roots. In some varieties the foliage can be used for greens a month after seeding.

Soil and climate

They do best in a cool and damp climate. In hot weather they become hard and fibrous and are likely to go to seed. If you live in a hot climate, you must sow either very early in the spring so that they can be harvested when very young and tender, before the hot weather of the summer, or late in the fall when they will grow happily into the winter and come to full maturity. Light fertile loam is ideal, but they will grow in most soils. They like neutral or slightly alkaline soil like all brassicas, so lime if your soil is acidic.

Soil treatment

Cultivate deeply. Like all root crops, turnips and rutabagas like a very fine tilth. If you get a lot of rain in your area, grow your turnips and rutabagas on ridges. You can encourage their capacity for fast growing by manuring a year in advance.

Propagation

In cool temperate climates, turnips for picking and eating young should be sown in late spring and then, if you are a turnip lover, twice more at monthly intervals. Sow main crop turnips for storing in late summer. Even early fall is not too late to get a good crop of "turnip tops," which are even tastier than spinach and contain lots of iron. Sow rutabagas in early summer. The same sowing will provide young sweet rutabagas for eating in late summer and main crop for storage. Sow both turnips and rutabagas thinly in shallow drills where they are to grow. An ounce (28 g) of turnip or rutabaga seed will sow 250 feet (75 m) of row.

THINNING THE SEEDLINGS
Turnip or rutabaga plants shouldn't be crowded together. Start thinning them out, using a hoe, when they are still quite small. Leave about nine inches (23 cm) between plants. If you're using the deep-bed method, a distance of about six inches (15 cm) between plants will be enough.

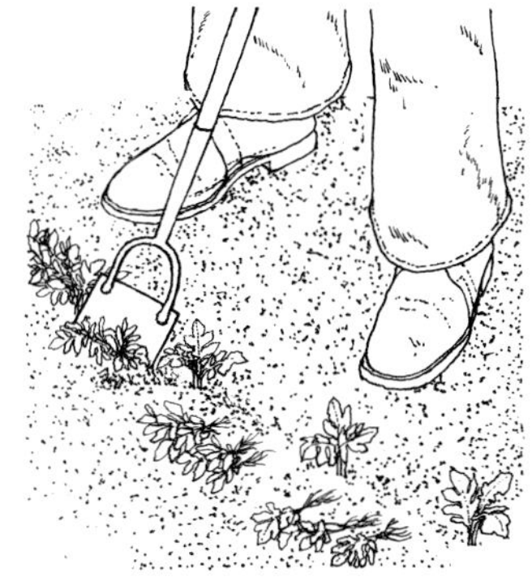

Care while growing

When the plants are still tiny, thin them out with the hoe so as to leave one plant about every nine inches (23 cm)—deep-bed method, one plant every six inches (15 cm). Leave a shorter distance for successional summer sowings that are to be eaten very young, and a greater distance for the winter main crop, which is intended for storing.

Pests and diseases

Turnips and rutabagas are subject to most of the pests and diseases that afflict cabbages and the same remedies can be applied (see Cabbages).

FLEA BEETLE When they are very young, turnips and rutabagas may get flea beetle (see p. 124). This shows as tiny holes on the leaves. A good shower of rain should cure it, or a good hosing with water if there is no rain. If this doesn't work, either derris or pyrethrum dust will kill them.

BORON DEFICIENCY Turnips and rutabagas are usually the first vegetables to suffer from boron deficiency. The core of your turnip or rutabaga will develop grayish-brown areas that will eventually rot and stink. A minimal amount of boron dissolved in water and added to the soil is sufficient to correct this deficiency.

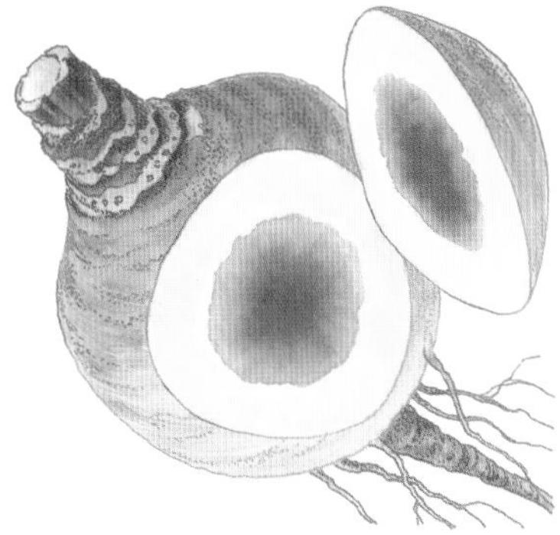

BORON DEFICIENCY
Turnips and rutabagas are good indicators of boron deficiency in your soil, since they are generally the first vegetables to suffer. The core of the turnip or rutabaga will turn a grayish-brown color and begin to rot and stink—sometimes becoming completely hollow in the center. An ounce (28 g) of boron is sufficient to restore a quarter of an acre (1,000 sq m) of land. Dissolve it in enough water to cover this area.

Harvesting and storing

The early successionally sown crop should be pulled during the summer when they are not more than three inches (8 cm) in diameter and are then at their sweetest. Main-crop turnips should be harvested before the first very hard frost and put in either a root store or a clamp (see p. 136). Rutabagas in all but the severest climates can be left out in the ground until they are wanted. So, eat your turnips before your rutabagas.

Kohlrabi

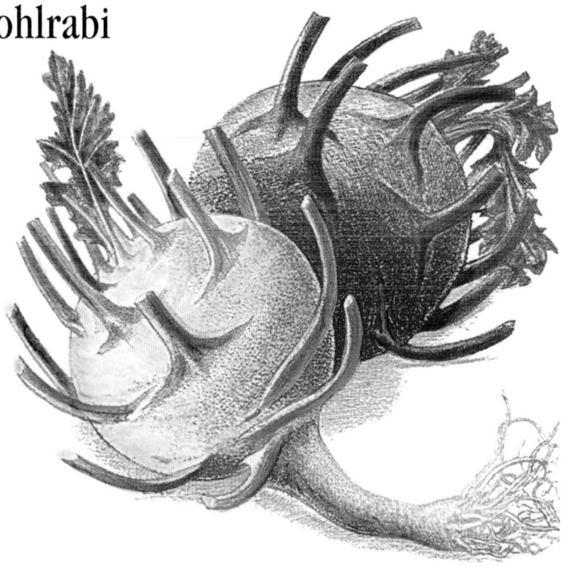

This strange-looking plant is merely a cabbage in which all the nutrients are stored in a swollen stem instead of in tight-packed leaves.

Soil and climate
Kohlrabi likes the same conditions as other brassicas but is even more dependent on moist soil; drought makes them hard and woody.

Propagation
It is better not to transplant kohlrabi, but to sow the seed out where the plants will grow. Sow thinly in two or three successional sowings between April and June.

Care while growing
Thin the plants you want to eat in the summer to six inches (15 cm) apart—deep-bed method, four inches (10 cm). Those you want to store through the winter, thin to ten inches (25 cm).

Pests and diseases
Kohlrabi can suffer from the pests and diseases that afflict cabbages (see Cabbages).

Harvesting and storing
Pick the plants very young and tender—about two-and-a-half inches (6 cm) across—and eat them raw or cooked. Store them as shown below.

STORING KOHLRABI
Pack kohlrabi in an unheated shed or basement between layers of straw.

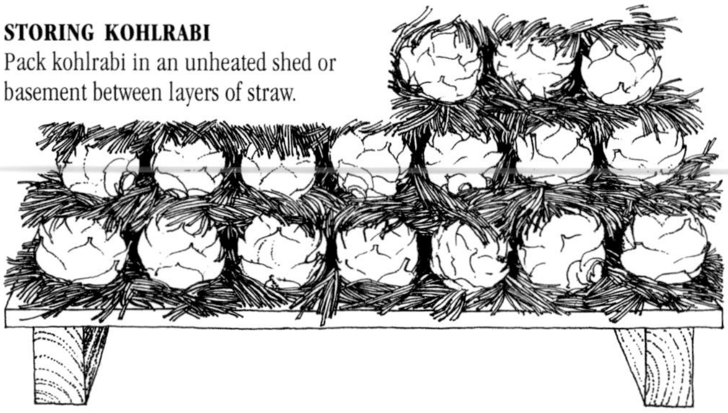

Radishes

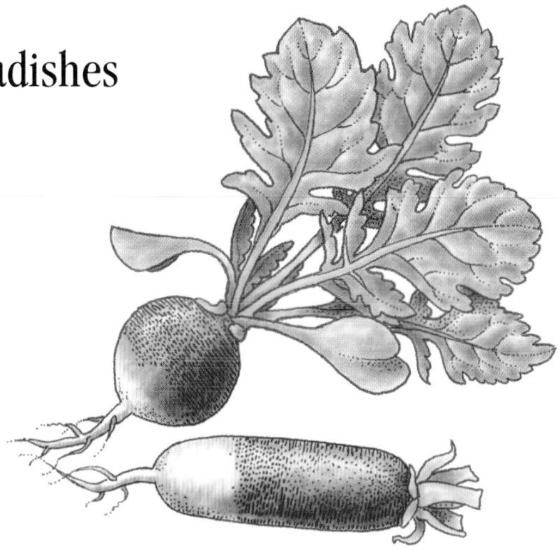

SUMMER RADISHES
Radishes grow in three to four weeks, are rich in iron and vitamin C, and are excellent for adding bite and crispness to salads. They are good for growing in odd vacant corners, and they do well in windowboxes as well. Winter radishes are larger and can be black, white, red, or red and white.

Soil and climate
Radishes like good, rich, damp soil and a cool, moist climate. Since they grow fast and are eaten quickly, it does not matter if they are grown in beds not reserved for *Brassicaceae*, for they don't have time to develop diseases. In hot countries they can be grown only as a winter crop. In cooler climates they can be grown in spring, summer, and fall.

Soil treatment
Like most *Brassicaceae*, they don't like acidic soil, so you should lime if it seems necessary.

Propagation
Just sprinkle the large black seeds thinly in shallow drills and cover them, or else broadcast and rake in. Sow very few at a time, but sow often—even once every two weeks—so you have fresh tender radishes whenever you want them. The seeds will keep for five years, so don't throw them away. If you want early radishes, you can sprinkle them among other crops that you are forcing in a hot-bed, or in a deep bed covered with transparent plastic or glass. In the deep bed (see p. 106) sow one inch (2.5 cm) apart in each direction.

Pests and diseases
FLEA BEETLE These pests are the only hazard. If heavy rain does not wash them away, give them a good hosing. If this fails, dust with derris or pyrethrum.

Harvesting
Just pull, wash and eat. If you have too many, pull them out anyway. Don't let them grow up hard and woody, or go to seed.

WINTER RADISHES
White radishes should be sown between late spring and midsummer; others toward the end of the summer. They will all be ready for harvest at the end of the fall. White ones must be dug and stored in peat. The others can be left in the ground until required during the winter.

Seakale

Seakale is a perennial whose young shoots should preferably be harvested in spring, but if you wish, it can be forced for eating fresh in winter.

Soil and climate
Seakale likes rich, deep, well-manured loamy soil and a cool, damp climate. Don't try to grow it anywhere hot and arid.

Soil treatment
Dig deeply—at least two spits—and incorporate plenty of rich manure into the soil.

Propagation
Seakale can be started from root cuttings, called thongs, or from seed. The former method is preferable, because plants started from root cuttings begin to yield the second year, a year earlier than seedlings. However it is said that a new race should be raised from seed from time to time. Get your thongs from a seed merchant or a fellow gardener. They are just bits of root about four inches (10 cm) long. Plant them six inches (15 cm) deep and 30 inches (75 cm) from each other in late winter—deep bed 15 inches (38 cm) apart. If you plant from seed, sow in shallow drills in March.

Care while growing
If you sow seed, thin to four inches (10 cm) apart and transplant to 30 inches (75 cm) apart the following spring.

Keep well weeded. You cannot eat green seakale because it is bitter. Therefore the plants must be blanched—that is, deprived of light completely so that they turn white. You can do this by covering them in situ with pails, boxes, or overturned flowerpots with the drainage holes blocked. If you want fresh seakale during the winter, you can force its growth. Either spread hot manure over the blanching covers so as to provide heat, or take the roots from their outdoor bed in the fall and plant them in loam in a hot-bed, or warmed frame, or even in a warm basement. Keep your seakale warm—the soil should be 55 to 60°F (13–16°C)—and dark, and you will get a good winter crop.

Pests and diseases
Small seedlings are occasionally attacked by flea beetle (see Turnips); otherwise they are not prone to attack.

Harvesting
Cut shoots when they are about a foot (30 cm) high in spring, unless you have forced the plants for a winter harvest. Like all perennials that are harvested for food, seakale must be treated with respect. After you have taken your just dues, let it grow up into the sunshine, green and strong, and build itself up for next year.

Cresses

WATERCRESS
If your garden has a corner that is persistently damp, watercress is the ideal crop. It has the distinctive hot flavor of the *Brassicaceae*.

Soil and climate
It does best in cool climates, but will grow perfectly well in a warm one, especially if it is standing in cold flowing water.

Propagation
It is possible to create a bed next to a flowing stream. Flood the bed by admitting water from the stream after sowing the watercress. You can grow it from seed, either planting the seed in the wet mud just above the water or sowing it indoors, in potting mix, in earthenware pots, which should be kept in a tray into which water flows constantly. You can bring it to maturity like this or else plant it out in a stream or damp bed. Another method is to buy really fresh commercial watercress from the store, put it in a plastic bag with some water, take it home, and plant it.

Care while growing
Pinch out the top shoots to make the plants bushy. If a plant flowers, cut it back.

Pests and diseases
Never grow watercress in water to which sheep or cattle have access. If you do, you might just get liver fluke.

Harvesting
Pick out side-shoots. The more you pick, the more grow.

CRESS
Like mustard (see p. 199), cress is eaten in the seedling stage, although if you are growing them together remember that cress takes a few days longer than mustard to germinate. Grow it on damp burlap or a damp peat* bed. Sow it thickly throughout spring and summer.

LAND CRESS
Also known as American cress, this is a relatively hardy salad plant, which should be pulled after about seven weeks of growth. Sow successionally through the summer for several months' salad supply. Sow half an inch (1.5 cm) apart and later thin to about six inches (15 cm) between plants. Find a fairly shady site that will not dry out, and protect the crop under glass as the weather becomes colder.

Solanaceae

Potatoes, tomatoes, peppers, and eggplants are all members of the *Solanaceae*. There is something a little exotic about this family, for it includes such dark and midnight subjects as deadly nightshade and tobacco, as well as such luscious tropical annuals as green peppers and chili peppers.

But there is nothing exotic about the potato. Even that great farmer and writer of the early nineteenth century, William Cobbett, termed it "the lazy root," for he thought it would supplant wheat, the cultivation and after treatment of which he considered to be the nursery of English virtues.

The other important member of the family is the tomato, which is so closely related to the potato that a hybrid has been that which has inferior potatoes on its roots and inferior tomatoes on its stems. Potatoes and tomatoes, like green and red peppers, came from tropical south and central America.

Most of the edible *Solanaceae* come from this area and they therefore require very rich, damp, and fertile soil, as similar to the rich leafmold of the tropical jungle as possible. Furthermore, none of the food-bearing *Solanaceae* are frost-hardy, which means, if you live in a cool climate, you must either start them off indoors, or not plant them until all danger of frost is past.

The *Solanaceae* have several pests and diseases in common and it is therefore advisable to grow them all in the same bed, or in the same part of the rotation. In this way your land is given a rest from solanaceous plants for the full cycle of four years, and there is no chance for disease to build up or for pests to accumulate. Certain eelworms, for example, can multiply to frightening proportions if tomatoes and potatoes are grown too often on the same land. Never touch any solanaceous plants when your fingers have been in contact with tobacco, because tobacco is a member of the *Solanaceae* and frequently contains virus disease.

One of the great values of most plants of this family is that they are rich in vitamin C. Potatoes are the richest source for most inhabitants of temperate climates; chili peppers are the richest source in many parts of the tropics. The reason why Indians eat hot curry is not to cool them down, but to provide themselves with vitamin C. All in all life would be much poorer were it not for this tribe of strange, soft-stemmed, potash-hungry, tropical-looking plants.

Potatoes

The potato is one of the few plants on which a person could live if he could get nothing else; and, unlike the others, it requires very simple preparation: no threshing, winnowing, grinding, or any of the jobs that make grain consumption a difficult technical operation.

Self-sufficiency from the garden in temperate climates is unthinkable without the spud, and I would recommend anybody, except those with the tiniest of plots, to devote at least a quarter of his land to it, and preferably as much as a third. Being a member of the family *Solanaceae*, it provides the soil with a rest from those families that are more commonly represented in our gardens. Without the potato break, we would find ourselves growing brassicas, for example, far too frequently on the same ground.

Soil and climate

Never lime for potatoes. They thrive in acidic soil: anything over a pH of 4.6. Scab, which makes them unsightly but doesn't really do them much harm, thrives in alkaline conditions but is killed by acidity. Potash is essential for good potatoes (but if you put on plenty of manure or compost, you will have enough of that) and so is phosphate. Nitrogen is not so important, although a nitrogen shortage (unlikely in a good organic garden) will lower your yield.

Unfortunately, the potato did not originally evolve for the climates of the northern hemisphere. It evolved in the Andes, and the wild potato is a mountain plant, although tropical. Its provenance makes it very frost-tender; the least touch of frost will damage its foliage and halt its growth.

Soil treatment

It is well worth digging the soil deeply the previous fall, and incorporating a heavy ration of manure or compost while you're at it; 8 cwt (400 kg) per 100 sq yd (84 sq m) is about right. Another excellent thing to do is to broadcast some green manure crop, such as rye, the previous fall after your root break. If you do this you should leave the crop undisturbed until a month or two before you want to plant your potatoes—unless, that is, your green manure is clover, in which case you should dig it in in the fall. In any case, dig the green manure crop well into the ground and at the same time dig in compost or manure. Or, and this works perfectly well, you can, if you have been short of time in the winter or if the weather has not enabled you to dig, actually dig the green manure crop when you plant the potatoes. Throwing any available compost or manure into the bottom of the furrow, plant the potatoes on top of this and fill in with the green manure.

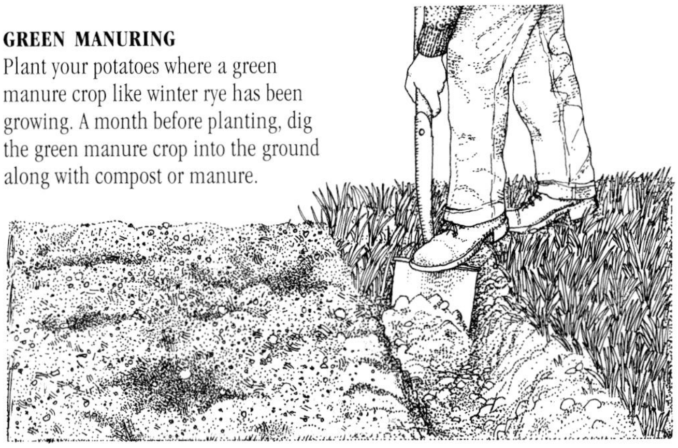

GREEN MANURING
Plant your potatoes where a green manure crop like winter rye has been growing. A month before planting, dig the green manure crop into the ground along with compost or manure.

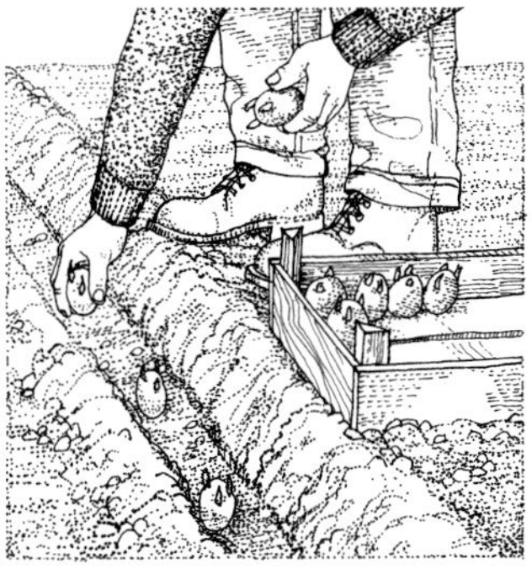

SOWING EARLY POTATOES
Make a trench five inches (13 cm) deep with a hoe, put in some manure or compost if you have it and put the seed potatoes in, rose-end up, about a foot (30 cm) apart.

COVERING EARLY POTATOES
Three to four inches (8 to 10 cm) of soil is usually enough covering for the early seed potatoes that you have planted. You will be earthing them up later on anyway.

Propagation

I know of nobody who grows potatoes from actual seed, although potatoes set seed in little green fruits that look just like small tomatoes. It is better to use sets, which are, in fact, just potatoes, although they are called "seed." If you plant a potato it will grow into a potato plant that produces between six and a dozen more potatoes. (The actual potatoes, by the way, are not *roots*—they are swollen underground *stems*.)

Potatoes grown in temperate climates near sea level are heir to certain diseases, which are the price they pay for growing in the wrong place. Among these are certain virus diseases that are transmitted by aphids. If you plant potato "seed" you will probably get a good crop of potatoes. But if you plant them where aphids abound, and you plant the new generation of sets from them the next year, the crop is likely to be slightly less. If you go on for a third year, and a fourth year, the crop will diminish even further. This is because there is a build-up, with every generation, of the virus diseases introduced by aphids.

The remedy is to get your "seed" from people who grow potatoes in places where there are no aphids. In practice ,"seed" potatoes must be grown above a certain altitude, or else on some windswept sea island where aphids wouldn't have a chance. The specialist seed growers carefully "rogue" the potato plants as they grow (that is, pull out any weak or diseased potatoes) and protect them from infection. The tubers that they lift are subsequently certified by the government agricultural authorities of the country concerned as being disease-free seed and fit for consumption.

All this does not mean that you cannot keep and plant your own tubers. Most people do, and you can even buy "once-grown" or "twice-grown" seed from your neighbors. And if you have land at over 800 feet (240 m) in the northeastern US, or on a sea-girt island, you can probably grow "seed" forever, both for yourself and to trade for other goods with your neighbors.

Seed potatoes should, ideally, be about 1½ ounces (40 g) in weight. You can cut larger tubers in half, provided you leave some "eyes" (small shoots) on each half, but I don't like doing this, as it can let in disease. Ideally, seed should be "chitted" before being planted. That is, it should be spread out, one spud thick, in a cool place, and in diffused light. Don't allow frost to get to it (frost will immediately rot spuds) and keep it out of hot sunlight. If the place is too hot and dark, long gangly shoots will grow off the spuds and tend to break off before you plant them. (If you can plant them without breaking these shoots off, though, the potatoes will grow very well.)

So the best thing to do with your seed potatoes is to lay them in chitting boxes in midwinter. These will stack one on top of the other and admit light and air to the spuds, and can be carried conveniently out to the garden for planting.

EARLIES " Earlies" are potatoes that grow quickly and can be eaten straight from the ground. They are not for storing. Plant them as early as you can, but remember that frost will kill them once they appear above the ground, unless they are protected with cloches or a thick covering of straw or compost. If they get frosted, you may be able to save them by hosing the frost off with warm water.

MAIN CROP Your store of "main crop" potatoes will go a long way toward keeping you alive during the winter. Plant them in late spring.

Chitting is well-nigh essential for early potatoes, but if you don't get around to chitting for your main crop, never mind—plant them just the same. You will still get a crop—it will just be later, that's all. And never plant any diseased tuber, or one that looks defective in any way. You will simply be spreading disease among your own crops, as well as those of your neighbors.

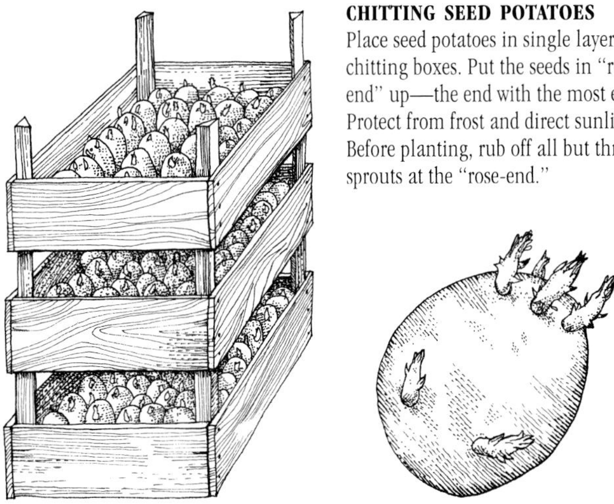

CHITTING SEED POTATOES
Place seed potatoes in single layers in chitting boxes. Put the seeds in "rose-end" up—the end with the most eyes. Protect from frost and direct sunlight. Before planting, rub off all but three sprouts at the "rose-end."

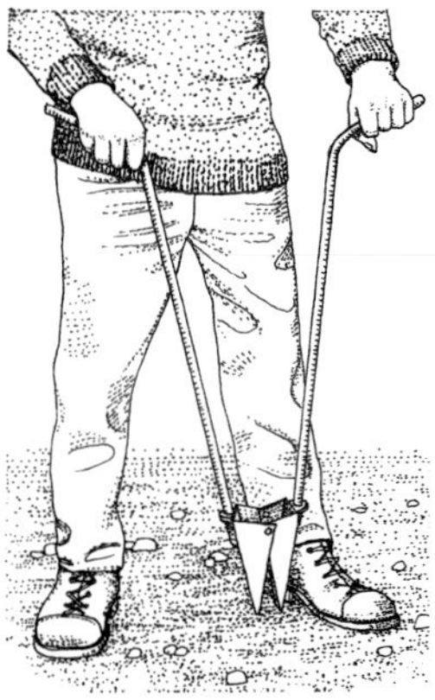

USING A POTATO PLANTER
If your soil is light, loamy, or sandy, you can save yourself some toil by using a potato planter. Ram it down into well-dug ground with your foot, drop a potato into the cup, then close the cup by pushing the handles of the planter together. Withdraw the planter, and the potato will be left buried in the ground. The method isn't quite as good for the potato as simple trenching, but it is easier and quicker. The potato planter is useless for sowing in heavy soil or clay, but it is ideal for the deep-bed method.

Don't put early potatoes in too deep: if they have four inches (10 cm) of soil over them when you have finished, that is usually enough. If the land has been dug before—that is, if you are not digging it for the first time since the previous fall—just make a furrow with the corner of a hoe, about five inches (13 cm) deep, put the spuds in, and cover with about four inches (10 cm) of soil. You will earth them up well enough later, and do not want the crop to grow inconveniently deep in the ground. If you have light, loamy, or sandy land, you would do well to get a potato planter, which is also the ideal thing for the deep-bed method (see p. 106).

For early potatoes, have the drills about two feet (60 cm) apart, but have them 30 inches (75 cm) for main crop. Put earlies in a foot (30 cm) apart in the rows: main crop about 15 inches (38 cm). Remember, main-crop plants have much longer to grow and produce much bigger and heavier crops.

Now there are other methods of planting potatoes. One excellent one is to plant the potatoes on compost, cover them with more compost, and then a thick mulch of straw or spoiled hay. Or you can use leaves or leafmold in this way with good results. If you grow early potatoes with this method, you can gently remove some of the mulch, take a few spuds, and let the plant go on growing to produce some more. All these mulch-cover methods do great good in that they enrich the soil for other crops after the potatoes are finished. As you rotate your potato crop around your garden, the whole property becomes enriched.

PLANTING ON COMPOST
Put a good layer of compost in the furrow, and plant the potatoes on top. Cover them with compost, then mulch with straw.

HARVESTING FROM COMPOST
Lift some of the mulch, pull a few potatoes, and replace the mulch. The plant will go on to produce the main crop.

A very effective method, which has the advantage that it can be done in a small space (even on a patio), is to grow potatoes in a barrel. Fill the bottom of a barrel with a thin layer of soil and plant a single "seed" potato in it. Keep adding more soil as the plant grows upward and you will find that more and more tubers will form in the new soil. Finally the green plant will be sprouting out of the top of the barrel. Wait for the plant to flower and then simply empty the barrel out. You will find a huge number of potatoes in it.

You can work the same principle to even better effect by laying an old car, truck, or tractor tire on its side, filling it with soil, and planting one or more potato sets in it. When the plants have grown, but before they flower, add another tire and fill it with soil. Allow the plants to continue growing. Keep adding tires until the plants reach about four feet (1.2 m). Then harvest by dismantling the whole structure. This is better than the barrel method because the plants have plenty of light throughout their growth.

GROWING POTATOES IN BINS
Growing potatoes in bins is particularly useful if your space is limited—even a patio will do. Take an old garbage can, and fill about a sixth of it with soil. Plant one or more potato sets. When the plants have grown, but before they flower, put another layer of soil on top. Continue building up layers of soil as the plants appear, until they reach about four feet (1.2 m). When the potatoes are ready for harvesting, simply empty out the bin and you'll find you have a surprisingly heavy crop.

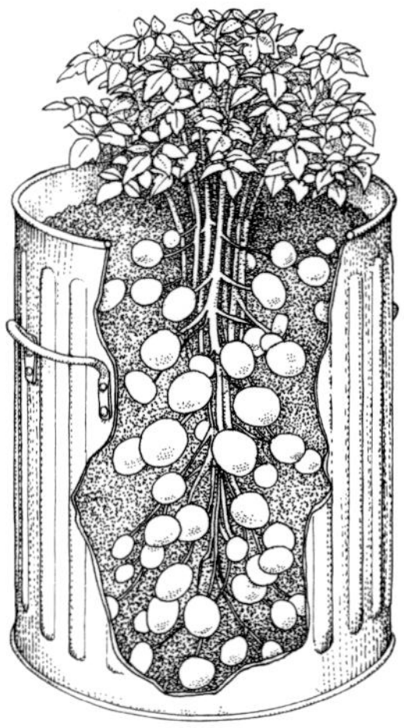

If you plant in a deep bed you should get an enormous crop. You can plant 18 inches (45 cm) deep. And in the very soft earth of the deep bed you can plant with a potato planter. Leave a foot (30 cm) between potatoes.

Care while growing

Potatoes need a lot of room under the ground, and they will turn green if exposed for more than a day or two to the light, in which case they become bitter and poisonous. This is because they produce a toxin called solanin. It is usual therefore to earth them up—that is, to draw or throw soil up around the plants so as to protect the growing tubers from the light and give them plenty of room for growth and expansion. Of course people who grow beneath a mulch don't have to do this, but they do have to make sure that there is plenty of mulch to cover the potatoes completely. Potatoes do not like weed competition, and in the very rich, deeply cultivated soil that they grow best in, weeds grow at an amazing rate. When you ridge up, you must be sure to kill the weeds. When weeds sprout up on the ridges between the potatoes, you must either hoe them out or pull them by hand. Throw them down in the furrows to rot and they will help by forming a mulch.

One tip about earthing up potatoes: for some reason the plants stand upright at night and in the early morning, but sprawl helplessly during the heat of the day. Potatoes are far

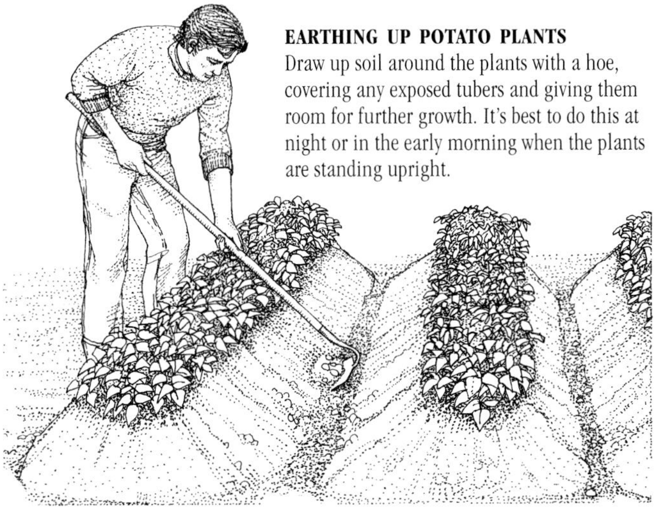

EARTHING UP POTATO PLANTS
Draw up soil around the plants with a hoe, covering any exposed tubers and giving them room for further growth. It's best to do this at night or in the early morning when the plants are standing upright.

easier to earth up early in the morning when they are standing up like guardsmen on parade. Like real guardsmen, when the sun gets hot they are inclined to faint. You may have to earth up several times; the final one should be thorough, the ridges patted down with the back of the space and made steep and even, for in this way they defy the spores of potato blight should this disease strike (as it very likely will).

When the tops of the main crop meet over the rows, they will suppress weeds and, after the final earthing up, you can relax for a while.

Pests and diseases

POTATO BLIGHT When the potato first came to Europe, blight did not affect it. In cool, damp climates, people living on peaty acidic soils—where the potato grew better than any other food crop—became entirely dependent on it. In Ireland, particularly, it became the mainstay of the poorer country people, to the exclusion of almost anything else. Then, in the middle of the nineteenth century, the blight struck. In one year it spread right through Ireland, blasted the tops of the growing crop, and later rotted the potatoes in their clamps, turning them into a slimy mass. Millions of people died.

No cure was found for blight until somebody noticed that potatoes growing downwind from copper smelting plants in South Wales did not get blight. So could copper prevent this dreaded disease? A mixture of copper sulfate and lime was tried—similar to the mixture already being used by the vignerons of Bordeaux against mildew on their grapes. It was found that if the foliage was sprayed with this at "blight times"—when the temperature and humidity of the air were above a certain point—the foliage was protected from the drifting spores of the blight. So now, to avoid blight I spray, very thoroughly above and below the leaves, with Bordeaux mixture*, and I do it about once every two weeks throughout the hot and humid weather of the summer. If you are in a dry windy area you may not get blight. Ask your neighbors. You can buy a commercial spray or else mix up your own Bordeaux mixture* (see p. 105).

And what if you do get blight? You will know by black patches that appear on the leaves, and thereafter develop borders of white powdery stuff which is, in fact, the spores of the fungus that causes the disease. You cannot cure it by spraying then, although you may protect healthy plants from its spread. But do not despair. Unless the attack is a very early one the spores will not spread down to your tubers, and if you have earthed up well, spores that are washed down by the rain will not sink into the earth and come in direct contact with the tubers. You must cut the haulms (foliage) off with a very sharp blade (sharp so as not to drag the potatoes out of the ground) and burn them. It is sad for an organic gardener to have to say burn anything, but—yes—burn them. Then leave your tubers undisturbed in the ground for at least three weeks after you have removed the haulms. If you lift them immediately they will come into contact with billions of spores on the surface of the soil. If you leave them be the spores will be washed down the steep sides of the ridges into the furrows where they will sink harmlessly into the soil below. Leave the spuds for as long as possible. In the moderate climate where I live I often don't lift them until I need them to cook—even after Christmas sometimes. They are safer there in the ground than they are if I lift them.

WART DISEASE This is becoming less common as it is becoming compulsory, under EC regulations in Europe at least, to grow only immune varieties. The disease manifests itself by wartish growths that can cover the surface of the potato. Burn all such spuds. I've grown wart-free potatoes for a lifetime, but when it does infest your ground, you should plant only immune varieties. Otherwise, don't grow potatoes for six years on that land and hope the disease will die out. 65 pounds (30 kg) of quicklime* per 100 sq yd (84 sq m) of infected land are said to kill it.

SCAB You will probably get this if you grow spuds in very alkaline soil or in soil that has been recently limed. It is not serious. If you want to sell potatoes it matters, for they don't look nice. If you just want to eat them, don't worry about it; simply peel the scab off. But plenty of manure or compost will prevent scab, so organic gardens simply shouldn't have it.

POTATO ROOT EELWORM This arrived in Europe from its home in South America before World War I and has been with us ever since. It attacks chiefly the monoculturists who grow potatoes on the same soil year after year—or at any rate too frequently. Don't grow spuds too often on the same soil. If you get it really badly you will have to give up growing potatoes there for at least ten years, although it is said that if you grow several crops of *Tagetes minuta* on the land, and compost it, or dig it in as green manure, it will suppress eelworm. And, if you grow a crop of *Tagetes minuta* the year before planting potatoes, it will fool the eelworm cysts, by its secretions, into remaining dormant during the tenure of the potato crop.

COLORADO BEETLE This is yellow with four black stripes on its back. It hibernates deep in the soil and emerges in early summer to lay its eggs on potato foliage. The grubs then eat the leaves and can easily destroy a whole crop. Potatoes

SPRAYING FOR BLIGHT
During hot and humid weather, you can protect your potato crop against blight by spraying over and under the leaves every two weeks or so with Bordeaux mixture*. If some of your crop is affected in spite of this, continue spraying the healthy plants, so that the disease is prevented from spreading.

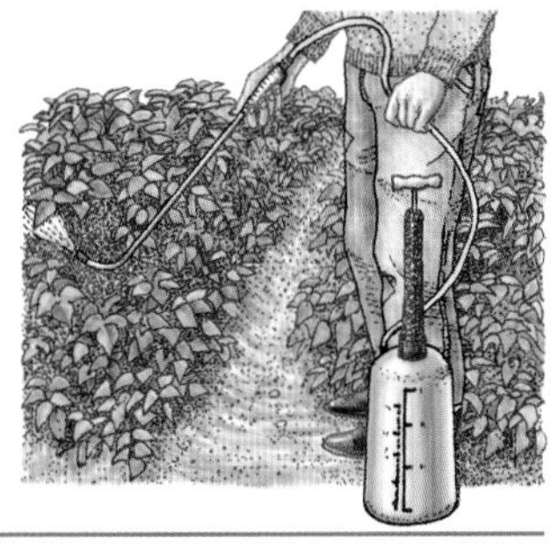

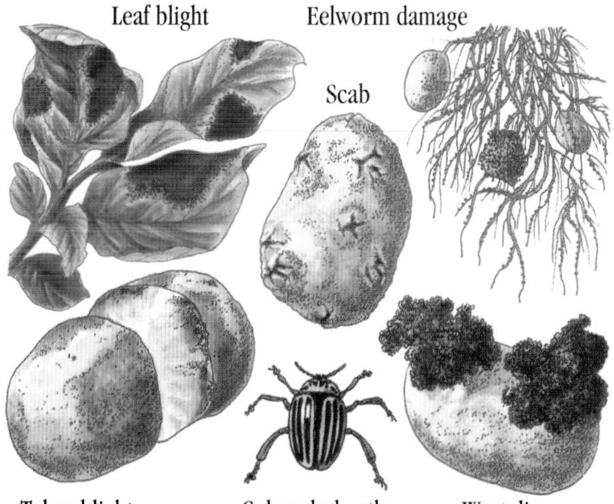

grown on a very large scale are the most susceptible. If you do see a beetle on the leaves, squash it, and immediately notify the authorities. The grubs can be sprayed with derris, pyrethrum, or, best of all, nicotine*. Deep dig in the winter to expose the beetles to attack by birds.

The other diseases of the potato (and there are over a hundred of them) should not be a problem provided you use only clean, healthy seed and grow on heavily manured or composted ground, preferably not more than once every four years. Don't suffer any "volunteers" to exist; that is, plants that have grown up from potatoes that you have inadvertently left in the ground. They will only cause a build-up of disease.

Harvesting and storing

You can harvest any time after the plants have flowered. Dig potatoes carefully with a fork, taking great pains not to spear any, and if you do spear any, eat those first, for if you store them with the others they may cause rot. You can scrape and eat early potatoes immediately: just dig them as you want them half an hour before a meal. Earlies have a lot of their carbohydrates in the form of sugar, because they are still busy growing and must have their energy in a still soluble form.

Main-crop potatoes, however, have ceased to grow, and the sugar has all turned into starch, which is really what potatoes are. So lift your main-crop potatoes as late as you like but before the very hard frost sets in, and preferably in dry weather. By then the tops will have wilted and dried. Leave the spuds lying on top of the ground for a day or two for the skins to "set," or harden, and the spuds to dry. Don't leave for longer, because potatoes left too long in the light will grow green and become bitter and poisonous, but two days won't hurt. If you have a lot of potatoes—say, a ton (1,000 kg) or more—you can clamp them. That is, pile them in a steep-sided heap, cover with straw or dry bracken, and then cover with soil. The colder your winters, the thicker the covering of straw you will require. In places with very cold winters you cannot clamp at all because the spuds will become frosted. Leave small chimneys of straw sticking out on top of the ridge of the clamp every two yards or so, and build straw-filled tunnels at similar intervals around the base. Beware, a thousand times beware, of rats. If they make their homes in your potato clamp, it's goodbye to your potatoes.

If you have less than a ton (1,000 kg), or live in a very cold climate, store your potatoes indoors. The requirements are: they should be in complete darkness; they should be ventilated; they should be as cold as possible but not subjected to frost, which rots them. So, a plastic or metal garbage can with a few ventilation holes knocked in the top and bottom will do, or else wooden crates, or barrels, but again, with ventilation. And in fact your potatoes will come to no harm if you just leave them in a pile in a completely dark corner of a frost-proof basement or shed. Sacks are no good, except for certain open-weave synthetic fiber sacks; burlap or canvas will rot, and so will paper.

The reason for the old-fashioned "root cellar" in the northern US is very hard winters. In most of Europe the clamp is the traditional way of storing potatoes in large quantities, and on the whole it is the method of storing I prefer.

DIGGING AND CLAMPING
Harvest your early potatoes at any time after the plants have flowered. Late potatoes can be left in until the plants have died down. Dig potatoes out carefully with a fork, making sure you don't spear any. If you do, don't store them because they will very likely cause rot. If your crop is large—say, over a ton (1,000 kg)—it can be stored in a clamp (right). Pile your potatoes steeply on a bed of straw, cover with more straw, and mound over with soil. Ventilation is important, so make small chimneys of straw every two yards along the top of the clamp, and insert straw tunnels at the same intervals at ground level.

Potato clamp

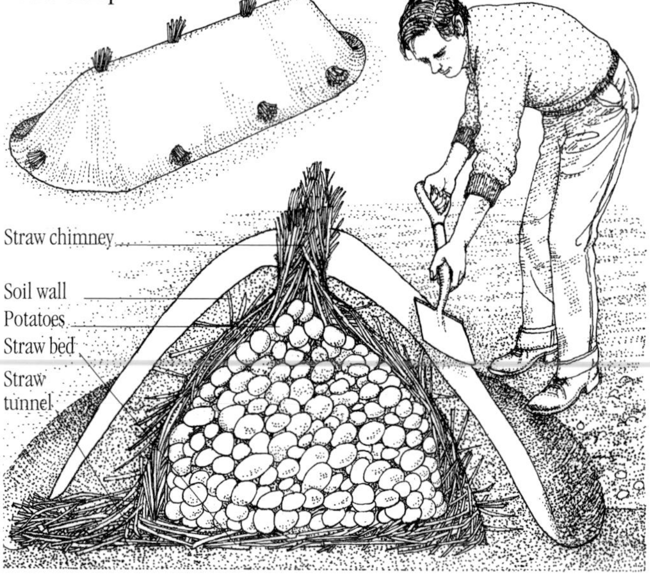

Tomatoes

Like the potato, the tomato is a native of tropical America. It is a perennial, but in temperate climates, where it is only half-hardy, it is grown as an annual. The tomato has enormous value for the self-sufficient gardener, for not only does it improve any dish to which it is added, it is also a very rich source of a whole range of vitamins and, most important, these do not seem to be damaged a great deal by cooking or canning. If you grow enough tomatoes in the summer to have them canned (or frozen) all winter, you will not suffer from vitamin deficiency.

Soil and climate

To grow tomatoes outdoors, you must have at least three-and-a-half months of warm weather with plenty of sunshine. If you don't live in this sort of climate, you must start them off indoors and plant out in the summer. If the summer turns out too cold or cloudy to bring them to ripeness, they can be ripened indoors or else made into green tomato chutney. In cool or cold climates, they grow well in greenhouses, and are far and away the most valuable greenhouse crop. But remember that, unlike cucumbers, tomatoes do not like a lot of humidity.

Tomatoes do well in any rich soil. In light soil they will give an earlier crop than they do in heavy soil. On the other hand, they grow well in heavy clay that has had several years of compost application.

Soil treatment

For each seedling, I like to dig a hole about a foot (30 cm) deep and as wide across, and fill it nearly to the top with compost, as though I were planting an apple tree. Then I fill the hole up with soil and, when the time comes, plant out in that. It is best to do this about six weeks before planting out. In fact, it is a good idea to prepare the stations at the same time as you sow the seed indoors. However you manage it, tomatoes need rich soil, plentifully endowed with rich muck or well-rotted compost. As the compost rots and settles, the ground sinks a little, which helps the tomatoes retain water.

Propagation

Except in very hot climates, sow the seed indoors in seed boxes during spring. In all but the coldest climates, an unheated greenhouse will do; a temperature of 70°F (21°C) is ideal. Sow thinly in either commercial seed-starting mix or a mix that you make yourself (see p. 92). Cover the seed boxes with newspaper during the night, but in the day make sure they are in full sunlight. Two or three weeks after sowing, prick the tiny plants out three inches (8 cm) apart in larger seed boxes or, better still, peat* pots. Again, use commercial potting mix or your own mixture. Don't water too much; keep the soil slightly on the dry side—just dry-to-moist. And always give them all the sun there is.

After a month, start gently hardening the plants off. Put them outside in the sun in the day but indoors at night, or else move them to cold frames and keep the covers on at night but off in the day. At the beginning of summer they should be sufficiently hardened off for planting out in their stations. If you have cloches, plant your tomatoes out two weeks earlier and keep them covered for that period.

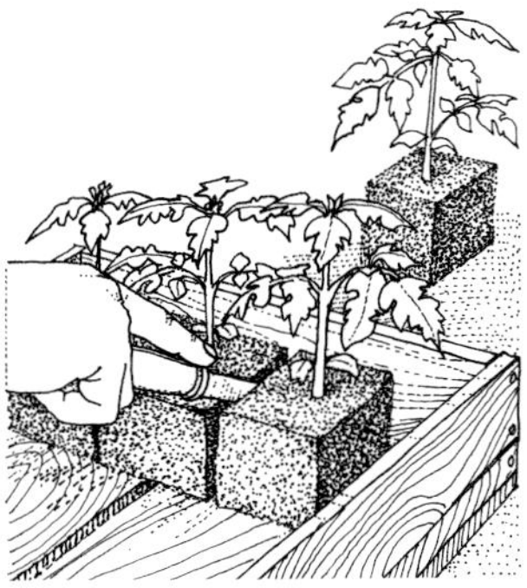

PLANTING OUT SEEDLINGS
Avoid disturbing the soil around your tomato seedlings. Two weeks before planting out, cut squares around each plant in the seed box, so as to keep the soil intact when it comes to planting.

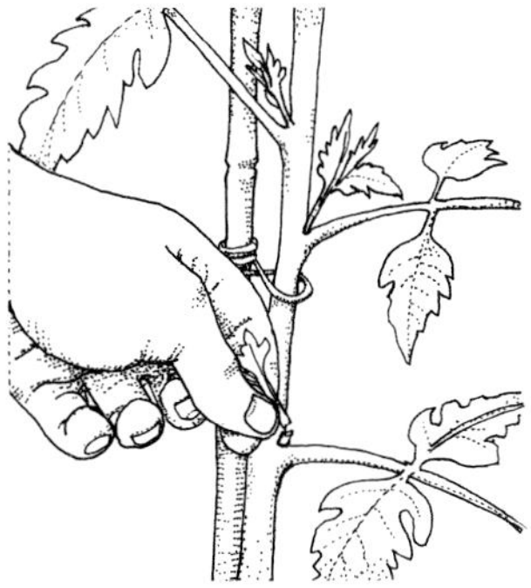

PINCHING OUT SIDE-SHOOTS
A straggly or untidy tomato plant may not set any fruit. With your fingers, pinch out the little shoots that emerge at the point where the leaf stalks meet the main stem.

Planting out

You must provide tomatoes with adequate vertical support, because by nature they are trailing plants. The best way is to use stakes, at least five feet (1.5 m) tall. Sink them about a foot (30 cm) into the earth. Position the stakes when you plant out the tomatoes, but don't damage the root clusters. Plant the tomatoes quite deeply so that the lowest leaves are just above the ground. Tomatoes tend to put out adventitious roots from their stems and this should be encouraged. If you have any long straggly plants, you can make them stronger by laying a length of the stem horizontally in the ground; the stem will send out new roots.

Spacing depends on what sort of plants you intend to grow. In cooler climates, plants should be kept small, and two feet (60 cm) apart in rows three feet (90 cm) apart is about right. In warmer climates, they can grow larger, so more space should be allowed. In hot climates you can let them sprawl, and then plants should be four feet (1.2 m) apart. If you use the deep-bed method (see p. 106), allow two feet (60 cm) between plants.

A method that is worth trying in a small garden is to plant out leaving only a foot (30 cm) between rows. Then stop the plants when they have set one truss only and allow no side-shoots to develop. This sounds ruthless, but you should get more ripe tomatoes than with usual methods.

When planting out, take great care to disturb the soil around the roots as little as possible. If you are using peat* pots, just soak the plants and plant the peat pots as they are.

If you use seed boxes, cut the soil in squares around each plant down to the bottom of the tray with a knife. Do this two weeks before you plant out, and try to retain that soil intact when you plant.

RING CULTURE Ring culture is another way of growing tomato plants from seedlings. It is best practiced in the greenhouse, though it can be done outdoors. The method is worthwhile if you are short of soil or short of space, or if your soil is disease-ridden.

Two weeks before you want to plant your seedlings, stand rings of plastic nine inches (23 cm) in diameter and nine inches (23 cm) high on a bed of clean gravel. Fill the rings with commercial potting mix or your own mixture (see p. 92). Two days before planting, water the gravel and the rings with water in which a complete organic compost or manure has been steeped—a thick soup, in fact, and lace it if you can with a little fish meal.

When you plant the seedlings, water the rings with ordinary water, and continue watering through the rings only for the next ten days. After that, when the roots have reached the gravel, water the gravel only, and ensure that it is permanently moist. Once a week, give the gravel a good dousing with your organic soup. Otherwise treat like ordinary tomatoes.

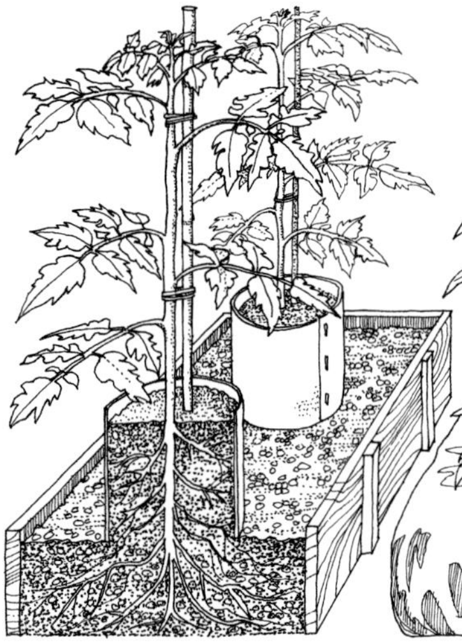

THE RING CULTURE METHOD
Stand rings of plastic on a bed of clean gravel, fill with potting mix, and plant seedlings. Feed the plants with organic compost or manure steeped in water. After ten days, water the gravel only.

PLANTING IN GROW BAGS
Plant four seedlings into a commercially prepared grow bag or in an old fertilizer bag filled with peat*. Take care with watering since the bags won't drain and the plants may become waterlogged.

PEAT BAGS This is a space-saving method that works indoors and outdoors, and is especially good for people who have just a patio or a balcony. Buy a specially prepared bag or simply fill an old fertilizer bag with peat*. Plant four seedlings in each bag and water carefully; there is no facility for drainage, so the plants can easily get waterlogged.

Care while growing

Tie the tomato plants loosely to their stakes with soft string and keep tying as they climb. Don't tie too tightly or you will cut the stems. An excellent alternative is to drop a tube of wire netting, about 15 inches (38 cm) in diameter, over each plant. The plants will climb inside the tubes.

Pinch out the little shoots that spring up at the base of each leaf stalk; otherwise you will get an untidy straggling plant that probably won't set any fruit. And don't let your plants get too high.

KEEPING TOMATO PLANTS LOW
Four trusses on each staked plant are enough. To stop them from growing higher, simply break off the tops above the fourth truss.

FALL RIPENING
When the nights draw in, take the plants off their stakes and lay them on straw under cloches. Tomatoes must be warm to ripen.

Tomatoes do not want too much water but they want some; if the ground dries right out, the fruit will crack. The very best thing to water the plants with is liquid manure. (Make it by half-filling a barrel or tank with farmyard muck and topping up with water.) Remember that tomatoes need warm roots, so nip the bottom leaves off and train the plants as upright as possible so that the sun can get to the soil around the roots. In fall, when the nights begin to draw in, take the plants off their stakes, lay them down horizontally on straw, and cover them with cloches. This certainly helps to ripen the fruit.

Pests and diseases

BLIGHT Outdoor tomatoes are just as susceptible to potato blight as potatoes are, so spray your tomatoes with Bordeaux mixture* (see p. 105). Spray once every two weeks during the warm summer weather, and if it pours with rain just after you have sprayed, spray again.

CUTWORM These pests bite plants off just above the ground. Protect the plants with small collars of cardboard (see p. 124) or sprinkle wood ash around them.

Harvesting and storing

Pick the fruit gently with the stalks on and take great care not to damage the skin. Red tomatoes must be eaten fresh, or must be canned or frozen immediately. (Tomatoes for cooking can be stored in a freezer.)

Green tomatoes, or tomatoes that are not quite ripe, can be covered with cloth or paper and kept in a cool place until they ripen. Keep them in the dark, never in the sun. A time-honored method is to lay a sheet of soft felt in the bottom of a drawer in a cool room, lay a layer of tomatoes on top, making sure none of them are touching each other, lay another piece of felt on the tomatoes, then more tomatoes, and so on. Lay the greenest at the bottom and the ripest on top. Be sure the tomatoes are all healthy or you may end up with a drawer full of mold. Green and ripe tomatoes can be stored as chutney (see p. 218).

Peppers

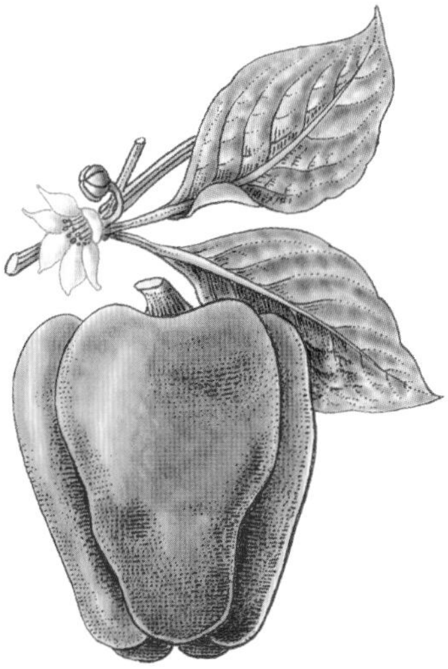

Green peppers, red peppers, and chili peppers are all capsicums. In their native South America they are perennials, but we grow them, like tomatoes, as annuals.

Soil and climate
They are slightly hardier than tomatoes. You can grow them outdoors in warmish temperate climates, but they are better started off under glass, and better still if they do all their growing under glass. They need at least 65°F (19°C) when they are flowering or they won't set fruit.

Soil treatment
They prefer light soil and benefit from compost.

Propagation
You can buy seed, but I think it is much better to buy some ripe red peppers of the kind you like best, break them open, and take the seed out. Sow the seed indoors at least six weeks before the last expected frost. Sow a few seeds in each pot and when they are about five inches (13 cm) tall, thin to the strongest one. Plant outside about three weeks after the last expected frost (two weeks earlier if you have warmed the ground with cloches) in beds prepared as if for tomatoes (see Tomatoes). Like tomatoes, plant them deeply.

Care while growing
Treat just like tomatoes but give them more water while they are young. Always water the roots, never the peppers; if they get too wet, they are liable to rot. Mulch heavily.

Pests and diseases
ANTHRACNOSE As long as you plant your peppers well away from your beans, they will not suffer from anthracnose. If they do get it, they will go bad. Burn them.
CUTWORM Protect your seedlings with cardboard collars (see p. 124) when planting out.

Harvesting and storing
Cut them off (don't break them) with an inch (2.5 cm) of stem on each fruit. If you have more than you can eat fresh, hang the vines up in a dry, windy place to dry. You may have to finish the drying process by hanging them over mild heat indoors. You can then just drape them, decoratively, in your kitchen or storeroom until you want them in winter.

Eggplants

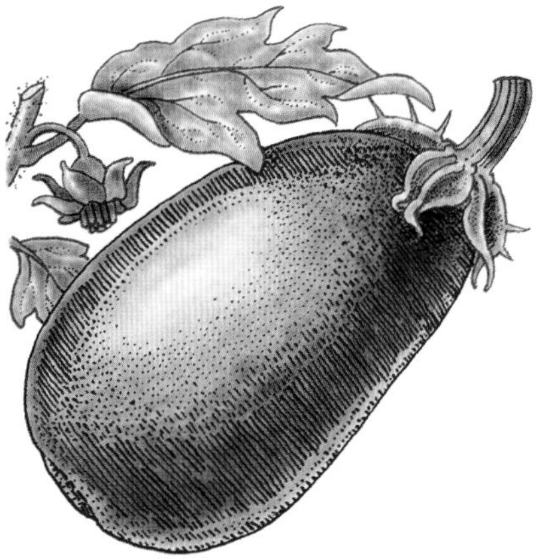

Even the most dedicated flower gardener has to appreciate the eggplant. As well as their luscious fruit, they have luxuriant purple flowers and large velvety leaves. They are not high in nutrition, but they make up for this with their unique flavor.

Soil and climate
In frost-free areas, the eggplant plant will grow as a perennial bush, but where there is the slightest touch of frost it must be grown as a tender annual. It needs a hot summer and deep rich soil with plenty of moisture, but it does not thrive in wet weather.

Soil treatment
Dig plenty of manure or compost into your soil. Eggplants like a pH of about 6.

Propagation
Sow the seed indoors about ten weeks before you plant out. Soak the seed overnight and then plant each seed in a peat* pot filled with good potting mix. If you have no peat* pots, sow an inch (2.5 cm) apart in seed boxes filled with potting compost. When the seedlings are two inches (5 cm) high, plant out into even richer potting mix four inches (10 cm) apart in seed boxes or in a cold frame.

When your plants are about ten weeks old, and the ground outdoors is warm to the hand, plant the peat* pots out three feet (90 cm) apart—18 inches (45 cm) apart if you have a deep bed (see p. 106). Take great care not to damage the roots.

Care while growing
Until the warm weather comes, keep cloches over them if you can spare them, and if you have a deep bed, use mini-greenhouses. Keep them well watered—with manure water (see p. 103) if possible—but don't overwater.

Pests and diseases
FLEA BEETLE These pests may attack when the plants are young. The leaves are quickly eaten away if they are not checked. Use derris dust to get rid of them.
MOULD The plants may get moldy in damp climates. The answer is to reduce the humidity in the air, so that the mold dries up. If your plants suffer much, try in the future to use fungus-resistant strains.

Harvesting
Cut them off—don't pull them—as soon as they have that lovely unmistakable bloom and before they are fully grown. The plants will then continue to fruit.

Apiaceae

Carrots, parsnips, celery, celeriac, Hamburg parsley and Florence fennel belong to the *Apiaceae* family (formerly the *Umbelliferae*). Members of this family also include several herbs such as caraway, angelica, and parsley.

Umbelliferous plants have numerous tiny flowers borne on radiating stems like umbrella ribs. They are a decorative family: the foliage of carrots is very attractive and will look good among your ornamental shrubs. Many members of the family are good to eat (beware the poisonous hemlock, though), and very many are eaten by animals. Cow parsley, for example, is well worth gathering for feeding to rabbits.

A feature that all members of the *Apiaceae* share is that their seed is very slow to germinate. So do not despair if you sow some *Apiaceae* seed and the plants take ages to show their heads. Just sow a few radish seeds along with the others; the radishes will be up in no time and will show you where the rows of your *Apiaceae* are.

The members of the *Apiaceae* family are closely related to all sorts of interesting plants, wild and cultivated, including such things as ginseng, whose ground-up roots are said to "relieve mental exhaustion," and sarsaparilla, which is used to make old-fashioned root beer.

Carrots

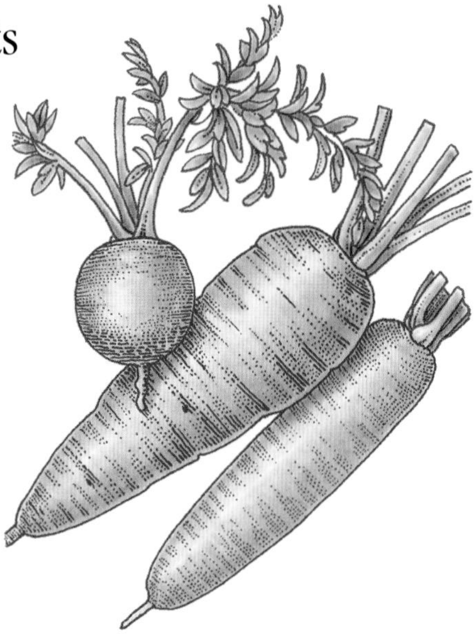

Like so many valuable food plants, the carrot is a biennial, and stores in its first year what it is going to spend in its second in the form of seed. We thwart its aims by gobbling it up in the first year before it has time to grow to maturity. Carrots have been bred to be either long and slow-maturing but heavy-cropping, or short, stubby and quick-maturing but lighter-cropping.

The most important constituent of carrots is carotene, which the human body converts to vitamin A. No other vegetable or fruit contains comparable quantities of this vitamin, which, among its other virtues, improves your eyesight; hence the parental exhortation to generations of children: "Finish up your carrots and you will be able to see in the dark".

Soil and climate

Carrots are a cold-climate crop. They can be sown very early in the spring in temperate climates, or in the fall or winter in subtropical ones. Carrots prefer to follow a crop that has been manured the previous year. A sandy loam is ideal for them. Heavy clay is not good for them, but if that is the soil you are blessed with, you can improve it enormously by copious and constant manuring or composting. Two pails of compost and two of leafmold applied to every square yard will convert even heavy clay into suitable soil for carrot-growing. They like deep soil, particularly the heavy-cropping long-rooted main-crop varieties.

Soil treatment

Do not apply fresh manure just before sowing carrots because it makes them tough, watery, and inclined to fork. If the soil is acidic, it should be limed, although carrots will flourish in a pH as low as 6. Like all root crops, they like both phosphate and potash; plenty of manure or compost should supply this, but be sure to apply manure at least six months before the carrots are planted. Rock phosphate and wood ashes are also worthwhile. The land should be dug deeply and raked down to a fine tilth.

Propagation

Draw little furrows about half an inch (1 cm) deep if you live in a moist climate but an inch (2.5 cm) deep in a drier one. Sprinkle the tiny seeds fairly thickly—four or five to an inch (2.5 cm)—as some of them don't germinate. With the deep-bed method (see p. 106), sow two inches (5 cm) apart each way. Pelleted seed (seed that has been coated with fertilizer so that each seed forms a little pellet) can be used to good

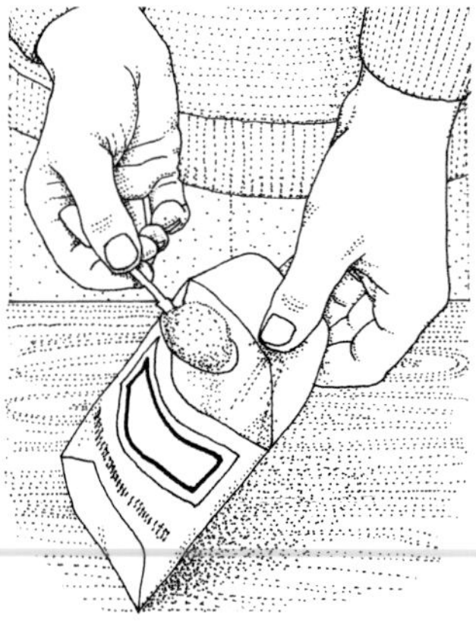

DREDGING CARROT SEEDS
Since carrot seeds are hardly visible, it is difficult to scatter them sparsely. Coat the seeds by shaking them up in their packet with a spoonful of slaked lime or ground limestone.

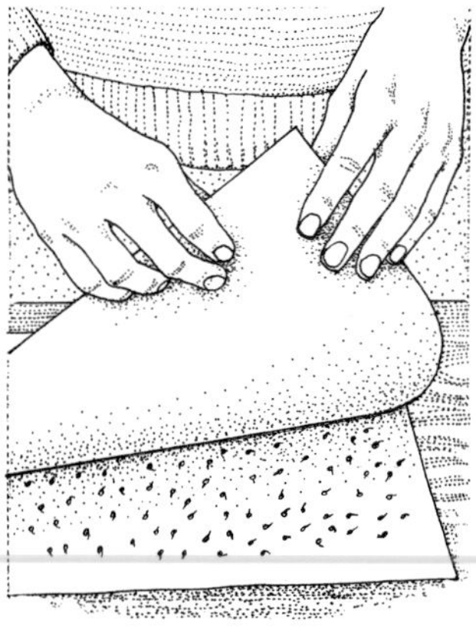

GERMINATING CARROT SEEDS
Carrot seeds need moisture if they are to germinate properly. Two days before planting, encourage them by placing them between two sheets of wet paper towel. But watch for mold.

effect with carrots. Sow it with a precision drill, far more thickly than you think you need to, as germination is even poorer than in conventional sowing. Cover the seed lightly with fine soil or, better still, with fine dry compost. In dry weather, soak the furrows well. Don't worry if nothing happens for a bit. Carrots take a long time to come up.

Try interplanting carrots with onions row by row. This is said to deter both carrot and onion fly, for the scent of the one masks the scent of the other.

Care while growing

You can just leave early carrots as they are in the ground, which will be very overcrowded, and pull them for what gardeners call "bunching." They are tiny like this but they taste very sweet.

Main-crop carrots, which you want for the winter, you will have to thin, and here your problems start. For as soon as you bruise a few carrots, and disturb the soil around their roots, you attract the carrot fly, which is said to be able to smell a bruised carrot for up to six miles. In areas where carrot fly is very bad, it is often best to do your thinning only on a wet day, preferably in light rain. Then, after pulling the thinnings, carefully step the ground down around the remaining carrots. Thin to an inch-and-a-half (4 cm) between plants first; later on thin to three inches (8 cm) apart. For a deep bed (see p. 106), thin once to two inches (5 cm). You can eat the thinnings.

Don't hoe near carrots. It is not good to loosen the soil and even worse to cut the carrots. You can hoe between the rows, but hand-weed only in the rows. Try not to let the carrot bed dry out. If you have to water the soil, give it a very thorough soaking; you want the water to go deep down and pull the carrots downward with it. Just watering the surface is no good at all.

Pests and diseases

CARROT FLY Carrot flies lay eggs on carrots; the larvae burrow into the root and spoil the crop. The carrot fly looks much like a small housefly; still, it is not the fly itself you are likely to notice, but the leaves of your carrots turning dark red as a result of the damage to the roots. The fly can be kept away with a dressing of soot, or you can mix an ounce (30 cc) of kerosene* in a gallon (4.5 l) of water and sprinkle that on, shaking it up well as you do so. Alternatively, mix a pint (0.6 l) of kerosene* with a bushel of sand and sprinkle that on at the base of the plants. If the carrot flies are bad, you may have to repeat the dressing every two weeks, and after every rain in the case of soot. If you have suffered from a very bad attack, dig the bed thoroughly in late fall, so that birds have a chance of getting at pupae in the soil. In an organic garden, you will have lots of beetles, and these may well eat up to half the carrot fly eggs before they hatch.

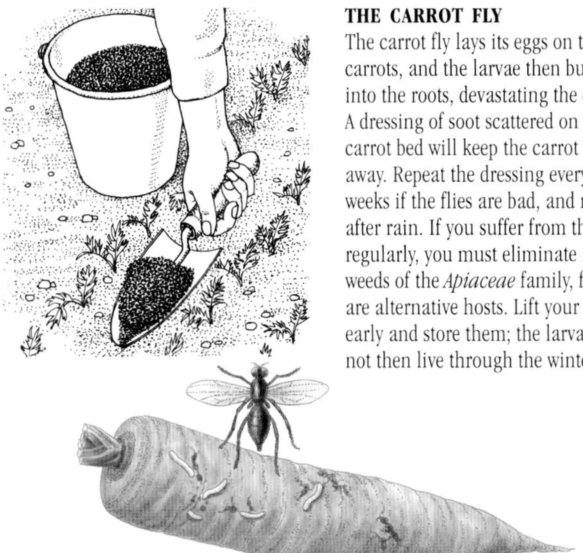

THE CARROT FLY
The carrot fly lays its eggs on the carrots, and the larvae then burrow into the roots, devastating the crop. A dressing of soot scattered on the carrot bed will keep the carrot fly away. Repeat the dressing every two weeks if the flies are bad, and repeat after rain. If you suffer from these regularly, you must eliminate all weeds of the *Apiaceae* family, for these are alternative hosts. Lift your carrots early and store them; the larvae will not then live through the winter.

CARROT DISEASE This can be bad in inorganic gardens but is unlikely to worry the good organic gardener. Brown spots appear on the roots and ultimately tiny red spores come to the surface of the soil; these are the spores of the mycelium that causes the disease. Burn all diseased roots and sprinkle the diseased soil with two parts of sulfur mixed with one part of lime and don't plant carrots there again for at least five years.

Harvesting and storing

If you pull carrots out of the rows at random you will attract carrot fly. So when you pick early carrots for eating fresh in summer, start at one end of the row. Main-crop carrots for winter storing can be left in the ground until well into the winter. In cold climates, lift them before the ground gets too hard and store them, not letting them touch each other, and twisting off the leaves first. Pull your carrots as early as possible if you have had a bad attack of carrot fly. The emerging larvae are then unable to pupate and produce a new generation of flies. If you do leave the carrots in the ground too long in wet weather, the roots are inclined to split; make sure that when you lift the carrots, you take care of the roots. Eat any damaged in lifting; don't store them with the others. And remember that if you wash carrots before storing, they will inevitably go totally rotten.

Store in a well-ventilated, cool place; just above freezing suits them best. Do not store year after year in the same root cellar because disease will gradually build up. It is best to store them in sand or peat*. You can use a variety of containers: a garbage can with plenty of holes punched into it to let in the air; a barrel; a dark corner of a cold shed (but beware of rats and mice); or even a box sunk in the ground outdoors, covered with a lid, some straw on top of that and then some soil. If you have a very large number, you can clamp them (see Potatoes).

STORING THE CARROT CROP
When harvesting your carrots, be careful not to damage the roots. If you do, then don't store with undamaged carrots. It is best to store in a cool but well-ventilated place. An old barrel makes a good container. Place the carrots so that they don't touch each other, and build them up in layers of sand or peat*. Make sure the barrel is ventilated—drill holes if necessary. Don't store carrots year after year in the same place because you may get a gradual accumulation of disease.

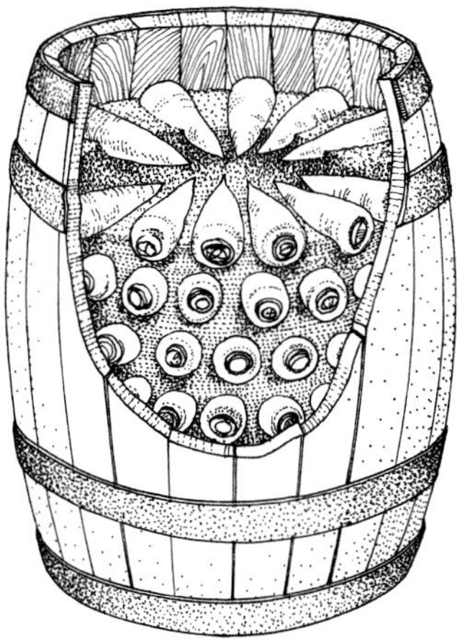

Parsnips

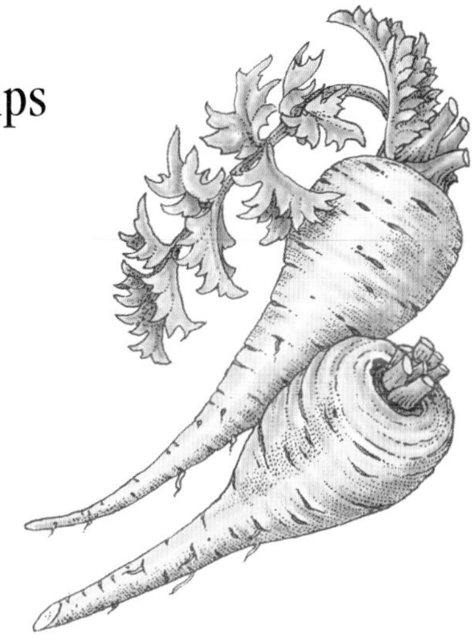

Parsnips are biennials like all good root crops and are even slower-growing than carrots. They are a good crop to grow in dry soil, because they are capable of forcing their food-storing taproots two feet (60 cm) down into the soil in their search for water.

Soil and climate

Parsnips will grow in fairly poor soil; they are so slow-growing that they do not need very rich conditions. On the other hand, in good soil they will grow better, more quickly, and produce more tender roots. And, of course, like every plant they flourish best in soil in which there is a high content of organic matter. They like soil about neutral: pH about 6.5. Very heavy soil is not good for them because it makes them fork. Stones and too much fresh manure also make them fork. A cold climate suits them best: without frost, they don't develop their full flavor.

Soil treatment

The deeper you dig, the better; for a really heavy crop dig in very well rotted manure or compost at least 18 inches (45 cm) deep—any less and your parsnips will fork.

Propagation

Traditionally, parsnip seed is the first of the year to be planted outdoors (not counting shallots, which are not a seed anyway). They were, and often are, sown in late winter—February in a temperate climate. However, in common with many other gardeners, I find it better to sow them later—well into the spring. Parsnips sown late are smaller, sweeter, and less woody, and they keep better. But, unless your garden really is an old-established organic garden in which the soil is largely humus, you should give late-sown parsnips a dressing of fish meal, bone meal, or some other organic fertilizer high in phosphate.

Again breaking with tradition, I like to sow parsnip seeds sparsely, but continuously, in drills, and thin the plants to about ten inches (25 cm) apart when they grow up. With the parsnip seed I sow radish seed. The radishes grow much more quickly than the parsnips and show you where the rows are so that you can side-hoe (the radishes "declare themselves," as gardeners used to say). The radishes also keep the crust of the soil broken, thereby giving the parsnips a better chance, and the leaves of the radishes shelter the young parsnip shoots from the sun.

THINNING PARSNIPS
Sow parsnip seed thinly in long furrows one inch (2.5 cm) deep. When the plants declare themselves twenty to thirty days later, thin them to about ten inches (25 cm) apart. Hoe frequently, and mulch if you can and provide plenty of water.

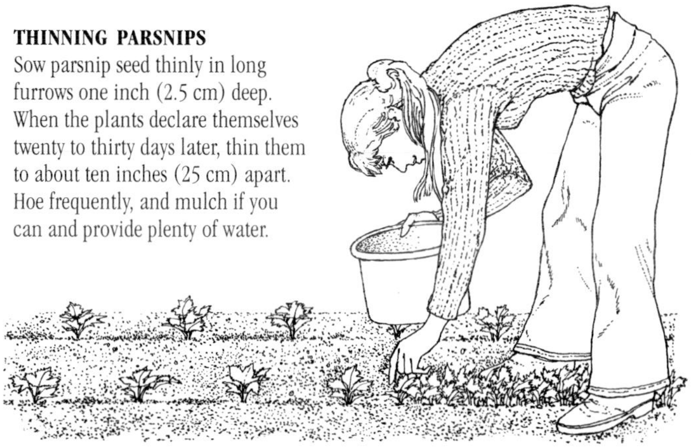

The drills should be about an inch and a half (4 cm) deep. After sowing the seed, push the soil back with your boot and walk along the row to firm it. Better still, cover the seed with fine compost and then firm it.

Care while growing

Young parsnips need plenty of moisture. Hoe from time to time, and mulch with compost if you have it.

Pests and diseases

CELERY LEAF MINER If you see tunnels mined in the leaves of your parsnips, look for the maggots, which will be living in blisters on the leaves, and squash them. To guard against this pest, spray with an ounce (30 cc) of kerosene* to a gallon (4.5 l) of water.

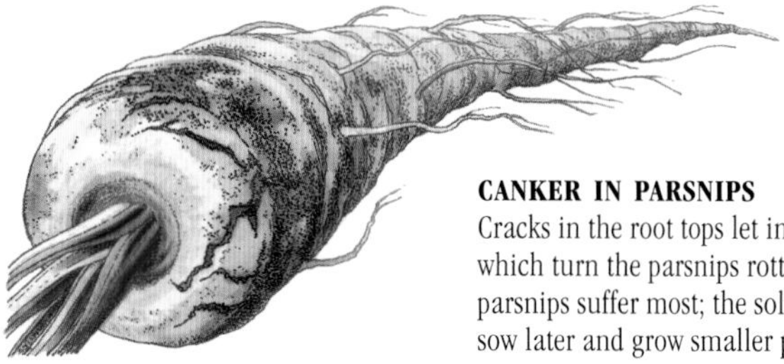

CANKER IN PARSNIPS
Cracks in the root tops let in fungi, which turn the parsnips rotten. Large parsnips suffer most; the solution is to sow later and grow smaller parsnips.

CANKER This is very common. The tops of the parsnip roots go rotten and brown. The worst attacks of canker occur in acidic soil and in soil that contains too much fresh manure. So, if your parsnips suffer badly, lime before sowing and refrain from adding fresh manure. A late crop is less likely to get canker than an early one.

RUST FUNGUS This appears as a rusty mold on the crown of parsnip plants. Mix two parts of lime with one part of sulfur* and sprinkle this on the soil around the plants. Don't grow parsnips on that bed for at least five years.

Harvesting and storing

Parsnips are completely frost-hardy but they don't like alternate freezing and thawing, so don't just leave them in the ground until you want them. Dig them up carefully and store in soil or sand in a very cold place. The very best thing is to make a pile outdoors. Do this with alternate layers of soil and parsnips—cover the whole pile with straw as an insulator, and cover this with patted earth to stop the straw from blowing away.

Celery

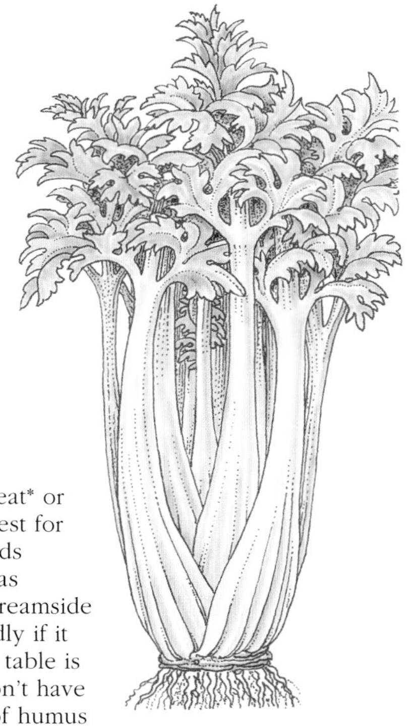

Celery, like parsnips, benefits from frost. In my view it should be eaten only in winter, when it is the most delicious vegetable available. During the rest of the year, it is tasteless and insipid, because it has not had the benefit of frost.

Soil and climate
Organic soil, such as peat* or soil rich in humus, is best for celery. Above all it needs constant moisture. It was originally a marsh or streamside plant, and it suffers badly if it dries out. A high water table is desirable, but if you don't have that, you need plenty of humus in the soil and you must water frequently if the weather is dry. Celery stands fairly acidic soil and does not need lime. Grow it in any climate where there is frost.

Soil treatment
I like to take out trenches about a foot (30 cm) deep and 15 inches (40 cm) wide, and dig plenty of compost or peat into the bottom. I do this in spring.

Propagation
Prepare a seed box by filling it with either a mixture of three parts sifted loam, one part of leafmold and one part of sharp sand, or with commercial potting mix. I prefer to do this in late winter but early spring is not too late. After the soil has been well soaked, sow the celery seeds sparsely and give them a light covering of soil.

Place the seed box indoors at a temperature of about 60°F (16°C) with either glass or old newspaper over the seed box. If you use glass, wipe off the underside twice a day to prevent moisture from dripping onto the seedlings. Keep the plants near the panes of the greenhouse or window so they don't get drawn sideways. Keep the soil just damp at all times; it is best to water with a fine spray, or you can stand the seed box in an inch (2.5 cm) of water and let the seedlings soak it up.

The plants will produce seed leaves before forming true leaves. As soon as the first pair of true leaves has appeared, prick out the plants into another box, which should contain three parts of loam, one part of leafmold, and half a part of rotted manure. Put the plants in carefully, two inches (5 cm) apart, and continue to spray them with water. Gradually harden them off by admitting more air, until late spring, when they can be put outside. Never let young celery plants dry out or your sin will be brought home to you months later when they suddenly run to seed before they are ready.

Although mature celery benefits from frost, the young plants will be damaged by it. So don't plant them out until you are sure there won't be another frost. Plant them out a foot (30 cm) apart—deep-bed method (see p. 106) six inches (15 cm) apart—in the bottom of the trench you have already prepared for them.

TRENCHING FOR CELERY
Dig trenches for celery, a foot (30 cm) deep and 15 inches (38 cm) wide. Work in three inches (8 cm) of manure, and cover it with three inches (8 cm) of topsoil.

PLANTING OUT SEEDLINGS
Plant out the seedlings a foot (30 cm) apart in the prepared trenches. Water the seedlings regularly until they are established, and continue to do so if there is no rain.

EARTHING UP AND TYING
In late summer, gather up the celery leaves and stems in a bunch. Tie the tops together, pack soil as tightly as you can around the plants, and remove the ties.

EARTHING UP AGAIN
Two or three weeks after the first earthing up, repeat the process; bank up the soil around the celery plants until only the leaves are still left showing.

If you have two or more rows of celery, plant them three feet six inches (105 cm) apart, because when you come to earth them up, you will need the space. Sow lettuces, radishes, and other quick-growing catch-crops between the rows. Harvest these before you need the soil to earth up the celery. Keep them well watered, especially for the first two weeks, if there is no rain.

Care while growing

There are several ways to earth up celery. If you have no help, tie the tops of the plants together, and pack the soil around the plants as tightly as you can, but without getting too much earth inside them. Fill the trench level in this way, then remove the ties on the tops of the plants. This should be done in late summer. Two or three weeks later, earth up again, using more lime to thwart slugs, and bank the soil well up around the plants. A further earthing up may be necessary in another few weeks. If you have enough peat*, use it in the later earthings up, but lay slates or tiles on the slopes of the bank to stop it from washing or blowing away.

The purpose of all this earthing up is to keep the stalks away from light. Like potatoes, they become bitter as they turn green from exposure to light, so the higher you keep them earthed up, the more crisp white celery you will have to eat. You can wrap the celery in collars of paper or plastic sheet before it is earthed up to prevent soil from getting inside the plants, but this method attracts earthworms and slugs.

Pests and diseases

LEAF MINER This is very common. It occurs when the maggots of the celery fly begin to mine tunnels into the leaves of the celery plant. Pick off any affected leaves and burn them. To control the disease, spray liquid manure* over the plants once a week. The smell deters the flies from laying their eggs.

LEAF SPOT Also known as "celery blight," leaf spot can destroy your celery if it is not quickly checked. The disease, which is spread through the seeds, will cause small yellowish brown spots to appear on the leaves. Spray immediately with fungicide or the whole plant will become affected. And only sow seeds that have been dipped in formalin*.

DAMPING-OFF DISEASE Young celery plants will suffer damping-off if they get too much water or too little air. The main symptom is a watery soft rot. Wipe off condensation if you're growing seedlings under glass, and make sure the seedlings are properly aired.

Harvesting and storing

Pull celery whenever you want to eat it, and if you are interested in flavor, do not start eating it until there has been a frost. Thereafter try to make it last until well into the winter. It is a very good idea to protect part of a row with cloches as soon as the really heavy frosts set in. This will keep the celery good until late in the winter. (So as not to waste cloche space, plant a double row of celery in the same trench.) If you haven't got cloches, you can use straw or bracken at night, but take it off on warmer days.

SELF-BLANCHING CELERY

To my taste, self-blanching celery is not as good as celery, but it is easier to grow. You raise it from seed in the same way as celery, plant it on the flat in late spring or early summer, and it is ready to eat in late summer. It won't stand frost at all.

Celeriac

The most delicious part of a stick of celery, in my opinion, is the crunchy base—the heart, as it were, of the celery. With celeriac you grow all "heart" and no stem. The plant consists mostly of a bulbous base, or swollen stem, with only small green shoots, which you can't eat, above it. Grow it exactly as you grow celery.

Soil and climate

Celeriac needs rich and mellow soil that has been well-manured. It grows best in a cool, moist climate.

Soil treatment

The deep-bed method (see p. 106) is ideal for celeriac, but, whether you use it or not, dig deeply and incorporate plenty of manure or compost.

Propagation

Sow indoors in late winter or outdoors in a nursery bed at the beginning of spring. Plant out in early summer.

Care while growing

Celeriac should be watered frequently and kept free from weeds. When you hoe, draw soil away from the plants. Don't earth up as you would for celery.

Pests and diseases

CELERY LEAF MINER Like celery, celeriac can be attacked by leaf miner, though usually to a lesser degree. Simply pick off affected leaves or spray the plants with liquid manure*.

Harvesting and storing

You can begin to harvest celeriac in the late fall. In a very cold climate, store celeriac in a root cellar, but if your winters are milder, leave them in the ground and harvest them as you want them.

Hamburg Parsley

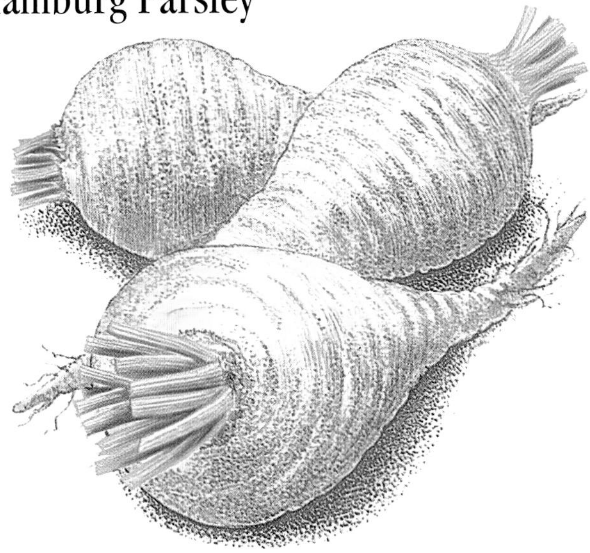

Hamburg parsley or "turnip-rooted" parsley grows as a root vegetable and can therefore be stored. It is not just an herb for flavoring things, although the leaves taste very similar to ordinary parsley; it can also be eaten raw or cooked like celeriac, which it resembles.

Soil and climate
You can get a good crop of Hamburg parsley wherever ordinary parsley will grow, namely, in moderate climates. The more sun the plants get, the better.

Soil treatment
Hamburg parsley tolerates poorer soil than most root crops, but dig deeply and work in plenty of well-rotted manure.

Propagation
Like all the *Apiaceae*, the seed takes a long time to germinate. Soak the seed before planting and sow sparsely in drills either in fall or early spring. A dressing of a high-phosphate fertilizer such as fish meal or bone meal is useful.

Care while growing
When the plants come up, you should thin them out to about four inches (10 cm) apart, either in the rows or the deep bed (see p. 106). Hoe the plants. They don't need to be watered much unless it is a particularly dry summer.

Pests and diseases
CANKER It is possible that the canker that can attack ordinary parsley may also attack Hamburg parsley. It gives a rusty appearance to the stems and causes the roots to decay. To prevent it, do not water too much. If you do get it, burn all your diseased plants.

Harvesting and storing
The leaves can be picked off and used as an herb, but the roots should be pulled as late as possible, because—and this is unusual—the largest roots taste the best. But don't leave the plant in the ground after the first frost. Store it in peat* or sand, in wooden crates or garbage cans in a root cellar or other similar cold storeroom.

Florence Fennel

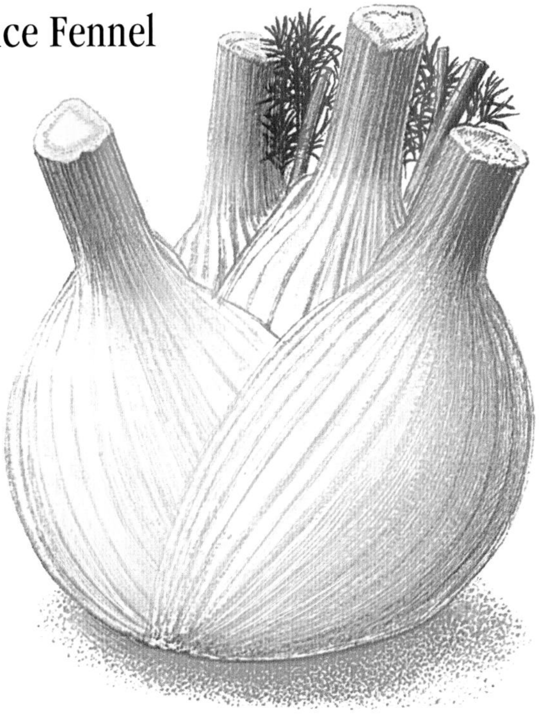

Unlike ordinary fennel, which is used only as an herb (see p. 196), Florence fennel can be eaten on its own because the leaf-stalks swell at the base of the leaves to form large "heads." Sliced raw, these can be added to salads. The stems can be cooked like celery, and the leaves and seeds can be used for flavoring. It is well worth growing for its unique taste, which combines the flavors of aniseed and licorice.

Soil and climate
Florence fennel will grow as a perennial only in hot climates, but it can be grown as an annual nearly anywhere. If you want really tender and delicious stems, grow it in the richest ground you have.

Soil treatment
Prepare the ground as for celery, digging deeply and making sure you incorporate a lot of manure.

Propagation
Sow seeds thinly in shallow drills, either in the late fall or during early spring.

Care while growing
Florence fennel needs very little attention. Thin to six inches (15 cm) apart in the rows or deep beds (see p. 106) and water when required. It is worth earthing up the heads so that the plants retain moisture.

Pests and diseases
Florence fennel is happily disease-free, as well as also being highly resistant to pests.

Harvesting
Cut the heads when they are about two inches (5 cm) across. The stems above them should be harvested before they get too old and stringy and stripped of their outer skin.

Liliaceae

Onions, leeks and asparagus are all members of the *Liliaceae*, or lily family.

As I have discussed elsewhere (see p. 19), the main subdivisions of the plants that provide us with nearly all our food are the two great classes: the monocotyledons (monocots for short) and the dicotyledons (dicots for short). Most of our vegetables are dicots. That is, they have two seed leaves—the first leaves they make—while the monocots have one. But the important difference between the two classes is nothing so trivial: it is that they have entirely different ways of growing. The dicots grow outward from the edges of their leaves. The monocot grow from the bases of their leaves, which means that they push their leaves up and out from the bottom.

The members of the lily family are monocots, as you can easily see by looking at their leaves. These don't have the network of veins that dicots have, but instead have parallel veins—each vein starting at the base of the leaf. Onions and leeks make use of this particular form of leaf growth to store nourishment in hard swollen bulbs, which are nothing more than many leaf bases compressed together to form a bulb. Asparagus is not so instantly recognizable as a monocot, but it is one nonetheless. The shoots, which are the edible part, grow from a "rhizome"—a horizontal stem that remains underground. The attractive, fernlike branches that sprout as the shoots grow taller are like leaves and these grow up from their bases like the leaves of the other monocots. The rhizome takes nourishment from the leaves in one year and uses it to produce the next year's shoots.

Humans, and other sapient animals, have always made use of the way plants store up in their first year's growth what they are going to use for flowering and fruiting in their second, and, although we don't eat lily bulbs, we do eat onions, leeks, and asparagus.

Onions

COMMON ONIONS

With careful growing, harvesting, and storing, there should be no part of the year in which your kitchen lacks that most indispensable ingredient, the onion. But you may well feel confused by the amount of conflicting advice that is available. Onions have been grown for centuries, and it sometimes seems as if each grower has his own special technique. In fact, the right timing and method depend largely on where you live.

Soil and climate

Onions need good rich soil. Sandy loam, peat*, and silt are all fine, but onions don't like clay, sand, or gravel. They grow successfully in widely different climates, although they prefer cool weather while they go about developing their leaves, followed by very much hotter weather during the period when they make their bulbs.

Soil treatment

Onions have very shallow roots and grow quickly, so they need plenty of nourishment in the top four inches (10 cm) of soil. Muck the onion bed well the previous fall with well-rotted manure or, even better, with a large amount of thoroughly rotted compost. They like plenty of potash and phosphate but not too much nitrogen. It is useful to add any of the following to your soil: wood ash, ground rock phosphate, soot, seaweed meal, a sprinkling of salt. The soil should have a pH of 6. If it is lower, then add lime.

Before planting out, firm the ground by treading or rolling—preferably both. And the soil must be dry, whether you are sowing seed or transplanting.

Propagation

If your neighbor is a successful onion-grower, it's a good idea to ask him for advice. There are four possibilities.

LATE SUMMER SOWING The idea of this is to get the onions to form bulbs by the early spring, before hotter weather causes them to bolt, and then allow them to mature through the summer. Sow the seed thinly in shallow drills, cover with half an inch (1.5 cm) of compost, and firm the ground. If the winter is particularly hard, put cloches out during the worst of it. In the spring, thin out the onions to about six inches (15 cm) apart. The thinnings can be used in salad.

WINTER SOWING In very frosty areas, it is best to sow onion seeds indoors in midwinter, for planting out in the spring as soon as the ground is dry enough. Sow seed thinly in seed boxes filled with either a commercial potting mix or a mixture of three parts sifted loam, one part leafmold, one part fine compost, and a sprinkling of sand. Keep moist but not wet, and cover with glass or paper. When the seedlings show, remove the covering and keep the seed box at about 65°F (19°C) near a window. When the second leaf is about half an inch (1.5 cm) long, prick the seedlings out into another box so that each plant is two inches (5 cm) away from the next. Use the same soil mix for the pricking-out box, but add one part of well-rotted manure. Harden the plants off gradually until they are in the open air, in a sheltered place, by the last frost. Plant the seedlings out in the spring. It is best to leave their permanent bed rough until the last moment so the soil can dry out. Before planting, rake it down as fine as powder and step it down firmly.

SPRING SOWING This is only good where there are cool, damp summers, and you won't be able to store the bulbs. Sow out of doors as you would onions sown in late summer and thin them out to four inches (10 cm) apart when they are big enough for the thinnings to be used as salad onions.

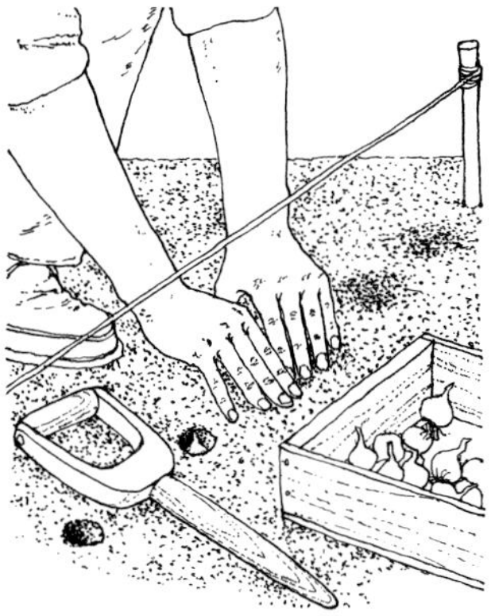

PLANTING ONION SETS
To grow onions from sets, it is essential to press the soil very firmly around each set. Make holes six inches (15 cm) apart; press the sets into the holes. Contrary to the usual advice (which is to leave the tops just showing), I suggest that you bury the sets completely, so that they are out of the reach of birds.

ONION SETS If you prefer them to seed, onion sets grow best in temperate climates where they should be planted in the spring. Make dibber holes every six inches (15 cm)—deep-bed method, four inches (10 cm)—along a line and push the sets into them, pressing down the soil around each set to hold it firm. I bury the sets completely—so that the topknot is just below the surface where the birds can't pull it out.

Care while growing

The most important thing is to keep your onions well weeded. In their later stages, they benefit from a mulch and this can be done with uprooted weeds. If your onions flower, pinch out the stems while they are still small.

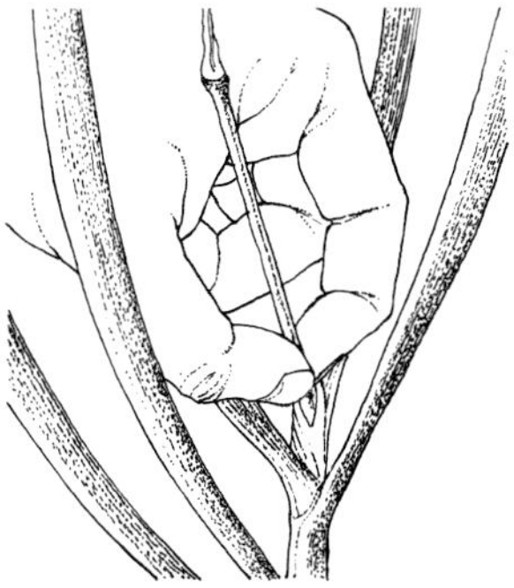

PICKING OUT FLOWER STEMS
If necessary, pinch out the flower stems while they are still quite small. If you don't do this, the plants will "bolt," that is to say, they will produce flower heads. These in turn prevent the bulbs from forming properly. When you are picking out the flower stems, take care not to loosen the bulbs.

Pests and diseases

ONION FLY The maggots of the onion fly are one of the nastiest pests we have to suffer, as they can stop onions from growing altogether. The maggots eat into the bulbs of seedlings. Spring-sown onions are the worst hit, but onions grown from sets aren't likely to be affected. To deter onion flies and their maggots, dust your rows of onions with flowers of sulfur* or soot, or sprinkle with an ounce of kerosene* to a gallon of water. Dust regularly, particularly when you thin the plants.

DOWNY MILDEW This occurs particularly during wet seasons, and causes grayish or purple streaks on leaves. If you find it, dust with Bordeaux powder*.

EELWORM These microscopic worms will cause the tops of your onions to wilt. Burn all affected plants and don't grow onions or allow chickweed, which also harbors them, to grow on that land for six years.

NECK ROT This disease attacks onions in storage. A gray mold forms on the onion skins and later the centers turn brown. Prevent it by drying your onions off well after harvesting and storing in a cool airy place.

ONION SMUT Onion smut shows as black blisters on stems and bulbs. If you do get it in your garden, water the rows with a solution made from a pint (0.5 l) of formalin* and four gallons (18 l) of water when you sow. However, you are unlikely to get onion smut.

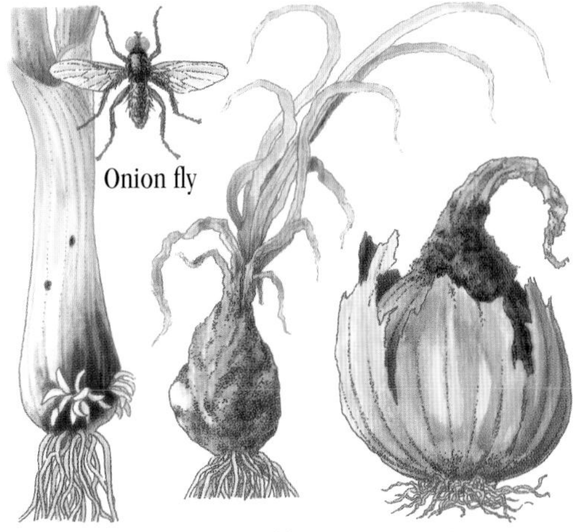

Onion fly damage Downy mildew Neck rot

Harvesting and storing

It is well worth bending over the tops of your onions when they will grow no more. This will start them ripening. In a hot, dry fall, ease the onions out of the ground and leave them a week or two in the sun to dry. But don't damage the skin or you'll let in neck rot. In wet places, lay the onions on wire netting after harvesting in order to keep them off the ground. But put them under cover if it's rainy. It is vital to dry onions well. When they have dried, the best thing to do is string them. Otherwise, put them in layers in some cool, airy place. They must not suffer severe frost, but it is better for them to be cool than hot. If you hang a string by your stove, use them fast before they rot.

BENDING THE TOPS
When the tips of the leaves start turning yellow, bend over and break the necks of the onions. This starts the ripening process. At the same time, loosen the onions with a fork to start them drying, but take care not to damage the skins. A few days later, ease them out of the ground and leave them in the sun to finish drying out.

It is vital to dry onions well. When they have dried, the best thing to do is string them. Otherwise, put them in layers in some cool, airy place. They must not suffer severe frost, but it is better for them to be cool than hot. If you hang a string by your stove, use them fast before they rot.

STRINGING ONIONS

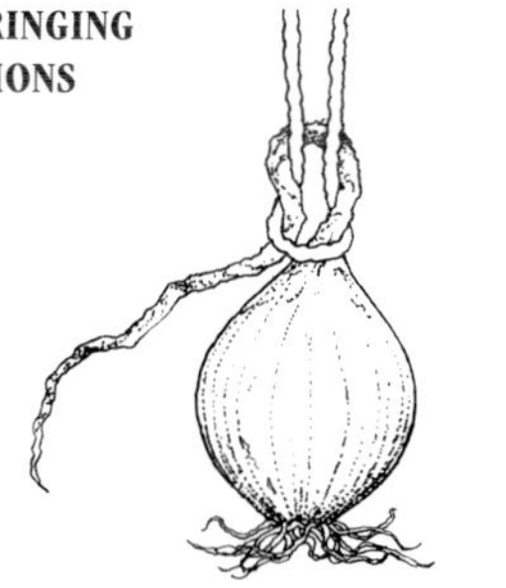

1 Knot together the ends of a three-foot (90-cm) length of string and hang the loop from a hook. Weave the first onion through the loop.

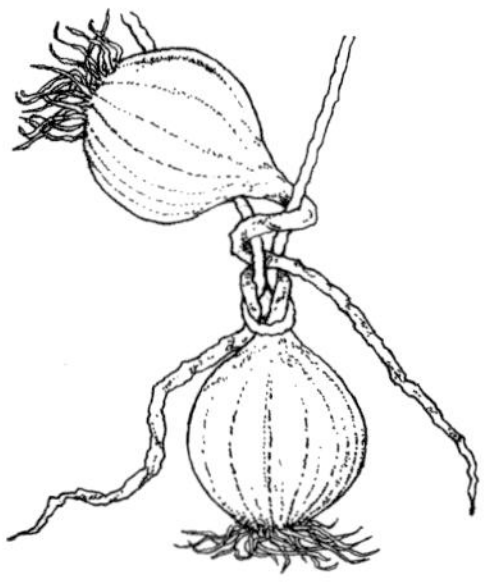

2 Use the leaves of the second onion to weave in and out of the string. The weaving must be tight and the second onion should finally rest on the first.

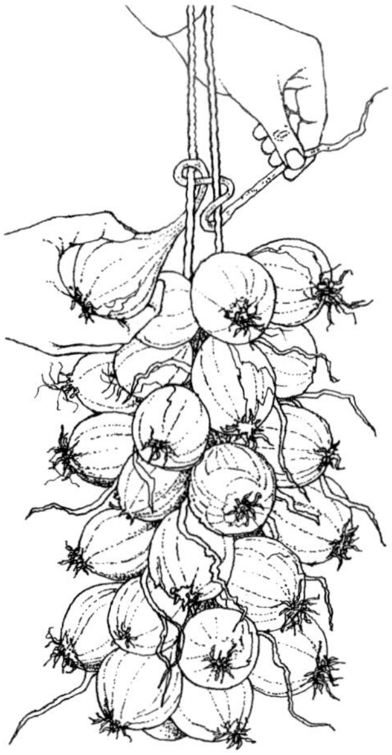

3 One by one, add onions to the original two, weaving first from the left, then from the right. As the bunch grows, check to make sure it is well balanced. When you finish, hang up the string in a cool, dry, airy place and the onions should keep until early summer. Remember that all onions for stringing need long, dry leaves.

SPRING ONIONS

To fill the gap after the last of your stored onions have gone rotten—if you haven't planted spring onions specially—buy bunches of green onions and plant them out three inches (8 cm) apart. They will quickly produce small bulbs. Alternatively, sow any onion seed, or the special green onion varieties in late summer, and pull them in the spring. But, of course, if you have sown onions outdoors in late summer, you will have spring onions anyway as a result of thinning them.

PICKLING ONIONS

Pickling onions are basically ordinary onions grown close together and in poor soil so that they stay small and mature early. Sow the onion seed thinly in the spring. Don't thin them. Pull them when the tops wither.

SHALLOTS

In England, it is traditional to plant shallots on the shortest day and to pull them on the longest, but they will grow if you plant them any time before early spring. Prepare the ground in the same way as for onions, and plant the shallots as you would onion sets, about six inches (15 cm) apart. When you harvest, hang them in a string bag to store.

Leeks

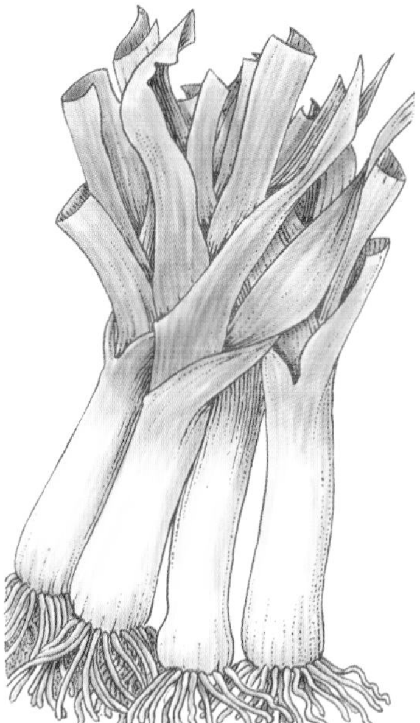

Cawl, the great soup that is the Welsh national dish, has leeks for its heart and soul—and not for nothing have the Welsh made this their national emblem. If you wish to grow only one vegetable of the *Liliaceae* family, begin with the leek. It will not let you down.

During the dreary hungry gap when all else fails, which stretches from late winter through to the middle of spring, the stalwart leek is there to relieve you from a diet of salted this and frozen that.

Soil and climate

Leeks like rich loam, but they can, and often do, put up with practically anything. Still, for leeks that you can be proud of, give them plenty of manure and plenty of compost. They need moist but well-drained soil, and they like a high pH—between 6 and 8. They are temperate climate plants but will grow in any climate except a tropical one. In my view, though, leeks grown in particularly dry weather don't seem to have the flavor of those grown under cloudy skies.

Soil treatment

For really fine leeks, treat your soil as you would for onions (see Onions). It is common to follow early potatoes with leeks—forming a part of the legume break in our recommended rotation (although they are not of course legumes). In this case, the heavy manuring the potatoes have had will be perfect, though the leeks will probably want some lime.

However, I generally plant leeks with the brassicas because they are then sharing the only bed that does not need digging up until late the following spring. While leeks are very closely related to onions, they do not fit easily into the onion bed, for the onions are strung and hung up in a cool shed before we even start to eat the leeks.

In hard ground, trench for leeks, digging one spit deep and mixing in plenty of compost or well-rotted manure.

Propagation

In very cold climates, sow seed indoors in late winter and keep at about 60°F (16° C). For summer leeks you should do this anyway, but I think it is a pity to eat leeks in the summer—"the fruits of the earth in their season" is what suited Adam and Eve before they tasted of the wrong fruit—and it is nice to come to leeks around Christmas with an unjaded palate. For winter leeks, in all but the coldest climates sow fairly thickly along the drills in the brassica seed-bed in early spring. Do not throw the leftover seeds away after sowing; each quarter-ounce (7 g) holds about a thousand seeds, and they keep for around four years. Plant out in the garden when they are four inches (10 cm) high.

It is common advice to snip an inch off the tops of the plants and at the same time snip the roots to about half their length. Then simply drop each plant into a hole made with a dibber and pour a little water in. This works perfectly well but, since I prefer not to mutilate plants more than I have to, I plant them as I do cabbages. Simply make a hole with your dibber, pour a little water in, and push the unmutilated leek down into the mud. I suggest you try half a row each way and see which grows the best. Plant them six inches (15 cm) apart—deep-bed method four inches (10 cm) (see p. 106)—in rows wide enough to work between—say 16 inches (40 cm).

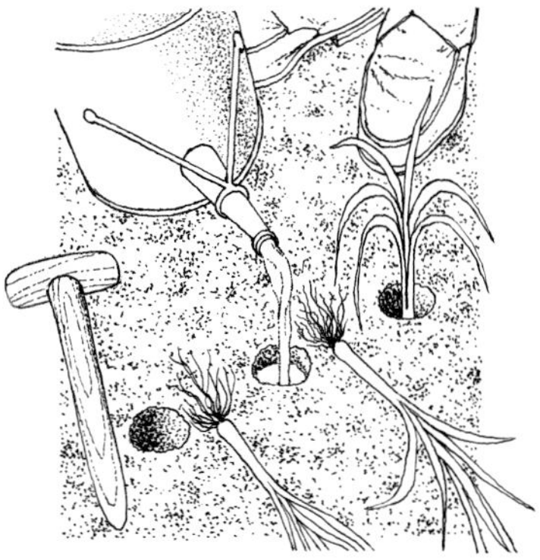

PLANTING LEEKS
Using a long dibber, make holes eight inches (20 cm) deep and six inches (15 cm) apart, in rows wide enough apart to allow space for maneuvering—say, 16 inches (40 cm). Pour a little water into each hole, and push the leeks down into the mud, roots first. To snip an inch off the top and the longer roots beforehand is a matter of personal preference.

Care while growing
Never let leeks go short of water. Hoe them and mulch if you can. Earth them up every now and then as they grow, in order to blanch them.

Pests and diseases
Leeks are said to suffer from all the same pests and diseases as onions, but I have never known a leek to suffer from anything—except somebody digging it out of the ground and eating it and that, surely, is a glorious death.

Harvesting and storing
When digging up leeks, avoid cutting off too much of the foliage because the leeks' prime food value consists of the vitamin A to be found in the leaves. So, as well as eating the stems, try and make use of the leaves, in soups and so on: they taste very savory.

In freezing climates, leeks can be dug up before the first severe frost and heeled into soil in a basement. In less severe climates they should be earthed up nearly to the tops of the plants. In temperate climates, leave them in the ground if possible until at least Christmas, and spare them even after that until late winter when the Brussels sprouts begin to look a sorry sight and you are tired of broccoli and kale. Then dig them up as you need them. However, if you wish to clear the bed for something else, it is perfectly all right to dig all your leeks up and heel them into the ground in a short row out of the way and near the kitchen.

Asparagus

If you ignore the common associations with luxury and decadence, then asparagus is in fact an excellent vegetable to grow, for it gives you a fresh bite of green stuff in the spring. It can be cut and cut again for a period of six to eight weeks, and it is impossible to conceive of anyone ever getting tired of so exquisite a taste. On the other hand, an asparagus bed takes up a lot of space, especially since you have to wait for three years from the time of sowing for your first taste.

Soil and climate
Asparagus came from the seaside and can put up with salt. It likes a mild, moist summer and a cold enough winter to make it go to sleep for half the year. Only a very severe frost will harm it after it has died down for the winter, but it is a good idea to cover the bed with straw if the winter is frosty.

The plant flourishes in a light, well-drained loam and it will grow well in sandy land if it has been well-mucked and composted and is not allowed to dry out. The soil should have a pH of 6.5, which may mean you have to add lime. It is possible to artificially create soil that is suited to asparagus—heavy clay, for instance, can be properly modified by adding plenty of compost and sand—and it is well worth doing this if you can, as asparagus is a perennial that will go on feeding you for years.

Soil treatment
The bed must be entirely free from perennial weeds, because once the roots of these are thoroughly intertwined with the roots of the asparagus, which spread out as far as five feet (1.5 m) you will never get them free. So the fall before you intend to plant, fork the planned bed over and over again and remove every last inch of couch grass root, ground elder, or convolvulus.

Then, if your land needs it, fork in lime and also if necessary rock phosphate or wood ash or some other form of potash, but if you have enough compost you needn't bother with this. Half a pound (200 g) of fish meal and half a pound (200 g) of ash per square yard is good insurance.

The old-fashioned method of preparing the bed, and one that I think can't be beaten, was to take out a trench five feet (1.5 m) wide and one foot (30 cm) deep, placing the topsoil to one side separately. Put six inches (15 cm) of good stable manure in the trench, six inches (15 cm) of rotted sod on top of this, and another few inches of well-rotted manure or compost. Replace the topsoil on this and you will have a bed for three rows of plants. An alternative is just to dig plenty of manure or compost into the soil in the fall and plant in that, in either single rows or rows about four-and-a-half feet (1.4 m) apart.

Propagation

Asparagus can be grown from seed. Collect your own seed by hanging up some of the female ferns when they are quite ripe, letting them dry, and then rubbing the seed out with your fingers. But if you grow them from seed you won't get any asparagus to eat for three years. If you can't wait that long, buy two- or three-year-old crowns from a nursery. Even one-year-old crowns save you some waiting time, they are cheaper than older plants and transplant more easily.

If you are sowing seed, do it either indoors in seed boxes or outdoors in a seed-bed, sowing in early spring and later thinning to three inches (8 cm) apart.

Whichever method you choose, when the plants are ready for their final bedding out, take out a trench or trenches nine inches (25 cm) deep and a foot (30 cm) wide. Make a cushion of fine soil in the bottom, slightly convex in section, and lay the plants two feet (60 cm) apart in the rows—deep-bed method a foot (30 cm) apart (see p. 106). Never let asparagus crowns dry out. If you buy crowns from a nursery, keep them moist until you put them in the ground, then give them a good soaking. Cover immediately with fine soil and compost.

Care while growing

Asparagus benefits enormously from heavy mulching. I like to mulch with seaweed, but otherwise mulch with any organic material that is nonacidic—don't, for instance, use pine needles or oak leafmold unless you mix thoroughly with lime first. If you use sawdust or anything very hard to break down, put on some high-nitrogen material, like blood meal. Don't hoe too deeply or you'll damage the asparagus roots. Your mulch is the best weed suppressor.

Many people cut the ferns down in fall as soon as they turn yellow, but I prefer to leave them to die down naturally and return their goodness to the plants.

Pests and diseases

ASPARAGUS BEETLE Use your fingers to pick off any beetles that appear on the asparagus shoots, and kill them. Do it early in the morning when the beetles can't fly. Otherwise dust with derris powder when the dew is on them.

ASPARAGUS RUST If you live in a very humid area, your asparagus may be affected by this fungus, so plant a rust-free variety. If you still get it, spray with Bordeaux mixture*.

Harvesting and storing

Don't touch the plants for three years if you grew them from seed, or for two years if they grew from roots. Then cut the shoots before they begin to open as you want them. Cut for six weeks after they begin to shoot the first year, and thereafter for two months. Midsummer day is the last day you should let yourself cut asparagus. From a five-year-old bed sporting two healthy plants, you will probably find that you can cut enough asparagus for one person to have one helping a week throughout the cutting season.

Reckon on ten years as the lifetime of a bed, and plant again in time to avoid a dearth after that. A good tip, when you scrap an old bed, is to take the plants out carefully at the end of fall and plant them on a hot-bed. You'll benefit by getting asparagus a month earlier than usual.

Asparagus can be canned or frozen (see pp. 220 and 227). My personal prejudice is against doing either because the fierce pleasure of eating fresh asparagus in the spring—surely one of the heights of gastronomic experience—is lost if the palate is jaded by having munched away at canned or frozen stuff the rest of the year.

TRENCHING FOR ASPARAGUS
Dig a trench nine inches (25 cm) deep and a foot (30 cm) wide. Leave the bottom slightly convex, so that it will hold water, and cover it with a layer of fine soil.

PLANTING THE CROWNS
Set out the asparagus plants on the layer of fine soil, two feet (60 cm) apart. Lay them with the roots spread out carefully so they can absorb all available moisture.

MULCHING THE PLANTS
Seaweed or fish meal is ideal for heavy mulching; otherwise use any nonacidic organic material. When the ferns turn yellow in fall, let them die naturally.

HARVESTING THE CROP
Three years from sowing or two from planting is the time to harvest your asparagus. Cut the shoots for not more than six weeks the first year, and two months subsequently.

Chenopodiaceae

Beets, spinach, seakale beet or Swiss chard, and spinach beet are the edible members of the *Chenopodiaceae* family. This very distinctive family originally came from the seashore, and its members share with the cabbage tribe the characteristic that their leaves have a tough, cutinous surface that is designed to limit transpiration (the escape of water from the plant) and thus preserve the moisture that the plants win so precariously from their salty surroundings. Seashore plants share with desert plants the need to conserve moisture, for their salty environment tends to draw it out of them by osmosis.

The beet family also has the peculiarity of producing its seed in little fruits that remain intact until they germinate in the ground. What we plant as the "seed" of beet and spinach is in fact a small fruit containing four or five seeds, each of which grows into a plant. Thus you find that when you plant these fruits, you get plants coming up in clusters. The clusters should be thinned to leave one plant from each for the best beet or spinach.

The family includes sugar beet, which is grown for sugar, of course (the roots can be up to 21 percent sugar), and mangolds, or "cattle-beets," which are grown specifically for feeding to cattle and other stock. The beet family is characterized by the fact that the swollen roots that we grow them for display rings when you cut them across. These rings are alternately storage tissue and conveying tissue. It is the storage tissue that contains the nutrients that the beet stores up in its first summer, so that it can leap ahead quickly in its second summer and produce seed before it dies. The conveying tissue transfers the nutrients from the roots to the rest of the plant.

Beets and spinach can send their roots down ten feet (3 m) or more into the soil. They also occupy the soil very fully, filling it with a mass of fibrous rootlets. This has a benign effect, breaking up the soil and loosening it, so that when the roots rot, they leave passages for water and air far down into the subsoil.

It is possible now to get fragmented seed of the *Chenopodiaceae*—where the fruitlets have been broken—or to get pelleted seed. Each of these makes it possible to sow singly, either by hand or with a precision drill, which saves thinning later. But the expense is really only justified if you are growing on a market-garden scale.

Beets

The western culinary imagination never seems to get much beyond boiling beets for hours and then soaking them in vinegar. But we have only to look to Russia and to the delicious borscht to see just what can be done with this sadly underused vegetable.

Soil and climate

Beets are really a cold-climate crop; in hot countries, they must be grown in the winter, early spring, or fall. Beets like well-drained soil, with plenty of humus, but not fresh manure. They won't grow at all if the ground is too acidic, and need a pH of about 6.5.

Soil treatment

In my suggested four-course rotation (see p. 73), beets should come in the root break. The land should by then have been heavily manured three years before for the potatoes, limed two years before for the *Fabaceae*, and probably well-mulched with compost one year before for the brassicas. The soil should be well-dug and without too many stones.

Propagation

The main crop of beets for storing should be sown about three weeks before the date of your usual last frost. But you can go on sowing successionally until midsummer. It's worth soaking the seed for a day or two before sowing. Sow the seed thinly three-quarters of an inch (2 cm) deep in drills a foot (30 cm) apart. When the plants appear, first single each cluster, then thin out to three inches (8 cm) apart for small beets and nine inches (25 cm) apart for large beets for winter storing—deep-bed method (see p. 106), four inches apart (10 cm). You can eat the baby thinnings.

Care while growing

Hoe several times to suppress weeds until the shiny leaves of the beet start doing the work themselves. Beets grown in dry conditions become hard and woody and they can often bolt. Mulching and watering should prevent this.

Pests and diseases

LEAF MINER The larvae of the beet fly will bore through the leaves. Pull off the affected leaves and burn them.

BEET BEETLE Beets may become infested with tiny black beetles. Water the plants with nicotine* water.

Harvesting and storing

The tops of beets can be eaten as well as the roots, but if you're going to store beets, wring the tops off—don't cut them since it makes them bleed. And avoid bruising.

By far the best way to store beets is to lay them gently in moist sand without allowing them to touch each other. If you lay them in the open, they will shrivel up.

Spinach

Spinach in reasonable quantities is good for you, but if Popeye had eaten a quarter of the spinach we saw him eat in the cartoons, he would have died of oxalic acid poisoning. If you like spinach, you can have it year-round with careful planning. In hot climates or even in more temperate ones, New Zealand spinach (which belongs to the *Tetragoniaceae* family) can be grown in the summer.

Soil and climate

High temperatures during the first couple of months of its growth will cause spinach to bolt to seed, so it is really a winter crop in hot climates. It likes good rich loam, and should be kept damp, but it will grow in most soils if there is enough organic matter present. It isn't a good idea to plant in soil where fresh manure has recently been applied.

Soil treatment

Spinach will go in odd corners as a catch crop, but it doesn't like acidic soil, so lime if necessary to bring the pH to between 6 and 6.5. If the soil is over 6.7, spinach will be one of the first plants to show manganese deficiency, since manganese gets locked up by too much lime. The ground should be well dug; alternatively, use a good layer of compost.

Propagation

In cool climates, the prickly-seeded variety of spinach is best in the winter and the smooth-seeded variety in summer. In hot climates the smooth-seeded variety should be planted in winter. Even in pretty cold climates you can sow smooth-seeded spinach in winter (though not, of course, where the ground is frozen) and successionally thereafter—say, once a month—until late summer. In wet winters it is a good idea to sow spinach on raised ridges for better drainage. In spring and summer, sow on the flat. Sow three-quarters of an inch (2 cm) deep and, when the plants declare themselves, thin to four inches (10 cm) in rows a foot (30 cm) apart. Allow three inches (8 cm) in both directions between thinned plants in the deep bed (see p. 106), and take care not to step on the bed when thinning.

Care while growing

Hoe and mulch if you can to keep down weeds. Water only if the weather is very dry; if the soil is allowed to dry out, the plants will bolt.

Pests and diseases

MOLD Spinach is virtually free from pests and diseases in good organic gardens, but in muggy, damp weather it can get moldy. This mold manifests itself as yellow patches on the leaves and gray mold on their undersides. If it appears, the best thing to do is scrap it and plant some more. If you plant some every month (a short row, of course), you won't have to go long without it. If it bolts, then the only place for it is the compost pile.

Harvesting

Simply pluck the leaves from your spinach as you need them. In fact, if you want the plant to go on bearing, you should pluck the outer leaves fairly often. Spinach tastes best if you simply rinse it quickly and put it, still wet, straight into a pan with a close-fitting lid. Don't add any more water; just leave it to simmer for five minutes or so.

HARVESTING SPINACH
You need a lot of spinach before you have enough for one helping. The secret is to have sown enough plants: as long as you spread the plucking over several plants, the more often you pluck the leaves, the better. Pick young leaves from the outside, taking care not to denude any one plant.

NEW ZEALAND SPINACH

New Zealand spinach is not in fact a member of the *Chenopodiaceae* family but it is similar enough to warrant talking about here. New Zealand spinach doesn't have the high oxalic content of true spinach, so its nutritional value is more available. It is less frost-hardy than true spinach and it should be grown in the summer. In very hot climates it will stand up to heat far better than ordinary spinach, but it does need protection from the sun.

Don't sow until all danger of frost is past. Grow New Zealand spinach in rows four feet (1.2 m) apart in good soil, or closer together in poor soil. When sown directly into a deep bed, it will grow particularly well. Soak the seeds in water for twenty-four hours before sowing; otherwise their hard cases make them slow to germinate. Plant three seeds to each station and later pull out the two weaker plants. Otherwise treat New Zealand spinach just as you would treat true spinach.

Swiss Chard

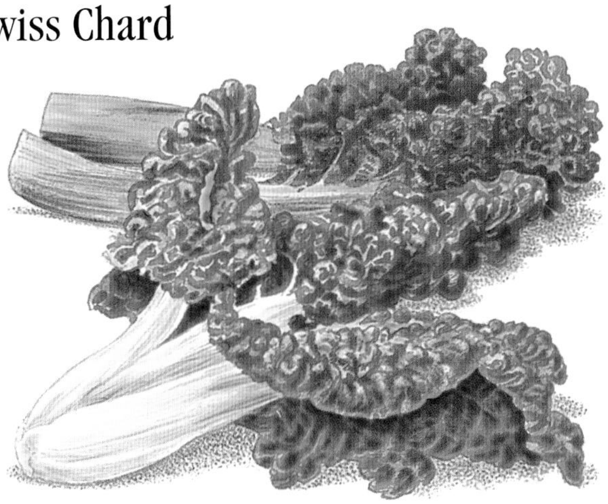

Swiss chard or seakale beet is just a beetless beet, as it were: it's a beet that puts down deep narrow roots instead of making one swollen root. It is an excellent crop for your garden, since it sends its tough roots three feet (90 cm) down into the subsoil and draws up what is good down there. Both the leaves and the midribs of the leaves can be eaten; you should cut the midribs up before cooking them, and they need cooking longer than the leaves.

Soil and climate
Swiss chard will grow in most climates except the very hottest and in any soil that is not waterlogged.

Soil treatment
Swiss chard, like the other beet crops, needs a pH of about 6.5, so lime if necessary. A small amount of well-rotted compost or manure is a good idea.

Propagation
Soak the seed as you would beet seed (see Beets). Sow seeds an inch (2.5 cm) deep and three inches (8 cm) apart—deep-bed method (see p. 106) also three inches (8 cm) apart. The plants need room, so make the rows 18 inches (45 cm) apart. The seed should be sown two or three weeks before the last expected frost, though in milder climates you can sow in late summer for harvesting and eating in winter and spring.

Care while growing
Swiss chard needs little attention, though it will probably be worthwhile to mulch it.

Pests and diseases
Swiss chard is hardy and resistant to pests and diseases.

Harvesting
When the leaves are seven inches (18 cm) long, start breaking the outer ones off and eating them. As the leaves get bigger, tear the thin leaf off the midribs.

RUBY CHARD
Ruby chard is exactly like Swiss chard except that its leaves and stems are deep red in color. It is said to thrive in heavy soils better than its rivals.

Spinach Beet

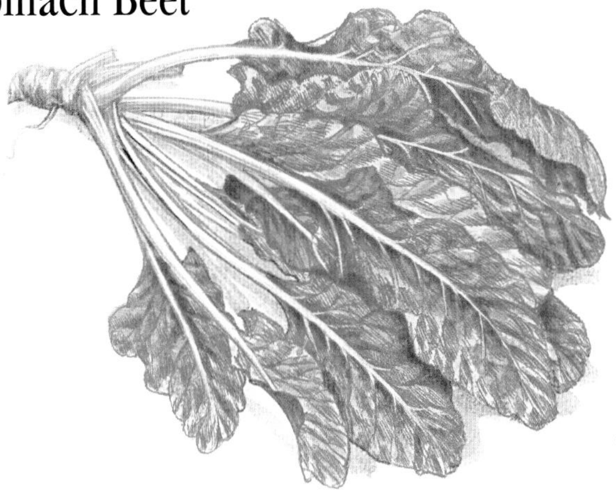

Spinach beet is a beetless beet, meaning that it does not make a swollen root as beets do. It is also known as "perpetual spinach" because you can get a year's supply of leaves from only two sowings, instead of having to sow successionally through the year, as with spinach. You can eat the leaves and the plant will go on to produce more. Spinach beet is hardier than spinach; it is unlikely to bolt in summer, yet can withstand frost. It tastes similar to spinach, but contains less oxalic acid.

Soil and climate
A cool moist climate is best for spinach beet, and it likes good deep soil. It grows very well, like the rest of the deep-rooting beet family, in deep beds (see p. 106).

Soil treatment
Dig deeply and add as much compost or manure as you can spare. You don't have to worry about forking roots with spinach beet, so there is no danger of adding too much fresh manure as there is with many root crops.

Propagation
If you sow in spring and again in midsummer, you can have leaves to eat throughout the year. Sow seeds an inch (2.5 cm) apart in rows 18 inches (45 cm) apart. When the seedlings are well established, thin to leave six inches (15 cm) between plants. If you have a deep bed (see p. 106), allow six inches (15 cm) between plants in all directions.

Care while growing
Keep the ground clear of weeds and pick off any flower stems that appear. Your crop will benefit from mulching.

Pests and diseases
Spinach beet is not greatly affected by pests or diseases, but watch out for slugs (see p. 104) in fall.

Harvesting
Pull the leaves off the plants carefully by twisting them downward. Pick from the outside, leaving newer, younger leaves to go on growing. Never denude a plant altogether. Don't allow any leaves to become large and old, for this will stop the plant from producing more. If they grow too old for eating, pull them off and put them on the compost pile.

Cucurbitaceae

Cucumbers, squashes, zucchini, pumpkins, and melons belong to the *Cucurbitaceae*, a family that has evolved to live in extreme climatic conditions.

Nothing is more fascinating in working with nature than observing how plants have evolved to fill nooks and crannies left in the complex ecosystem of the larger and grander species. Thus in the huge tropical rain forests, where mighty trees strive against each other to reach the light, you will find soft, apparently defenseless fast-growing creepers using their mighty rivals to support themselves, making use of speed in growing and flexibility of habit to carry on a kind of guerrilla existence down below.

Other members of the *Cucurbitaceae* are adapted to deserts, and these again make use of speed of growth, sacrificing strength and rigidity and all sorts of other virtues to this end. They shoot away quickly from a seed that has been lying dormant perhaps for years and store away the water of a flash rainstorm in quick-growing fruit. The Tsava melons of the Kalahari and Namib deserts of Africa are prime examples of this. As soon as it rains, these spring up all over the formerly waterless and barren deserts, and the San people, and other desert dwellers, are able to leave the water holes to which they have been confined and roam where they will—secure in the knowledge that they will find water wherever they go, in the Tsava that lie all around. Meanwhile, the Tsava benefit from their depredations, for as animals and men eat their fruit and suck in their water, the seeds get dispersed, to lie dormant perhaps for years again until the next rain.

Melons and cucumbers seem to be made to comfort people in hot dry climates—in Persia (present-day Iran), a dish of sliced cucumber in vinegar is often offered to the thirsty guest.

Cucumbers

In temperate climates, nearly all the cucumbers that have smooth edible skins can only be grown in greenhouses (see p. 209). There is one smooth-skinned strain*, optimistically called "Burpless," that is hardy and can be grown outdoors, but otherwise the methods described on this page and the next apply only to ridge cucumbers. Ridge, or outdoor, cucumbers are not fully hardy, but can be grown with success in temperate regions as long as all risk of frost is avoided. They are much easier to grow than frame cucumbers, and very prolific in favorable conditions.

Soil and climate

Cucumbers like soil that is nearly all compost or well-rotted manure. They will not stand wet feet, although they like plenty of moisture all through their growing period. Ridge or outdoor cucumbers can be grown in most parts of Europe and North America: they are fast-growing and complete their life during the heat of the summer. Although they are easily injured by frost, they mature fast enough to be unaffected by winter weather.

Soil treatment

The soil needs deep digging to ensure adequate drainage, and a large quantity of ripe manure or compost. Half manure and half soil is ideal. Dig it in a foot (30 cm) down in each station. Like tomatoes, cucumbers must have full exposure to the sun. The two crops like much the same soil, so it is a good idea to alternate cucumbers with tomatoes in a sunny bed up against a south wall or fence. Remember that cucumbers are good climbers, and take up far less room and do better if they are allowed to climb. If you can train them up a fence or a trellis on a wall, so much the better.

Propagation

You can sow cucumber seeds outdoors in their stations during the first week of May, with glass jars inverted over them for protection. Remove the jars when the plants get too big for them. Cloches are even more useful, because they can be left on longer and can be improvised with transparent plastic and wire.

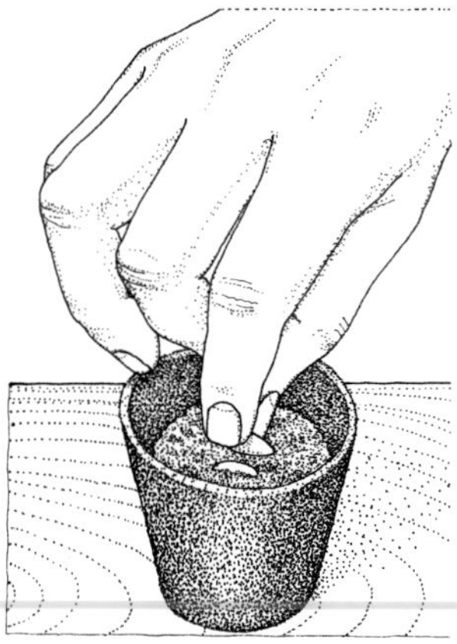

SOWING SEED INDOORS
To get early cucumbers, start them off indoors in peat* pots or soil blocks. Sow two seeds on edge in each pot or block. Do not press the soil down; cucumbers are that rare thing, a vegetable that dislikes firm soil. Water them and keep them warm, but out of direct sunlight. You can then transplant them, at about the time of the last expected hard frost, without disturbing the roots. Sow more seed outdoors at the same time.

The alternative is to start your cucumbers indoors in peat* pots or soil blocks. Then plant them out, long after the last possible frost, in their prepared stations. If your land is heavy and wet, it is best to plant them on ridges or heaps, because if you plant them on the flat in badly drained land, they are likely to die from damping off.

If you are not letting them climb, they can do with six feet (1.8 m) of space all around them; four feet (1.2 m) is the absolute minimum. If you are planting them in a deep bed (see p. 106), allow two feet (60 cm) each way, or else stagger them among other crops. I sometimes plant mine right on top of an old compost pile. They love it. When you plant them out, sow some more seed outdoors as well—this will give you a second crop just when your first is running out.

Care while growing

Cucumbers must have water all the time: many people sink a flowerpot into the ground near each plant and pour water straight into this so that the water quickly reaches the roots. Keep them well weeded.

TRAINING CUCUMBERS

Cucumbers thrive if encouraged to climb. Train them up a fence or trellis (left); when they reach the top, pinch out each growing point (below).

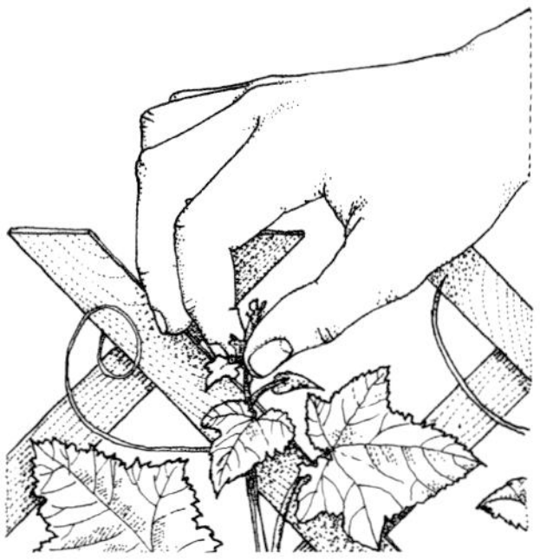

When they have six or seven true leaves—not seed leaves—it is best to pinch out the growing points to make them branch and straggle. Do not remove the male flowers from ridge cucumbers. Do this only with smooth-skinned cucumbers—the ones that you can only grow in the greenhouse. Keep the fruit off the ground by placing a bit of slate, plastic sheet, tile, or glass under each one.

Pests and diseases

MILDEW OR BLIGHT This can develop in very hot and muggy conditions. It produces white powdery patches on leaves and stems. Avoid by planting resistant varieties.

CUCUMBER BEETLE These are yellow and black striped beetles. The adults attack the leaves, and the larvae the stems and roots. Repel them with nicotine spray* (see p. 104).

CUCUMBER MOSAIC VIRUS Leaves become mottled and shriveled. Pull and burn affected plants immediately.

Harvesting

Pick and eat your cucumbers as soon as they are ready. Above all, don't leave any to grow old on the vines.

PICKLING CUCUMBERS

Treat pickles in exactly the same way as ridge cucumbers. They don't need quite as much space between plants as cucumbers do; two feet (60 cm) is about right. Pick them when they are only two inches (5 cm) long.

Pumpkins

Pumpkins have much in common with squashes, but the two can be distinguished botanically: pumpkins have soft rinds and hard stalks, squashes hard rinds and soft stalks.

Soil and climate

Pumpkins like humus if they can get it, but they will grow on unmanured ground as long as it is well drained. They need plenty of water.

Propagation

You can sow pumpkin seed outdoors, under polythene cloches or inverted jelly jars in spring, or else without protection in early summer. You can also sow seed indoors in peat* pots or seed boxes shortly before the last frost, and plant the seedlings out at the start of the hot weather. Allow four feet (1.2 m) between plants, or 30 inches (75 cm) if you use a deep bed (see p. 106).

Care while growing

Let each side-shoot grow a male flower—a flower without a pumpkin attached to it. Then, when the next female flower—a flower with a tiny green pumpkin—appears, cut the shoot short just above the female flower. This will encourage the growth of several smallish pumpkins instead of one enormous one. Place upturned saucers under your pumpkins to prevent them from suffering from rot (see Squashes).

Harvesting

Color is no indication of ripeness. It is best to wait until the first frost before cutting off the pumpkins. Leave about two inches (5 cm) of stalk on each one.

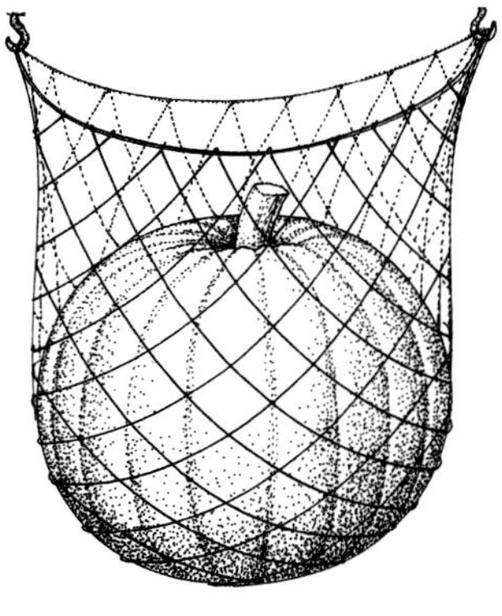

STORING PUMPKINS

Pumpkins need to be stored at a warmer temperature than any other vegetable: 50–60°F (10–16°C) is ideal. Never hang them up by their stalks, as eventually these will shrivel up, and the pumpkins will drop and bruise. By far the best way of storing them is to hang them in nets in an airy place.

Squashes & Zucchini

There is a bewildering variety of squashes, and zucchini are simply young squashes. The best thing to do is to try several different ones, see which you like, and then plant one good one for eating fresh and another that is good for storing. But bear in mind that the huge wheelbarrow-size squashes that praise the Lord at harvest festivals in English churches are better for that purpose than for any other.

Soil and climate

The best place to grow a squash is on a compost pile. They will not grow very well on unmanured ground and like as much humus as they can get. They don't like poorly drained land, but they do need plenty of moisture during their short, quick growing season. Work in plenty of compost or manure, and lime if the pH is below 6.

Propagation

In warm climates, sow the seed straight outside in the bed at the beginning of summer. In countries with shorter summers you can sow them inside in peat pots or seed boxes two weeks before the last frost and plant them out toward the beginning of summer. Otherwise, sow the seed outdoors in situ, under cloches or inverted jelly jars, and remove the cloches when the warm weather begins. Allow four feet (1.2 m) between stations—30 inches (75 cm) for a deep bed (see p. 106). Sow two, or even three, seeds, two inches (5 cm) deep in each station, and when they declare themselves, remove the weakest plants, leaving only one. It is said to be best to sow the seeds on edge and not lying flat. I find they grow whatever I do.

POLLINATING THE FLOWERS
Dust a little pollen from the male flower into the open female flower. To tell them apart, look for a squashlike swelling, directly behind the female flower; this is absent in the male.

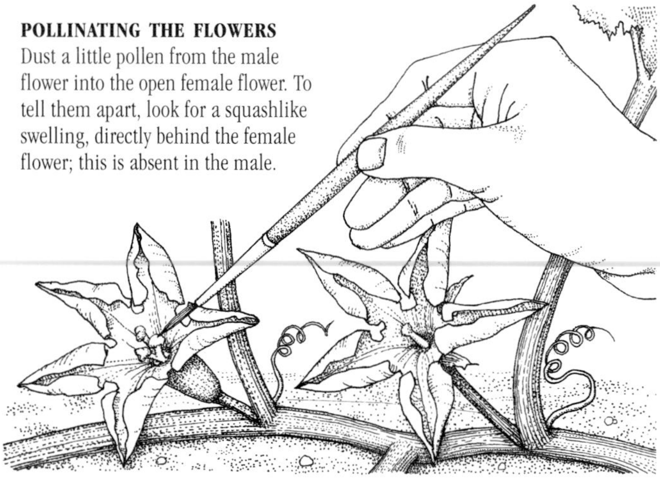

Care while growing

Never let them go short of water. Mulching is very good for them. Though insects pollinate some flowers, it is important to pollinate by hand as well. Trailing varieties must be trained up fences or a wigwam of sticks.

TRAINING UP TRIPODS
If trailing varieties of squash are allowed to grow along the ground, they take up a great deal of room; the shoots can be several yards long. To save space, train them to run vertically. Grow them up trellises or chicken wire; alternatively, train them up tripods. Tie three poles, seven feet (2 m) long, together at the top, and train one plant up each pole. When the shoots reach about five feet (1.5 m) high, pinch out the growing points.

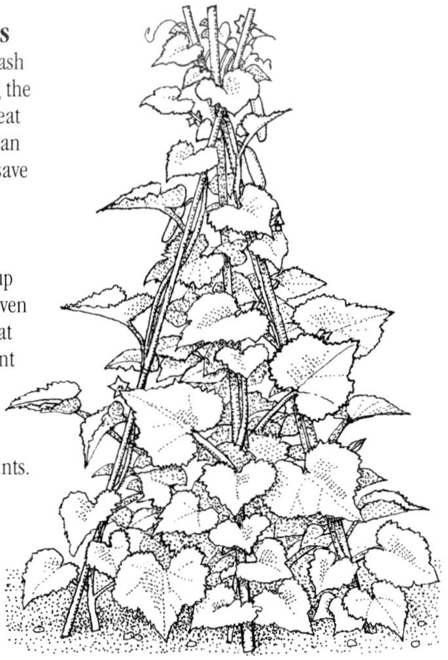

Pests and diseases

MILDEW White patches caused by lack of moisture in the soil, and a high degree of humidity in the air, appear on leaves. The solution is to water them more.

VINE BORER This pest hollows out the squash stems; the leaves go limp and die. Burn affected stems and prevent the pest spreading by earthing up the remaining ones.

SUPPORTING SQUASHES
As the squashes appear and grow to their full size, there is a danger of rot coming up from the soil and affecting them. So keep the squashes just off the soil, by laying a piece of slate, plastic sheet, tile, wood, or glass under each one.

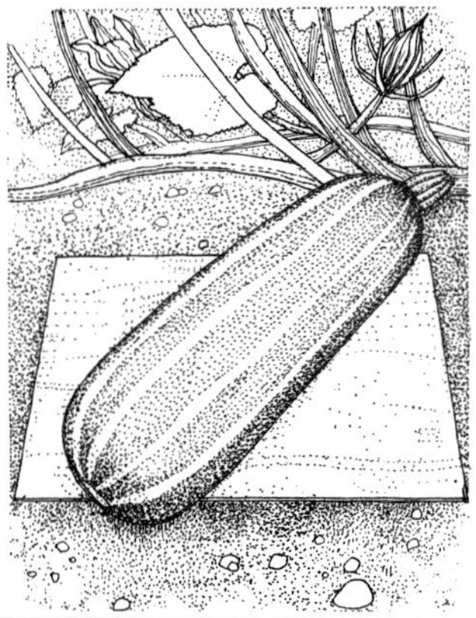

Harvesting and storing

Pick zucchini when they are between four and six inches (10 –15 cm) long. If you leave them to mature, they will turn into squashes. Keep picking to make more grow. Don't harvest until the frost is really threatening. Cut the stems off several inches from the vegetables; don't break them off, as the wound lets in rot, and be certain not to bruise them. Store like pumpkins (see Pumpkins).

Melons

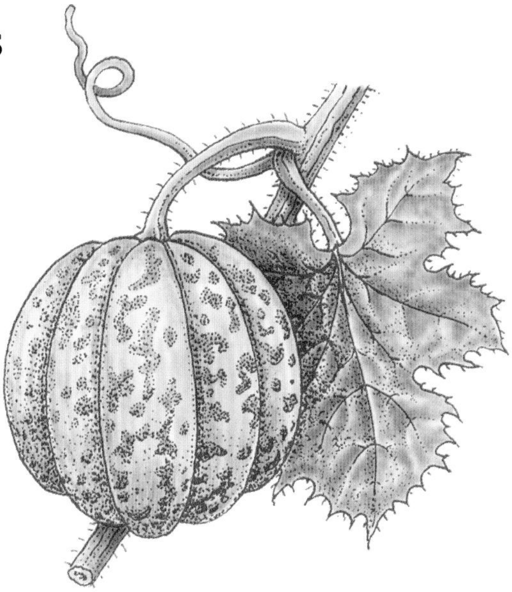

There are three main types of melons: the true cantaloupe, which comes from Europe; the netted melon; and a group that includes honeydews and casabas. My own preference is perhaps for the cantaloupe, but I have never tasted a melon—as long as it was ripe and freshly picked—that was not a memorable experience.

Soil and climate

Wherever you receive four months of summer, which culminates at melon-ripening time with days of 80°F (27°C) and nights of not less than 50°F (10°C), you can grow melons if the soil is right.

They like lightish soil with plenty of humus, not heavy clay, and they like it alkaline: a pH over 7 is best. You will probably have to lime for them.

Soil treatment

Unless your soil is already half manure or compost, dig in plenty, enough to give you a covering of at least four inches (10 cm) before you work it in. Or else dig out "hills" or stations, spaced about four feet (1.2 m) apart, digging to one spit (a spade's depth) deep and several spits square. Dump in a big forkful of manure at each hill, and then cover this with soil again so that each station forms a shallow mound.

Propagation

In hot climates, just sow the seeds two weeks after the last frost, edge-upward, six to a station, and an inch-and-a-half (4 cm) deep; keep them well watered. If you have a deep bed (see p. 106), you should sow cantaloupes 18 inches (45 cm) apart; they will do better if you intercrop them with other plants. Thin to two or three plants in a station when the seedlings are established.

In cool climates you must sow indoors in late spring or else outdoors toward the beginning of summer under cloches or upturned glass jars. If you sow them indoors, don't transfer them outside until all frosts are past and summer has definitely begun.

A method of getting them off to a flying start is to shoot the seeds first, by putting them between wet paper towels in a warm place until they begin to sprout. Do this about a week before you would normally sow the seeds, indoors or out. Don't let them dry out, but don't drown them.

Growing melons in a cool climate is a race against time. Fruit that sets after the middle of summer should ripen before the frost kills it off.

Care while growing

Treat just like outdoor cucumbers. As soon as the little melons are as big as your fist, balance them on a glass jar or an old tin can to keep them off the ground. They are more likely to ripen like this.

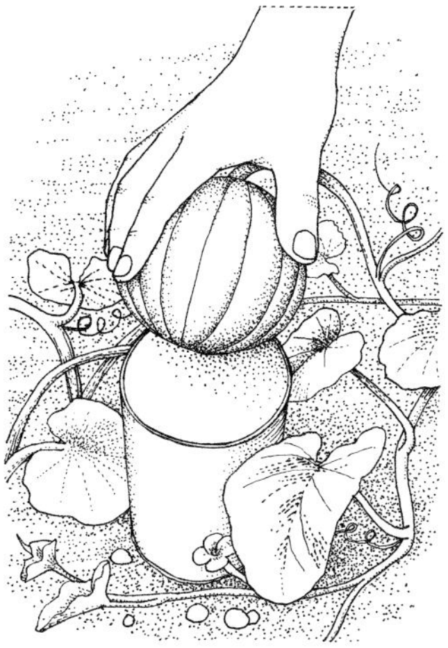

PROTECTING MELONS
You should care for your melons in the same way as for outdoor cucumbers. They need plenty of water and must be kept weed-free. When the little melons have grown to about the size of your fist, balance them on an old tin can to keep them off the ground. This will keep out rot and also makes them more likely to ripen.

Pests and diseases

CUTWORM These pests eat through the stalks of young seedlings. If you get them in your garden, put paper or cardboard collars around the small plants when you plant them out.

ANTHRACNOSE This is a fungus disease that causes brown spots on the leaves and ultimately causes the melons to go moldy. You can avoid it—along with some other fungus diseases—if you don't grow cucurbitaceous plants more than once on the same ground in four years.

BOTRYTIS, OR GRAY MOLD Very wet conditions may bring this on. Although the plants need constant moisture, there is no need to soak the ground. If botrytis strikes, all you can do is to destroy all affected fruits and vegetation.

Harvesting

Melons generally begin to crack around the stems when ripe, and the fruit will break easily from the stem. Tapping helps when you have had experience: a ripe melon makes a unique hollow sound. Some experts get on their hands and knees and sniff the melons to see if they are ripe. You don't store melons; you just pick them and eat them.

WATERMELONS

These need really warm weather for at least four months and a terrific amount of space because the vines straggle over large areas. Unless you have both of these, and sandy or very light soil, don't consider them. But if you have these conditions, go ahead and grow them—there is nothing so refreshing on a hot day. When you think a watermelon is ripe, knock it with your knuckles and heed Mark Twain's advice: a ripe melon says "punk," an unripe one says "pink" or "pank."

Asteraceae

Lettuce, chicory, endive, salsify, scorzonera, dandelions, globe artichokes, cardoons, Jerusalem artichokes, and Chinese artichokes are all members of the *Asteraceae* (formerly the *Compositae*).

The *Compositae* derive their name from the fact that their flowers, which look just like single flowers, are in fact clusters of many small flowers packed together and cunningly disguised as single flowers. The family contains two important groups, which are best described as the salad group and the thistle group. The salad group—lettuce, chicory, endive, salsify, scorzonera, and dandelions—are fast-growing, soft-stemmed plants, which gardeners should usually treat as annuals. They have white milky sap in their stems and a slightly bitter taste. This is the family from which most of the ingredients that make up the basis of good salads come.

The thistle group embraces a lot of the world's thistles, of which globe artichokes and cardoons are of especial interest to gardeners. Closely related to these are a number of other plants that look very different, but to a botanist are quite similar; these include Jerusalem artichokes, Chinese artichokes, and sunflowers. The vegetables in this group generally have to be cooked, but it is surprising what you can eat if you are hungry. Not long ago I was slashing down spear thistles in my meadows when somebody came along and told me that I could eat them. She proceeded, somewhat painfully, to peel the stalk of one of them (a young one) and yes, you could indeed eat them. You would have to be very hungry to make a meal of them though—unless you found a better way of getting the prickles off.

Lettuces

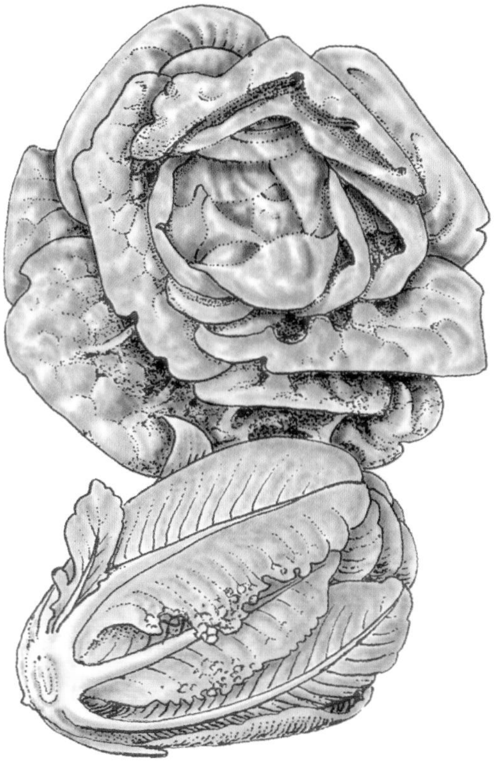

Lettuces have been cultivated for such a long time that several distinct and firmly fixed varieties have developed. The three common types are cabbage lettuce, cos lettuce, and butterhead lettuce. The cabbage variety is green on the outside with crisp, white leaves that form a closed heart; the cos has tall green leaves and forms a loose, elongated head; and the butterhead has tender, green leaves in a flattened, soft head.

With careful planting and selection of varieties, you should be able to grow lettuce for a large part of the year, at least in mild climates. For the purposes of cultivation, lettuce falls into three categories: winter lettuce, spring lettuce, and summer lettuce. Winter varieties are bred to grow through mild winters. They do especially well if they are covered with cloches. Spring varieties are very fast-growing so that they can come to harvest in early summer. Summer lettuces grow big and lush and form the vast bulk of the lettuce crop. Lettuces are excellent for small gardens and windowboxes, and grow well indoors too.

Soil and climate

Lettuces like cool moist conditions. They will grow well in shade, and are inclined to "bolt" or run too quickly to seed in hot sun. Thus they do best in cooler and moister climates, and should only be grown in winter in hot regions.

To grow lettuces well you need good rich soil. The ground should be well-drained but humus-rich to retain water. Lettuces will not grow well in heavy ground, so if you have, for example, clay soil, you must temper this for some years with plenty of manure or compost. The deep-bed method (see p. 106) is ideal for lettuces.

Soil treatment

The best plan is to give the lettuce bed a heavy dressing of really well-rotted manure or compost: a pound (500 g) to the square foot (900 sq cm) is not too much. The soil should be pH 6 or 7, so lime if necessary.

Propagation

You can sow lettuce seed direct in the bed, or in a seed-bed or seed boxes for transplanting. Seed will only germinate in fairly cool, moist conditions, so in very hot countries it is a good idea to put the seed between two sheets of wet paper towel and keep it in a refrigerator for five days before planting. Allow ten inches (25 cm) between plants and one foot (30 cm) between rows—deep-bed (see p. 106), eight inches (20 cm) all round.

Winter lettuce should be sown outdoors in early fall. Spring lettuce can be started indoors in seed-boxes (or peat* pots) in late winter, or else sown out of doors in fall and allowed to stay more or less dormant during the winter, protected by cloches or even some straw or leaves. Summer lettuce is best sown direct outdoors, and sown successionally right through the summer. Lettuces are also a good crop for windowboxes or indoor culture.

Care while growing
Lettuces should be hoed, and watered if necessary. Mulching is also beneficial. Shade the plants if there is a very hot sun, and in winter protect them from heavy frost.

Pests and diseases
CUTWORM Cutworms sometimes gnaw into the stems of seedlings near the ground. If your lettuces suffer badly from this, place collars around the seedlings when you plant them out. This will also discourage slugs, who love to eat lettuces.
LETTUCE ROT In some gardens there is a rot that comes from the soil and may attack the lettuces, spreading through the plants. This can be prevented by a layer of sand on top of the ground around each lettuce. With proper rotation, however, this should not be a problem.

PREVENTING LETTUCE ROT
Lettuce rot comes from the soil and spreads through the plant, via the lowest leaves that are touching the soil. Avoid it by spreading a layer of clean sand over the soil around each lettuce. Also, never plant lettuces in the same place two years running.

Harvesting
Lettuces will not store, and lettuces that have been stuffed inside a refrigerator for days are worthless. So pull them when you want them—making sure that the roots come out with the lettuces. Don't let them grow to seed (unless you want seed), but pull them before they have a chance to bolt, and give them to your poultry or put them on your compost pile when they become overripe.

CELTUCE
Celtuce or stem lettuce is grown for its thick stems rather than its leaves. It needs soil with lots of manure or compost. Sow successionally in shallow drills from spring to midsummer, and thin the seedlings to about twelve inches (30 cm) apart. Water the plants well, as they tend to get tough if allowed to dry out. Otherwise treat them as you would lettuce. The stems will be ready for cutting three months after sowing. You can eat the leaves too; these should be harvested as they form.

CORN SALAD OR LAMB'S LETTUCE
In my view, corn salad, or lamb's lettuce, which is a member of the family *Valerianaceae*, is rather tasteless, but its virtue is that it provides salad during the winter. It grows wild in cornfields, and this gives it an extra hardiness when it is cultivated in gardens. Sow it in early fall and treat exactly as you would lettuce. When the seedlings have three leaves, thin them to six inches (15 cm) apart. If the winter gets very hard, cover the plants with a mulch of leaves until early spring, when you can begin picking again. Harvest a few leaves from each plant as you need them.

Chicory

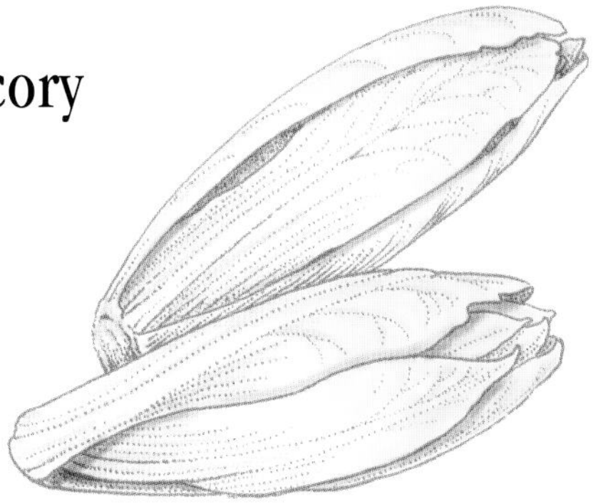

Chicory produces green leaves that are eaten as salad in summer, and, more importantly, shoots that are forced indoors to provide winter salad. Witloof is the best variety for winter forcing.

Soil and climate
Chicory prefers a coolish climate, but will grow virtually anywhere. Rich soil with a pH of 6–7 is best.

Soil treatment
Dig the bed at least two spits (spade-depths) deep, so that the long, straight roots will come out easily when you need to dig them up in the fall. Chicory grows well without mature compost or manure, but if you have some to spare it is worth digging it in well, especially in deep beds.

Propagation
Grow chicory from seed sown half an inch (1 cm) deep and three inches (8 cm) apart in rows two feet (60 cm) apart, or, if you have a deep bed (see p. 106), in stations six inches (15 cm) apart in both directions. Sow successionally through spring and summer for salad; in June for winter forcing.

Care while growing
Weed your chicory bed and keep the soil loose. If you sowed chicory for forcing, thin them to six inches (15 cm) apart and eat the thinnings as salad. Chicory suffers very little from pests and diseases.

Harvesting and storing
Pick leaves for summer salad as you want them. Dig your roots for forcing after the first hard frost, and force them as described below.

FORCING CHICORY
After the first hard frost, dig up a few chicory roots for forcing. Cut off the tops to within an inch (2.5 cm) of the crown. Plant them in a pot, or box of soil in a dark basement, with the temperature not less than 50°F (10°C). Shortly, new sprouts, or chicons, will grow; if you break them off carefully, a second crop will follow. Never pick them until you need them—even an hour in the light makes them droop.

Endives

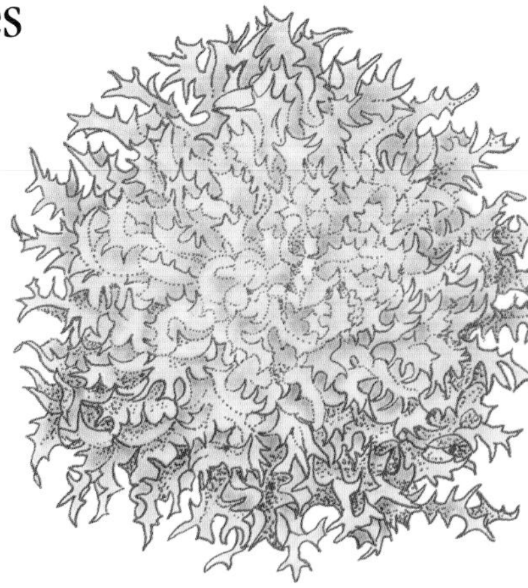

Endive is one of the many minor salad plants, which can be eaten either raw or cooked. It has more flavor, but lacks the delicious crispness of lettuce. Curly endive comes to harvest in late summer; broad-leaved, or Batavian, endive is picked in fall or winter.

Soil and climate
Endives will grow in practically any soil. A neutral pH between 6 and 7 is best, and the more humus in the soil, the better. Endives prefer a cool moist climate to a hot dry one.

Propagation
Sow curly endive seeds thinly in shallow drills at the beginning of summer. Make two more sowings at three-week intervals. Sow broad-leaved endive in late summer. Thin the seedlings when they appear, to nine inches (23 cm) apart or six inches (15 cm) in deep beds (see p. 106).

Care while growing
Endives don't like too much sun, so shade them if necessary and keep them moist. In cooler climates, this shouldn't be necessary. Three or four months after sowing, start blanching by covering the plants with something that will keep out the light. Whitewashed cloches are ideal. Alternatively, you can pull them up and plant them in earth in seed boxes and put them in a dark place.

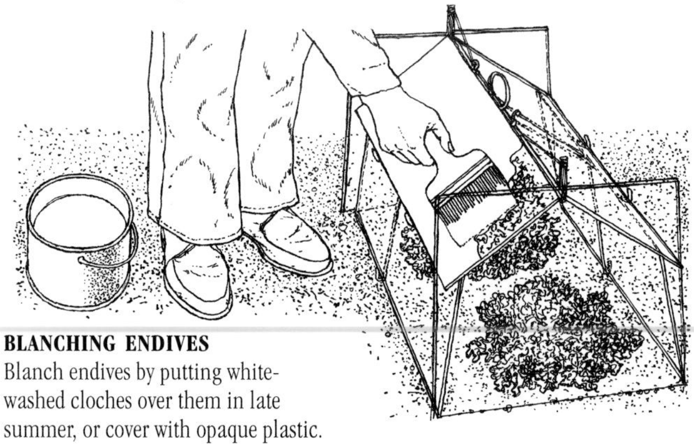

BLANCHING ENDIVES
Blanch endives by putting white-washed cloches over them in late summer, or cover with opaque plastic.

Harvesting
Simply eat endives once they are blanched; this will be about three weeks after you covered them.

Salsify

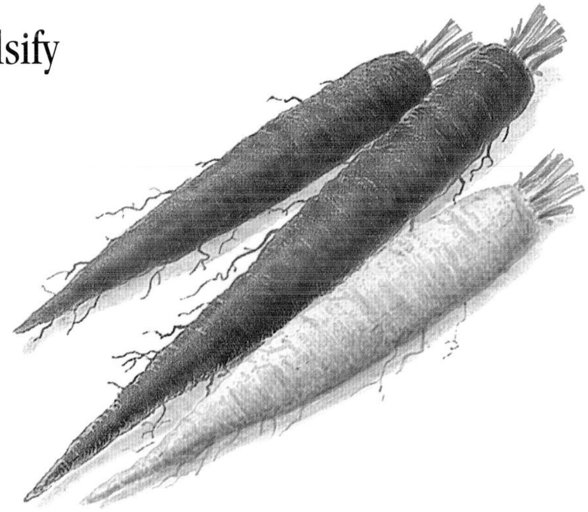

Salsify or "vegetable oyster" is grown mainly for its long roots, though the leaves may also be eaten. There are in fact two kinds—true salsify, and Spanish salsify, which is also known as scolymus—but the differences are minimal.

Soil and climate
Salsify likes deep, rich loam. If you have to grow it in heavy clay, take out a trench a foot (30 cm) wide and 18 inches (45 cm) deep and fill it with well-rotted manure.

Soil treatment
A bed dug deeply with well-rotted manure or compost is best. Fresh manure will cause the salsify root to fork. The soil should have a neutral pH, between 6 and 7.

Propagation
Sow seed in early spring an inch (2.5 cm) deep and two inches (5 cm) apart in rows about a foot (30 cm) apart. In a cold climate, you can sow it anything up to a month before the last frost, but you must cover it with glass. Thin to four inches (10 cm) when the plants appear. Sow closer and thin to three inches (8 cm) in a deep bed (see p. 106).

Care while growing
Keep the salsify weeded, and mulch in fall before hard weather sets in. It rarely suffers from pests or diseases.

Harvesting and storing
The roots should be allowed to grow to an inch-and-a-half (4 cm) in diameter and eight inches (20 cm) long before you harvest them. They are improved by frost—even quite a hard frost—so you can leave them in the ground quite late. But in very cold climates, harvest them in the fall and store them in damp sand in a cool basement. Since salsify is a biennial, it will go to flower and seed in the spring of its second year. If you still have some in the ground at this stage, cut the flower stalks before they get too hard; you will find they taste very good if you eat them like asparagus.

SCORZONERA
Very similar to salsify, scorzonera should be grown in the same way. One minor difference is that you can, if you want, treat it as a biennial, digging up and eating the roots in their second year of growth.

Dandelions

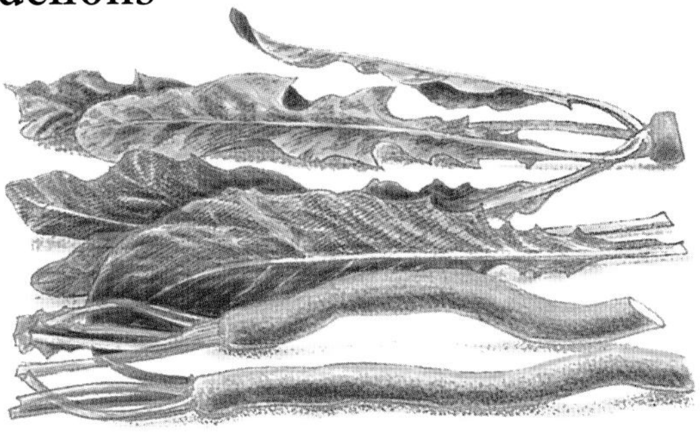

The wild dandelion has been tamed and improved to make a very useful vegetable. The leaves can be used, sparingly, in salads and are especially useful for salads early in the year before the first lettuces are ready; mulched over, the plants will survive the winter and grow away fast in the spring. Cooked dandelion leaves make an excellent vegetable, better tasting and far more vitamin-rich than spinach. The roots may be dried, ground, and used as a substitute for coffee. The coffee tastes quite good, but doesn't give you the lift that caffeine does. We might all be a lot healthier if we went over to it. The flowers make fine wine, but it is sensible to eat the leaves of your cultivated dandelions and gather flowers from wild dandelions.

Dandelions are perennial plants; if you look after them they will last for years.

Soil and climate
Dandelions thrive in any soil and in any climate except those that are very hot.

Soil treatment
Dig well and incorporate compost or manure.

Propagation
Dandelions grow easily from seed that can be bought from seed catalogs. Sow successionally from the middle of spring until midsummer. Sow thinly in rows 18 inches (45 cm) apart and thin to a foot (30 cm) apart in the rows when the leaves are two inches (5 cm) long. On the deep bed (see p. 106), broadcast and subsequently thin to about a foot (30 cm) apart each way.

Care while growing
Water if the bed dries out, and keep it well weeded. Cover the plants with a light mulch in cold winters. Pick off any flowering shoots (use wild dandelion flowers for wine).

Pests and diseases
Cultivated dandelions are closely related to their wild ancestors and are therefore very resistant to disease. Keep the slugs away though (see p. 104).

Harvesting
Dandelions are perennials so you must not cut too many leaves off during the first summer of their lives, because the roots gain strength from the leaves. Cut them hard the second year. You can dig dandelion roots up in the fall and force them to grow shoots in the same way you can force chicory (see p. 159).

Globe Artichokes

Eaten as it should be—that is, when still young and tender—the globe artichoke is, in my view, supreme among vegetables. But more than that, it is as beautiful a plant as you are likely to have in your garden. A thistle *par excellence*, I've actually seen a globe artichoke grow to a height of ten feet (3 m). You will, however, need a fair amount of space to grow globe artichokes, so they are better left out of the very small garden. On the other hand, they are decorative enough with their fernlike, grayish leaves not to be grown only in the vegetable garden. They are not related either to Jerusalem or to Chinese artichokes.

Soil and climate
Low-lying black alluvial soil is ideal for globe artichokes. It should be moist but not waterlogged. If you don't have this—and there is a good chance you won't—any rich, moist soil will do. They are not really winter-hardy in cool temperate climates, since hard frosts cut the leaves right down. Even so, the roots will survive and the plants will shoot up in the following spring. Some protection, such as straw or leaves, helps them to survive the winter.

Soil treatment
The bed for your globe artichokes should be dug deeply. Work in large amounts of organic material. The pH should be around 6.5, so test the soil.

Propagation
Globe artichokes can be grown from seed and it doesn't take as long to get heads from them this way as many people seem to think. Sow the seed in a hot-bed in late winter, plant out in spring, and you should be eating artichokes in early fall. You can also sow seeds in their permanent position in the spring, but you will have to wait to harvest blooms until the next year.

A popular alternative to sowing seed is to grow the artichokes from suckers or offsets. If you uncover the old plants at the roots in the spring or fall, according to your climate, you will find a number of small shoots getting ready to grow. Cut some of these out carefully, taking with each shoot a "heel," which is a bit of the mother plant. But don't take so much as to harm the mother plant. Plant these suckers straight into the ground at about the same depth they were before. Do this in early spring in cold climates or in fall in hot ones. They should give you heads by the following summer.

Spacings vary according to variety and the kind of soil the plants are to grow in—in very rich soil plant them farther away from each other because they are likely to grow much bigger. Generally, four feet (1.2 m) between plants is about right. Make it five feet (1.5 m) in a deep bed (see p. 106), because the soil will be rich.

You can have good heads to eat for six months of the year in cooler climates if you protect some old plants very well in the winter. These will give you blooms in late spring and early summer. If you planted good suckers in early spring, these will give you a crop in late summer. And you'll get heads in the fall from suckers planted, say, six weeks after the first planting. In very hot climates you should get most of your crop in the winter and spring.

Since globe artichoke plants tend to lose vigor after a few years (though I have had good crops from an eight-year-old plant), try replacing a quarter of your total crop every year with new plants. That is, each fall dig out the oldest quarter of the crop, and replace them in the spring with new plants. But take the suckers from the old plants before scrapping them and plant them in sand or soil indoors. In the spring these can be used for replanting.

Care while growing

Heavy mulching with compost or manure is always a good idea with globe artichokes. And where the winters are cold, it is best to cut the plants down to the ground in the fall and to pile hay, straw, or leaves over them. But if you do this, uncover the plants on mild days to let them dry out. In summer droughts, soak the ground around the plants regularly and thoroughly.

Pests and diseases

BOTRYTIS This causes a gray mold on leaves and stalks and may attack plants in very warm wet seasons. If it does, remove and burn all affected plants.

ARTICHOKE LEAF SPOT In hot, muggy weather, leaves may go brown and die. Use a weak solution of Bordeaux mixture*.

Harvesting

Ignorance of the following rule makes artichokes less popular than they should be: harvest them when they are very young! Don't wait until they are huge prickly things, as hard as wood and as sharp as needles. If you cut them when they are still tight, green, and small, you can eat practically the whole thing. If they are too old, only the bottom tips of the leaves and the heart will be edible. Simply cut through the stalk about an inch below the globe.

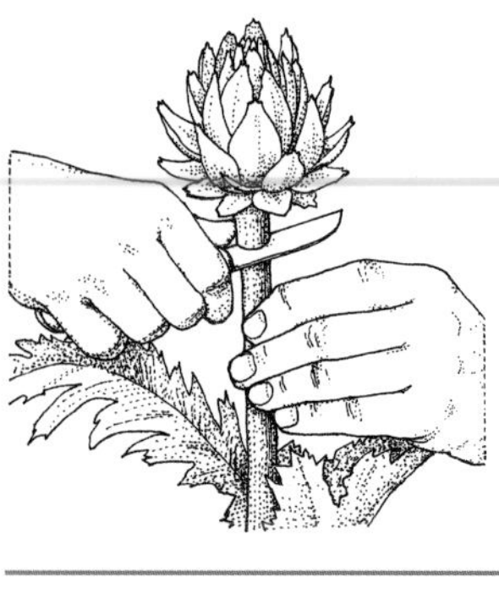

HARVESTING GLOBES
If you harvest your globe artichokes while they are still very young, you can eat the whole head, instead of just the heart and the base of the inner leaves. Harvest the globes by cutting through the stalk about one inch (2.5 cm) below the globe.

Cardoons

Cardoons are thistles that are very closely related to globe artichokes. They have been specifically bred for their stems, which should be eaten blanched. They taste delicious but take up a lot of space.

Soil and climate

Although strictly speaking perennials, cardoons are always grown as annuals, so they are not as fussy about soil as globe artichokes are. Cardoons grow in a wide range of climates, although they don't like it too wet.

Soil treatment

Dig holes about one foot (30 cm) across and three feet (90 cm) apart, and fill these with compost. Or you can dig out trenches as for celery (see Celery).

Propagation

Sow three or four cardoon seeds in each hole or at three-foot (90-cm) intervals along the trench in the spring. Alternatively, sow in peat* pots a little earlier and plant out toward the end of spring. Pull out all but the strongest plants when they appear.

Care while growing

Cardoons should be kept watered and weeded, and, since they need blanching, they should be earthed up like celery (see Celery). But before earthing up you should wrap the plants in straw, or put a length of drainpipe over them to protect them. They are rarely affected by pests or diseases.

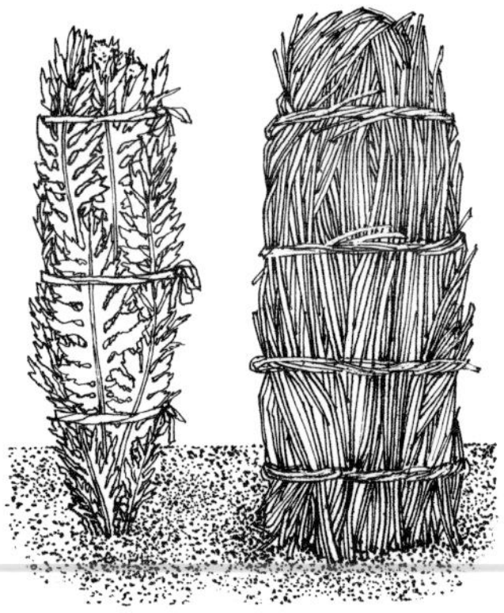

BLANCHING CARDOONS
About three weeks before harvesting, bunch up the cardoon plants and tie them together. Then wrap them completely with straw, or anything else that will keep out the light. Leave the tips of the leaves showing. When the blanching process is complete, you will have cardoons all through the fall and winter.

Harvesting

You can begin to harvest your cardoons in fall and continue well into the winter. Three weeks before you intend to harvest a batch, blanch the plants, which may be three feet (90 cm) tall. Gather them up and tie them together, then wrap each one with straw.

Jerusalem Artichokes

It is most unfortunate that the Jerusalem artichoke was so named, since people constantly mix it up with the globe artichoke. Strangely, they are members of the same family, but only a botanist could see any resemblance between them. The Jerusalem is in fact more closely related to the sunflower; the plant looks very similar except that it has small flowers and tubers on its roots. The tubers are pleasant to eat and are especially good for diabetics because they contain a special form of sugar and no starch.

Soil and climate

Jerusalem artichokes grow best in light or sandy soil. They do very poorly in boulder clay. In light land they will grow like weeds if you let them—as high as seven feet (2 m)—and they will smother any plant that tries to compete with them. They grow in practically any climate.

Soil treatment

If you have to grow them in heavy soil, dig it well, make sure it is free of perennial weeds, and incorporate as much manure or compost as you can. Sandy soil is little trouble, although the more manure, the bigger the crop.

Propagation

Simply dig a hole with a trowel and plant the tubers six inches (15 cm) deep. I put them in in late winter. Even a tiny scrap of a tuber will produce a plant. Plant 18 inches (45 cm) apart each way—deep-bed method (see p. 106), 15 inches (38 cm). They make a smother crop.

Once you have had a crop of Jerusalem artichokes it is very difficult to get rid of them. They will come up again year after year unless you hoe them out constantly.

Care while growing

In light land you don't need to do anything. In heavy land hoe between them because Jerusalem artichokes are not vigorous enough to beat weeds. Alternatively, mulch heavily. Pests and diseases rarely affect them.

Harvesting and storing

Dig them up in late fall, or else leave them in the ground until you want them. In climates with hard frost, dig them up after the tops have died back and store them. The tops are fine for mulch, and can be woven as a windbreak.

CHINESE ARTICHOKES

Strictly speaking a member of the *Lamiaceae* family, the Chinese artichoke is also grown for its tubers, in exactly the same way as the Jerusalem artichoke. Regular watering and fertilizing with soluble manure will add extra flesh to the tubers, which should be lifted in fall when the leaves die.

Sweet Corn

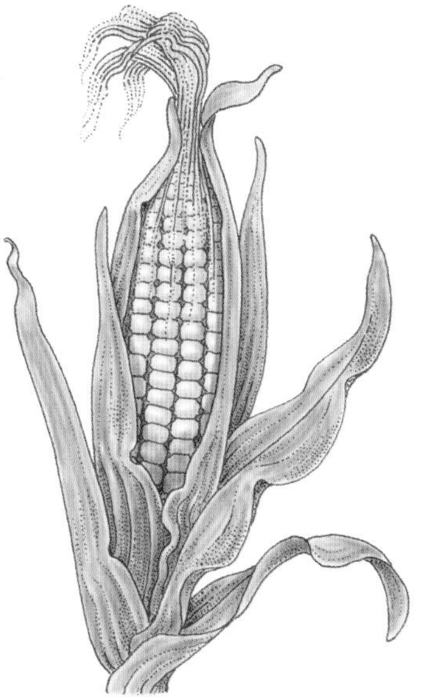

Sweet corn is a member of the *Poaceae*, or grass family; this includes all the world's cereal crops and a lot of other species besides.

The plants, which I have seen as tall as 12 feet (3.6 m), send a mass of fibrous roots deep into the soil. This takes a lot out of the soil, but when in due course the plants are returned in the form of compost they put most of the goodness back again. There are small varieties bred for gardeners that only grow to four feet (1.2 m). These are useful in a small garden because they don't cast so much shade.

Soil and climate

Sweet corn likes deep, well-drained, humus-enriched loam. Clay is too cold for it in northern climates; it must have an early start, as it needs four months to come to maturity. It will grow in light sandy soil, but only if there is plenty of humus in it. Hungry old gravels and sands just won't grow it at all. It needs good soil.

As for climate, sweet corn prefers four months of hot weather without much cloud cover. However, you can grow it in fairly cloudy countries, if you start it early enough.

Soil treatment

Sweet corn needs plenty of humus thoroughly mixed into the soil (not just dumped on top at the last minute) because the roots go deep as well as spreading wide. A couple of inches (5 cm) of well-rotted manure dug well in a spit (a spade's depth) deep is ideal. Alternatively, plant sweet corn after a heavily manured main potato crop. Soil must be neutral: pH 6.5 to 7.

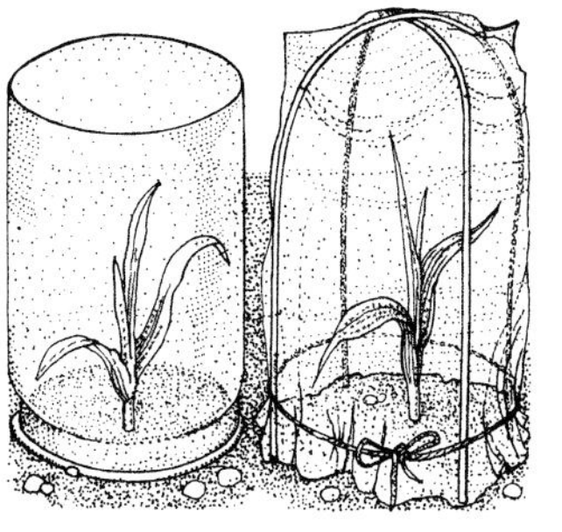

PLANTING SWEET CORN
So that the roots are not disturbed by transplanting, start sweet corn off outdoors under protection. An overturned jelly jar makes an admirable cloche; alternatively, improvise one out of polythene and wire. Remove the cloches before the plants become too big.

Propagation
It is better to sow sweet corn in wide blocks instead of in long thin lines. This is because it is wind-pollinated and in a thin line some plants may get missed. To give it a flying start, sow it outside two weeks before the last probable frost under cloches, overturned jars, or plastic umbrellas. Remove these, of course, before the plants get too big for them. Or better still, especially in cool climates, start it off indoors in peat* pots, soil blocks, or even small flowerpots. Sweet corn doesn't take to being transplanted but this doesn't mean it can't be done. In the cool climate where I live, I like to sow it in peat pots in the greenhouse in early spring and transplant it—pots and all—a month, or six weeks, later.

Sow the seed or transplant the plants about a foot (30 cm) apart each way, in blocks or in the deep bed (see p. 106).

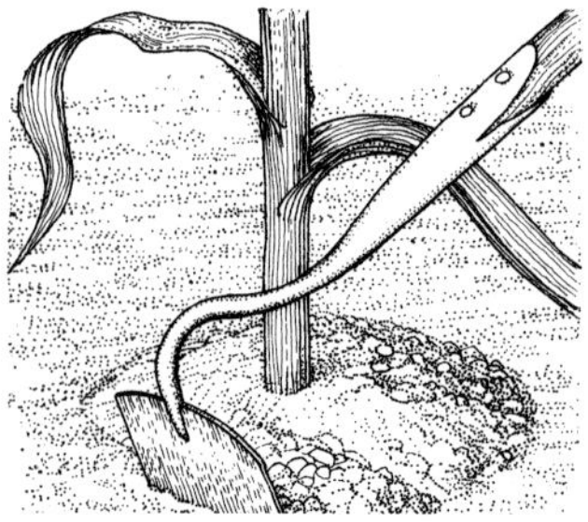

EARTHING UP SWEET CORN
If you intend to earth up your sweet corn stalks while they are growing, sow the seed 18 inches (45 cm) apart instead of 12 inches (30 cm) apart. Pulling up the soil around the plants in this way gives them support while they mature and causes them to put out more roots higher up.

Care while growing
Sweet corn doesn't like to go short of water. Mulching does it nothing but good. Earthing up is beneficial too, because the plant will put out more roots higher up its stem.

Pests and diseases
EARWORM These bore into the tips of the ears of corn. If you see them when you are picking the ears, destroy them.
SMUT Smut is a fungus that causes large gray boils to appear on the kernels. Burn the affected plants and don't leave rotting sweet corn on the ground; bury it or compost it. Otherwise it may develop smut.

HARVESTING SWEET CORN
Pick the corn by jerking the cobs sharply downward and breaking them off; if you cut them off you risk damaging the plant.

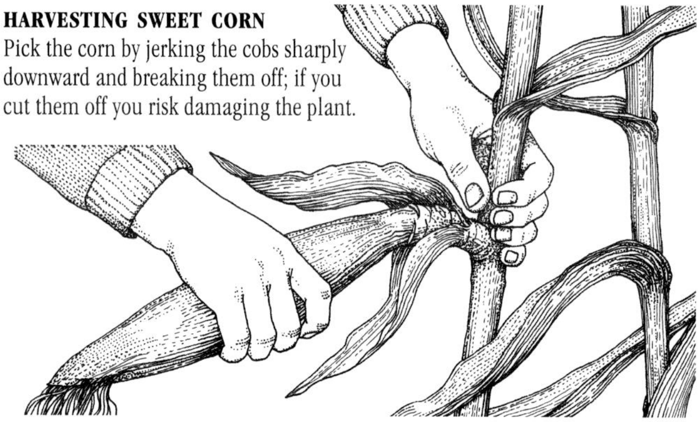

Harvesting
The ears are ready to pick as soon as the white, silky tassels go brown; otherwise, test for ripeness by opening the husk and pressing your fingernail into a grain. If it is firm but still slightly milky it is ready. Rush the corn to the saucepan! The moment you pick it, the sugar starts turning to starch and the flavor is lost.

Okra

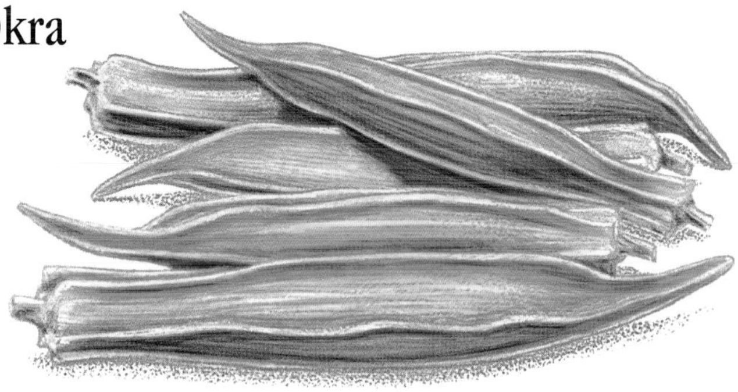

Okra, familiarly known as "lady's fingers," is a tropical plant of the *Malvaceae*, or mallow family, whose most renowned member is the cotton plant. The pods of the okra plant make the most delicious vegetable with a very subtle and distinctive taste. They are much used in Indian curries. When the seeds are well developed they can be shelled and cooked like peas. The whole plant is extremely attractive with large yellow and red flowers. Three or four plants will keep a family well supplied with okra.

Soil and climate
Okra can be grown in greenhouses in cool climates. Outdoors it needs a lot of summer sun, and it is not much good trying to grow it in countries with cool cloudy summers. But where outdoor tomatoes really thrive and crop reliably, okra can be grown. It likes a light soil with plenty of humus but not too much fresh manure, since this will produce too much leaf and not enough fruit.

Soil treatment
Okra grows especially well by the deep bed method (see p. 106) and in cooler climates can be grown under portable deep-bed mini-greenhouses (see p. 111). Otherwise cultivate deeply and work in some well-rotted compost. The soil should have a pH between 6 and 7.

Propagation
Okra can be started indoors, but only in peat* pots because it doesn't like being transplanted. If you sow the seed outdoors, wait until the soil is thoroughly warm. You can help here by warming the soil yourself under a mini-greenhouse or under cloches (see p. 96). Dwarf varieties should be sown 30 inches (75 cm) apart; larger varieties three and a half feet (1 m) apart.

Care while growing
Okra should be watered occasionally, but not swamped.

Pests and diseases
CATERPILLARS Pick caterpillars off and step on them.

Harvesting and storing
You can harvest about two months after sowing. Harvest when the okra is still quite young, a few days after the flowers have fallen. Pick pods every other day whether you need them or not, so that the plants keep producing more. Pods can be frozen or canned, and in Italy I have seen them laid out on a rack in the sun and dried. They keep well like this, but I prefer them fresh. Keep picking them as long as they grow, which can be right up to the first frost.

Rhubarb

Rhubarb is a vegetable because we eat its stems, not its fruit. It is thought of as fruit merely because our whim is to eat it as a dessert, like most other fruit. These days there is so much frozen fruit, as well as fruit from abroad, that there is no longer an endless gap between the last of the stored apples and the ripening of the raspberries. In the old days that gap was filled with rhubarb. But rhubarb is still well worth growing, for it is a good fruit substitute. Its stems contain oxalic acid, which scours pans clean and sets your teeth on edge.

Soil and climate

Rhubarb likes a cold climate (it comes from Mongolia) and is no good at all in a hot one. Unless it enjoys frost in the winter it does not have the dormant period that it needs, and its stalks, instead of being red and edible, are green and inedible. It likes quite acidic soil, so don't give it lime, but otherwise it will grow in any well-drained soil, and it seems to thrive in that milieu of nettles, old rusty cans, and broken bottles at the end of many old country gardens.

Soil treatment

Put rhubarb in a part of the garden devoted to perennials because, properly treated, it will continue to grow and yield for years. Clear its bed of perennial weeds, dig it deeply, and put plenty of manure on it. It pays to dig a deep pit, discard the subsoil, and fill in with manure and topsoil.

Propagation

Rhubarb seldom breeds true with seed, and so the usual method is to use divided roots. Commercial growers generally dig up their beds every four years, divide the roots, and replant on fresh ground. You can either buy root cuttings or get them from a neighbor who is in the process of dividing his rhubarb. Just plant the bits of root the right way up three feet (90 cm) apart, and up the plants will come. Nothing can stop them.

PLANTING RHUBARB CROWNS
It is possible to raise rhubarb from seed, but this is not the most reliable method of growing it. It is best to get hold of some rooted crowns. Dig a deep hole, fill it with compost, put back the topsoil, and plant the roots the right way up and three feet (90 cm) apart. They are sure to grow.

Care while growing

Plentiful mulching is good for rhubarb. In the winter, when it dies down, you can bury it deeply in a mulch of manure, leafmold, compost, or what you will. As long as it is provided with ample organic matter in this way, you need not dig it up every four years: it will last almost indefinitely.

CUTTING OFF FLOWERS
Flowering rhubarb will not produce succulent juicy stems. The flower stalks divert all the nourishment away from the plant; cut them off when they appear.

In the spring draw the mulch away from around the plants to let the sun warm the soil. Then, when the plants begin to grow well, you can cover them with old pails and in the winter, cover the pails with fresh long-strawed manure, so that the heat from the manure will force the rhubarb on, and you will get stems to eat early in the spring.

You can cover the crowns with oil drums, painted black to absorb the heat of the sun, open at the bottom end, and with a six-inch- (15-cm-) diameter hole cut in the top. The oil drums should then be pressed firmly into the ground around the plants; they should be adequately weighed down to prevent them from blowing over.

Pests and diseases

RHUBARB CURCULIO This is only found in North America. It is a rust-colored beetle about an inch (2.5 cm) long. It bores into every part of the rhubarb plant, but can easily be picked off. The beetle lives in dock plants, so suffer not docks to exist near rhubarb.

FORCING RHUBARB
Cover each plant with an old bucket, and when winter comes, insulate the bucket with long-strawed manure. If you lift two-year-old roots in late fall and force them indoors, you can have rhubarb to eat through the winter. Otherwise, start forcing the plants in the same way outdoors in very late winter, and you will reap the benefits in early summer.

Harvesting

Spare the plants altogether for their first year, and thereafter only harvest the big, thick stalks: let the thinner ones grow on to nourish the plants. Never take more than half the stems of a plant in one year. Don't cut the stems, as this lets in rot; break them by pulling them back from the plant, then forcing them downward an inch or two. This does not hurt the crown. Stop harvesting altogether in July.

You can make jam with rhubarb (see p. 222) but the best thing you can do with it is to make wine (see p. 224).

Mushrooms

Mushrooms, which are fungi and not vegetables at all, are an obvious choice for the self-sufficient gardener who has space to spare indoors. Mushrooms have a higher mineral content than meat (twice as high as any other vegetable), and contain more protein than any other vegetable except for certain types of beans. Another good thing about growing mushrooms is that the compost you need for growing them can all end up in your garden outside.

Climate

In warm weather you can grow mushrooms outdoors or indoors without artificial heating, using the method I will describe. In the winter, keep the temperature over 60°F (16°C). Never leave mushrooms in direct sunlight.

Soil treatment

To grow mushrooms you need boxes that ideally should be two-and-a-half-feet (75 cm) long, nine inches (25 cm) wide, and nine inches (25 cm) deep.

You can buy suitable compost and this is really the best thing to do for growing small quantities of mushrooms. However, to make enough compost for 60 square feet (6 sq m) is not difficult. Get four bales of wheat straw (no other straw will do) and shake it out into layers, soaking it thoroughly with water. Leave it for a day or two, but throw on more water from time to time, because it must be saturated. You should also have: seven pounds (3 kg) of gypsum (from a home improvement store), 28 lb (12.7 kg) of poultry manure, and 14 lb (6.3 kg) of mushroom compost activator.

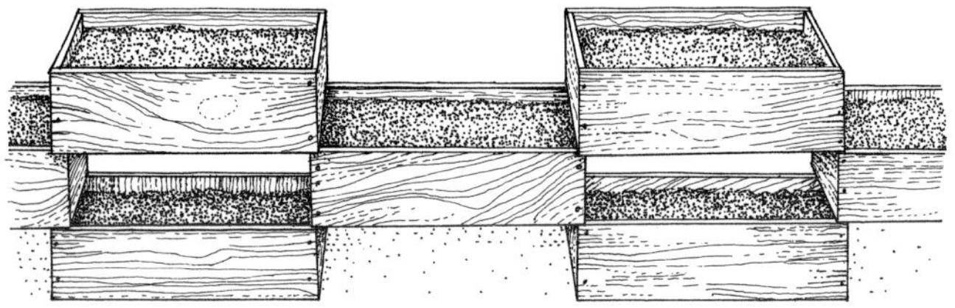

STACKING MUSHROOMS IN BOXES Allow at least six inches (15 cm) between the top of one box and the bottom of another. There should also be perhaps a dozen half-inch (1.5-cm) holes in the bottom of each box. I like cedarwood boxes best, but you can use fiberglass trays.

When the straw is thoroughly wet, spread some out 12 inches (30 cm) deep over an area five feet (1.5 m) square. Shake over this layer a trowelful each of the poultry manure, gypsum, and activator. Add another foot (30 cm) of straw and on this another sprinkling of the other goodies, until all the materials have been used up. The pile should be about six feet (1.8 m) high. If it is outdoors, cover it with an old carpet, paper, or plastic.

By the fourth day the temperature of the pile should be 160°F (71°C). Leave it another two days, then turn it so the outsides are in the middle. If any part of the pile appears dry at this turning, sprinkle water on it, just enough to moisten it but not enough to wash away the special ingredients. When you turn the pile, shake out the straw thoroughly and rebuild very carefully. The success of your crop depends on this care.

After another six days, turn again. Be even more sparing with water, but if there are any dry patches or gray patches, sprinkle them lightly. Then, after four more days, turn yet again. If the compost appears too damp, apply more gypsum. Six days later the compost will be ready for the boxes.

Propagation

When it is ready for use, the compost should be fairly dry and springy; it should consist of short pieces of rotted straw but should not be sticky. Fill each box, tamping the compost down well with a brick, until the final topping-off is level with the top of the box.

By now you will have bought some spawn. There is "manure" spawn, which comes in lumps that you break into small pieces, and "grain" spawn, which you simply scatter on the compost. I suggest that beginners use manure spawn, because it is easy to use and reliable.

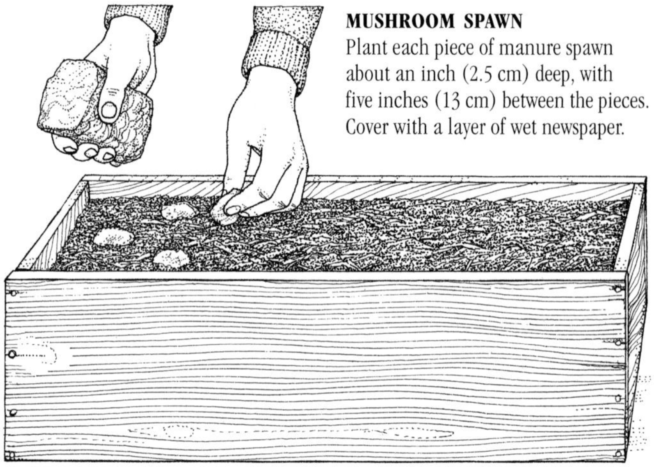

MUSHROOM SPAWN Plant each piece of manure spawn about an inch (2.5 cm) deep, with five inches (13 cm) between the pieces. Cover with a layer of wet newspaper.

Care while growing

During the next week or two, do not let the temperature fall below 60°F (16°C); 70°F (21°C) is even better. On the other hand, beware of overheating; 90°F (32°C) may kill the spawn. After three weeks you should see the white threads of the mycelium growing in the compost. At this point you must apply "casing." Mix some well-moistened horticultural peat with the same bulk of freshly sterilized loam (the loam should be from permanent grassland). Put an inch-and-a-half (4-cm) layer of mixed peat* and loam on top of the compost and press it down gently. Mushrooms should appear about three weeks later. Give them a little water. Keep the temperature between 60°F (16°C) and 64°F (18°C).

Harvesting

When you harvest mushrooms, twist them out. When the crop seems over, try to persuade it to go on cropping a little longer by watering it with a weak salt solution. Eventually dump the spent compost on your compost pile, wash the boxes with formaldehyde solution, and put the boxes out to weather for several weeks before you use them again.

CHAPTER SIX

The Cultivation of Fruits

Containing the planting, growing, and harvesting instructions for members of the families Rosaceae, Rutaceae, Grossulariaceae, Moraceae, Ericaceae, Oleaceae, and Vitaceae.

Rosaceae

Apples, pears, quinces, cherries, peaches, nectarines, apricots, plums, damsons, raspberries, blackberries, and strawberries all belong to the useful and beautiful family of the *Rosaceae*. It is a huge family that includes agrimony, burnet, mountain ash, 500 species of hawthorn, and, of course, the mighty rose.

Most of the fruits grown in temperate climates belong to this family, which splits into several subdivisions: among them are plants that have stone fruits, like cherries and plums; those that have berries, like strawberries and raspberries; and those that have what botanists call pomes, like apples and pears.

All the species are insect-pollinated, which is why they have such enticing flowers. They also depend on birds and animals to scatter their seed—suitably manured—which is why they have attractive and edible fruits. And so, with the help of other living things, the cycle renews itself, and the *Rosaceae* continue to enhance our lives.

Apples

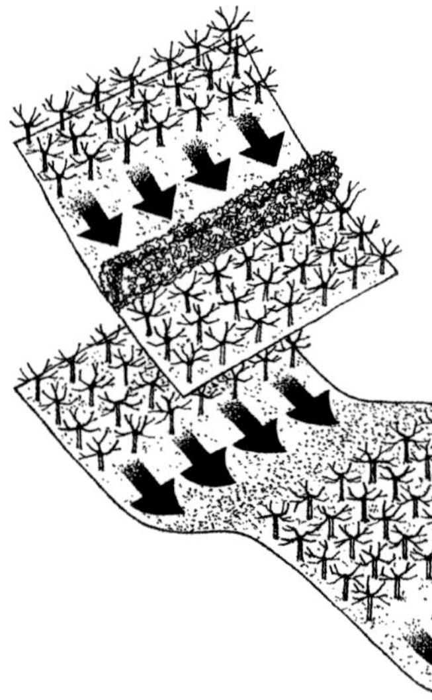

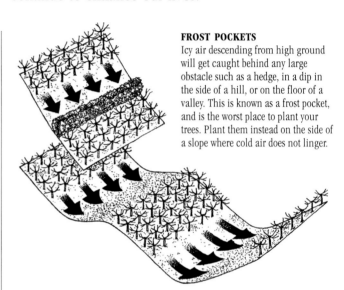

FROST POCKETS
Icy air descending from high ground will get caught behind any large obstacle such as a hedge, in a dip in the side of a hill, or on the floor of a valley. This is known as a frost pocket, and is the worst place to plant your trees. Plant them instead on the side of a slope where cold air does not linger.

Apples are far and away the most important hard fruit crop of temperate climates. If you grow early and late varieties and a variety bred for keeping, you should have apples to eat throughout the year. The gap will come in the summer when you should have plenty of soft fruit. How many trees you should have and whether they should be standard, half-standard, dwarf, espalier, or cordon is discussed on page 76.

Soil and climate

Apples prefer a good, deep, well-drained loam, though they will do well on a heavy loam. They do not thrive in gravel, very sandy soil, heavy stubborn clay, or shallow soil above chalk or limestone. But if you have unsuitable soil you can always dig a big pit where you want to plant a tree and bring in some good topsoil from outside. And of course any soil can be improved—be it too light or too heavy—with plenty of compost or manure.

In places where peaches, outdoor grapes, figs, apricots, citrus fruit, and such things grow well and freely outdoors, it is better not to grow apples. The apple is a cold-climate tree and does better for a period of winter dormancy. It doesn't mind very cold winters (certain varieties will even grow in Alaska) but it doesn't like late frosts once it is in blossom. Late frosts are generally the sort that creep over the land on a still, clear night. Therefore, take care not to plant apples in frost pockets—those places where cold, frosty air gets trapped after it has flowed down from high ground. The floors of valleys and dips in the sides of hills—especially if they contain some obstruction, such as a thick hedge—are likely to be frost pockets. If your land is not flat, plant your apple trees fairly high on the side of a hill, or on a gentle rise, where cold air will not linger. But don't plant them where it is too windy.

Soil treatment

It is a good thing to clear-cultivate land—thoroughly dig it over—before you plant fruit trees on it. Ideally, after clear-cultivating you should grow a crop or two of green manure (see p. 86) and dig, plow, or rototill this into the soil. But you may be in too much of a hurry to do this—I always am—in which case simply clear-cultivating is perfectly adequate. It improves drainage and kills perennial weeds. Firm the soil well after digging by rolling or treading it. Then leave it for two weeks to settle. Make absolutely sure the land is well drained.

If you have heavy land, be very careful. You might easily dig a hole for a fruit tree, fill it with magnificent free-draining loam and compost, and find that you have done nothing more than dig a pond. Water sinks freely into the loam, fills the hole, then cannot get away because of the surrounding clay, and the tree dies for lack of air to its roots. You can get around this problem by filling the bottom foot (30 cm) of your hole with loose stones and then laying a line of drain pipes to connect each hole to a ditch or a low-lying piece of ground. Such an arrangement will enable the water to run away.

Soil should be around neutral for apples, about pH 7. So if your soil is acidic, lime it. But take care not to make it too alkaline, for although stone fruit trees need plenty of lime, apples don't like too much.

Propagation

Most varieties of fruit tree will not breed true from seed. Seeds are the product of sexual reproduction and therefore each seed will have characteristics of both father and mother in it. To establish new varieties you have to grow trees from seed (indeed there is no other way), but once you have found a good variety the only way to reproduce it faithfully is by vegetative rather than sexual reproduction. In other words, you use hard-wood cuttings instead of seed.

Unless you are interested in the propagation of fruit trees for its own sake, it is best to buy them from a nursery. Most of the fruit trees that you buy will consist of two different cultivars, or varieties, of the same species of tree grafted together. Nurserymen select root stocks for such qualities as hardiness and degree of vigor, (vigor to a nurseryman means the size of the tree when it is fully grown) and then select other varieties for good fruiting qualities, and graft the latter on top of the former. Planting a tree is described on page 98.

If you were to plant a Cox's orange pippin seed straight into the ground, it would give you Cox's apples (that is, if it survived at all), but the tree would not have anything like the hardiness and vigor that would result if you grafted a Cox on to the rootstock of a crab apple tree. It is the rootstock that decides the growth habit of the tree. So by employing dwarfing varieties as rootstocks you can grow dwarf, or smaller, fruit trees than would grow from seed.

There is one organization that is eminent throughout the world for the breeding of new rootstocks and that is the East Malling Research Station, in Kent, England. There, thousands of new varieties of apples and pears are grown, and watched and evaluated, and successful ones are selected for widespread vegetative reproduction. The result of this is that all over the world, apples and pears are grown on Malling rootstocks.

Most apple trees are now grown on dwarfing rootstocks. The different stocks are distinguished by the letter M (for Malling) with a number after it. The most dwarfing rootstock is M9. It produces early maturing trees, large good fruit, and very dwarf trees, but it only grows well on good soil. M26 and MM106 are good semi-dwarfing rootstocks, and M25 is a very good rootstock for large trees. M2 and M111 are good for growing large trees on poor soil.

You can plant seeds to grow your own rootstocks and cut your own "scions," which are healthy fruiting twigs of the current season's growth, usually about 18 inches (45 cm) long. You can then join the two together by grafting or budding. This is interesting to do if you have orchard space to spare, and it can be a profitable sideline. The techniques of grafting are described in detail on p. 99.

Plant standard apple trees 16 feet (5 m) apart. You can grow large trees in a circular deep bed (see p. 110): goblet or dwarf varieties thrive in an ordinary deep bed, six feet (1.8 m) apart, with other plants growing in between the trees and all along the side of the bed.

Care while growing

Maintain mulch around the tree at all times. Bear in mind that mulches quickly disappear; the earthworms pull the organic material down into the soil, where it rots and does a lot of good. So replace the mulch as often as necessary.

For the first four years of the tree's life, keep the ground around it free of grass; in an orchard this means all the ground between the trees. You may well grow strawberries on this ground, for these will not interfere with the nourishment of the apple trees. But the very best thing you can do for the young trees is to keep the ground clear-cultivated all summer, and then sow a winter green manure crop in the fall. A mixture of half winter rye and half winter vetch is ideal. Rototill, or dig this in shallowly, in the spring.

PRUNING Pruning fruit trees is a science in itself, and the best way to learn about it is to watch an experienced pruner. The basic techniques are described on p. 100.

The idea of pruning is to shape the bush, and control the number of fruiting spurs so that you get plenty of good fruit and not too much inferior fruit. There are two main forms of pruning: winter pruning and summer pruning. They are quite different and have different purposes.

Winter pruning, which is principally to shape the tree, encourages growth but may delay fruiting: the more you prune a tree in winter the faster it will grow. But a tree putting all its energy into growing can't produce fruit. So once a tree has reached its adult size (usually after about four years for standards), restrict winter pruning to a minimum. Summer pruning, which consists of shortening the current year's growth, helps to stop the tree from growing too fast or too big, and encourages earlier fruiting.

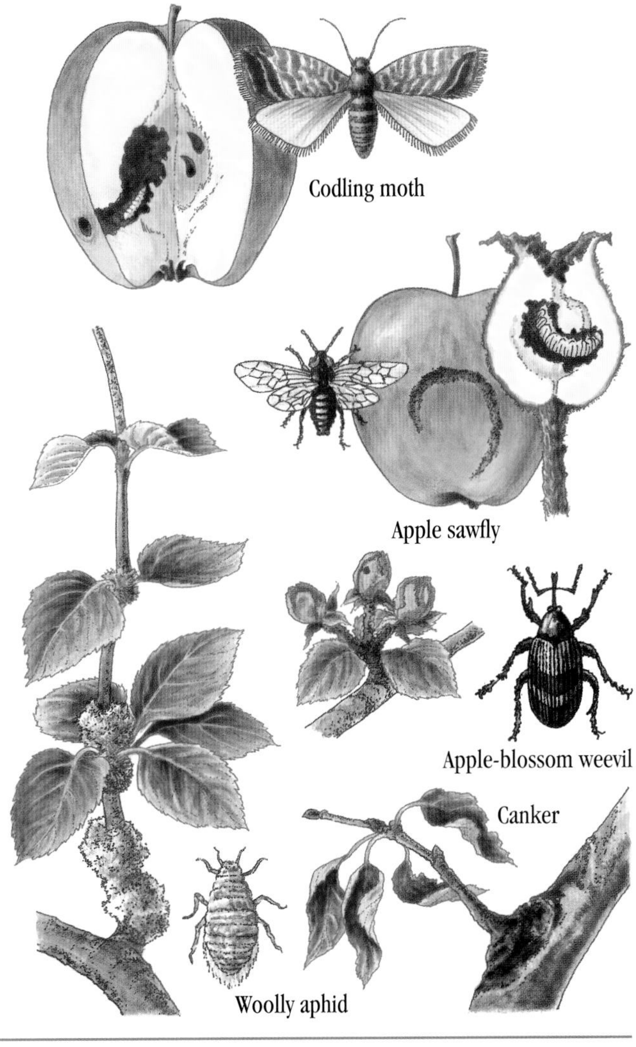

Pests and diseases

Most commercial fruit-growers combat pests and diseases by constantly spraying with deadlier and deadlier fungicides and insecticides. The pests are poisoned, but so are their natural predators. As the pests build up a resistance to various pesticides, the pesticides have to be applied in larger and larger quantities: to spray twenty-four times in a season is not unusual.

My own belief is that a thorough spraying with a proprietary winter wash* (see p. 104) in late winter, well before the buds even think of opening, is all the spraying you should need to do. But apart from spraying, there are several other things you ought to do.

First and foremost, be hygienic. Do not leave prunings, dead dropped fruit, or other rubbish lying on the ground. If you can't eat dropped fruit, put it on the compost pile. Burn all prunings. When the leaves fall leave them to be dragged down by the worms unless they have mildew on them, in which case you should burn them. Don't just leave diseased or cankered trees; pull them out and burn them.

If you have a lot of trouble with pests, wrap a band of paper, or cloth, covered in grease around the trunk of each tree. Any pests that walk up the trunks will get stuck.

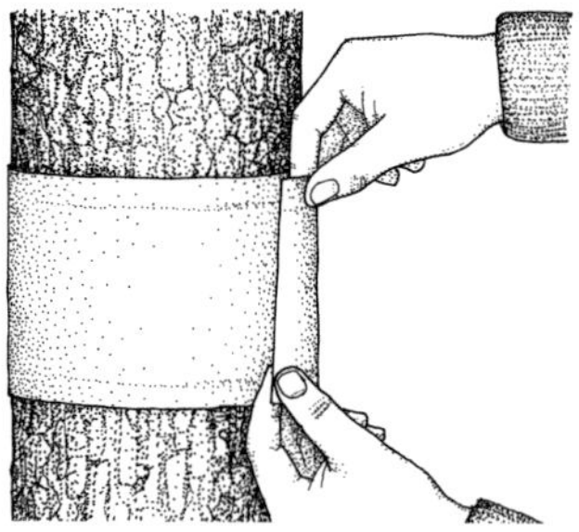

GREASE BANDING
There are various organic ways and means of keeping down pests, but if you find that you are still having a lot of trouble, you can cover a band of paper or cloth with grease and wrap it around each tree. All the pests walking up and down the trunks will get stuck and die.

If you have hens, keep them under your fruit trees, because they eat a lot of harmful grubs. In midsummer, carefully examine all your trees, remove any deformed or diseased fruit and put it on your compost pile.

MILDEW If your trees suffer from either downy or powdery mildew, which causes a whitish down on the leaves, then you should burn all the leaves when they drop in the fall or put them right in the middle of a compost pile.

CODLING MOTHS Codling moths lay their eggs on the blossoms and eventually the caterpillars burrow into the fruit. The solution is to wrap corrugated paper or old burlap around the trunks and major branches in midsummer. The caterpillars will take refuge in these to pupate, and when they do so you can burn them in the fall. In the old days, people used to build bonfires in their orchards in midsummer simply so that the codling moths would fly into the flames.

SCAB Scab is a fungus that makes brown patches on fruit. As long as the scabs are small, they are not important. Winter washing* and careful hygiene are the only cures.

APPLE SAWFLY Yellowish maggots tunnel into the fruit, sometimes leaving them completely inedible. This happens in late summer, and the apples are covered in ribbonlike scars. Trap the maggots at this time in glass jars, covered with wire gauze with holes too small to admit bees. Fill the jars with water mixed with sugar, honey, treacle, or else molasses, and then hang them from the branches on the sunny sides of your trees.

WOOLLY APHID OR AMERICAN BLIGHT Woolly aphid is a pest that attacks apples and leaves, causing growths that look like cotton batting on the leaves. Painting such patches with denatured alcohol* will kill the grubs. You can also grow buckwheat near your fruit trees. This will attract hoverflies, which lay their eggs near the woolly aphids. When hatched, the hoverfly will crawl under the "wool" and eat the aphids.

APPLE-BLOSSOM WEEVIL Apple-blossom weevils lay their eggs in the blossoms. This often causes the blossoms to turn brown and die. The adult weevils eat the leaves. If you get these, put on the burlap or paper trap for codling moths a month early. It will then trap both pests.

CANKER Fruit trees are mainly attacked by canker in wet climates. Rot develops on branches or trunks. Prune off the affected branches and the affected parts of trunks down to the clean wood and paint the wounds.

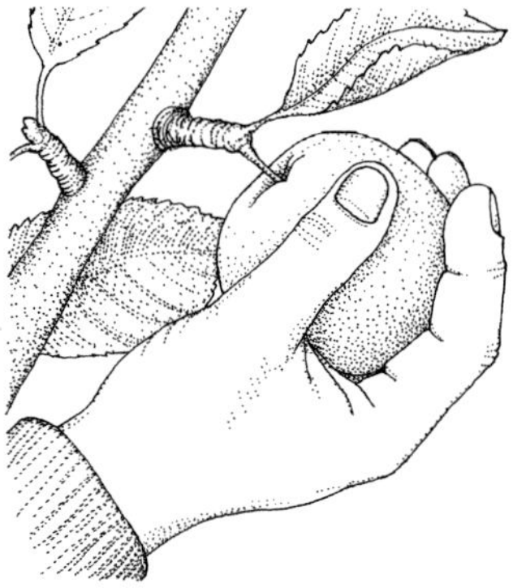

PICKING APPLES
Don't bump or bruise apples in the process of picking them. The time to pick them is when they come off at once as you twist their stalks sharply upward. Pick over the same tree several times if necessary, so that each apple is picked when it is just ripe.

Harvesting and storing

Summer, or early, apples should be picked and either eaten or preserved pretty fast. Late apples are the ones for storing.

Don't try to store any damaged, unripe, or overripe fruit, or fruit from which the stalks have come out. A temperature of 40°F (4°C) is ideal for storage. Frost is fatal and so are excessively high temperatures. Ventilation must be good but not too vigorous; you don't want them in a draft. Too dry a place is also to be avoided; it is a good idea to throw water on the floor if the air seems too dry. Places with thick walls and stone, dirt, or tiled floors are better than attics. Either lay the fruits out in a single layer so they are not touching each other, or better still, wrap each fruit in newspaper or even oiled wraps. Don't store any fruit together with strong-smelling substances.

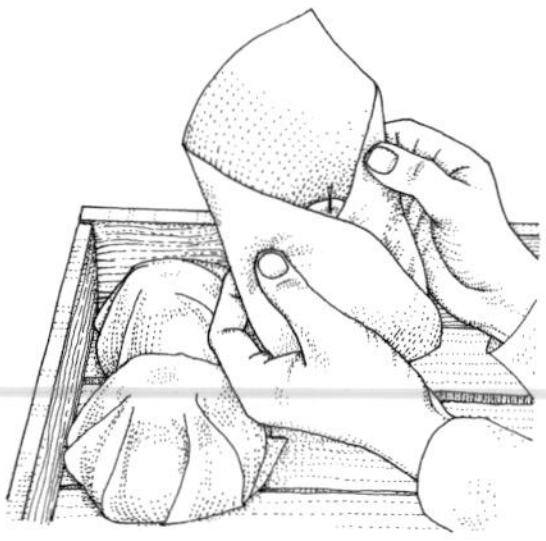

STORING APPLES
Late apples are the ones to store. They should be kept in a well-ventilated room, free from drafts and very dry air. A temperature of 40°F (4°C) is ideal. The very best way of storing is to wrap each fruit in oiled paper, and place them in boxes or crates, in a single layer, not touching each other.

A new way of storing apples and pears is in thin polythene bags. The apples need not be individually wrapped. Seal the bags and store them at an even temperature; 40°F (4°C) is ideal. Make pinholes in the bags, one for each pound of fruit inside.

Pears

If you have space to spare after putting in three apple trees, a pear tree is a good fourth choice, but remember that most varieties need a partner nearby for pollination. The culture of pear trees is very similar to that of apple trees (see Apples), although pears are rather more fussy and they need extra care and attention.

Soil and climate
Pears suffer more from frosts than apples do, because they blossom earlier and the frost can kill or severely injure the blossoms. They also need a dormant period in order to fruit. They prefer heavy soil, but it must be well drained.

Soil treatment
Before planting any fruit tree, you should clear-cultivate the soil (dig it over thoroughly), and pears are no exception to this rule. Soil should be around neutral, pH 6.5 to 7.5.

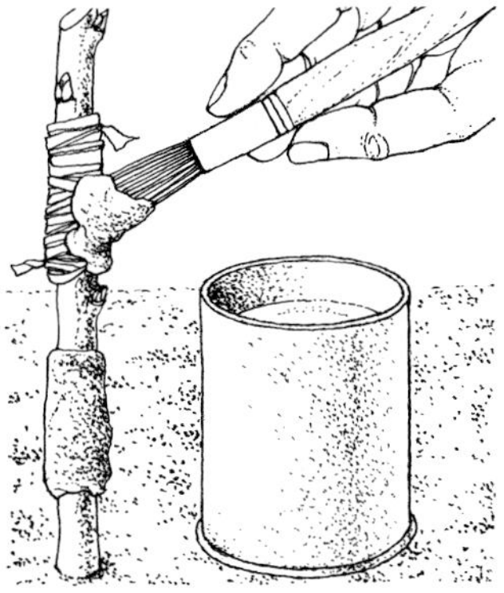

"DOUBLE WORKING" PEARS
Pears are generally planted on quince rootstocks (the quince is closely related to the pear, but hardier and smaller). However, some varieties of pear will not take on quince rootstock and must first be grafted on to a compatible pear variety, which is itself grafted on to the quince.

Propagation
Pears will not breed true from seed. To produce trees that will fruit well and are also hardy and vigorous requires grafting (see p. 99). The difference between apples and pears is that pears sometimes have to be "double worked."

East Malling Research Station (see p. 169) is a worldwide source of pear rootstock. Malling Quince A is most commonly used. Malling Quince C is the best rootstock if you want dwarf trees. Pears are often self-sterile and have to be planted with mutually fertilizing varieties. Plant pear trees in the way described on page 98. They can also be grown in a circular deep bed (see p. 110), or six feet (1.8 m) apart in an ordinary deep bed as long as they are kept small.

Care while growing
Except that pears can bear heavier pruning than apples without being stimulated into rampant growth, the pruning procedure for apples and pears is identical (see Apples and p. 100). And tip-bearing pears should be treated like tip-bearing apples.

If a pear tree ceases to produce new growth—and this can happen in a tree that is still very much alive—cut back into two- or even three-year-old wood in order to stimulate new growth.

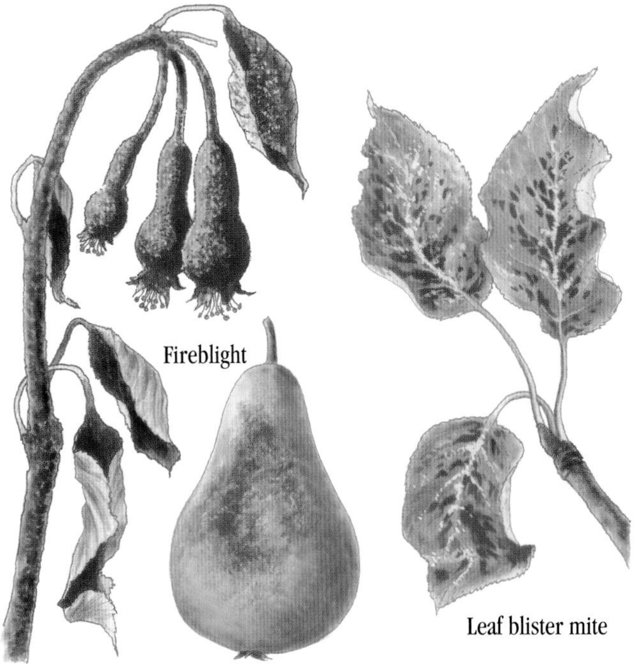

Pests and diseases
Pears can get all the apple diseases and the same steps should be taken (see Apples). There are also some pests and diseases that are peculiar to pears:

FIREBLIGHT Fireblight must be tackled as soon as it is discovered. It attacks at blossom time, causing the blooms to blacken and shrivel, and subsequently every part of the tree blackens as though it has been on fire. Cut out all affected parts at least six inches (15 cm) back from the site of infection with a sterilized knife and burn them immediately. Several varieties of pear are resistant to fireblight.

LEAF BLISTER MITE These tiny mites attack leaves in spring, causing green or red blisters to appear. Pick off and burn the leaves immediately.

PHYTOPHTHORA ROT This is a disease caused by a fungus. Brown patches appear on the skin and the flesh rots. Burn all rotten fruit and spray with Burgundy mixture*.

Harvesting and storing
Pick pears a little before they are completely ripe, as soon as they come off the tree easily when you lift them away. Take exaggerated precautions not to bruise them. Store them like apples as near 30°F (–1°C) as possible, but before you eat them bring them into room temperature and wait for them to ripen. Eat them when they are slightly soft. There is one day in the life of every pear when it is perfect, and with pears, perfect is perfect.

Cherries

It is only really worth planting a cherry tree if your garden or orchard fulfills two conditions. First, there must be ample space to spare after allowing for vegetables, soft fruit, and staple tree fruit—apples, pears and plums; a cherry tree can cover an enormous area of ground, often about 500 square feet (45 sq m). Second, your garden must be relatively free of birds. If it isn't, the birds will eat the cherries, in which case the best thing to do is grow your cherries against a wall and hang a net over them.

There are two kinds of cherries: sweet and sour. Generally speaking, sweet cherries are for eating fresh, and sour cherries are for cooking, canning, and jam-making. Sour cherries have the advantages that they are less attractive to birds and can be grown anywhere in the garden, while sweet cherries need a sunny position or a south-facing wall.

Soil and climate

Sweet cherries thrive on lightish well-drained loam. They even do quite well in gravelly soil, although they send their roots deep and need a good depth of soil beneath them. Sour cherries will do better in clay than sweet cherries will, but they also prefer light deep soil. Both sorts prefer a pH of 6 or 7, but will tolerate more lime than apples, so a pH of 8 will do. They will grow in temperate climates and there are even sour varieties that will fruit in very severe climates. The blossoms of most varieties, however, are frost-tender and should not be grown in frost pockets (see p. 168).

Soil treatment

Clear-cultivate the soil (that is, dig it thoroughly).

Propagation

Cherry scions are mostly grafted on wild cherry rootstock. Malling 12/1 (see p. 169) is the most common. The simplest thing is to buy the cherry tree you want, already grafted, but if you want to do your own grafting, the appropriate methods are described on p. 99. As nearly all cherries are not self-fertile, it is a good idea to have two varieties grafted on one tree. Choose varieties that flower at the same time. Plant the trees just like apple trees (see p. 98). In an orchard they should be 45 feet (13 m) apart. If cherries grow in a border up against a wall, it can be dug as a deep bed (see p. 106). Otherwise cherry trees are too large to grow in a deep bed.

Care while growing

Prune cherry trees as illustrated below. It is an advantage to apply material high in nitrogen—about an ounce (28 g) for each year of the tree's growth until it is five years old. Simply sprinkle it on the ground near the base of the tree. Thereafter apply five ounces (140 g) per year. One ounce (28 g) is found in one pound (500 g) of cottonseed meal or a half pound (225 g) of blood meal.

Keep the soil bare under cherry trees for the first five years, but don't dig deeply. Hoeing or mulching is sufficient. After the fifth year, clear all weeds away, plant some daffodils, tulips and crocuses around the tree, grass the land down, and leave it. The other alternative is to run chickens under your cherry tree. If you do this make sure you have enough chickens to produce about 25 pounds (11 kg) of manure in a year.

Pests and diseases

BLACK CHERRY APHID These aphids cause severe leaf curl, which is sometimes accompanied by black patches on the leaves. If your cherry trees are badly affected, spray with tar wash or Burgundy mixture* (see p. 105).

SILVER LEAF DISEASE If left unchecked, silver leaf disease may kill the tree. It is caused by a fungus that lives on dead wood, so you won't get it as long as you prune well in early summer and cover all wounds with paint.

Harvesting

Harvest sweet cherries when they are quite ripe and eat them immediately. With sour cherries, pull the fruit off, leaving the stalks; otherwise you will tear the tree.

PRUNING A CHERRY TREE
Start with a "maiden" tree. In the spring, shorten all its branches by six inches (15 cm). The next spring, cut out all main branches but five. A year later, prune all but two secondary branches on each main branch. Every spring thereafter, cut out all dead or inward-pointing branches.

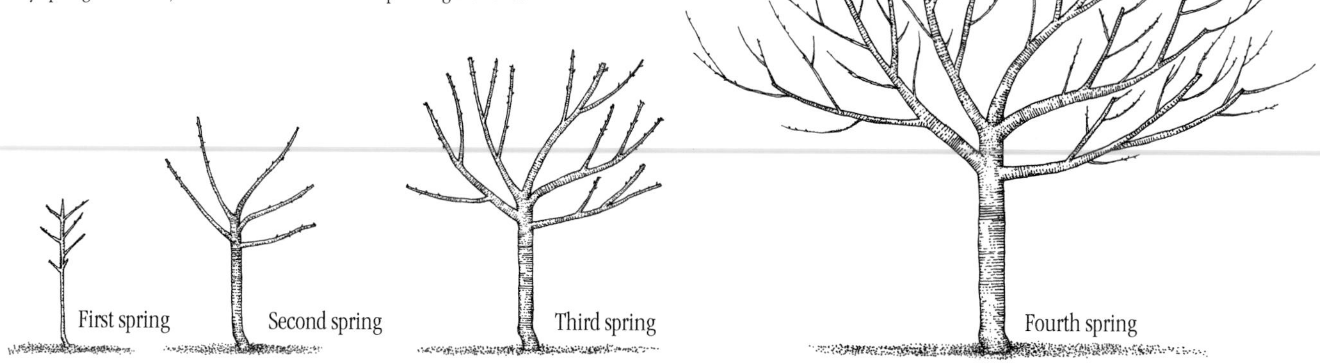

Peaches & Apricots

Peach and apricot trees are very similar and should be grown in the same way; they share the same pests and diseases. Nectarines differ from peaches only because they have smooth skins; botanically they are identical. A peach or apricot is only worthwhile in a cool climate if you already have more than enough apples, pears, and plums. In warmer, but not subtropical, climates they can be treated as a staple. They can be grown in greenhouses (see p. 212).

Soil and climate

Peaches and apricots will grow in sand, or very sandy or gravelly soil, provided there is plenty of humus in it. They like a hot summer and a fairly cold winter. Ideally the winter temperature should go below 40°F (4°C) for some weeks to give them a dormant period, but it should not get too cold. Peach and apricot trees must have a sunny position and must not be grown in frost pockets (see p. 168). Land sloping down to a lake, river, or estuary is fine. In cool temperate climates they are a very uncertain crop to grow outdoors, but fan-training them up south-facing walls can be successful. If you have to grow them in the open, strangely, a north-facing slope is better than a south-facing one. This is because they won't come into flower so soon on a north-facing slope and will therefore miss the last frosts.

Soil treatment

Dig the soil well and dig in plenty of humus, but not humus too rich in nitrogen. Too much nitrogen makes peach trees rampant, sappy, and more susceptible to frost damage. Peat* and leafmold are very good. A pH of 6 or over is ideal.

Propagation

Plant peach trees in the early spring, except in very mild climates, because cold weather could injure them in their first year. Choose a variety that is known to be suitable for your area (ask a specialty nursery), and plant the trees as you would apple trees (see p. 98). Peaches can be grown in a circular deep bed (see p. 110). You can buy peach trees ready grafted, or you can do your own grafting (see p. 99). St. Julien A is a good root stock if you want a small tree; Brompton is best for large trees. Special hardy varieties of peach are now bred that do not need grafting.

Care while growing

The fruit only grows on the previous year's wood and it is wise to remember this when pruning. When you plant a sapling, cut the tree back to about two feet (60 cm) above the ground, cutting just above a branch. Prune it hard again in the early summer; cut back all branches to within an inch (2.5 cm) of the trunk (not flush with it). New branches will sprout that first summer beside the stubs of the old ones. Rub off all of them except three, which will form the framework, or "scaffold," of the tree. Do this as soon as the tiny branches show themselves.

The aim now is either to let the most upright of the new branches go upward and make a trunk, or, better still, to let all three grow up and away from each other and form an upside-down tripod. All your subsequent pruning, which must be done each year in early summer, should maintain this shape. Cut out all inward-pointing shoots and cut back all shoots that have died at the tip, until you reach clean white wood with no brown in the middle. Protect all wounds meticulously with tree paint.

In cold climates, give all peach trees nitrogen in the very early spring, about an ounce (28 g) for every year of the tree's growth. This controlled amount of nitrogen enables the tree to grow and fruit vigorously in the summer but to stop growing long before the freezing winter, when new, sappy wood would suffer from frost. Give dressings of compost or manure, in late fall.

Fruit should be thinned so as to give one fruit every nine inches (25 cm) of wood. This is best done in two stages: toward midsummer, thin to four inches (10 cm) apart and then, about four weeks later, when the fruit is the size of a walnut, thin to nine inches (25 cm).

Peach leaf curl

Pests and diseases

PEACH LEAF CURL This is very common in Europe. The leaves curl and crinkle. Spray with Bordeaux mixture* (see p. 105) in midwinter and again a month later. Spray once more in the fall just before leaf fall.

LEAF SPOT This is a bacterial disease that causes brown spots on the leaves, and it can be serious. If a tree gets it, give it plenty of manure and it should get over it.

Harvesting and storing

When peaches have turned yellow and are slightly soft to gentle pressure, it is time to pick. Turn the fruit slightly and it will come off. It can be stored for up to two weeks in a cool basement, or frozen (see p. 227) or canned (see p. 220). Apricots should be picked and eaten when soft and ripe, or they can be picked a little earlier when still firm, and then dried. To dry them, split the fruit in two and remove the stones. Leave them in trays in the sun, split side up, for up to three days.

Plums & Damsons

Plums and damsons are nice easy crops to grow compared with apples, pears, and peaches. They are fairly hardy, don't get many diseases, and yield very heavily in some years.

Soil and climate

Plums like deep soil, and will flourish in deep loam or clay as long as they are well drained, but they do not thrive in dry, shallow soil at all. Damsons are slightly more tolerant of shallow soil than plums. Plums flower early and are therefore susceptible to spring frosts, so don't plant them in a frost pocket (see p. 168). Like other temperate climate fruits, plums need to lie dormant through a cold winter.

Soil treatment

A neutral soil is best, around pH 7, so lime if your soil is acidic. Clear-cultivate (thoroughly dig) the land before planting plums, and then, ideally, grow a crop, or even two crops, of green manure. Dig or rototill these into the soil. The land must be well drained. If it is not, fill the bottom foot (30 cm) of the hole you dig for each tree with stones, and bury a line of drain pipes to lead the water away to a ditch or lower piece of ground.

Propagation

Plums and damsons should always be grafted (see p. 99), and nearly always will be if you buy from a nursery. "Myrobalan B" is a good rootstock for large, heavy-cropping trees that will tolerate clay. "St. Julien A" and "Common Plum" are best for small trees. Plant plum trees as you would apple trees (see p. 98). Plums are not self-pollinating, so you must plant at least two, and preferably several, compatible varieties. You must get advice from a nursery on this.

Allow 24 feet (7 m) between standard trees if they are planted on "Myrobalan" stock and, say, 15 feet (4.5 m) between trees planted on semi-dwarfing stock such as St. Julien. Plums can be planted in the circular deep bed (see p. 110). Plant plum trees in early winter; but in areas with exceptionally cold winters, plant in early spring.

Care while growing

Plums benefit from quite rich feeding. It is an advantage to have hens or other poultry running under them. Otherwise apply heavy dressings of compost, or stable or cow manure.

PRUNING Plums can be pruned to form all the tree shapes described on p. 101. Prune when you first plant the tree, and thereafter confine pruning to early summer only, because silver leaf disease may develop if you prune in winter.

In some years plum trees bear a considerable weight of fruit, and because their branches tend to be quite slender, those that carry a lot of fruit may need support. There are two ways of doing this. You can build a T-shaped wooden scaffold, which should be firmly rooted in the ground next to the trunk and secured to it with a plastic strap. Ropes from the top of the "T" can be tied around the drooping branches. The other solution is to use a forked branch as a support. Protect the branch with burlap to save it from chafing.

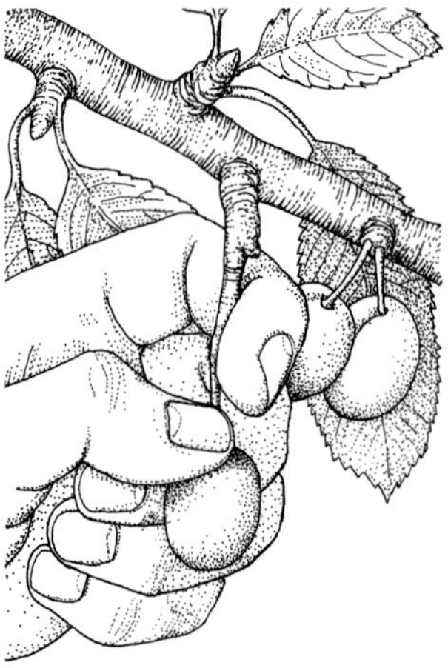

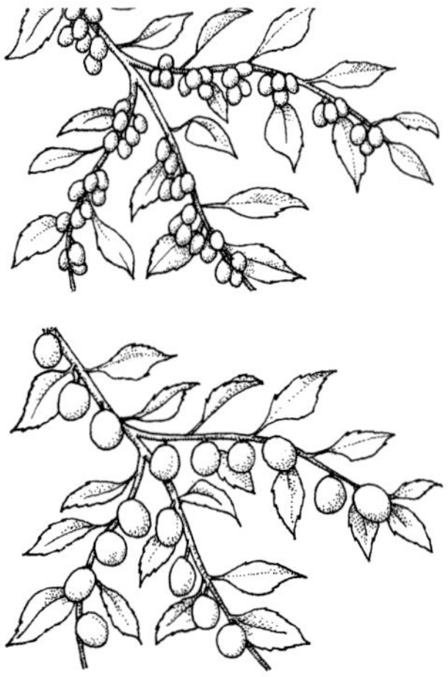

THINNING PLUMS
If your crop is allowed to become too heavy on the tree, the plums may be small and tasteless. Therefore you should thin them when the plums reach about half their final size. Don't pick off the stalk when you pinch off the fruit. Leave at least two or three inches (5–8 cm) between remaining plums.

Pests and diseases

Plums suffer from the same pests and diseases as apples (see Apples), as well as a few of their own.

SILVER LEAF DISEASE The symptom of this is silvering of the leaves, but the disease attacks and can kill the whole tree. It is caused by a fungus that grows in dead wood. As long as you rigorously prune out all dead wood in early summer, burn the prunings, and cover the wounds with paint, your trees won't suffer from silver leaf.

BACTERIAL BLIGHT This shows first as black streaks on the young shoots. Later on, black spots appear on leaves and fruits, which become inedible. There is no cure for this, beyond the pruning and burning of all diseased wood. Some varieties are more resistant to this than others. "Myrobalan" rootstock confers a degree of immunity.

HEART ROT Leaving sawn-off stumps of branches on a tree can cause heart rot; the stumps heal slowly, so bacteria can get in and kill the wood under the bark. You won't get it if you cut all branches flush with the trunk.

Harvesting and storing

For jam or jelly, or for canning (which suits plums admirably), pick the fruit as soon as the bloom appears on their skins, but before they get soft. For eating fresh, pick them when quite ripe, which is when they give a little and come easily off the tree. In hot, dry climates, plums to be turned into prunes can be left on the tree until they are quite dry and ready to be shaken off. They should then be dried in the sun on trays. In damp climates, they can be dried artificially (see p. 216).

Quinces

Quinces are so closely related to apples and pears that the latter are often grafted on to quince stock, because quince stock is hardy and produces small trees. Quinces are not grown as much as they should be; they have a very special and delicate flavor and quince jelly is one of the world's great gastronomic experiences.

Soil and climate
Quinces will grow in any soil or climate that apples will (see Apples), although they are a little more tender. They prefer a warm summer and a fairly cold winter. Heavy soil suits them best, but it must be well drained.

Soil treatment
Dig the land thoroughly, and if you are not in a hurry, grow and dig in one or two crops of green manure. The soil should be neutral, pH 7. Quinces don't like too much nitrogen but need phosphate and potash.

Propagation
If you don't buy a seedling from a nursery, the best way to propagate is from cuttings made from the suckers that quinces put out every year. In the fall, cut lengths about nine inches (23 cm) long, and bury two-thirds of their length in sandy soil. After a year, move these to their permanent positions.

Care while growing
You can prune quinces to all the tree shapes (see p. 100) or you can leave them strictly alone, in which case you will get a spreading bush shape. Quinces are not prone to being attacked by pests and diseases.

Harvesting and storing
You can leave the fruit on the tree until after the first hard frost. Then make jelly (see p. 223) right away, or if you don't have time to make jelly immediately you can store the quinces in cool moist conditions for up to three months.

MEDLARS
Medlars are hardier than quinces and can therefore cope with colder conditions. They do best if they are grafted (see p. 99) on to thorn, pear or quince rootstock. Otherwise treat them like quinces. The fruit is unusual in that its five seeds are visible and it is only edible—and it is actually very good to eat—when it is half rotten.

Raspberries

Raspberries are one of the best soft fruit crops the self-sufficient gardener can grow. They are hardy, and will stand neglect, although they shouldn't have to. They are easy to grow and heavy-yielding.

Soil and climate
They prefer the soil to be slightly acidic, so do not lime under any circumstances. Lime can cause chlorosis (yellowing of the leaves). They do need good soil though, so if your soil is light and sandy, put in plenty of manure. Raspberries prefer sun, but if you have a garden where sun is at a premium, grow your raspberries in a shady area. They will stand colder climates than most other fruits.

Soil treatment
In the fall, dig a trench two spits deep and fill it with soil mixed with compost or manure. They need a lot of potash, so incorporate wood ashes with the soil if you have any; otherwise mix in some other potash fertilizer. They have both shallow and deep roots, and need a lot of humus.

If you only want one row, you have no problem: but the roots spread far and wide, so if you want more than one row then have them quite far apart: six feet (1.8 m) is usual in commercial gardens but four feet (1.2 m) will do if you are trying to save space.

PROPAGATING RASPBERRIES
Like strawberries, raspberries propagate by "walking." Raspberries "walk" by pushing out roots that send up suckers to form new plants. Just let your plants send up suckers, cut off the roots connecting them to the parent plant with your spade, then lift the suckers and replant them.

Propagation
I strongly advise you to buy raspberry stock that is certified disease-free. Such plants will give higher yields and last far longer than the plants your neighbor offers you when he has to get rid of his suckers in the fall. The certified plants you get will consist of one cane, or whip, with a heel of root attached to it. Plant roots a foot (30 cm) apart in rows four feet (1.2 m) apart. Put the root down about three inches (8 cm), cover with soil, and firm well. Immediately cut the cane down to nine inches (23 cm) above the ground.

As long as you start with disease-free stock, there is no reason why you should not then multiply your own raspberry plants in subsequent years. Like strawberries, raspberries "walk," but they do it in a totally different way. In the deep bed (see p. 106), raspberries should be planted in three rows with 18 inches (45 cm) between the rows. Their shallow roots make intercropping inadvisable. Don't plant raspberries where raspberries have been before.

Don't plant them immediately after potatoes or tomatoes either, for these plants get some of the same diseases.

Care while growing

Don't let them fruit the first summer: remove the blossoms, otherwise the plant will be weakened by fruiting. By the second summer they should bear well. Keep weeds down near the plant by heavy mulching—say, within a foot (30 cm) of them. Lawn cuttings, leaves, or compost are all good. Hoe between the rows. Don't allow grass or weeds to establish themselves; raspberries will not flourish in grass. So be sure that the mulch is thick enough every spring. The raspberry

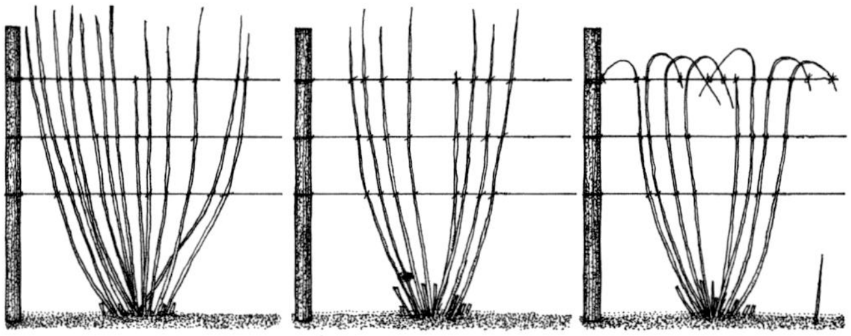

PRUNING RASPBERRY CANES
After cutting out all the old wood, thin the canes, leaving only the best six or eight to fruit the next summer. When these grow over the top wire, shorten them to six inches (15 cm) above it; or else bend them over in an inverted "U" shape and tie them to the wire.

rows are a good repository for wood ashes. Training is simple but necessary. You must have a fence, with three wires, the top one five feet (1.5 m) from the ground, the others at regular intervals below it. Tie the canes to these. Some people have three pairs of wires and simply shove the canes in between each pair. This works but individual tying is better.

PRUNING In the fall, after the leaves have died off, cut off all the canes that have fruited close to the ground and prune as shown above.

Remember that raspberries act like biennials, although actually of course they are perennials. The wood made in one year fruits the next year and then dies down. So cut out the wood that has fruited every year and keep the wood that grew that year, because that will fruit the next.

Pests and diseases

There are several virus and fungoid diseases that raspberries get. If you see any discoloration or other sign of disease, cut out and burn the affected part.

MOSAIC DISEASE This is the worst of various virus diseases and makes the leaves curl and show red and yellow mottling. Completely dig out the bushes that have this and burn them. If you don't, aphids will spread the virus to other plants.

IRON DEFICIENCY If you see yellowing between the veins of the leaves, suspect iron deficiency. This is especially likely if you have very alkaline soil.

Harvesting and storing

Eat as many as you can ripe and raw with cream. Store the rest; they freeze well (see p. 227) and can well (see p. 220). When rain falls on ripe fruit, pick the fruit immediately after the rain stops and can or freeze it; if you don't, the raspberries will go moldy. Don't leave moldy fruit on canes, because the mold will spread to the others.

Blackberries

In most temperate parts of the world, wild blackberries grow nearly everywhere and it is fun hunting for them. Nevertheless, for a regular supply it is worth keeping a few bushes.

Soil and climate

There are several species of blackberry, and cultivated varieties have been developed from them that will grow happily in every climate from the very coldest temperate region to the subtropics. They prefer rich, well-drained soil (pH 7), and a sheltered site.

Propagation

You can propagate blackberries from cuttings, suckers, by layering (see p. 95), or by division of roots—digging up a piece of plant with roots on it, and replanting. The simplest method of all is to propagate from tip cuttings—cut the tip off a cane, push it into the ground, and it will root. Wrap all planting material in moss or wet newspaper and store it in a plastic bag until you need it.

If you want to grow blackberries from seeds you must "stratify" them; this means that over the winter you must keep the seed in a box full of sand at a warm room temperature for three months, then store them at 40°F (4°C) for another three months.

Plant cuttings, layers, roots or seedlings in late fall or early spring. Allow six feet (1.8 m) between the bushes. It is a good idea to plant along a fence, and dig the bed to form a deep bed (see p. 106).

Care while growing

Blackberries fruit on last year's wood, so prune in winter by cutting all wood that has just fruited unless you have "Himalaya" or "Evergreen" varieties; these fruit for several years on the same wood and therefore it should not be cut out so ruthlessly. As a general rule, leave about ten strong newly grown canes to fruit the following year. They are very greedy plants and need rich mulching to grow and fruit well.

Pests and diseases

ORANGE RUST This shows up as bright orange spores under the leaves. Look for these if your plants give out spindly shoots with narrow leaves. Root out and burn infested bushes.

Harvesting and storing

Blackberries are ready for picking when they almost fall off the bush into your hand. Put them in shallow boxes and store in a refrigerator, or freeze (see p. 227) for eating in winter.

LOGANBERRIES

Grow loganberry bushes in a sheltered place; although they flower later than blackberries, severe spring frost will damage the canes. Plant the bushes ten feet (3 m) apart. Unlike blackberries, loganberries only fruit in fall for a two- to three-week period.

Strawberries

Strawberries are fun to grow, if a little awkward. Most gardeners would agree with the remark "Doubtless God could have made a better berry but doubtless God didn't."

Strawberries are a "walking" plant, because they are perennials that don't have an elaborate root system. Therefore they exhaust the ground on which they grow within a year or two. To escape from it and find fresh ground, they send out runners that meander over the ground until they find somewhere to send down roots.

STRAWBERRIES IN BARRELS
Strawberries grow well in pots and tubs of all kinds. A barrel makes an ideal container. Drill several staggered rows of holes three inches (8 cm) wide and 15 inches (38 cm) apart. Drill the rows at eight-inch (20-cm) intervals. Drill several holes in the base and put a layer of gravel in the bottom. Then insert into the center a vertical wire mesh tube four inches (10 cm) in diameter. Fill it with gravel. Then fill the barrel with potting mix up to the first row of holes. Set one plant next to each hole, with the crown emerging. Repeat all the way up the barrel, watering each layer as you go. Finally, set four or five plants in a circle at the very top.

There are several varieties of what are called "remontant" or "perpetual" strawberries. These fruit later than ordinary strawberries and continue fruiting into late fall. It is a very good idea to plant a few, so as to give yourself a treat in the cold weather. If you force ordinary strawberries in the spring under cloches, plastic tunnels, or mini-greenhouses (see p. 111) and have remontants as well, you can have strawberries from early summer to late fall.

Soil and climate

Strawberries are a woodland plant, and you should bear this in mind when choosing a site for them and looking after them. It means that they tolerate shade, although they fruit far better in sun; they like plenty of humus (they will grow in almost pure leafmold as they do in the wild); and they don't object to fairly acidic conditions. They do better on light soil than clay, but granted plenty of humus they will thrive in any well-drained place. They are a temperate-climate crop and develop a far better flavor in a cold climate than a hot one. It is best to move on to totally fresh ground every three years with new plants.

Soil treatment

Dig the land one spit deep, incorporating plenty of compost or any well-rotted organic manure. Strawberries do well on the no-digging system (see p. 83) as long as the bed has had enough compost put on it. They are also potash-hungry, so if you have wood ashes to spare, use them on your strawberry patch. Farmyard manure can be rich in potash.

ENCOURAGING RUNNERS
Start with disease-free strawberries from a reputable source. From then on you can propagate from runners. Every year you should remove the blossoms from a proportion of your healthy plants, so that they are encouraged to send out plenty of strong runners.

Propagation

The first time you plant strawberries, get virus-free stock, from a reputable source, certified healthy. Unless you want to grow new varieties of strawberry, in which case you should grow them from seed, the best thing to do is to multiply them from runners. There are a few varieties that do not make runners, and these are multiplied by dividing up the crowns themselves.

Most varieties of strawberry will make runners that will root themselves whatever you do, but you can encourage them by removing the blossoms from a few of your plants. You have merely to sever the runners from the main plant, dig out the little mini-plant on the end of it, and transplant it. But an even surer way of doing it is to bury small pots of soil in the ground near the parent plants and peg the ends of runners down on these pots. When the runners have rooted properly, sever them from the parent, dig up the pots, and transplant to their new positions. In this way you can establish a new strawberry bed every year and scrap one every year, after it has fruited for three seasons. Every fall you will have a newly-planted bed, a year-old bed, a two-year-old bed, and a three-year-old bed, the last of which will be ready for digging up. Always plant your new beds as far from the old ones as you can, to hinder the spread of disease.

You can plant or transplant strawberries at any time of the year (if the winters are mild enough) but it is traditional to plant in late summer, as you can then harvest a crop the next year. In regions with dry summers, plant in early spring. Plant 15 inches (38 cm) apart with 30 inches (75 cm) between rows. Plant so that the crown is at ground level but the roots are spread out widely and downward. Water the

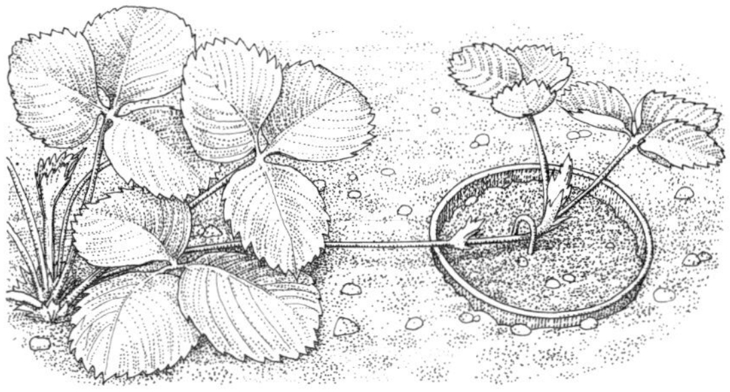

NEW STRAWBERRY PLANTS FROM OLD
Bury pots of soil in the ground near the parent strawberry plants, and peg down the ends of runners on top of the pots. Eventually you will be able to sever the old plants from the new plants, dig up the pots, and then transplant them to their permanent positions in your garden.

new plants well. Strawberries do very well on the deep bed (see p. 106). Plant and space them as for a conventional bed.

Care while growing

It is very easy for a strawberry bed to become infested with weeds. The plants straggle relentlessly and make most methods of weeding very difficult. Hoe for as long as you can hoe and then weed by hand. If you have planted between the end of one summer and the beginning of the next, let the crop fruit in the year after that but not before: during the plants' first summer, pick off the flowers.

Prick over the ground with a fork in the spring and, as the crop begins to spread, put plenty of straw under the straggling stems. This suppresses weeds and keeps the fruit clean and healthy. But keep an eye out for slugs.

If you suffer much from birds, you will *have* to use a net. You can either have a net low over the strawberries, in which case you will have to remove it every time you want to pick a strawberry, or else a fruit cage (see p. 184), which is expensive unless you make it yourself.

PROTECTING THE PLANTS
When the fruit begins to form, put a good mulch of straw under the plants. This keeps weeds down and keeps the fruit clean and disease-free. If birds are a nuisance, make a net. Set up posts and invert a glass jar over each one, before putting the netting over the framework; the jars stop the netting from catching.

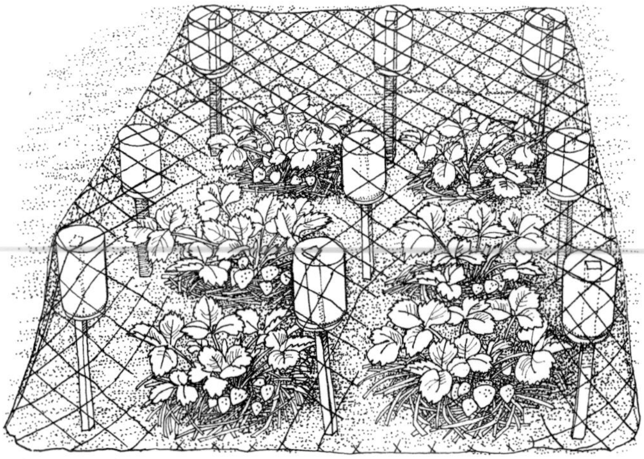

Pests and diseases

Don't try to chase your strawberries on with nitrogen, because it makes them soft and open to disease.

POWDERY MILDEW This white powder will make strawberries turn a dull brown color. Spraying with sulfur* at regular intervals can help.

APHIDS These are a menace because they spread virus diseases, principally strawberry crinkle and strawberry yellow edge, both of which show in the leaves and weaken the plant. To prevent it, spray the plant centers hard in April with a nicotine spray* or with derris. Don't use the nicotine when the berries are nearly ripe. Remove any stunted or discolored plants and burn them; these diseases are incurable.

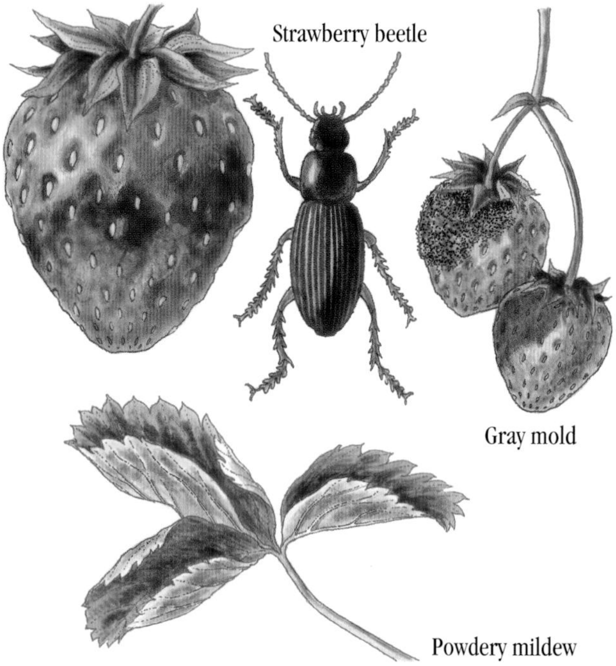

STRAWBERRY BEETLE This little pest feeds on strawberry flesh. Keep the bed well weeded, and it will be discouraged from settling near your strawberries.

GRAY MOLD This is also called botrytis. It appears first as a gray spot on the flowers and then on the strawberries themselves, where it grows to form a gray fur that rots the fruit. Dust with flowers of sulfur* at the very first sign.

ROT If berries rot after rain, remove them to the compost pile. Pick all ripe berries immediately after rain.

Harvesting

Pull the fruits off the plant with their stems intact. Leave the stems on until just before eating; when the stems are removed, vitamins and other nutrients are lost. Store them in the shade for a few hours, or in a refrigerator for a day or two. Strawberries can be frozen, but go soft when thawed.

After you have harvested your crop, remove the straw from under the stems and clear the bed of dead leaves, surplus runners, and weeds.

Rutaceae

Oranges, tangerines, kumquats, lemons, limes, and grapefruit are all members of the *Rutaceae*. Because it includes the genus *Citrus*, the *Rutaceae* family is as important to people who live in the subtropics as the *Rosaceae* family is to people who live in temperate climates. Citrus plants are very aromatic, broad-leaved evergreens.

Citrus fruits grown in the tropics do not taste as good as those grown in what is known as a Mediterranean climate. At the other extreme, they can't stand frost, although oranges can bear marginally lower temperatures than lemons, which are damaged and sometimes killed at temperatures below 26°F (–3°C). This means that citrus fruits growing out doors tend to be limited to the Mediterranean seaboard in Europe; to Florida, southern California, and part of Arizona in North America; and to South Africa, subtropical South America, and Australia. But this does not prevent citrus from being a viable greenhouse crop: in the 18th century, conservatories known as orangeries were common among the wealthy in England.

My own feeling is that if I only had the space under glass to grow one citrus tree, I would grow a lemon. One orange tree provides only a small proportion of the fruit needed by a family through the year, whereas one lemon tree would fill a family's requirements.

Oranges

What the apple is to temperate regions, the orange is to the subtropics. It is heavy-yielding, delicious, will keep well, and has the enormous advantage that it can just be allowed to hang on the tree, for as long as six months. It is a rich and reliable source of vitamin C. In temperate climates, oranges must be grown under glass. This is described on p. 212. They can be grown in tubs, which must be brought indoors in the winter, but the yield from trees grown like this is fairly low.

Soil and climate

Orange trees can stand winter frosts as low as 20°F (–7°C) but fruit and young growth are injured at temperatures below 25°F (–4°C). Oranges like lightish land: sandy loam is ideal; heavy clay is unsuitable. Deeply drained soil is necessary, for they will not thrive on a high water table. Slightly acidic soil is best for them: they will tolerate a pH anywhere between 5 and 7, although they prefer it to be around pH 6.

Soil treatment

Make sure the site is well drained. Dig deeply, and incorporate phosphatic material and potash into the soil. Rock phosphate, granite dust, wood ashes, compost, or farmyard manure all make a reserve for the roots to draw on in times to come.

Propagation

Nearly all orange trees are grafted on rootstocks, because the best fruiting varieties are not the most hardy and vigorous. It really is best to buy them ready-grafted, because grafting oranges is a delicate process, but if you want to graft your own, the techniques are described on p. 99. When you buy your trees, you should take note of the rootstocks, as these will affect the type of fruit that your trees produce. The most common rootstocks are listed below:

TRIFOLIATE This is best for most gardens. It is very hardy, can put up with cold better than most other varieties, and is a dwarfing stock.

CLEOPATRA This is the best rootstock for both tangerines and small oranges.

ROUGH LEMON This rootstock thrives on sandy soil. It produces fruit early, but the trees are rather short-lived.

SWEET ORANGE Good in well-drained sand, this rootstock is useless in clay, where it gets foot-rot. It produces juicy, smallish fruit.

SOUR ORANGE Strange as it may seem, the sour orange rootstock is good for sweet orange fruiting stock, because it is hardy and disease-resistant.

There is a huge selection of fruiting stocks available. They are all either sweet for eating, or sour for making marmalade. The most common sweet varieties are "Jaffa," which are large and juicy, "Valencia," which are good to eat and have a long fruiting season, and "Washington Navel," which is the best one for hot, dry climates like the southwestern US.

You can plant orange trees at any time of the year as you would apple trees (see p. 98). Good nurseries send the trees out with their roots "balled": that is, wrapped in burlap with soil inside around the roots. Plant orange trees with extreme care so as not to disturb the soil around the roots.

Place the roots, wrapping and all, in the hole; pour some topsoil around the ball, then carefully withdraw the wrapping. A big tree should be 25 feet (7.6 m) from its neighbors: one on a dwarfing stock such as Trifoliate, 20 feet (6 m). Water well after planting and make sure the tree is well watered for two weeks. After that, continue to water regularly—about once a week, depending on the soil.

A modification of the deep-bed method may be used for oranges and all other citrus fruits. Deep-dig a circle for each tree. The diameter of the circle should correspond to the dripline—that is, where the branch extremities of the adult tree are expected to be. Keep the circle of soil raised, heavily mulched, and don't step on it.

PLANTING ORANGE TREES
Your orange tree should arrive from the nursery with its roots already "balled"; that is to say, with a ball of soil wrapped around the roots with burlap. Place the tree in the hole prepared for it with the wrapping still around the roots. Pour some topsoil into the hole before gently removing the burlap.

Care while growing

In regions with high rainfall, some watering in dry periods may be necessary for the first three years; after this, none is needed except in cases of real drought. In dry areas, where there is little rainfall, trees need a good soaking every two or three weeks; this means 20 to 30 gallons (90–140 l) per tree. More water than this may wash the nutrients down out of reach of the roots. Watering "little and often" encourages foot-rot.

You should also feed the soil by mulching heavily with organic material once a year. If low-nitrogen material is used, such as hay or straw, some source of more concentrated nitrogen—blood meal or cottonseed meal—should be added to help rot the material down. If rotted compost is used, nothing need be added.

Pruning is minimal with orange trees. Trees should come from the nursery already pruned so as to leave a suitable "scaffold" of four or five branches. Small sprouts that come from the trunk under this scaffold should be rubbed out by hand when they are tiny. Old weary trees can be stimulated into new life by pruning some of the old wood away; choose wood in the center of the tree, which does not get much sun. Cut out any frost-damaged branches, but not until the summer after the frost. It is important not to stimulate trees into excessively rank growth by cutting out too much wood. Sometimes upper branches grow so long that the lower growth is shaded. You cannot remedy this by pruning. But in a group of several trees, the solution is to take out one or two, so that more light reaches the rest.

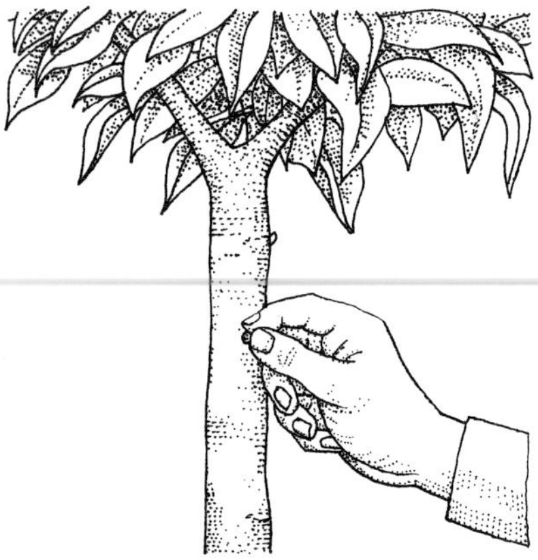

RUBBING OUT SHOOTS
You will usually receive your tree with a good scaffold of four or five branches already established; it is therefore unlikely to need much pruning. If you notice any new shoots emerging from the trunk, rub them out by hand while they are still very small.

Pests and diseases

The many pests that trouble orange trees in inorganic orchards are seldom present in organically managed ones. Eelworm, for example, never becomes a serious problem in an orchard planted on organically rich soil, because predators thrive on unsprayed trees.

FOOT-ROT In long periods of wet weather, orange trees are susceptible to foot-rot, which rots the bark near the soil line and in extreme cases can kill the tree. You can prevent this by observing a few simple rules: keep the mulch at least a foot (30 cm) away from the trunk of the tree; keep that circle free of fallen leaves and debris; don't water right up against the trunk, and don't water too *often*; always keep the junction of trunk and roots clear of soil.

Harvesting and storing

The delightful thing about oranges is that you can leave them on the tree and just pick them when you want them. Pull tight-skinned oranges off with a twist: loose-skinned ones should be snipped off with a little stalk left on them. They can be stored under refrigeration—at 30°F (–1°C) with a humidity of 80 or 90 percent—but you are unlikely to need to do this, because oranges have a very long harvesting season. And remember that a green orange is not necessarily unripe. Orange-colored oranges will sometimes turn green again when the weather gets warmer. They still taste the same.

TANGERINES, MANDARINS, AND SATSUMAS

Tangerines, mandarins, and satsumas are all classified as *Citrus nobilis*. Tangerines have a deeper-colored skin than mandarins and the name "satsuma" was originally applied to a particular variety of tangerine. The terminology has now become confused and the names "mandarin" and "satsuma" are often used to apply to the whole group. The fruits are in general smaller than ordinary oranges, the skins are looser, and the sections separate more easily. The advantage is that the trees are smaller and more hardy than orange trees and are therefore suitable for small gardens, roofs, and patios. Cultivate the trees as you would orange trees, bearing in mind that most varieties are not as productive as sweet orange trees.

KUMQUATS

A kumquat tree is an attractive proposition, particularly in a small garden, and they can be grown on roofs and patios in tubs. Kumquats belong to the *Fortunella* genus, but this is so closely related to *Citrus* that crosses between kumquats and oranges can be made. Certainly the orange-colored kumquats look exactly like tiny oranges. The fruits are rarely bigger than one and a half inches (4 cm) in diameter, but are very juicy and good to eat; also, the peel is spicy and makes splendid marmalade or candied peel. Kumquat trees are very decorative and rarely grow taller than ten or twelve feet (3–3.5 m), and their other advantage is that they are hardier than virtually any other citrus fruit (especially if grafted on to Trifoliate rootstock). You grow them in exactly the same way as oranges.

Lemons & Limes

Apart from ordinary lemons, you can also grow the Meyer lemon, which is a hybrid particularly suitable for small gardens. It is hardy—sufficiently so to survive temperatures of 15°F (–9°C)—and quite small. Outdoors it makes a bush about six feet (1.8 m) tall. It grows well in tubs, on roofs, and on patios. Growing lemons in a greenhouse is described on p. 212. Limes are used in much the same way as lemons, but contain more acid and more sugar.

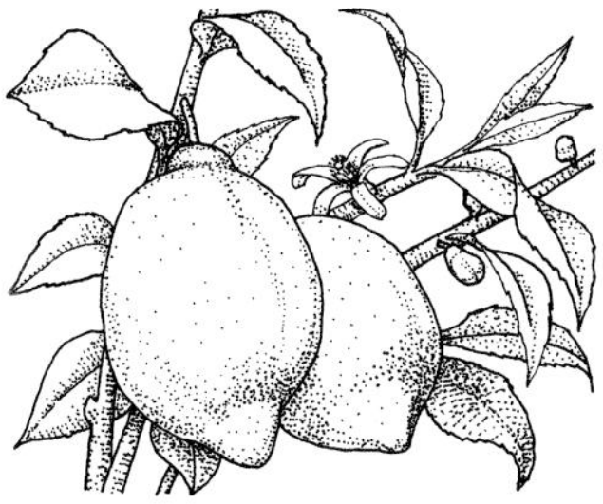

MEYER LEMON BUSHES
This hardy little lemon hybrid will flourish, given a sheltered spot, full sunshine, and plenty of compost. Flowers, immature fruit, and ripe fruit are all to be found on a thriving bush at one and the same time.

Soil and climate

Lemon trees are slightly more tender than orange trees and prefer heavy soil. As they fruit all year, the crop can be badly damaged by winter frost. This also applies to limes, which are less hardy still. Lemons and limes are very much a subtropical fruit. They will tolerate most soils, provided that the water table is below the depth of their roots, which go down no more than four feet (1.2 m). You should incorporate plenty of phosphatic material (see p. 88) into the soil.

Propagation

Buy lemons and limes already grafted. Plant them as you would any other tree (see p. 98). The most usual rootstocks are the same as for oranges. You can plant lemons in a circular deep bed (see p. 110).

Care while growing

Lemons need a little more pruning than oranges, enough to stop them from straggling and becoming vulnerable to bad weather. Shorten any outward-straggling branches to inward-pointing buds so as to keep the tree compact. You can do this at any time of the year. Limes do not need any pruning. If your trees are thriving but not bearing, protect them from the wind and feed them a lot of extra compost; this extra attention should make all the difference.

Harvesting

Both lemons and limes bear year-round in suitable climates, so just pick them when you want them.

Grapefruit

Grapefruit evolved in the West Indies as a sport, or mutation, of the shaddock, which is a coarse and rather unattractive fruit, but the grapefruit, as we all know, is delicious. It is also a rich source of vitamin C.

Soil and climate

Grapefruit must have deep, well-drained soil and like it to be slightly acidic: a pH of 6 is best. As for climate, they can stand as much cold as oranges, 20°F (–7°C), but they need more heat to ripen perfect fruit. In temperate climates, grapefruit must be grown under glass.

Soil treatment

Well-drained soil is most essential. Deep digging—four spits deep if possible—is important, and you should incorporate some phosphate and potash into the soil. Compost or manure buried below the roots can only do good.

Propagation

Grapefruit are generally grafted on to sour orange rootstock, although on poor sandy soil it is better to use lemon. Plant the young trees (see p. 98) at any time of the year; because they are evergreens, one time is as good as another. You must plant them very carefully, as you must other evergreens (see Oranges). Plant trees 25 feet (7.6 m) from their neighbors. You can plant them in a circular deep bed (see p. 110).

Care while growing

Grapefruit need plenty of water. In high-rainfall areas they need watering for the first three years and then probably not at all. In dry areas they need a good soaking—say, 25 gallons (100 l) per tree every three weeks. Don't put water actually on the trunk. Heavy mulching can only do good, provided you keep the mulch two feet (60 cm) away from the tree. Prune them in exactly the same way as oranges; they suffer from the same pests and diseases (see Oranges).

Harvesting and storing

Grapefruit will stay happily on the tree for months, but when the fruits begin to turn yellow, test the occasional one so that you know when to pick them. When they are just right, pick them, wipe them with a clean damp rag, let them sit in a cool place in a breeze for a few days, then put them in the refrigerator.

Grossulariaceae

Black currants, red currants, white currants, and gooseberries are members of the family *Grossulariaceae*. They belong to the important genus *Ribes*, all of whose members are shrubs that display familiar small round berries. Currants and gooseberries are exceptionally hardy and are cultivated almost as far north as the Arctic circle. They are very popular in Europe, but less so in North America, because they can be alternative hosts to white pine blister rust and are for this reason prohibited in several states and countries. Personally I would rather have black currants and gooseberries and prohibit white pines, because I think they are both magnificent fruits, and black currants are probably the best source of winter vitamin C available to mankind. White currants, which are actually more yellow than white, have a fine, distinctive flavor when eaten raw. Red currants are grown primarily for making into red currant jelly, though they are good eaten raw or cooked.

Gooseberries

Gooseberry bushes are good plants to grow in smallish gardens because they yield a lot of fruit from a small area. They can be trained as cordons (see p. 101), in which case they take up hardly any space at all.

Soil and climate

Gooseberries will thrive in almost any soil but have a slight preference for heavy soils. They like a cool climate, and are very tolerant of shade. They can therefore be planted in places where there is too much shade for most plants.

Soil treatment

Dig deeply and incorporate manure or compost in the top spit, over quite a wide area, because the roots are shallow but spread a long way laterally. A pH of 6 to 8 is suitable for them. Incorporate some lime if the pH is less than 6.

Propagation

Plant new bushes during fall or winter. Bush plants should be five feet (1.5 m) apart and cordons a foot (30 cm) apart in the row. In a deep bed (see p. 106), gooseberries should be planted four feet (1.2 m) apart in a line down the middle of the bed.

Care while growing

When the bushes are two or three years old, cut half the length off each leader to a suitable bud. If the plant is droopy, cut to an upward-pointing bud: if it is upright, cut to an outward-pointing bud. Cut all lateral growths back to within three inches (8 cm) of the stem. In each subsequent year, cut out a good proportion of the old wood.

Every summer, shorten all laterals, keeping about five leaves on each. At that time you can examine the bushes for mildew, and cut out any shoots that are infected. Gooseberry bushes should be grown on a "leg," a short main stem. You must keep the ground under and between the bushes clear of weeds. Don't dig, for fear of injuring the shallow roots, but hoe, scuffle, or rototill very shallowly.

Pests and diseases

AMERICAN GOOSEBERRY MILDEW The first symptom is a white felt that covers the young leaves and shoots. The berries themselves acquire a brownish covering. The best prevention is not to give the bushes too much nitrogen. If you do get it, pick off and burn all affected shoots, and spray, in midsummer, with a mixture of half a pound (228 g) of soft soap, one pound (500 g) of washing soda*, and five gallons (23 l) of water. You can spray with this again in the spring when the bushes flower and once again when the fruit is set.

GOOSEBERRY SAWFLY These are small caterpillars with green and black spotted bodies and a yellow tail. They produce three generations in a season and can strip your bushes of leaves. Spray hard with derris, pyrethrum, or quassia*.

RED SPIDER MITE The tiny red mites cluster on gooseberry leaves, causing them to turn bronze with a white area underneath. Ultimately the leaves will dry up and die. The answer is to knock them off the bush with a jet of water.

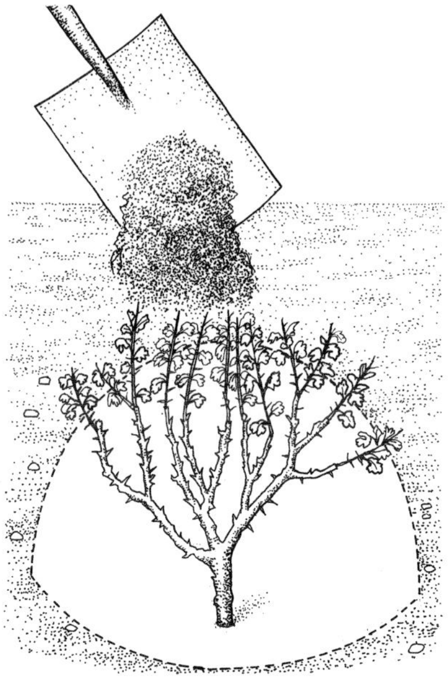

"MOUNDING" GOOSEBERRIES
Cut an old bush back in early spring to within twelve inches (30 cm) of the soil. This encourages new shoots to grow. Then in midsummer build up a mound of soil and compost around the bush, so that only the tips of the canes are visible. By fall the canes will have put out roots. You then gently remove the soil, cut out the canes with the strongest roots. and transplant them.

Harvesting

Strip the fruit off by pulling the branches through a hand protected by a thick leather glove. The fruit falls off and can be caught in a sheet. It can then be separated from the leaves and other flotsam that get stripped off by rolling the whole lot down a board. The fruit rolls, the rest does not. If you don't eat them fresh, can them or make jam.

Currants

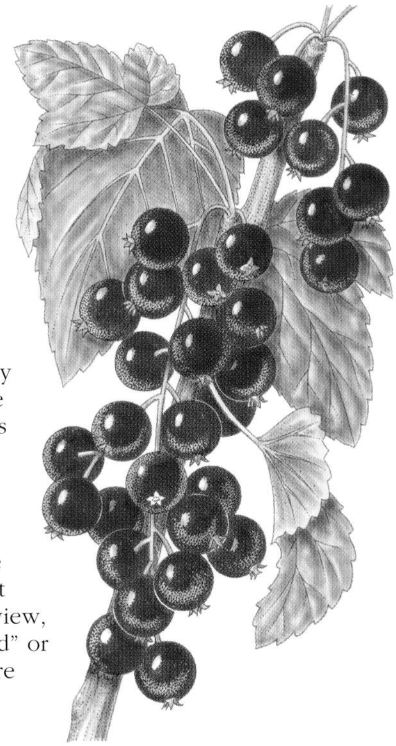

BLACK CURRANTS

Black currants are one of the best and most reliable sources of vitamin C in cold, moist climates. Alaska is not too far north to find them growing, and they thrive in Scotland and Sweden. They are hardy and easy to grow, store well, and make all sorts of delicious preserves and wines. They are heavy-yielding and do not take up too much room. You do not have to wait too long to start picking, either. In my view, of all fruits, either "hard" or "soft," black currants are the most rewarding to grow.

Soil and climate

A fertile heavy clay-loam with plenty of organic matter in it is perfect for black currants, but you can grow them on practically any soil if you add enough compost or farmyard manure. I have grown them with great success on heavy boulder clay and on sand, but with both I had to mulch heavily every year with organic material. A great advantage of black currants is that they are hardy enough to be planted in frost pockets. They like a cool and moist climate, because hot winds dry out their leaves. But they can be grown in hot dry countries as long as they have some shade, such as the north-facing wall of a house. You can also intercrop with apple trees in an orchard, although in this case make sure that the bushes are not starved of moisture as well as being protected from heat.

Soil treatment

Black currants are shallow-rooted. Nevertheless, prepare the ground by digging deeply, because they benefit from soil that is well drained and aerated. Also incorporate plenty of organic material before planting. Dig in ground rock phosphate if you can get some cheaply, hoof and horn, or anything else that is going to last a long time and release nutrients slowly. I always give black currants plenty of manure, but I don't use well-rotted compost, because I reserve this for the things that really cannot do without it. For all my soft fruit, including black currants, I use long-strawed stable or cow-shed manure.

Make sure the ground is free of perennial weeds: once the bushes are in the soil, it will be hard to destroy any weeds that are left.

Propagation

To start off it is best to buy bushes from a reliable nursery so that you are sure they are healthy. After that, you can multiply your own stock for the rest of your life, because black currants

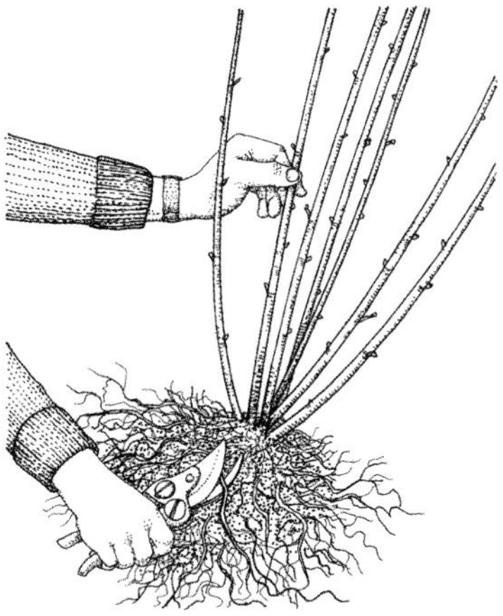

TRIMMING BACK ROOTS
When you are planting your black currant bushes, the roots should be wet. Spread them out well over the shallow holes you have dug in readiness, and cut back first any roots that are broken or torn. Then trim back any very thick roots, but leave all the fine, fibrous ones intact.

grow very easily from cuttings. Because currant buds start growing very early in spring, plant the bushes in late winter; if your winters are not harsh, plant in fall, so the roots can get established before the ground freezes.

When you plant your new bushes, dig wide shallow holes four feet (1.2 m) apart. If the roots of the bushes are dry when you get them, soak them in water for several hours before planting. Spread the roots carefully, first snipping off any very long or broken ones. If you use the deep-bed method (see p. 106), plant a row of bushes along the middle of the deep bed at intervals of four feet (1.2 m), preferably alternating currants with gooseberries. Use the space at the edges of the bed to grow annual vegetables. As soon as you have planted the bushes, snip off all the branches to outward-pointing buds, leaving at least three or four buds on every shoot.

When you come to propagate from cuttings you must use the current year's wood. Black currants fruit on the previous year's wood, so it is not expedient to cut too much new wood off the bushes. But when you are pruning older bushes, you will inevitably cut out a certain amount of old wood on which some new wood has sprouted.

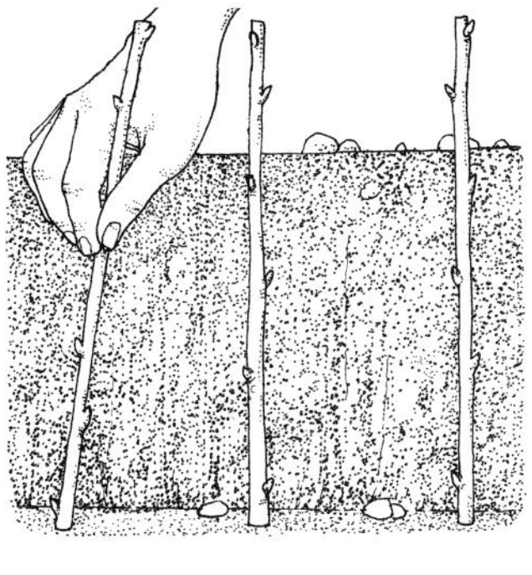

PROPAGATING FROM CUTTINGS
Prune off cuttings of new wood from eight to twelve inches (20–30 cm) long, and in the fall, plant them deeply in good light soil with two buds above ground. Keep them from drying out, and protect them with straw mulch; plant out the following spring.

Care while growing

Prune the bushes annually in early winter. The thing to remember is that they fruit only on the previous year's wood, so you cannot expect any fruit the first year. Therefore, preserve all new wood (which is yellow or light brown) every year, if you possibly can, so that it can fruit the following year, but cut out all the wood that has already fruited. You can tell which is the older wood because it will still have the little stalks of the berries on it.

Mulch with plenty of manure, and wood ash when you have it, and keep the ground clear of weeds.

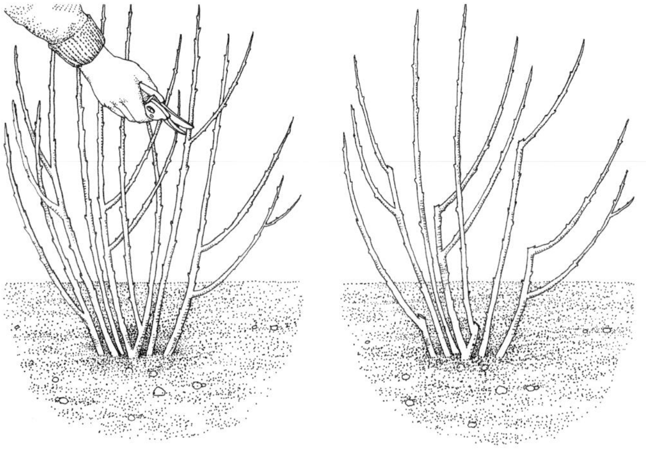

PRUNING BLACK CURRANTS
Every year in early winter, cut out as much old wood that has already fruited as possible, while preserving the new. Cut off the old wood just above a good new low-growing shoot.

Pests and diseases

LEAFSPOT FUNGUS This distressing disease can cause all the black currant leaves to turn brown and drop off in midsummer. Rake up all affected leaves and burn them or put them in a hot compost pile.

CURRANT MAGGOT The black currants themselves are sometimes attacked by maggots. This is more common in North America than in Europe. Currants that ripen before their time should be examined and any with maggots destroyed.

BIG BUD This very common disease causes the new buds of the plants to swell unduly in midsummer. Simply pick off all such swollen buds and burn them.

REVERSION The big bud mite carries the reversion virus. The leaves of the bushes change shape and look rather like nettle leaves. The bushes flower earlier and the flowers are brighter than usual, but the crop is poor and dies fast. There is no cure for reversion, so the moment you notice it, you should root up the bush and burn it.

CURRANT SHOOT BORER/CURRANT CORE BORER Apply a winter wash* in January with a tar-oil spray to prevent these pests; the spray will also keep off aphids. If you do see a cane's leaves wilting at the tip, cut the branch back until you find the tunnel, and kill the borer.

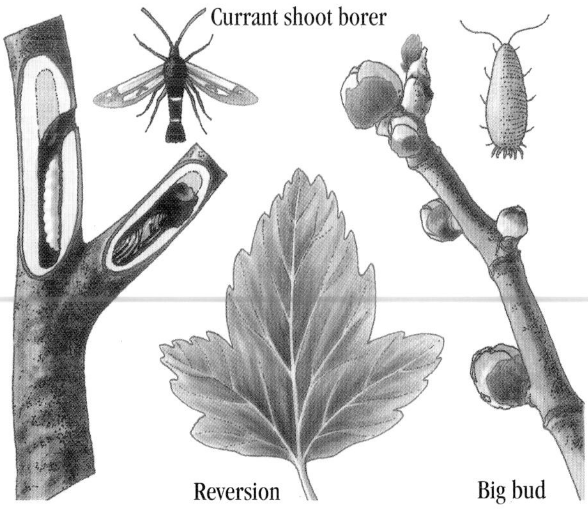

CORAL SPOT/DIEBACK These two fungus diseases are caused by too much nitrogen. Cut back any wood that seems to be dying, or that shows the distinctive red spots of coral spot on the branches. Cut right back to sound white wood, and burn the affected branches. Then stop feeding the bushes with high-nitrogen manure such as stable manure; mulch instead with waste vegetable matter, spoiled hay or straw, and a little well-rotted compost.

PROTECTING BLACK CURRANTS
If birds are likely to strip you of your entire black currant crop, you must protect the bushes. An excellent way of doing this is with a fruit cage of wire mesh supported on a frame.

Harvesting and storing

You can just leave currants on the bushes if there are no birds; likewise, if you grow them in a fruit cage there is no hurry to pick. But don't leave them too long, or they will fall off the bush. Currants freeze well, can well, and make fine wine, jelly or jam. They are one of gardening's greatest rewards eaten raw with cream.

RED AND WHITE CURRANTS

You should treat red and white currants in exactly the same way as black currants, but there is one important difference. The fruit on red and white currant bushes is borne on two- or three-year-old wood. This means that you prune for the first time when the bush is two years old, by cutting out all the wood except seven or eight good shoots.

Each year after that, cut out all new shoots, except for three or four that suit your plan for the shape of the bush. In the third year and every year after that, cut the oldest shoots right back to the ground.

The aim is to have a few one-year-old, a few two-year-old, and a few three-year-old branches on every bush, with plenty of short fruiting spurs on each branch. The other important thing is a good shape—open in the middle, not too spread-eagled, and yet not too bunched up.

Prune the fruiting spurs as you would on an apple tree (see p. 169), because the fruit is borne on spurs like apple spurs. The principle is to snip back side-shoots to one or two buds to encourage spur-making.

Bush trees should be trained on "short legs" (main stems a few inches high), but red and white currants are also excellent for cordon training or espaliers (see p. 101).

Moraceae

Figs and mulberries belong to the family *Moraceae,* whose other members include: hemp; hops; the rubber trees of Southeast Asia and their diminutive, the popular household rubber plant; and a number of tropical and semitropical trees with exotic-sounding names like the breadfruit tree, the snakewood tree, and the trumpet tree. Figs and mulberries are unusual members of the family in that they thrive in temperate climates. All they need is plenty of sun. And figs do better on poor soil than on rich soil, where in order to make them fruit, their roots must be confined artificially. Figs and mulberries are both delicate fruits that do not travel or store well. They should therefore be eaten fresh from the trees; otherwise, figs must be dried or canned, and mulberries must be made into jam or wine. Both trees are attractive and long-lived and grow to about 30 feet (9 m) high.

Figs

Figs are a Mediterranean fruit that was said, in ancient Greece, to be the food of the philosophers. Whether this is so or not, they are quite an experience to eat. They will also grow in much colder climates than is generally thought, as long as they get lots of sun and plenty of water.

Soil and climate

In temperate climates, figs will flourish on the worst soil you have, provided it is well drained and in full sun. They grow very well against a south-facing wall and will tolerate clay, lime-rich soil, sandy soil, or rubble.

Soil treatment

Figs like plenty of humus, so you can mix compost with their earth. Give them a little lime as well. In heavy clay, poor gravel, or sandy soil, you will have no problems; this is the sort of soil found in the warm countries where the fig is native. But in all other soils it is best to confine the roots. This can be done by growing the trees in a concrete box, or in any other sturdy receptacle, buried in the ground. They will also grow in barrels or huge pots, indoors or on a patio. Allow for drainage from the box or container.

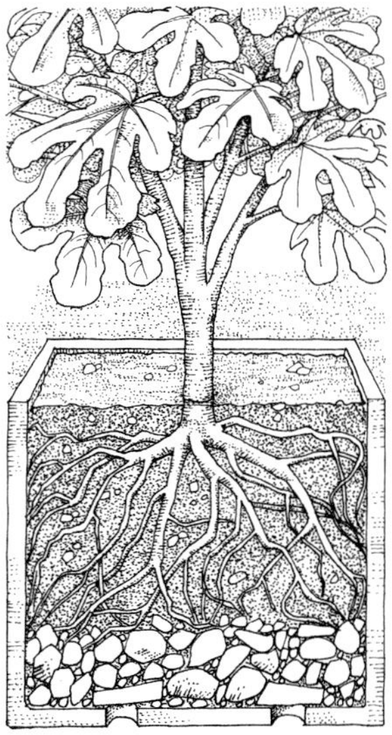

CONFINING FIG TREE ROOTS
If you plant a fig tree in good soil, it will grow well and get very large, but it may not set any fruit for as much as half a century. This is because the roots spread so much farther than the top of the tree that a severe imbalance is caused: nourishment at the root tips never reaches the leaves, so branches straggle and growth is weak. The solution is to confine the roots. Grow the tree in a concrete box buried in the ground. Allow for drainage from the box by making one or more drainage holes in the bottom. Shield the holes well with tiles or broken crockery, so that water can get out but the roots stay in.

Propagation

Figs will grow from suckers, cuttings, or layers (see p. 95). To grow them from cuttings, cut lengths of ripe wood about a foot (30 cm) long from an existing tree in late fall. Plant these cuttings in a shallow trench of good light loam, so that they come out of the ground at an angle of 45 degrees. Leave one growing point only above the surface. Plant the cuttings at nine-inch (22-cm) intervals.

Cover them with loose soil during the winter, so that they are completely buried. In spring, scrape away the surface soil and expose the cuttings, cover them with cloches, and water them whenever the soil is dry. They must not dry out. When the weather has really warmed up, remove the cloches, mulch well, and keep them well-watered until the fall. Then plant out the cuttings in their final position, taking great care not to damage the roots.

If an old tree throws up suckers from its roots, you can dig these out in late fall, keeping the roots intact, and plant them in their permanent position. You can also layer a low branch by pegging it down to the soil. Once it has rooted, transplant it.

Care while growing

Cut out a branch from time to time to keep the tree open, if it seems to be overcrowded. And in early summer each year, nip off the first half-inch (1 cm) of all the leading branches to make them bush out instead of allowing them to grow long and straggly. When the fruit begins to swell, water copiously. Two or three gallons (9–14 l) a day are in order if the weather is dry.

Pests and diseases

COTTON ROOT ROT Figs get this disease if they are planted after cotton. It is incurable; trees wilt and die.

SOURING If disease-carrying insects get into the open end of the fruit, the fruit will shrivel and taste sour. Pull off any diseased or shriveled fruit and put on the compost pile.

Harvesting and storing

Eat your figs straight from the tree when they are ripe. Any that you can't manage to eat should be dried. This can be done in hot sun, on racks, or in a drying box (see p. 216).

Miscellaneous

Mulberries

The name mulberry in fact refers to a multitude of fruits, from a white or red mulberry tree 50 or 60 feet (15–18 m) tall, right through to a white mulberry shrub grown to feed silkworms. The white mulberry grown so freely in the mountains of Persia (present-day Iran) is an insipid fruit, but the wine-red mulberry grown in Europe and America is a splendid fruit and deserves to be cultivated a lot more.

Soil and climate
Mulberries will grow in any garden soil with a neutral pH. Most varieties are very hardy in temperate climates, apart from the black mulberry which only grows in hot regions.

Soil treatment
Dig deeply and incorporate compost or manure.

Propagation
If you can get a tree from a nursery, just plant it as you would an apple tree (see p. 98). Allow it plenty of room to grow; trees should be about 30 feet (9 m) apart. They can be planted in circular deep beds (see p. 110). After you have planted one or two, the trees will proliferate, because birds drop seeds far and wide; otherwise you can propagate from cuttings (see p. 95).

Care while growing
There is nothing difficult about growing mulberries. Just give the trees a good mulching now and again. They are rarely attacked by pests or diseases. When the trees are established, sow grass around them, because this will facilitate harvesting.

Harvesting and storing
Mulberries are highly perishable, so eat them when they are ripe. From a commercial point of view this is a disadvantage; the fruit does not keep for any time at all and must be eaten very quickly. Wait for the fruit to fall to the grass below the tree and gather it immediately. If the tree is situated so grass cannot be grown beneath it, spread hay or straw on the ground during the fruiting season. A word of warning: mulberry juice stains very badly, so wear old clothes when harvesting.

Mulberries and cream are delicious; mulberry wine (see p. 224) can be superb. Birds adore mulberries; if you grow them near cherries they will eat the mulberries and leave the cherries; if you grow them in or near a chicken run, the great weight of fruit that a mature tree produces every summer will fall to the ground and feed the hens, and you will get all you want too.

Blueberries

Blueberries are the fruit for those who have sandy, acidic, waterlogged soils in cold climates. "Blueberry" is often misused as a collective name for several species of the *Ericaceae* or heath family, which includes such fruits as bilberries and cranberries. All these edible fruits grow wild, in cold mountain climates where no other fruit will grow. Only the blueberry proper can be cultivated with success, however, and there are several improved cultivars.

The bushy shrub has attractive white or pinkish flowers and brilliant fall leaf color. It can grow as high as 15 feet (4.5 m). The bushes are slow to mature; after three years they will probably provide you with at least some fruit, but it may take as long as eight years before they bear a full crop. Mature bushes bear very heavily.

Soil and climate
Wild blueberries grow on very acidic soils with a high water table. They do not possess root hairs, so they cannot suck moisture from damp soil particles as other plants can. They therefore need water within reach of their roots. Also, they cannot absorb nitrates and must therefore have their nitrogen in the form of ammonia. This means they must have acidic soil because ammonia-forming bacteria cannot live in anything else.

Ideally they should be planted in light loamy soil with plenty of humus and some sand, and the pH should not be above 5; 4.5 is ideal. They must have a cold climate with at least 100 nights at a temperature as low as 40°F (4°C), but they should be planted in full sunlight.

Soil treatment
Blueberries must have plenty of organic material. They will not grow in purely mineral soils, no matter how much artificial fertilizer is put into it. If the pH is above 5 you should lower it by digging in plenty of leafmold, sawdust, or peat* some months before you intend to plant.

Propagation
Blueberries do not root easily and it is better to buy plants from a nursery. Plant them in spring six feet (1.8 m) apart in rows eight feet (2.5 m) apart, in shallow holes that have been filled with an equal mixture of topsoil and organic matter. After planting, mulch heavily with sawdust, and cut half the length off each branch. Blueberries will grow from layers (see p. 95) as well. Nick the underside of each branch before pegging it down for layering.

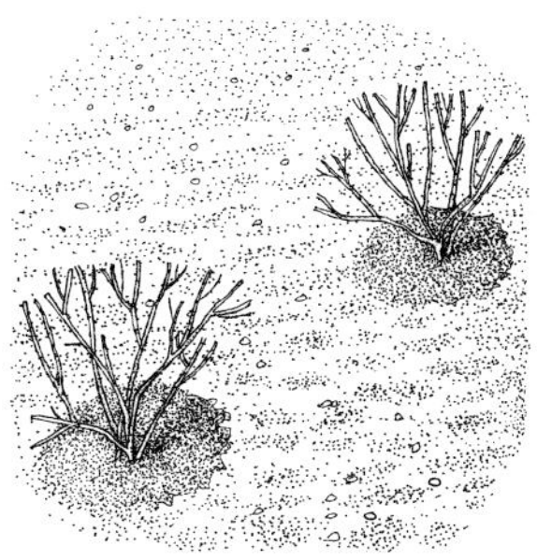

PLANTING BLUEBERRY BUSHES Immediately after the bushes are planted, give a four- to six-inch (10–15-cm) mulch of sawdust (above left). At the same time, cut all the branches back by half (above right); this gives the roots a chance to catch up with the top growth.

Care while growing

Keep the soil constantly moist under the mulch during the first year. For the first four years, don't allow the bushes to fruit at all. Strip off all flower clusters. From the fifth year on, remove all fruiting buds except one for every three inches (7 cm) of branch. Cut out some of the main branches, aiming to leave one for every year of the tree's age. Cut out all small weak laterals every summer. The bushes do not reach full maturity until they are ten to fifteen years old. From then on they will yield as much as 30 pints (17 l) per year. When strong new shoots grow up above the top of the bush, cut them down to the level of the bush to encourage the growth of laterals. Every year the trees must be heavily mulched with organic material. Don't let any lime or sea-sand come near them.

Pests and diseases

TENT CATERPILLARS These pests are the most harmful insects to attack blueberries. They spin tents of silk over the leaves; either pick off the eggs during the winter or remove caterpillars and eggs in the spring.

CANKER This causes reddish-brown wounds on the stem that kill the buds nearby, and in severe cases can girdle the trunk and kill the whole cane. Canker can be prevented by growing the blueberries in an airy place, by keeping the bush open by pruning, and by immediately removing any cankered material.

HARVESTING BLUEBERRIES Leave the fruit on the bushes until it has become really soft. Then test each one by rolling it gently between thumb and finger: the ripe berries will come off easily and the unripe ones will stay on.

Harvesting and storing

Let the fruit stay on the bushes until it really begins to soften, which will be about ten days after it turns blue. This is when the sugar content is at its highest, and consequently when the flavor is strongest and sweetest. If you pick the fruit any earlier, they will be rather tasteless. Then roll the berries gently so that the ripe berries come off and the unripe ones stay on. Store in a refrigerator, or freeze (see p. 227).

Olives

Not only can a man live on olives, bread, and wine alone (and many a man has), but from this fruit is expressed the best edible oil in the world. Olive trees belong to the *Oleaceae* family, and have stunningly beautiful flowers.

Soil and climate

The olive will thrive in practically any soil; it grows in Mediterranean countries where there is no true topsoil at all. But as to climate it is very specific. It needs a cold winter—around 45 to 50°F (7–10°C)—but never below 10°F (–12°C), because this will kill it; even 18°C (–8°C) will do it some damage. Although they don't suffer from late frost, they need a hot summer; it can scarcely be too hot. If you are not between latitudes 30° and 45°, either north or south of the equator, it is not much good trying to grow olives; nor can they grow above 2,500 feet (800 m).

Propagation

The easiest thing to do is to buy a sapling, and plant it like any other tree (see p. 98). Otherwise olives are best propagated from cuttings in a mist propagator (see p. 96). It is best to use softwood cuttings of the current season's growth. Take the cuttings in early fall. Small cuttings should be planted vertically, larger ones horizontally below the soil.

Care while growing

In the first three years after planting, the tree should be shaped to make four or five good strong branches for the scaffold. Cut out all other branches, crossing branches, and branches that grow inward. Let the new laterals on the main scaffold branches grow. By the fifth or sixth year the tree should begin to fruit. If in any one year the tree bears a terrific lot of fruit, thin it; otherwise it will strain itself and not give any fruit next year. In countries with very dry summers, irrigate plentifully during fruiting.

Pests and diseases

OLIVE KNOT This causes swellings on any part of the tree. Cut out such swellings and paint the wounds with tree paint.

SPLIT PIT Heavy watering after a drought when the fruit is swelling causes the stones inside the fruit to crack, ruining the fruit. Keep watering regularly while the tree is fruiting and you won't get it.

Harvesting and storing

Pick the biggest olives by hand from the tree in fall and use them for pickling (see p. 218). Fruit from which you wish to press oil should be left on the tree until late in the winter, when it will be quite shriveled. You then beat the branches with poles and catch the olives on tarps spread on the ground below the trees.

Grapes

"Without wine all joyless goes the feast," sang the poet, and certainly since ancient times the vine has had a notable effect on the development of civilization and culture.

Like the olive, the vine, which belongs to the family *Vitaceae*, grows out of the subsoil. There is a theory that the early Mediterranean mercantile civilizations came about because the overcropping of wheat, and the grazing of goats, caused the topsoil to waste away in those countries. The inhabitants were forced to farm their subsoil, which they did with such crops as vines and olives. They were then forced to trade wine and oil for wheat. This meant they had to become potters (because they had to make *amphorae* to carry wine and oil), shipbuilders, sailors, and merchants. This in turn sped up their industrial and mercantile development.

Some variety of grape is native to nearly every temperate region of the world, and to several subtropical ones as well. Grapes of the Mediterranean species, *Vitis vinifera*, will grow and ripen to wine status in southern England and Wales only with difficulty, and will thrive in North America only in California and Arizona. Americans are lucky in having two native species that will grow in cooler climates: *V. labrusca*, also known as the fox grape or Concord, and *V. rotundifolia*, the Muscadine or southern fox grape. In Britain, *V. labrusca* will grow much better than *V. vinifera*.

Soil and climate

Vines grow well on poor, dry, stony soil. They will grow on limestone soil, and certain varieties will even flourish on chalk, although this is not ideal.

Stony soils on slopes make good vineyards. Many of the best French vintages come from alluvial gravel terraces. I have grown grapes successfully on soil composed largely of decayed fossil seashells. Rich clay soil is bad for grapes, causing them to lose their fruit or ripen it too late. It is fortunate for humankind that the vine thrives on soil that is little good for anything else.

The climate most suitable for grapes is the Mediterranean type. The winter must be cold enough to give them a dormant period, but not so far below freezing as to harm their dormant vines. Most varieties can take temperatures as low as 27°F (–3°C) or even 17°F (–8°C). In cases where temperatures are lower than this, the pliant vines can be bent down and covered with soil to protect them from the elements.

But more important than winter temperatures are the warmth and sunshine that grape vines must have in the summer, both for the fertilization of the flowers in midsummer, and for the ripening of the fruit in late summer. Dessert grapes do not need as long a ripening period as wine varieties; grapes that are pleasant to eat may still not contain enough sugar to make good wine. Late spring frosts are not a problem, because the vines start growing late enough to miss them.

Soil treatment

Clear the soil completely of perennial weeds; incorporate rock phosphate and potash, and dig deeply. If the pH is much below 6, lime the soil to bring it to about 7. Good drainage is absolutely essential.

Propagation

You can, of course, buy year-old plants from a nursery. But most vines are grown from cuttings, although it's often hard to stop them from growing from seeds, which rarely produce strong, heavy-yielding vines. If you have existing vines, make cuttings by separating your winter prunings into two bundles: the ripe, reddish-brown wood and the tender new wood. Tie the ripe wood in bundles, marking the top end with a tiny scratch, label the bundles with the variety, and bury them in moist sand. Feed the new wood to rabbits or goats, or put it on the compost pile. Take the bundles out in March and select the pieces that are about as thick as pencils; make cuttings by chopping them into foot-long (30-cm) lengths, with a bud near the bottom of each. The best cuttings are made from canes with about three or four buds to the foot. Make a long, deep nick with a spade in the sandiest soil you have and plant the cuttings the right way up. The top bud should be just above soil level. Step the cuttings in hard.

During the summer, most of these cuttings will root, and by the following spring, they will be ready to replant. Most experts tell you to dig a wide hole for each cutting and spread the roots carefully over a mound of soil. I suggest that you simply snip off all the roots of each new plant to about two inches (5 cm) so that it looks like a shaving brush, then make a hole with a crowbar, about six inches (15 cm) deep, drop the plant in, and stamp the soil down firmly. I know this works because I have done it in England and seen it done in Italy. You will get excellent results this way because the new vine is forced to put out plenty of new fine roots.

GRAFTING Most European vines are grafted on to American rootstocks, because *Vitis vinifera* cultivars, which Europeans prefer, cannot be grown on their own roots; they are attacked by an aphid called *phylloxera**. American rootstocks have a high degree of immunity to this insect. In Britain this does not apply, because it is perfectly possible to grow *Vitis vinifera* on its own roots: there is no *phylloxera** in this country as yet.

Grafting grapes is quite simple and should be done in winter. Wood of the rootstock should be cut to one-foot (30-cm) lengths with three or four buds on each. The scions should be cut to two- or three-inch (5–8-cm) lengths with one bud. Cut the scion and the stock as you would for any other grafting (see p. 99), and tie them together with raffia or tape. Cover the joint with wax.

When this process is complete bury them shallowly in layers in moist clean sand. Put the box containing them in some place where the temperature does not fall much below 70°F (21°C): a heated greenhouse is ideal. As soon as warm weather comes, plant them out at a slant in a holding-bed, with one bud of the scion just above the soil.

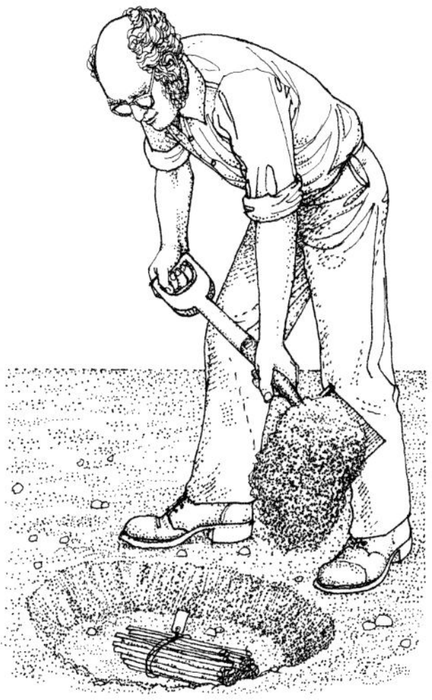

TAKING GRAPE CUTTINGS
In winter, tie ripe prunings in bundles and bury them in damp sand. Take them out in the spring and chop the best into sections a foot (30 cm) long. The best cuttings have three or four buds per foot.

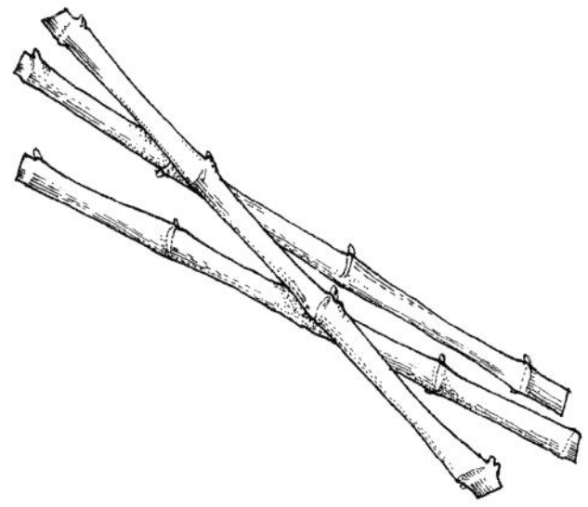

PROPAGATING FROM CUTTINGS
Take a spade and make a long, deep nick in sandy soil. Plant the cuttings, leaving the top bud just above soil level. Step them in hard. Most of the cuttings will root and by the next spring will be ready to plant out.

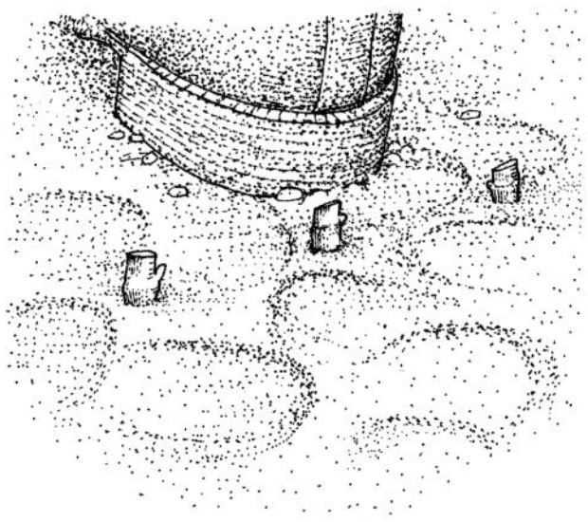

Soon after midsummer, scrape the soil away and cut off any roots that have grown from the scion with a sharp knife. Do this again at the same time the next year. Do not allow the scion to put down roots.

Plant out stocks and scions in the vineyard in the second or third year. Plant them with the joint just above the ground, but then heap some soil over the joint to cover it. After a year, hoe the soil away, since the joint will now be strong enough not to need this protection.

If you wish to change the scion of an unsatisfactory vine, you can try approach grafting. This is a very simple method. Plant a cutting of the desired scion in a pot. When it has taken, place the pot near the growing vine and slice off a short piece of bark with a little wood from the stems of both vine and scion. Put the two cut faces together, and bind and wax them. When the graft has taken, cut off the scion plant below the graft and the rootstock plant above it.

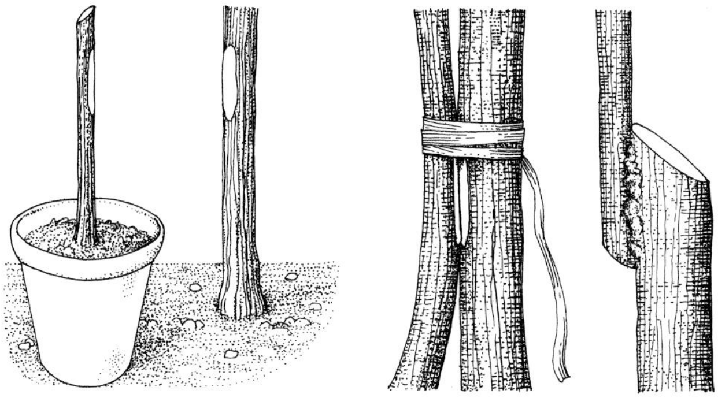

APPROACH GRAFTING
You can improve the quality of a vine by changing the scion. Plant the new scion in a pot near the vine. Chop off a little slice of wood from both vine and scion (above left). Put the two cut faces together, and bind and wax as usual (above right).

Care while growing

For the first three or four years it is most important to keep the ground beneath and between vines free of weeds. At first you can do this by deep digging or plowing. Then, as the roots spread, shallower cultivation is better, because this will not damage them. A rototiller is useful, but shallow scuffling with a hoe will do as well. Heavy green mulching is also effective: comfrey or alfalfa are good for this. Moderate feeding with manure or compost from time to time is beneficial.

PRUNING Training and pruning are subjects of labyrinthine complexity and endless argument: only the benign fermented juice of the grape itself serves to prevent such arguments from becoming vitriolic. The best thing to do, I suggest, is to copy your grape-growing neighbors. But as a general rule, the colder your climate is, the smaller you should keep your vines. In Italy you may find great straggling vines growing up elm trees. In Britain you must keep outdoor vines very small indeed, because the climate will certainly not allow you to grow many ripe grapes otherwise.

The thing to remember when pruning vines is that grapes only grow on this year's shoots, sprouting from last year's wood. Old wood will not fruit, and neither will new shoots springing from two or three-year-old wood. Therefore there must be just enough of the last year's wood to produce the current year's fruiting spurs. And it is these fruiting spurs that will send out new fruiting spurs next year. You can keep some of the present year's canes free of fruit, by stripping young fruit off them, and use them as the next year's base for new fruiting canes. But this method is expedient only in climates where grapes grow freely.

GUYOT METHOD In practice, in cold climates, you will probably need to use the Guyot method, which works as follows. Plant the vines four feet (1.2 m) apart in rows six feet (1.8 m) apart. Erect a two-wire fence along each row with the bottom wire 15 inches (38 cm) from the ground and the top wire a foot (30 cm) higher. Set a light stake four-and-a-half feet (1.3 m) long and tie it to both horizontal wires.

In the third winter after planting, cut all the canes except two down close to the ground. Tie the two remaining canes to the upright stake and pinch them off when they get a few inches taller than the stake. Do not allow them to fruit, and pinch the laterals off when they are a few inches long.

The following winter, cut one of the two vertical canes right off (it was only spare), bend the other one over, and tie it along the bottom wire. Come summer it will send out fruiting branches. When they are long enough, tie these to the top wire. Prune any that are not going to bear and snip off the ends of the fruiting branches, leaving four to five leaves above the flowering bunches.

Now new shoots will come from the stool (stem) of the plant. Keep two of them and cut off all the others. Cut the tips off when they are taller than the stake—say, five feet (1.5 m) high. The next winter, cut off the horizontal that bore the fruiting branches, bend down the better of the two vertical

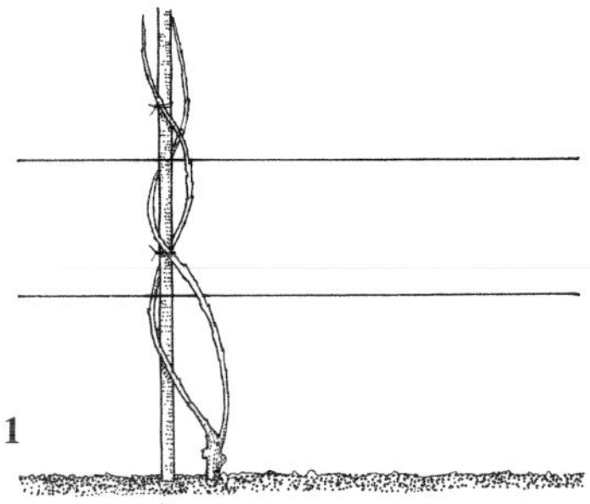

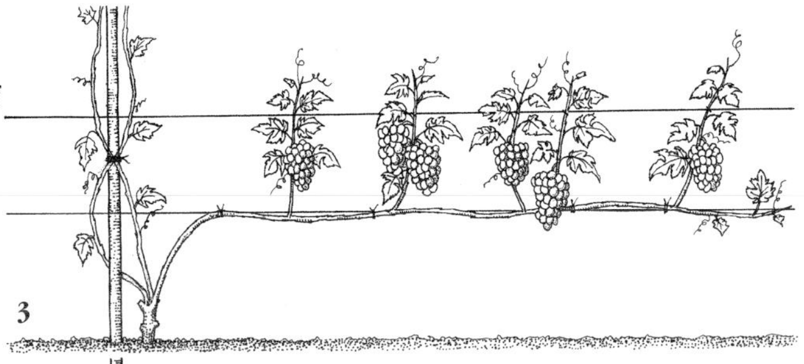

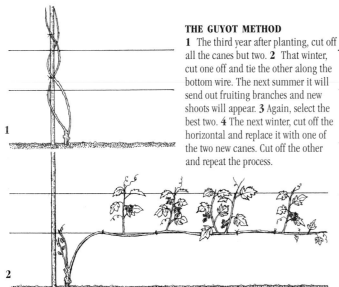

THE GUYOT METHOD
1 The third year after planting, cut off all the canes but two. **2** That winter, cut one off and tie the other along the bottom wire. The next summer it will send out fruiting branches and new shoots will appear. **3** Again, select the best two. **4** The next winter, cut off the horizontal and replace it with one of the two new canes. Cut off the other and repeat the process.

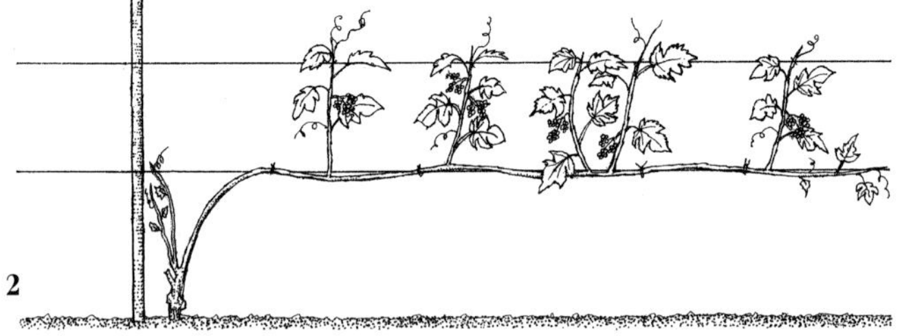

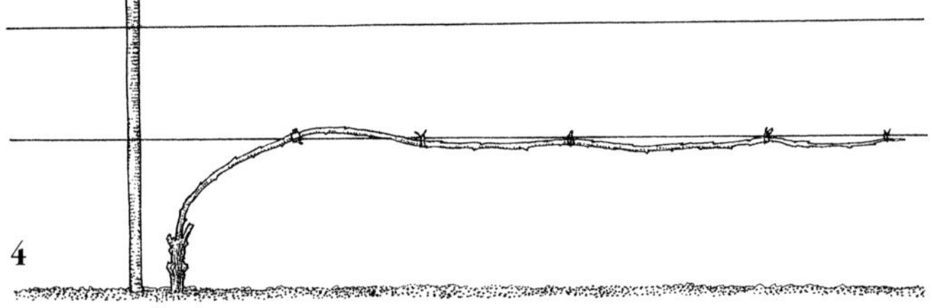

canes to take its place, and cut the other cane right off. Next year, repeat the process. In this way, every summer you always have one horizontal cane bearing fruiting wood, and two of the current year's canes being kept in reserve to fruit the following year.

If you are training vines up walls—one of the best ways of growing them—you can practice exactly the same method in a modified but more extensive form (see below).

It is a good idea to grow vines on south-facing walls if you can: they are more decorative than any ornamental creeper and far more useful.

Pests and diseases

POWDERY MILDEW OR OIDIUM This very common complaint originated in North America. A fine dusty film forms over the vine. To prevent it, dust with sulfur* every three weeks from the flowering stage until the grapes start to ripen.

DOWNY MILDEW This causes a much thicker layer of white down than powdery mildew. To prevent it, spray with Bordeaux mixture* (see p. 105) every three weeks. You will not get the disease under cloches or in greenhouses, because it is spread by droplets of rain.

BLACK SPOT OR ANTHRACNOSE This can appear after periods of wet weather. It causes well-defined black spots on the leaves.

TRAINING UP WALLS
A modification of the Guyot method works well. Instead of cutting the vine right back each year, let it establish a framework of old wood and then allow horizontal branches to develop as if the top of the old wood were at ground level. If the wall is high, plant two vines or more; let the permanent wood grow tall on some and keep it short on others. Vines fruit at their extremities, so if you try to make one vine cover the whole wall, you will only get fruit at the top.

Routine spraying for downy mildew should prevent it. If it does not, increase the strength of the mixture to one pound (500 g) of copper sulfate*, 14 ounces (400 g) of lime, and six gallons (27 l) of water.

VINE MITE This produces blisters on the tops of the leaves. Sulfur dusting* for powdery mildew will also control this.

BIRDS It is quite possible to lose your entire grape crop to the birds. If you suffer from them badly, you must enclose your vines with netting.

WASPS Wasps can decimate a crop of grapes. Prepare a bait made of some sweet stuff, include a few squashed grapes, and mix some poison in with it. Track down and destroy nests.

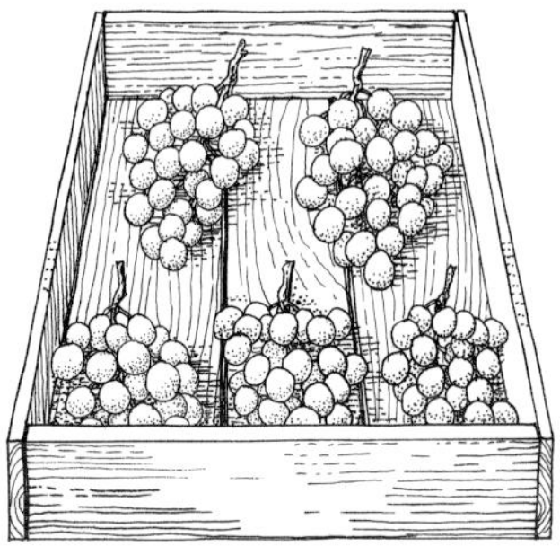

STORING GRAPES
Spread out the bunches of grapes in single layers as far as possible, and leave them after picking until the stems begin to shrivel (above left). Then store them in shallow trays and keep them in a cool, slightly humid place (right).

Harvesting and storing

Leave the fruit on the vines until they are fully ripe, because the riper they get, the sweeter they taste and the better wine they make. They are ripe when the stem of the bunch begins to turn brown.

Snip the bunches from the vines with pruners. Spread out the bunches in single layers and leave at 50°F (10°C) until the stems begin to shrivel. Then store in shallow trays in a cool, slightly humid basement or storeroom at 40°F (4°C). Grapes will keep fresh for several months when stored in this way. For instructions on making wine, see p. 224.

CHAPTER SEVEN

The Cultivation of Herbs

Containing the sowing, growing, and harvesting instructions for the many useful herbs that can be nurtured in the kitchen garden.

Angelica

BIENNIAL

Angelica takes up a lot of space; the plants can reach six feet (1.8 m) high and are quite imposing. If your space for herbs is in any way limited, this is one you can think of doing without. The leaf stalks can be candied or crystallized, and the roots and stems can be cooked with stewed fruit to provide natural sweetness.

Soil
Angelica is best grown at the back of the herb garden, in deep rich soil and partial shade.

Propagation and after-care
The seed does not keep very long; to make sure of good results I pick the seed fresh and ripe in the fall, seal it up in an airtight container, and sow it early in the spring an inch (2.5 cm) deep in moist soil. The seeds will take a long time to come up. Seedlings should be given six inches (15 cm) of space toward the end of the first year, two feet (60 cm) in the second, and anything up to five feet (1.5 m) after that. Angelica usually flowers in its second year, so strictly speaking it is biennial. However, sometimes it does not flower until its fourth or fifth year. After it has flowered, the plant will die.

Harvesting
If you want to harvest the leaves, cut them in early summer when the oils are strongest. As well as eating the leaves as a vegetable, you can dry them very successfully. Leaf stalks for candying should be picked at the same time as the leaves. As for roots, dig them up in the second fall, because they become too woody after that.

Anise

ANNUAL

Anise seed can be baked into bread and cake, and used to flavor cheeses, puddings, candies, and cordials with its delicate licorice scent. An ounce (28 g) of seed in half a pint (300 ml) of brandy, allowed to stand in the sun for two weeks, makes a fortifying drink.

Soil
Anise likes warm, well-drained soil, and a sunny position.

Propagation and after-care
Sow seeds in spring in situ, thinning when the plants are established to eight inches (20 cm) apart. Take care when thinning, because the herb is easily damaged. The more sun the seeds get, the more quickly they will mature.

Harvesting
You should be able to harvest in midsummer, when the seed heads have turned gray-brown. Cut the stalks, tie them in bunches, and hang them up to dry them out. Thresh them when they are thoroughly dried. Save some of the seed to sow the following spring.

Balm

PERENNIAL

Balm, or lemon balm as it is sometimes called, adds a subtle flavor to fruit salads or cooked fruit and is good when added to poultry stuffing. Its pleasant smell recommends it to makers of *potpourri* and the scent lasts for a very long time. If you have it in your garden it will attract bees, which is a good thing because they will pollinate your vegetables.

Soil
Balm likes shady places and rich moist soil, but it needs a little sun to prevent it from getting stringy and blanched.

Propagation and after-care
Sow seed in spring or early summer indoors or in a cold frame. It will take three or four weeks to germinate. Plant out when the seedlings are four inches (10 cm) high. Alternatively, sow seed outdoors in midsummer, then lift and replace the seedlings early the following summer.

If you have an existing clump or can buy or beg one, divide it up and plant the portions in the fall or spring. Balm divides easily into clumps. There is no difficulty in looking after it.

Harvesting
Harvest some leaves just before the buds flower, and then cut the plants right down in the fall and cover them with compost or leafmold.

Balm bruises easily, so be careful when picking and don't expect too much during the first year's growth. Dry it in a dark airy room and store in sealed jars in the dark.

Basil

ANNUAL

In cool climates basil must be sown every year, because frost will kill it. In warm climates you may be able to turn it into a perennial by cutting it right back in the fall, so that it shoots up in the spring. Basil grows very well in containers indoors.

Basil leaves have a strong flavor and used in large enough quantities will dominate even garlic. In France, many cooks steep basil leaves in olive oil and keep this for dressing salads.

Soil

Basil needs dry, light, well-drained soil. A sunny but sheltered position is what it likes best.

Propagation and after-care

Sow seed indoors in early summer. Wait until the soil is warm before planting out the seedlings, eight inches (20 cm) apart in rows a foot (30 cm) apart. Water the plants well, to keep the leaves succulent.

Harvesting

Pick off leaves as soon as they unfurl and use them fresh. Cut the plants down for drying in late summer or early fall; basil takes longer to dry than most herbs.

Bay

PERENNIAL

Bay has a hundred uses in the kitchen, and because it is evergreen there is no storage problem. Freshly dried leaves to put with pickled herrings, with stews, casseroles and soups, should always be available.

Soil and climate

Bay will do well on any average soil. It likes some sun, but needs sheltering from harsh winds. Bay is susceptible to frost, so in cold climates you should grow it in tubs that can be moved indoors in winter. Add compost occasionally, and some bone meal or other material that contains phosphate.

Propagation and after-care

You can buy a young tree and plant it in the winter, or you can propagate it easily from hardwood cuttings or half-ripened shoots.

Harvesting

Pick leaves fresh all year round. You must dry them before you can eat them. Dry them in layers (see p. 216) in a warm shady place. Never dry them in full sun. If the leaves begin to curl, press them gently under a board. After two weeks of drying, put them into airtight containers, preferably glass jars, because the leaves exude oil.

Borage

ANNUAL

Both the flowers and the leaves of borage are used in many different cool drinks, as they contain viscous juices that actually make the drinks cooler. You can sprinkle the blue flowers over salads, or use them for a tisane. The plant is very decorative.

Soil

Borage will grow on any piece of spare ground, but it likes sun, and prefers well-drained loamy soil.

Propagation and after-care

Borage can only be propagated from seed. Sow the seed in spring, in drills one inch (2.5 cm) deep and three feet (90 cm) apart. Cover the seed well with soil. The plant will self-sow.

Harvesting

Eight weeks after sowing, begin cutting the young leaves and keep cutting from then on. Pick the flowers when they appear. You may get two flowerings in one season. Dry the herb quickly at a low temperature.

Burnet

PERENNIAL

Use fresh young burnet leaves chopped up in salads or as a flavoring in sauces. Add the leaves to cream cheese, where they enhance the cool taste. Dried leaves are well worth adding to vinegar, and they can also be used to make a fragrant tea.

Soil and climate
A dry light well-limed soil suits burnet best. The plants need full sun in order to flourish. But burnet is hardy and will do well in most climates.

Propagation and after-care
Sow the seed in early spring, and later on thin the seedlings to a foot (30 cm) apart. It is a good thing to sow seed each year for a continual supply of fresh young leaves. If you want leaves specifically for drying rather than just fresh, you can propagate burnet plants by division.

Harvesting
Pick the young leaves frequently, use them fresh or dry them carefully.

Caper

PERENNIAL

The flower buds of this herb are pickled in vinegar a few hours after being picked, and then become what we know as capers. These are used to much effect in rice dishes, salads, stuffings, and sauces for meat and sea food. Caper grows wild in Mediterranean climates where it flourishes, but it is a difficult herb to grow in temperate climates (where nasturtium may be considered an acceptable substitute).

Soil
Caper does best on poor dry soils; the plant needs full sun and grows well on slopes.

Propagation and after-care
In subtropical areas, grow capers from cuttings or division, planting out the established bush into a well-drained mixture of gravel and sand. When you plant the bush, sprinkle enough water to set it, and thereafter hardly water it at all. You can grow caper successfully in rock gardens, if you simply drop the seeds with a little sand into crevices between rocks. In temperate climates try growing caper under glass, in a well-drained sandy loam, planting out the cuttings in early spring. It also makes an attractive pot plant on a sunny windowsill, but in these circumstances it is unlikely to produce enough buds for culinary use.

Harvesting
Pick off the flower buds as soon as they are fully developed. Leave them in the dark for a few hours before pickling them.

Caraway

BIENNIAL

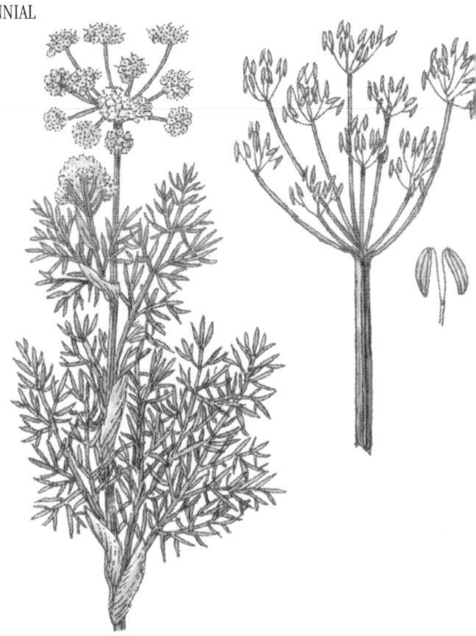

Caraway seeds have long been used in cakes, breads, cheeses, sweets, and sauces. Ground seeds can also be added to rich meats such as roast pork or to spicy stews like goulash. You can use the leaves of the plant in salads, and the roots can be cooked and served as a vegetable.

Soil and climate
Caraway is adaptable and will accept most soils, as long as they are not too wet, but it prefers clay loam and a sheltered location. It is winter-hardy and is best suited to a cool climate.

Propagation and after-care
Sow seed as soon as it ripens on an existing plant; alternatively, sow bought seed in late spring. Thin seedlings to a foot (30 cm) apart and keep them weed-free. Then leave them until the following year, when they will flower and seed. Caraway plants need protection from wind, so that the seed heads don't shatter before the seed is ripe.

Harvesting
When the seed turns brown, snip off the flowerheads and dry the seed in an airy place (see p. 216) before threshing.

Chervil

BIENNIAL

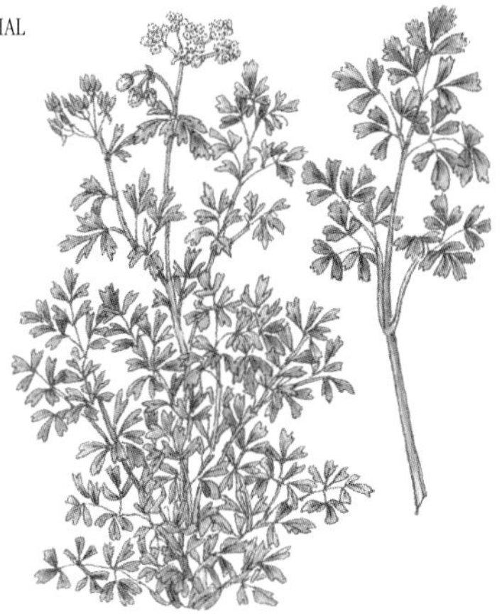

Chervil is a most important herb, a top priority if you are only choosing to grow a few. It is a substitute for parsley but its leaves have a far finer flavor; it is the basis of French cookery with *fines herbes*, the fundamental ingredient of a magnificent soup; it is a fine constituent of salads, one of the very best flavorings in the world for omelets, and it takes its place proudly in many a noble sauce. The habit of boiling chervil is a barbaric one; it must be added to cooked dishes at the last moment so that its delicate flavor can emerge unimpaired.

Soil
Chervil will grow in any soil except heavy clay or wet ground. It needs some shade in summer but full daylight in winter, so ideally grow it in the partial shade of a deciduous tree.

Propagation and after-care
You should sow some seed in the spring for summer use, and then some more in high summer for cutting in winter. Many chervil lovers sow successionally all through the summer. Chervil does not take to being transplanted, so sow it where you mean it to stay, in drills ten inches (25 cm) apart, thinning later to about eight inches (20 cm) between plants. From then on it will self-sow quite rapidly. I often cloche a summer sowing; alternatively, chervil grows well indoors in containers.

Harvesting
You can eat chervil from six to eight weeks after sowing. Cut the leaves off with scissors before the plant flowers. Chervil is tricky to dry, because it needs a constant low temperature, but luckily you can get it fresh all year.

Chives

PERENNIAL

Chives are mini-onions and like onions are members of the *Liliaceae*, but you eat the leaves of chives and not the bulbs. They are perennial, easy to grow, and you can go on and on forever snipping the grasslike tops and flavoring all sorts of food with them.

Soil
Chives will grow in most conditions, but they will do best on good soil with plenty of humus and they prefer a warm, shady position. They will grow well in containers indoors. They like a pH of 6 or 7.

Propagation and after-care
You can sow seed in the spring, but you will get quicker and better results if you plant seedlings or mature plants. You can buy clumps or bum them from a neighbor. Simply divide the clumps up and plant them. The spring or summer is the best time for this. You must keep them moist so it is best to plant them near a pond or water tank, or even a faucet. The plants die down in the winter but you can keep some going for winter use by planting them in a container indoors and putting it on the kitchen windowsill. Every three years or so dig up your chives and replant in fresh soil.

Harvesting
Chives are ready for cutting about five weeks after the seeds were sown. Just clip the "grass" off as you need it to within two inches (5 cm) of the ground. You can clip the tips of the leaves as much as you like without damaging the plants because like all the *Liliaceae*, chives are monocotyledons (see p. 18). Thus clipping the tips off has no effect whatever on the growing point down below. The more you cut them the better they will be.

Cilantro/Coriander

ANNUAL

Don't be put off by cilantro's intense smell, because it is a staple of Mexican cooking, and its seed, called coriander, is essential to Indian cooking: you can use the seeds crushed or whole in curries and mixtures for stuffing vegetables like squash, tomatoes and peppers. If you coat coriander seeds with sugar you can add them to your homemade marmalade, or your children can eat them as sweets.

Soil
Rich soil suits cilantro best. It also needs a sunny, well-drained site.

Propagation and after-care
Sow the seed in late spring in drills 12 inches (30 cm) apart; later thin to six inches (15 cm). The plants will very probably reach a height of two feet (60 cm) or more.

Harvesting
When the seeds begin to turn brown, cut the plants near the ground and hang them up to dry. Thresh the seeds when they are thoroughly dried and store them in jars. Never use partially dried coriander seeds; they have a very bitter taste.

Dill

ANNUAL

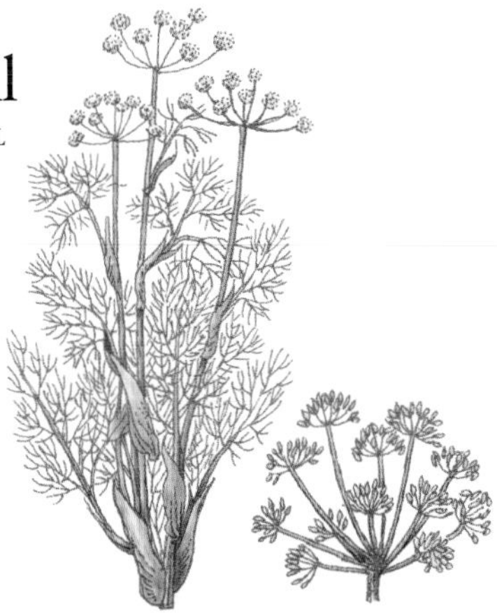

Dill seeds are mildly soporific and are much stronger than the herb derived from the leaves. They are traditionally used in the making of dill pickles. The slightly bitter taste of the seeds is absent from the leaves, which bring out the taste of fish or chicken.

Soil

Dill accepts almost any soil as long as it is well-drained. It must have sun but should not be allowed to dry out.

Propagation and after-care

Sow seeds flat on the bed in the spring, pressing them slightly into the soil. Sow successionally all through spring and summer for a continuous supply of dill leaves. Thin the plants to nine inches (23 cm) in rows a foot (30 cm) apart. As long as you keep them well watered, the plants will grow fast, producing great numbers of leaves before flowering. Very dry weather and inadequate watering will cause them to flower before the leaves are fully grown. Fennel is a bad neighbor for dill as cross-pollination can often take place.

Harvesting

You can start cutting the leaves when the plant is about eight inches (20 cm) tall and keep on cutting right through to late fall. The best time to cut for drying is just before the plant flowers. If you want to use the seeds for pickling, cut them when both flowers and seeds are on the head. Seeds you want to use for flavoring or for sowing the following spring should be left on the plant rather longer, until they go brown. Dry the seed heads before threshing them. The drying temperature must not rise above blood heat.

Fennel

PERENNIAL

Fennel looks very like dill, but has a quite different, stronger flavor. The leaves are much used for flavoring oily fish such as mackerel or herring, and should make part of the stuffing for "a pike with a pudding in its belly." You can use them raw in salads as well. The seeds are nice to chew and can be added to liqueurs.

Soil

Fennel grows well in any garden soil provided it is not acidic, too heavy, or too wet. It prefers rich chalk soil and a sunny location.

Propagation and after-care

Sow the seed in fall for a crop the following year. Sow three seeds in a station and leave 18 inches (45 cm) between stations. If you want seeds and not leaves, sow in early spring under glass. Another approach is to treat fennel as a biennial by digging up roots in the fall and storing them through the winter indoors in sand. The following spring, divide the roots (see p. 95) and plant 12 inches (30 cm) apart in rows 15 inches (38 cm) apart.

Harvesting

Cut leaves through the summer; harvest the seeds when they are still green and dry them out of the sun in thin layers, moving them as they sweat. Drying fennel leaves can be done if you use great care and a low temperature; it is best to use them fresh.

Garlic

PERENNIAL

Garlic can be added to almost any dish. It can be eaten cooked or raw, and it can even be chewed by itself. So grow plenty and use it with abandon.

Soil and climate

Garlic is a southern European plant, but it will grow in cooler temperate regions. It needs the same sort of good rich soil as onions need, with plenty of manure or compost incorporated. Plant it where it will get plenty of sun.

Propagation and after-care

Buy garlic bulbs, from the farmer's market or supermarket if they are cheaper there than at the seed merchant. Pick off the individual cloves and plant them. You can plant them in the fall, or in the early spring. The sharp end is the top end of the clove—plant each one in a hole deep enough to leave the top just covered with soil. Plant four inches (10 cm) apart in rows as close together as you can manage, or four inches (10 cm) in all directions in a deep bed (see p. 106). Keep the cloves weeded. They don't want too much water.

Harvesting

Fork the garlic out of the ground when the stems dry up and dry them out for a few days in the sun if possible, or under cover in some place where the rain won't reach them. Drying is essential if you want to store garlic. Tie the heads into bunches and hang them up in an airy, cool, dry place; use them as you need them, but keep some to plant the following year.

Horseradish

PERENNIAL

The roots of horseradish make a hot-tasting herb. Either grate the roots and use them as they are, moistened a little with vinegar, or make a sauce by mixing them with oil and vinegar or grated apples and cream. Horseradish goes well with roast beef, cold meats, and smoked fish.

Soil and climate
Horseradish likes deep, rich soil and will grow in any climate that is not too hot. In hot climates, horseradish must be grown in shade.

Propagation and after-care
Just plant three-inch (8-cm) pieces of root, about as thick as your finger. Contrary to normal practice, I prefer to put them in nearly horizontal and only two inches (5 cm) below the surface. You can plant horseradish any time of the year, and once you have it, you have it forever. The problem is how to stop it from spreading across the garden. You can confine it inside slates or tiles dug deeply and vertically into the soil. Another method is to set a twelve-inch (30-cm) land drain pipe into the soil on its end, fill it with loam and compost, and plant a piece of root in it. The plant will grow very well, produce clean, tender roots, and be very easy to harvest. And it won't spread. If you don't confine the roots you must dig it out of the ground where it is not wanted.

Harvesting
All parts of the root are edible. Just dig them up and grate them. In cold climates you can store the roots like carrots, in a container of moist sand.

Hyssop

PERENNIAL

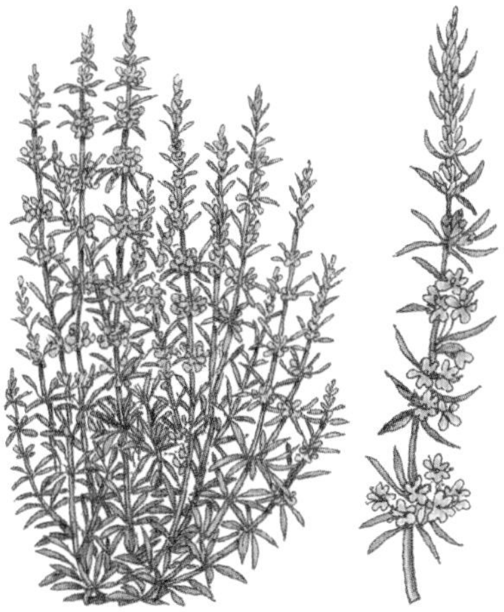

Hyssop is a member of the *Lamiaceae* (mint) family and has a pungent and rather bitter taste. The leaves and the ends of the stalks contain the flavor and will go with a variety of dishes. Hyssop is a good plant for encouraging bees into the garden, where they do a great deal of good by pollinating vegetables, especially beans.

Soil and climate
Hyssop prefers chalky soil, well-drained and containing plenty of lime. The plant thrives in warm weather, but will also manage to withstand winter in cool temperate climates.

Propagation and after-care
You can sow seed in drills a quarter-inch (0.5 cm) deep and transplant the seedlings in midsummer to the open bed when they are about six inches (15 cm) high. Plant them in rows two feet (60 cm) apart.

Harvesting
Once the plants are mature, about 18 inches (45 cm) high, cut back the tops frequently so that the leaves are always young and tender. Cut leaves and stalks for drying shortly before the plants flower.

Lovage

PERENNIAL

All parts of lovage except for the roots can be used in cooking. The bottoms of the stems can be blanched and eaten like celery. The leaves have a strong, yeasty, celery-like flavor as well, which means they can be used to flavor soups and casseroles when celery is not available. The seeds taste the same as the rest of the plant but the flavor is more concentrated.

Soil
Lovage is a hardy herb and likes rich, damp soil and a shady site.

Propagation and after-care
Plant seeds in midsummer, in drills an inch (2.5 cm) deep. Transplant the seedlings in fall or spring to positions two feet (60 cm) apart. By the time the seedlings are four years old they will have reached their full size and should be spaced about four feet (1.2 m) apart. Lovage grows immensely tall; one large plant will be enough to keep a family adequately supplied through the year.

Harvesting
If you want the very large, aromatic leaves for flavoring, water the plants especially well. If they have enough water you will be able to take plentiful cuttings at least three times a year. If you only want leaves, don't allow the plants to flower and seed. Lovage can be dried successfully in a cool oven, at a temperature of less than 200°F (100°C), with the door left a little ajar.

Marjoram (Pot)

PERENNIAL

Pot marjoram is the only type of marjoram that is truly winter-hardy in cool temperate climates. It is a plant that tends to sprawl, throwing out long flowering stems.

Soil

Pot marjoram prefers dry, light soil, with a modicum of sun.

Propagation and after-care

You can grow it from cuttings established under glass and planted out in the spring, or by putting in bits of root in spring or fall. Keep it moist until it is well established. The alternative is to sow seed in spring, in drills half an inch (1 cm) deep and eight inches (20 cm) apart. Thin to 12 inches (30 cm) apart when the seedlings are big enough to handle.

Harvesting

Harvest leaves and stems in late summer. Pot marjoram dies down in winter, but it is a good idea to pot it and bring it indoors each winter. If you do this the plant will grow through the winter and may well last years longer than it would if you left it outside. Seeds for sowing next year ripen in late summer or early fall.

Marjoram (Sweet)

ANNUAL

Sweet marjoram is the only annual of the three marjorams; it has a delicate aromatic flavor, and goes well with game and poultry stuffings.

Soil

Sweet marjoram needs medium-rich soil, with a neutral pH; it wants a good helping of compost and a warm, sheltered spot.

Propagation and after-care

Sow seeds in pots under glass in early spring. Plant out in early summer 12 inches (30 cm) apart. A combination of warmth and humidity is vital to the good growth of the seedlings while they are still young.

Harvesting

Pick leaves and stems toward the end of summer, before the buds open. Use them fresh or dry in thin layers in the dark (see p. 216) and you will get a strong-smelling green herb.

Mint

PERENNIAL

As well as common mint (also known as spearmint), you can grow apple mint, which combines in one plant the flavors of apple and mint; Bowles mint, which is the best type of mint for mint sauce; or peppermint, used to best advantage in peppermint tea. All these mints are slightly and subtly different, but you grow them all in the same way.

Soil

Mint likes moist soil—next to a stream is ideal. It needs sunlight to make it grow with full flavor, although it will stand partial shade.

Propagation and after-care

The best way of establishing mint is to get some roots from somebody who is being overrun by it. In the spring, lay them horizontally in shallow drills three inches (8 cm) deep. Don't harvest much mint that first summer. In the fall, cut the plant right down and cover the roots with compost. If you are overrun by it, simply hoe it out.

If you want to force some mint for using in winter, dig up some roots in the fall, plant them in a seed box in good potting mix, and keep them indoors or in a greenhouse under slight heat, say 60°F (16°C). Mint grows well in containers indoors.

Harvesting

Cut fresh leaves whenever you want them. If you want mint for drying, harvest it in midsummer just before it flowers, but don't cut it after a shower of rain; wet leaves will just turn black and go moldy. Peppermint leaves for tea should be dried and stored whole.

Mustard

ANNUAL

Mustard is grown extensively by gardeners for digging in as a green manure crop, just before it flowers. It grows quickly, makes a bulky crop, and deters the potato-loving eelworm. Mustard can be grown in the herb garden for seed, however, and it is this that makes the mustard that goes in mustard pots. The seeds are ground very finely and the resulting powder kept dry until it is needed, for mixing with water or vinegar. Seeds can be used whole for pickling or for adding to casseroles. Young mustard shoots cut two or three weeks after sowing form the mustard ingredient of the traditional British salad, mustard and cress.

Soil
The seed needs good rich soil, with a pH no less than 6.

Propagation and after-care
The culture of mustard for seed is very easy. Sow in early spring. Broadcasting very thinly will do, but it is better to sow thinly in rows two feet (60 cm) apart and thin to nine inches (23 cm) when the seedlings are established.

Harvesting
Pull the plants out of the ground before the pods are fully ripened, when they are a yellow-brown color. Hang them up in bunches to dry, and thresh the seeds out when the pods are well dried. Grind with a mortar and pestle.

Nasturtium

ANNUAL

Nasturtium is a great asset in the organic garden, because it seems to keep pests away from other plants, especially peas, beans, and soft fruit. People who love pepper but who find it upsets them should turn joyously to nasturtium, for it is an excellent substitute. The leaves spice up salads and add taste to a bland cream cheese spread. You can use the flowers and seeds in salads, and you can pickle the seeds while they are still young and green to use them like capers.

Soil
Nasturtium is an easygoing plant and will grow anywhere, given plenty of sun and light, sandy soil. Poor soil is best if you want a good crop of flowers; but if leaves are your priority, add plenty of compost to the soil.

Propagation and after-care
Sow the seeds *in situ* in late spring. Water them sparingly. The seedlings need little attention. Nasturtium will also adjust quite admirably to being grown in containers.

Harvesting
Cut the leaves in midsummer, just before the plants flower. Chop and dry them, before shredding and storing. The nasturtium flowers do not dry well and are best eaten fresh.

Oregano

PERENNIAL

Oregano, or wild marjoram, is a favorite ingredient in Italian cooking. Its strong spicy flavor suits strong-tasting, oily dishes; if you use it in more delicate dishes, use it in moderation.

Soil
Oregano prefers chalky or gravelly soil, and a warm, dry location. Hillsides are ideal locations.

Propagation and after-care
Sow seed in early spring, thinning later to between eight and twelve inches (20–30 cm). The final distance between mature plants should be 20 inches (50 cm). Hoe the seedlings well. Like pot marjoram, you can grow oregano from cuttings. It is slow to grow, and needs a hot spell to bring it on really strongly.

Harvesting
Pick the oregano leaves and stems in late summer. Seeds for sowing ripen in early fall. Use fresh or else dry in thin layers in the dark.

Parsley

BIENNIAL

Parsley will enhance almost any dish you care to mention, from the blandest poultry to the spiciest sausage. Its great virtue is that it never overpowers the natural taste of food, just brings it out more fully. Broad-leaved, or French parsley, which is grown in the same way, is more substantial and can be used as an ingredient in salads.

Soil

Many people think that parsley is difficult to grow, but I find that as long as it is given rich enough soil, with plenty of humus, this is not the case. The soil needs to be well worked so that the roots can penetrate deeply. Parsley thrives in containers, but again the soil must be rich and well drained.

Propagation and after-care

You can grow parsley from seed, but the seed is extremely slow to germinate. (A good tip to speed germination is to put the seed between two layers of wet paper towels in your refrigerator for about two weeks.) Sow seed in spring, and exercise patience. Put the seeds in drills half an inch (1 cm) deep; later thin to three inches (8 cm) and eventually, when the plants are mature, to eight inches (20 cm).

You can sow in late summer, for winter forcing. In the winter put a cloche or two over some of your parsley patch. Parsley usually goes to seed in the second year, so you should sow it fresh every year to ensure succession.

Harvesting

Pick a few leaves at a time from each plant. If you want bunches, you can pick off whole plants close to the soil, once the stem is eight inches (20 cm) high. For drying you should pick leaves during the summer and dry them quickly. Parsley is the only herb to need a very high drying temperature—between 100 and 200°F (39–93°C). Dry it in an oven with the door open.

Rosemary

PERENNIAL

This woody shrub originates from the dry, chalky hills of southern France. It is very ornamental and can grow to more than five feet (1.5 m) so it is useful grown in rows to divide vegetable beds. It goes best with rich meats, such as lamb, mutton, or pork. Its piney flavor is pleasant but pervasive, so exercise some restraint.

Soil

Light, sandy, rather dry soil suits rosemary best. It needs plenty of lime, shelter, and a southern exposure.

Propagation and after-care

Sow seeds six inches (15 cm) apart in shallow drills in spring. When seedlings are a few inches high, transplant them to a holding-bed, leaving six inches (15 cm) between plants. When they are well established, plant out three feet (90 cm) apart. Seeds do not always germinate, so taking cuttings—before or after flowering—is more efficient. Cuttings should be six inches (15 cm) long. Remove the lower leaves and bury two-thirds of their length in sandy soil in a shady position. By fall they will be ready for planting out. Protect them during their first winter by cutting them back to half their length in late summer; this enables the new shoots to harden off before the onset of cold weather. Then mulch with leafmold and cover securely with burlap.

Harvesting

Pick sprigs in small quantities from the second year onward. You can do this at any time of the year, but late summer is best for drying. You can use rosemary flowers for flavoring as well as stalks and leaves. Pick the flowers just before they reach full bloom.

Sage

PERENNIAL

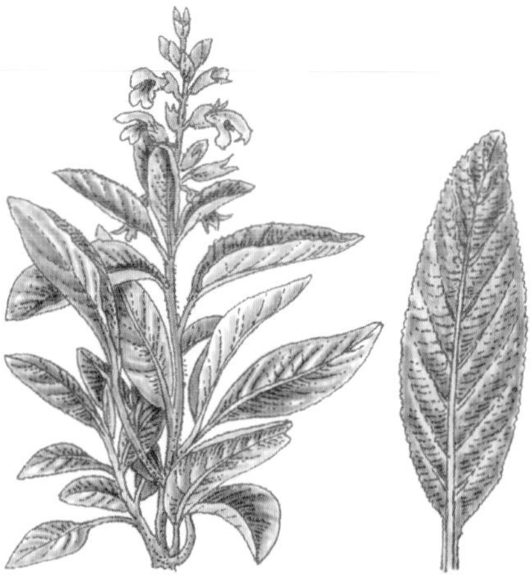

Sage is strong-tasting stuff—too strong to mingle well with other herbs—but it goes well with spicy sausages, fresh garden peas, or as a flavoring for cream cheese. For cooking use narrow-leaved sage; for drying, broad-leaved. A mature bush is about two feet (60 cm) high and is both a useful and an attractive plant to grow in a small garden.

Soil

Sage likes well-drained chalky soil, so lime well if the ground is at all acidic. It does not like damp ground nor too much water.

Propagation and after-care

Narrow-leaved sage can be grown from seed sown in late spring. Transplant seedlings 15 to 20 inches (38–50 cm) apart in early summer.

Broad-leaved sage is always grown from cuttings. Take cuttings with a heel of stem on them, plant out in spring, and water well at first. Sage will last for several years but it is just as well to establish a new bush from time to time.

Harvesting

If you want leaves rich in oils, it is best to wait until the second year before harvesting. Cut narrow-leaved sage in early fall. Broad-leaved sage will not flower in temperate climates; it should be cut in midsummer and again a month later to stop it from going woody. The leaves are tough and take longer to dry than most herbs.

Savory (Summer)

ANNUAL

Summer savory is an annual and tender in cold climates. It can be used fresh or dried. In spite of its strong aromatic smell, it has a more delicate flavor than winter savory. You don't need to grow large quantities, as it grows fast and you only need a little at a time.

Soil and climate

Summer savory will accept poor chalky soil, but thrives on rich humid soil as long as it has not been freshly manured or composted. It will grow in all but the coldest climates.

Propagation and after-care

Sow in late spring in rows a foot (30 cm) apart, and thin seedlings to six inches (15 cm). It is said that the seed should lie just on top of the soil to germinate, but I find it germinates quite well just below the surface where it is less likely to be eaten by birds. Work the soil well before sowing and keep it damp afterward. Summer savory will often seed itself and shoot up again in the fall.

Harvesting

You should be able to cut summer savory twice from the one sowing; once in summer and again in the fall. Cut shoots just before the flowers open. Harvest the seeds for sowing next year when they go brown. To dry summer savory, lay it on frames (see p. 216), cover with a fine-meshed net, and place in a dark cupboard with a low temperature.

Savory (Winter)

PERENNIAL

Winter savory is more hardy than summer savory. It grows a foot (30 cm) high and is bushy, making it an ideal plant for filling the low gaps in garden hedges. Winter savory's strong flavor makes it a good accompaniment for baked fish or lamb.

Soil

Winter savory grows well on poor soil and likes well-drained chalky land. It needs plenty of sun.

Propagation and after-care

Sow seed 12 to 15 inches (30–38 cm) apart in drills in late spring. Don't cover the seeds because they need light to germinate. You can also propagate by planting out cuttings two feet (60 cm) apart in the spring.

Harvesting

You can cut shoots in the second year from early summer onward. As with most herbs, harvest before flowering so that you get the maximum content of volatile oil; this also stops the stalks from going woody. Winter savory leaves become very hard when dried, so you should grow the herb indoors during the winter and pick it fresh.

Sorrel

PERENNIAL

Sorrel has a refreshingly acidic taste. It is a close relative of the dock and looks rather like it. It can go raw into salad, and is very good cooked with spinach, omelets, veal, or fish. Sorrel and lettuce soup can be quite exquisite.

Soil

Sorrel likes light, rich soil in a sheltered place, with plenty of sun, but it will grow very adequately in the shade.

Propagation and after-care

You can sow seed in spring and thin the seedlings out later to six inches (15 cm) apart. Alternatively you can propagate by dividing roots (see p. 95) in spring or fall. When the plant flowers in early summer, cut it back before it goes to seed.

Harvesting

You can start harvesting four months after thinning, or whenever a plant has formed five strong leaves. Cut the leaves with a knife or pull them straight off the plants; cook with them in an enameled pot because an iron one will turn them black. You can dry leaves in the dark and store them in airtight jars.

Tarragon

PERENNIAL

Tarragon is traditional with chicken, good with fish, and excellent in soups; and tarragon vinegar is an excellent salad dressing. There are in fact two varieties, which are often confused: Russian and French tarragon. Russian tarragon is the tougher and the taller of the two plants; French tarragon is stronger-flavored and needs to be checked in its summer growth to stop it from becoming too bushy in the winter.

Soil

Tarragon does not like "wet feet," so good drainage is vital. Try to plant it on a slope so that the roots never get waterlogged. Tarragon likes being exposed to the elements, and puts up with fairly poor soil; it does not even mind stony ground.

Propagation and after-care

It is best to buy young plants or divide mature ones, planting them out to two feet (60 cm) apart, after the last hard frost. Every four years transplant cuttings so that you have plants with a full flavor. Do this in spring or fall. You can grow tarragon in pots that you bring indoors in the winter, or cut the outdoor plants right down every fall and cover them well with compost or other litter.

Harvesting

Pick fresh leaves all through the growing period; this will encourage new ones to grow. If you want to dry them, cut the plants down to just above the ground before they flower. You may manage to cut three times during the growing season of an established plant. Dry in the dark at a fairly low temperature.

Thyme

PERENNIAL

This hardy perennial is native to southern Europe. Common thyme has a sharp bittersweet taste. Shoots, leaves, and flowers can all be used—fresh or dried—in soups, stews, and meat dishes of all kinds. Less hardy than common thyme, lemon thyme has a beautiful scent and taste. The leaves are delicious, chopped fine and sprinkled sparingly, on salads or meat. Otherwise lemon thyme is mainly used for flavoring. It is a good plant to grow if you have bees; it gives honey a delicious fragrance, but bees will only collect the nectar in hot weather.

Soil

Thyme likes light, well-drained soil that has been well limed. It does best in a sunny position, and is excellent in rock gardens.

Propagation and after-care

If you grow thyme from seed, sow it in late spring in drills a quarter-inch (0.5 cm) deep and two feet (60 cm) apart. It is more usual to propagate by division from an established plant, or by cuttings taken in early summer. Keep the beds well watered and weed-free. Cut back a little before the winter and in subsequent springs cut the shrubs well back to encourage new growth. Lemon thyme trails and in exposed positions should be protected during the winter with straw or leafmold.

Harvesting

Cut once in the first year, but from the second year onward you can cut twice. Cut early if you want to, but flowers can be used with leaves, so you can cut during the flowering period. Cut off shoots about six inches (15 cm) long, rather than stems from the base of the plant.

Tree Onions

PERENNIAL

Tree onions are delightfully pungent, and can be used for pickling as well as in stews, or chopped up raw in salads. Also known as Egyptian onions, they differ from other onions in that the onions grow at the top of the stems. The parent bulb stays in the ground to produce another crop the next year, although if you want you can eat the underground bulbs as well as the ones that form on the flower stems.

Soil

Tree onions like a sunny and well-drained spot. They should be started off with a heavy mulch of either compost or well-rotted manure.

Propagation and after-care

You can plant bulbs in fall and spring. Plant them in clumps, six inches (15 cm) apart in rows 18 inches (45 cm) apart. Mulch from time to time with compost. The stems may grow as tall as five feet (1.5 m), so when the little onions begin to form, use sticks to support the weight of the plants.

Harvesting

Pick the bulblets from the top of the plants as and when you need them.

CHAPTER EIGHT

Growing in the Greenhouse

Containing advice on the choosing and equipping of a greenhouse, and instructions on sowing, growing, and harvesting of greenhouse crops.

Growing in the Greenhouse

The primary function of a greenhouse is for propagating seeds and growing tender crops like tomatoes and cucumbers. It is therefore well worth having some form of greenhouse in all but the smallest garden. For very small gardens, seeds can be propagated on a convenient windowsill.

Now, there is a huge variety of greenhouses and you should think very hard about what best suits your needs and your pocket. I suggest you start by getting hold of as many suppliers' catalogs as possible, and by having a look at the greenhouses of as many of your neighbors as you can and asking their opinions.

TYPES OF GREENHOUSE

Starting with the smallest form of greenhouse, I suggest you consider the window-greenhouse. You can buy one of these and install it over a convenient south-facing window. Alternatively, consider taking a window out of its frame, building a wooden platform out from the house at the base of the window, and erecting a glass casing above it so as to form a protrusion from the house. This greatly increases the size of the platform that the original windowsill provides, and also pushes the seed boxes or pots right out into the light where the seedlings and plants do better. And you have the advantage that, if the room behind is heated, the window-greenhouse is heated too. A small window-greenhouse will grow enough tomatoes to keep the average family well supplied.

A lean-to greenhouse is a very common and sensible arrangement. It is best if there is a door leading from the house straight into the greenhouse, and it is even better if there is also a window connecting the two. Either or a window or a door can be left open, in winter for warmth to get from the house to the greenhouse, and in the summer for the delicious aroma of tomato plants to enter the house.

One advantage of a lean-to greenhouse is that you save half the cost of a free-standing greenhouse. Disadvantages are that it is often difficult to effect the join between walls and roof of the greenhouse, and the wall of the dwelling house. Also you often end up with too shallow a pitch in the roof of the lean-to, which can cause leaves, rubbish, and water to collect. The best things about a lean-to are that it acts as a sun-trap, is nice to work or sit in, and helps to warm the dwelling house in winter. You do not get as much light from the sky as you do with a free-standing greenhouse, but the increased warmth from proximity to the dwelling house more than compensates for this.

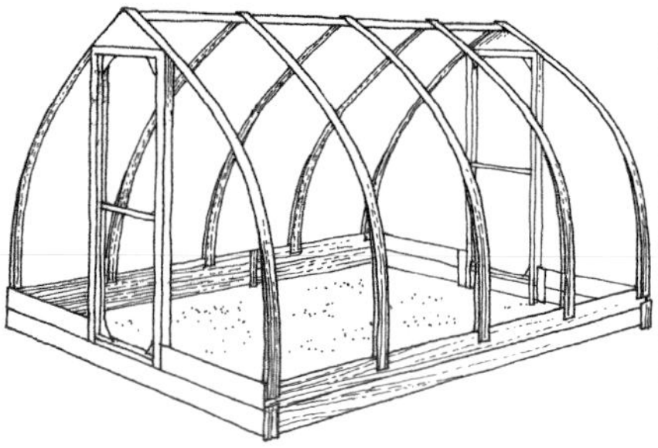

PLASTIC GREENHOUSE
You can build this kind of greenhouse yourself, using plywood and plastic sheeting. Flexible sheeting is cheaper than rigid plastic and easier to cut up, but it must be renewed every three years or so.

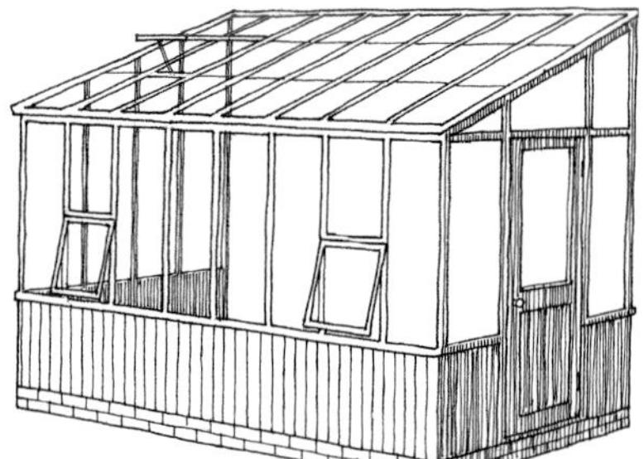

LEAN-TO GREENHOUSE
A lean-to greenhouse attached to a south-facing house wall is ideal in a small garden. A door or window connecting house and greenhouse allows heat to pass both ways.

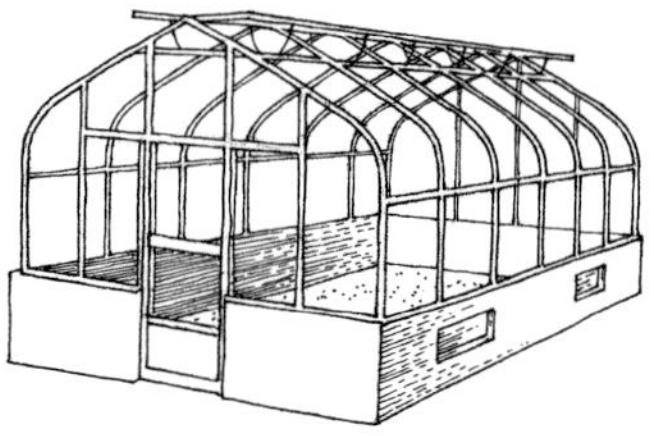

CURVED-EAVES GREENHOUSE
Greenhouses with curved eaves are very popular. The transparent fiberglass walls are erected in fewer pieces than conventional greenhouses.

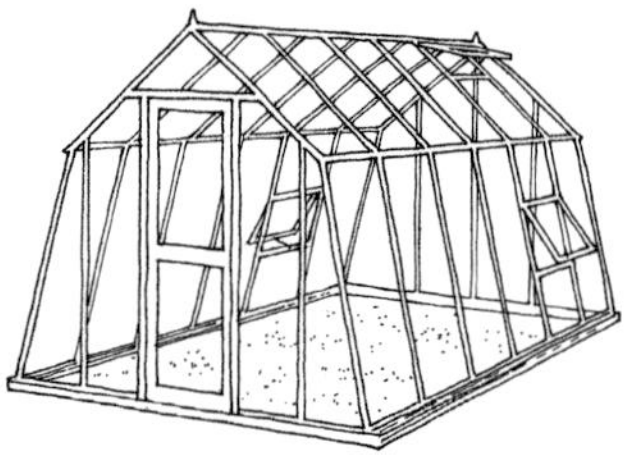

DUTCH STYLE GREENHOUSE
The sloping sides enable the greenhouse to attract more light, and the whole construction is very stable. You cannot grow very tall plants close to the walls.

A free-standing greenhouse can be a very ambitious affair, as large as you have space and money for. It gets more light than a lean-to, but has much worse heat insulation. If I built one I would build an insulated wall of stone or brick to the north side and paint that black to absorb sunlight during the day, and I would build it with a steeply sloping roof.

Lean-to and free-standing greenhouses can be built in all shapes and sizes and can be made from many materials. You can buy a greenhouse ready-made from a specialist firm, you can build one yourself, or you can build a basic structure yourself and buy components—panes of glass in wooden or aluminum frames—and fit them into it. Nowadays, it is probably cheaper, as well as quicker, to buy your greenhouse ready-made. The most common shapes are described above.

Basic materials for greenhouses

As far as the framework of your greenhouse is concerned, your choice is really between wood

FREE-STANDING GREENHOUSE
Most free-standing greenhouses are all glass. They allow in maximum light, and are ideal for growing fruit trees or large plants directly in soil, on the greenhouse floor. An alternative is to have low brick or wooden walls; this cuts down heat loss drastically, but means that you must have raised staging. Many greenhouses have a low wall on the north side, and glass all the way down on the south side.

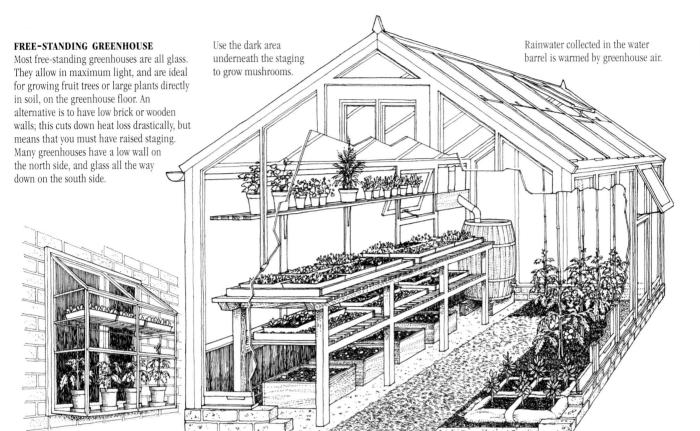

Use the dark area underneath the staging to grow mushrooms.

Rainwater collected in the water barrel is warmed by greenhouse air.

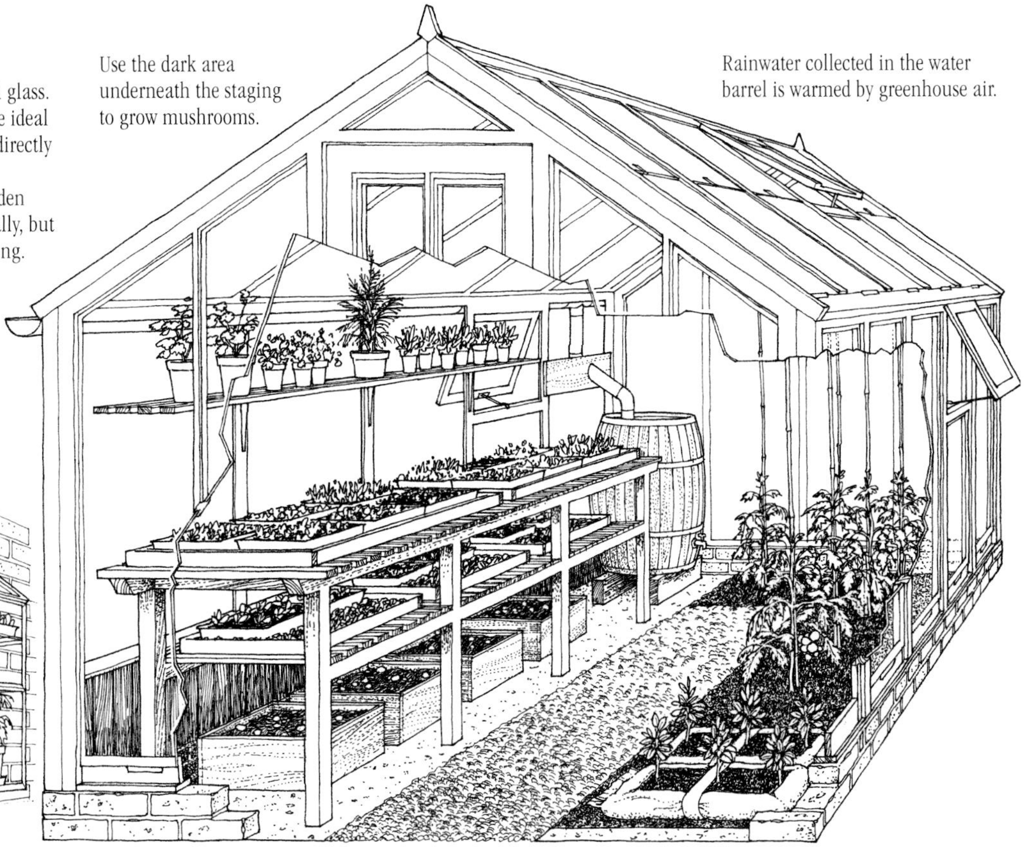

WINDOW-GREENHOUSE
These are designed to replace ordinary house windows. They are extremely efficient, because they get heat from the sun and from the house.

STAGING
If you want to grow plants in pots or seed boxes you will need staging (see p. 206). This is best made of wooden slats supported on frames made of seasoned wood.

CREATING HUMIDITY
In hot weather, spray the gravel path occasionally with water, to create humidity. For the same reason, rest your seed boxes on large trays full of moist gravel.

GROWING IN SOIL
Tomato and pepper plants are very suitable for growing directly in greenhouse soil, or in grow bags. The soil should be made up and brought from outside.

and aluminum, although a third choice—plastic piping—is currently being developed and may prove cheaper than the others.

Aluminum will not rust or rot, but it is generally agreed that it reacts more strongly to hot and cold than wood and therefore cools the greenhouse in winter; my own belief is that this is not a significant factor. Aluminum looks ugly, is hard to work yourself, and is fairly expensive. If you decide on aluminum, you really do need to buy your greenhouse ready-built.

Wooden greenhouses must be made from a weatherproof wood like cedar, redwood, or teak. It is pointless using soft wood or any wood that requires constant painting to stop it from rotting. Wood has a tiny disadvantage in that it obscures more light than aluminum, but it looks nicer and you can work it yourself. And a cedar greenhouse should last as long as you will.

The choice of transparent sheeting lies between glass, and three kinds of plastic: fiber-reinforced, which is a transparent form of fiberglass; PVC or acrylic modified plastic, which is a fairly stiff plastic; and polyethylene, which is flimsy.

Glass lets in a lot of light, looks good, lasts a very long time, is seldom broken by wind, and can easily be mended, but it costs a lot and requires a strong framework to carry its weight.

Fiber-reinforced plastic comes in large sheets; it is easy to fit and does not need extensive framing. It also takes some of the heat out of a hot sun, which is a very good thing. However, it does not admit as much light as glass and this is a serious disadvantage in winter. It is also inflammable, and will only last twenty years.

PVC is cheaper than either of the above, and transmits light well, but it will only last five years and can be ripped by a gale.

Polyethylene is very cheap—about a tenth the cost of glass—and transmits light very efficiently. But it will only last one or two years, and is very easily ripped by gales.

Transparent plastics are becoming very popular on both sides of the Atlantic, and as long as plastics remain cheap, relative to glass, they are well worth using. Glass is, of course, better in the long term, but it represents a substantial capital investment nowadays.

Heated or unheated?

The other great decision to make about your greenhouse is whether to heat it or not. My own feeling is that for the person genuinely gardening for self-sufficiency a heated greenhouse is a luxury that defeats its own object. It is very easy to put more calories of energy into a heated greenhouse than you get out as food produced.

A heated greenhouse is fine for the specialist who wants to grow flowers out of season, or the commercial grower who wishes to supply a luxury winter market, but for the person who is genuinely trying to be self supporting at low expense, it is not really worthwhile unless he can provide himself with cheap energy such as water or wind power, or has a good source of wood for burning.

There are useful crops that can be grown in an unheated greenhouse all year round anyway, and there are excellent ways of storing summer crops so they don't have to be forced in the winter. In the summertime, even in quite a cold climate, you can use an unheated greenhouse to grow, or start growing, most of the crops that grow their whole lives outdoors in Mediterranean climates—tomatoes, cucumbers, melons, peppers, eggplants, and so on. And in winter your unheated greenhouse will enable you to grow lettuces, radishes, spinach, and a few other cold-climate crops. Surely it is better to eat canned tomatoes and eggplants in the winter than to try forcing such things unnaturally and at great expense. A little heat occasionally, when the temperature is very low in the winter, just to prevent the cold-climate crops you are growing at the time from dying, is quite justifiable, but this is a very different matter from running a heated greenhouse all through the year.

Interior fixtures

Inside your greenhouse you will need some staging. You can, of course, grow plants in the soil on the floor of the greenhouse and not have staging at all; for the big plants like tomatoes and cucumbers, and for all fruit trees, this is the best way. But, for your seed boxes, and for the other vegetables that you will grow in pots, you need staging. Using benches in tiers is undoubtedly the best way to get the greatest possible number of crops into your greenhouse.

Your benches should be between two-and-a-half and three feet (75–90 cm) wide: not wider than three feet because that makes them awkward to work on. In a greenhouse ten feet (3 m) wide, two rows of benches—one on each side—are enough; against the north wall have a three-tiered bench, on the south a single tier so that it does not obstruct too much light.

If the greenhouse is much wider than ten feet (3 m), you can consider having another bench down the middle; this should be a double tier. If your greenhouse is as narrow as seven feet (2 m), just have one bench of three tiers against the north wall. If you take it that your path or paths should be 20 inches (50 cm) wide, you can easily work out the best arrangement for a given width of greenhouse.

To support your benches, consider using old galvanized water or gas pipes. These are strong, easily cleaned, and very permanent. Plate glass is best for the actual benches, if you can afford it. It is easily cleaned, lets light down below, and does not allow water to drip. If you use wooden slats for the benches you should place glass, plastic, or slate underneath the lowest tier to stop water from dripping on the plants below.

If you use wood for anything in a greenhouse, be careful about using creosote*. The fumes can kill plants. Old creosote is probably safe.

Paths should be gravel, crushed rock, or concrete. I prefer the two former. If you sprinkle gravel or crushed rock with water and rake it occasionally, it creates a cool moist atmosphere in hot weather.

If you have dark spaces under your benches you can grow mushrooms there. They provide a great deal of good protein in a very little space and don't mind the dark.

Greenhouse soil

Ideally, the soil of a permanent greenhouse should be "artificial." That is, it should not be the original soil of the site but a soil made up and brought in from outside. A good mixture is as follows: one part sphagnum-moss peat; one part coarse sharp sand; two parts good garden topsoil. If you mix a pail of vermiculite or perlite into a wheelbarrow of this mixture, so much the better. Vermiculite and perlite are fragmented rock products that keep the soil open and loose; they have no nutrient value in themselves.

Many growers pasteurize all their soil before bringing it into the greenhouse and then pasteurize it every year. If you grow the same crop year after year in the same soil, you have to do this in order to avoid a build-up of disease. I prefer to dig out the soil my tomatoes have grown in, put it outside, and bring in fresh soil.

Soil for seed boxes

You really should pasteurize—and I mean pasteurize rather than sterilize, because sterilization kills all life in the soil and this is not the organic gardener's aim—all soil for seed boxes, unless you buy professionally-made seed-starting mix, which is in fact a very sensible thing to do; you use so little that the expense is minimal. A three-cubic-foot (0.08-cu-m) bag will fill 18 seed boxes, 20 by 14 inches (50 x 35 cm) to a depth of one-and-a-half inches (4 cm).

If you do want to pasteurize your own soil, put it in an oven pan, cover with tin foil, and bring it to 180°F (83°C)—no more than that or you will kill useful bacteria as well as harmful ones. Alternatively, you can drench the soil in boiling water and then let it drain quickly, or you can cook the soil in a pressure cooker for twenty minutes at five pounds (2.3 kg) of pressure.

Before filling seed boxes, soak the seed-starting mix in water: one gallon (4.5 l) of water to six pounds (2.7 kg) of mix. Leave it for a day before putting it in the boxes or pots. If you are short of seed-starting mix, put two inches (5 cm) of sand and peat mixed in the bottom of each container and then add just half an inch (1.5 cm) of mix on top.

It is best to cover your benches with something that will absorb water—cinders are ideal before you put seed boxes or pots on them. This keeps the plants from drying out, and cinders discourage slugs and snails.

Greenhouse temperature

Even in a heated greenhouse, the temperature should vary between night and day. In a general greenhouse, with many different things in it, 65°F (19°C) by day and about 45°F (7°C) at night is ideal. If the temperature outdoors does not go much below 25°F (–5°C) you may be able to maintain these temperatures without any artificial heat, especially if you have a lean-to or a free-standing greenhouse with a black-painted north wall. If you can keep the air stirring in the greenhouse, with an electric fan, for example, this will help a lot. Heat rises, so the hot air tends to go up into the ridge, and its heat is lost. Stirring the air forces the heat down again and in this way the temperature remains even. It would be worth experimenting with a very small windmill fitted with a direct drive to a fan in the greenhouse.

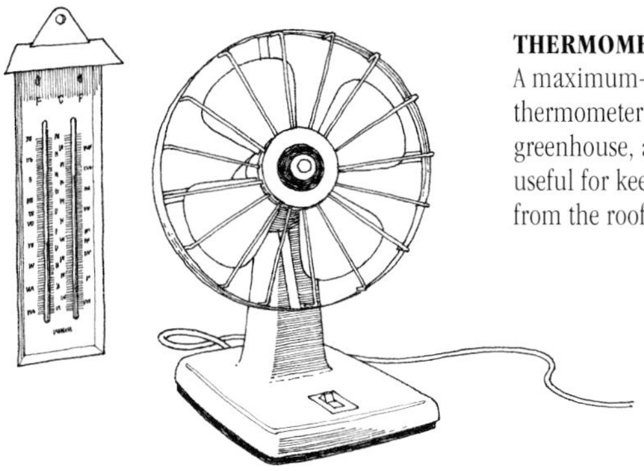

THERMOMETER AND FAN
A maximum-minimum thermometer is helpful in a greenhouse, and a fan is very useful for keeping hot air away from the roof.

A good way to keep heat in is to double-glaze. This can be done temporarily for the winter, by attaching plastic sheeting to the inside of the greenhouse. You can also keep the temperature up in winter by keeping the wind off your greenhouse. A screen of evergreens planted on the side facing the prevailing wind can be very effective protection for a free-standing greenhouse.

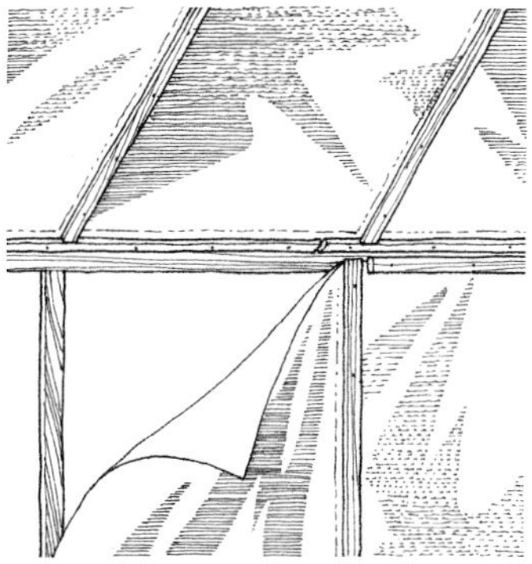

DOUBLE-GLAZING
The surest means of keeping your greenhouse temperature up in winter is to double-glaze. It is cheap as long as you use plastic sheeting and attach it with pins to the inside of the greenhouse. You will want to remove it when the weather warms up anyway.

If you need some artificial heat in winter, an electric heater with a thermostat is ideal, but it is expensive to run. The alternative is a special greenhouse kerosene heater. These give off fewer fumes than household kerosene heaters, and

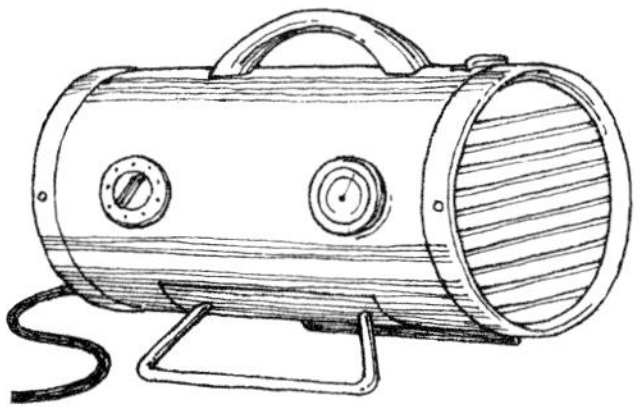

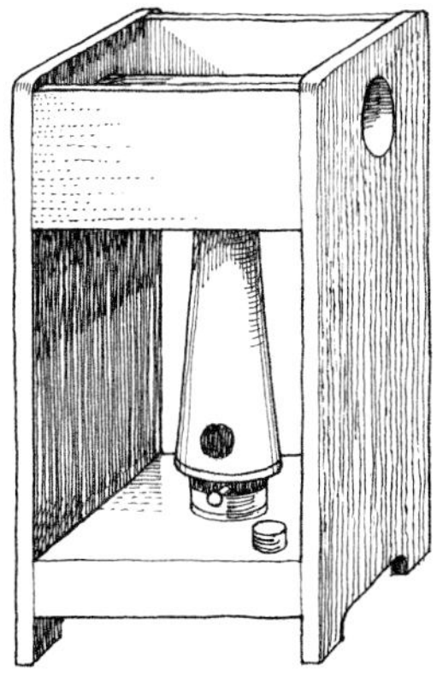

ELECTRIC AND KEROSENE HEATERS
Electric heaters can be controlled thermostatically, but they are expensive to run. Kerosene is cheap, but the burner must be adjusted regularly.

fumes are bad for plants. The disadvantage of these is that you must constantly check the temperature and turn the heater on and off, because the greenhouse must not get too hot.

Keeping cool in summer may be as hard as keeping warm in winter. Plain honest whitewash

WHITEWASHING GLASS
Whitewash painted onto a glass greenhouse keeps out the sun's hottest rays in summer. You can whitewash the whole greenhouse, or just the side that receives most sun; how much you do will depend on what you are growing and the climate in your part of the world.

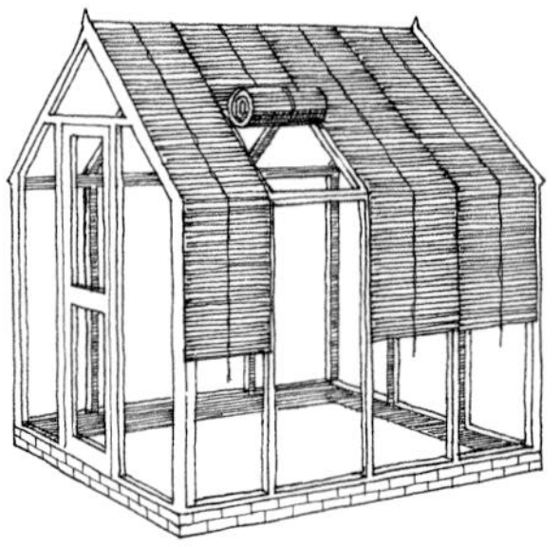

GREENHOUSE BLINDS
Bamboo blinds are a convenient and attractive means of keeping greenhouses cool. Roll them down whenever the temperature threatens to get too high.

on the panes is very useful. It washes off naturally when the fall rains come, and you don't need it any more. (Don't use whitewash on plastic. You may have to wash it off artificially, and this may scratch the surface.) Screens of bamboo or other sticks, or screens of plastic can be used either inside or outside the greenhouse, but they are expensive and don't last long. A good idea, especially in sunny climates, is to plant a screen of deciduous trees between the greenhouse and the sun. The leaves shade the house in summer and die away in winter when you don't want shade.

TREES FOR SHADING
Deciduous trees planted on the sunny side of your greenhouse will keep it cool in summer. The leaves will drop in fall, when the greenhouse needs sun again. Don't plant the trees too close to the greenhouse, because their roots will take nourishment from the greenhouse soil.

Ventilation

Ventilation is very important. Never let the air become "dead" as greenhouse growers call it, meaning stagnant; you must keep it "buoyant." This is difficult, given that you have to maintain a temperature as well, but constant attention to opening and shutting ventilators achieves a lot. Thermostats are fairly cheap and can be paired with ridge ventilators.

WATERING GREENHOUSE PLANTS

Watering plants in the greenhouse is difficult and requires great judgment. To water too much is counterproductive; at worst it kills plants. But to leave plants until they droop from drought is obviously disastrous as well. Watering little-and-often is bad policy. Plants need a good soaking every now and then, and dry periods in between.

You can check whether a plant needs watering by sticking a skewer into the soil. If it comes out clean and dry, the plant needs watering. If particles of earth cling to the skewer, the plant is all right as it is. Tapping the sides of a clay flowerpot is another indication. If it rings hollow, water. If it doesn't don't.

Don't, as a rule, water foliage in your greenhouse. Put the water on the soil only. Water in the morning—never in the evening; plants don't want to go to bed sodden and cold. Don't use freezing cold tap water; if you can manage it, 70°F (21°C) is best for most greenhouse plants. A rain barrel in a greenhouse is an excellent idea; the water then reaches the temperature of the air.

GREENHOUSE CROPS

If I had a criticism of the divine plan, it would be that tomatoes flourish in very different temperature and humidity conditions from cucumbers, and these two crops are far and away the most valuable things that a gardener can grow under glass. There are three things you can do: set yourself up with two greenhouses; divide your greenhouse up with a partition; use your greenhouse to grow cucumbers, eggplants, melons, and other vegetables that like humidity, and grow tomatoes under mini-greenhouses (see p. 111) outdoors. (You have no problem, of course, if you are blessed with a climate in which it is possible to grow tomatoes reliably without any protection at all.)

If you divide your greenhouse up, I would suggest that you partition off a small area of the greenhouse where you will grow a few very early tomatoes in the winter and cucumbers in the summer. The main part of your greenhouse will then be for that most important of all crops, tomatoes, in the summer. In winter it will be devoted to lettuces. The information about growing individual crops, which follows, is intended to supplement the information provided in the Cultivation of Vegetables and the Cultivation of Fruit sections (see pp. 113–190).

Tomatoes

SOIL TREATMENT Prepare the greenhouse soil by forking in at least half a wheelbarrow load of well-rotted compost per square yard. Some wood ashes, fish manure, or other high-potash fertilizer are worth adding if you have them.

PROPAGATION Sow seed in the last week of January. Sow it in the greenhouse if you have some heat, otherwise sow it indoors. Plant seeds carefully in a seed box in good seed-starting mix. It pays to buy this from a reputable retailer, since you need such a tiny amount and the subsequent crop is so important to your family economy.

However, you can make your own (see p. 92). If you have no propagator (see p. 92), keep the seed box at about 70°F (21°C) by day and 65°F (19°C) by night, by covering the tray with glass and putting the newspaper over the glass. It is important to wipe the underside of the glass every day to prevent water from dripping onto the seedlings below.

PLANTING TOMATO SEEDLINGS
Tap the pot gently all around. Scissor the plant between your fingers and upend the pot into your hand. Pull the pot away leaving the root ball intact. Plant gently and water right away.

After eight or ten days—as soon as the little seed leaves are fully out—prick out into three-inch (8-cm) peat* or earthenware pots. After about three weeks, when the plants in the pots are well grown, plant them out in the bed, leaving 15 inches (40 cm) between plants. Give each plant a stick or string to climb up. In the case of peat* pots, just plant the pots; with flowerpots, tap the plants gently out of the pots, keeping the ball of soil as intact as you can, and plant carefully. Water the plant at once.

You can, of course, grow the tomatoes in pots, or other containers, without ever planting them in a bed. In this case, you should use commercial potting mix in ten-inch (25-cm) pots.

CARE WHILE GROWING Keep the temperature in the greenhouse between 65 and 73°F (19–23°C) by day and don't let it drop below 50°F (10°C) at night. For most of the year you should be able to achieve this without artificial heat. For early winter-sown tomatoes, you may need some form of heating. Keep the greenhouse well ventilated; tomatoes don't want a stale and humid atmosphere. Water very well—on the soil, not the plant—whenever the leaves begin to wilt, but don't water too much. A good soak about once a week is perfectly adequate.

Pinch out side-shoots. When the fruit begins to ripen, remove some leaves to let the sun in if this seems necessary, but don't keep hacking the leaves out, for they are what make the plants grow. Don't overfertilize. Once every two or three weeks, it's a good idea to give them a pail of compost or manure soup, or comfrey tea (see p. 103).

A very good method of growing tomatoes in greenhouses is in grow bags (see p. 138). These are plastic bags that you can buy ready-filled with peat* or specially prepared soil mix. Apart from the fact that you get far more from them than you pay out for the bags, the peat or soil mix ultimately adds to the fertility of your garden. Ring culture (see p. 138) also works well in a greenhouse.

HARVESTING Pick tomatoes as they ripen. You should be picking from midsummer until well into the beginning of fall. In early fall, pick the remaining green tomatoes and put them in a drawer to ripen.

Cucumbers

SOIL TREATMENT The bed of the greenhouse should be dug well and plenty of compost or manure—strawy manure is best—should be incorporated. For each plant, make a mound of compost mixed with loam and sand. Mounds should be about six inches (15 cm) deep and a foot (30 cm) wide. Allow two feet (60 cm) between mounds.

PROPAGATION Sow the seed from midwinter onward in small peat* pots or flowerpots—one seed to each pot. Use commercial seed-starting mix, and stand the pots where the temperature will never go below 70°F (21°C). If your greenhouse is unheated, keep them in your house. If they are in flowerpots, it is a good thing after two weeks to pot them on into bigger—say, six-inch (15-cm)—pots, using commercial potting mix. Don't jam the soil around them too hard, soak well after repotting, but thereafter water only when the soil inside is dry.

When the plants are six inches (15 cm) high, plant them carefully in the middle of the prepared mounds in the greenhouse. The temperature when you do this must be at least 70°F (21°C).

Arrangements must be made for the vines to climb, so give each plant a vertical wire or push in a substantial cane. Take these up the sides of the greenhouse to the roof, and put in horizontal wires at 18-inch (45-cm) intervals. Now the temperature must never fall below 70°F (21°C) and it is best if it rises to about 90°F (32°C) in the day. When the plants are young they need hardly any ventilation. If you do open the ventilators a little during the morning, close them early in the afternoon. This is why cucumbers don't coexist happily with tomatoes, which need much drier air.

CARE WHILE GROWING By about midsummer you may have to whitewash the glass above the cucumbers so that they don't receive much direct sun. At this time of year you need high humidity but you must not keep the roots constantly wet. A good watering twice a week is enough. But syringe the plants with warmish water once a day, and keep the floor and walls of the greenhouse moist.

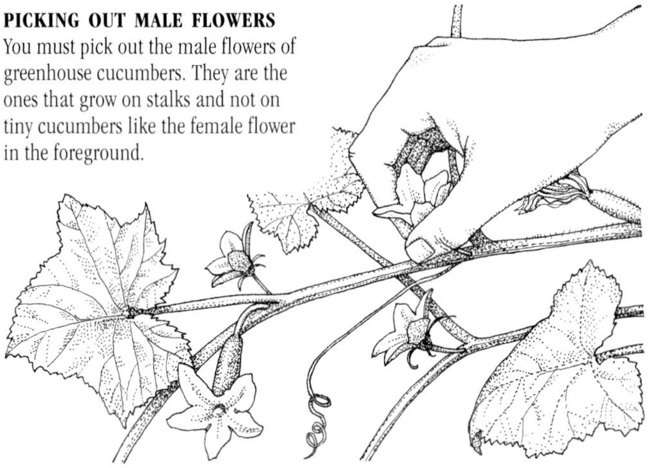

PICKING OUT MALE FLOWERS
You must pick out the male flowers of greenhouse cucumbers. They are the ones that grow on stalks and not on tiny cucumbers like the female flower in the foreground.

As the plants grow, you must train them. Tie the vines loosely to the wires or canes. Stop the main stem by pinching out the growing point when it gets to the roof or when it is six feet (1.8 m) high. Nip out all male flowers (the ones that grow on little stems and not on the mini-cucumbers) so that they will not pollinate the fruiting females and produce bitter fruit with large seeds. Nip out any female flower that grows on the main stem. Stop all laterals (branches from the main stem) at two joints and each sublateral (branches from a lateral) at one joint.

A good dose of compost, manure, or comfrey tea from time to time will be all to the good.

HARVESTING Cut your cucumbers when they reach an appropriate size, and eat them as soon as you possibly can. Never leave them to grow old and shriveled on the vine.

Melons

SOIL TREATMENT Melons will grow very happily in any greenhouse, or part of a greenhouse, where cucumbers are growing. They need the same temperature and humidity conditions as greenhouse cucumbers, and the soil should be treated in the same way. It is well worth preparing mounds for melons, as you would for greenhouse cucumbers.

PROPAGATION Propagate melons as you would cucumbers, by sowing one seed to a flowerpot or peat* pot in midwinter, and make sure that the temperature does not drop below 70°F (21°C). Plant the seedlings out when they are six inches (15 cm) high. You will need to give them vertical and horizontal wires or canes to climb up.

CARE WHILE GROWING Protect your melons from direct sun in the middle of the summer. Water them fully about twice a week, but don't water around the stem of the plant at soil level, because this is likely to cause collar rot—a disease that rots the base of the stem. Instead, sink a flowerpot into the ground six inches (15 cm) away from the plant and water into that, taking care not to splash the stem of the plant. Give the whole plant a light syringeing with water once a day.

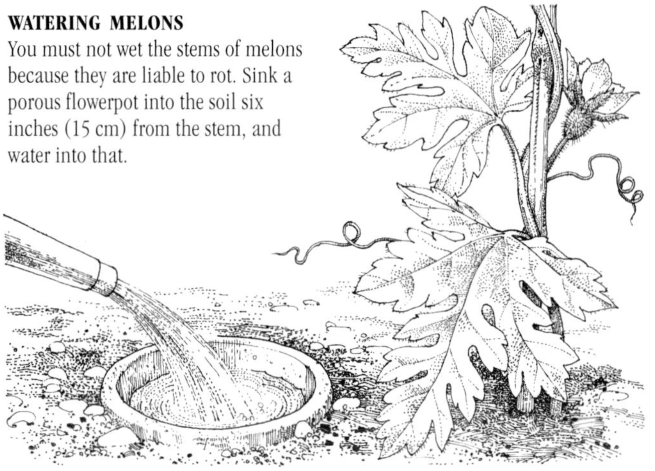

WATERING MELONS
You must not wet the stems of melons because they are liable to rot. Sink a porous flowerpot into the soil six inches (15 cm) from the stem, and water into that.

As your melons climb, attach the plants loosely to the wires or canes. Large melons may need to be supported in nets or cloths. Fix these to the sides of the greenhouse or to your supports.

HARVESTING Melons are ripe when the skin around their stems begins to crack and they come away easily from the vine. Pick them when they are ripe and eat them as soon as you can.

Eggplants

SOIL TREATMENT Eggplants like rich soil, so dig in plenty of compost.

PROPAGATION Eggplants are best grown in summer, unless you have a heated greenhouse. Sow seeds in peat* pots, because eggplants do not like being transplanted. The seeds need plenty of heat: 75 to 85°F (24–30°C). If the temperature falls below 70°F (21°C), they very likely won't germinate. Keep them, therefore, in the part of the greenhouse where you grow your cucumbers. When the seedlings are four inches (10 cm) high, transplant them into their permanent bed.

CARE WHILE GROWING Keep your eggplants well watered, preferably with manure water (see p. 103).

HARVESTING Cut eggplants as soon as the skins are shiny and deep purple. The plants will then produce more.

Peppers

SOIL TREATMENT Dig well and incorporate compost.

PROPAGATION If you have an unheated greenhouse, you must grow peppers in the summer if you live in a cool climate. In a warm climate they will thrive in an unheated greenhouse even in winter. It is best to start the seed off in a propagator or in a seed box with a pane of glass over it. The temperature should be about 80°F (27°C). When the seedlings are two inches (5 cm) high, transplant them into peat* pots. When the plants are four inches (10 cm) high, plant the pots in their permanent positions. This should be in the part of the greenhouse set aside for cucumbers, because peppers need plenty of warmth and moisture.

CARE WHILE GROWING Keep well watered; be sure to water the roots and not the peppers. Water on the peppers will cause them to rot.

HARVESTING When they reach the right size,cut them from the vine leaving a stalk an inch (2.5 cm) long on each pepper.

Okra

SOIL TREATMENT Dig deeply and work in plenty of compost. Manure is not good for okra; it causes the plant to put its energy into making leaves rather than fruit.

PROPAGATION Unless you have a heated greenhouse, sow your okra seed in early summer. Okra seed is stubborn, so soak it in water for 24 hours to get it started, and then sow it in peat pots in a propagator (see p. 97). If you don't use a propagator, still use peat* pots, because okra seedlings prefer not to be transplanted. Plant out when the plants are two inches (5 cm) tall.

CARE WHILE GROWING Like cucumbers, okra plants thrive on heat and moisture; hot, dry air is bad for them and can cause the buds to drop off. Give okra plants a good watering twice a week, and syringe them with water daily.

HARVESTING Pick the pods when they are young—about two inches (5 cm) long. The plants will then go on bearing.

Lettuces

SOIL TREATMENT Rake a good measure of compost into your soil or, if you grow lettuces in containers, make sure the soil contains peat* or compost.

PROPAGATION Sow the seed in a seed box in late summer if you want lettuces for eating in winter. Keep the seed box moist, at about 60°F (16°C), and covered with glass and newspaper. Remove the newspaper when the seedlings first appear. When they are half an inch (2 cm) high, prick them out, using extra seed boxes, to give them more room. Remove the glass and keep them at about 55°F (13°C).

Water occasionally, but take care not to overwater. When the lettuces are three inches (8 cm) high, plant them out into the greenhouse bed or into suitable containers. This final planting out should occur in mid-fall when you will still be eating your outdoor lettuces.

If you want early spring lettuces, sow some seed in mid-fall for planting out in midwinter. You may well find yourself planting next year's tomato crop among these spring lettuces. This does not matter: you will clear away the lettuce long before the tomatoes need the room.

CARE WHILE GROWING Keep the lettuces watered, but do not wet the plants—only the soil. If you let the soil dry right out, the poor lettuces will flop on the ground and will very likely suffer from gray mold (see p. 157). The ideal temperature for the winter lettuce house is 65°F (19°C) by day and 55°F (13°C) by night. They are, after all, a cool climate plant, and will survive out of doors all winter in a temperate climate if they are protected with cloches.

HARVESTING Pick lettuces young or old, when you want them, but remember that, if they are left too long, they will bolt.

Radishes

Radishes are easy to grow in an unheated greenhouse. Simply broadcast the seed in the soil in the greenhouse and rake it in, or sow it in seed boxes. They can be harvested within a month and present no problem at all.

Peaches

SOIL TREATMENT Before planting, incorporate plenty of humus into the soil, but avoid using an excess of nitrogenous material because this encourages unnecessary growth. Keep the soil moist but don't let it get sodden—spraying it on sunny days will maintain humidity. When the fruits are ripening, apply liquid manure*.

PROPAGATION Plant the tree in potting mix in a ten-gallon (45 l) tub; keep it in the greenhouse until all danger of frost is past and then put it out for the summer, in a sunny but very sheltered place. Alternatively, if you have a fairly large greenhouse, you can grow a fan-trained peach tree: train it along wires parallel to the wall and eight inches (20 cm) away from it.

CARE WHILE GROWING Hand-pollinate the flowers on the tree with a small brush. Later, when the fruits are about half their final size, thin them so that there are nine inches (25 cm) between fruits. You must prune fan-trained trees by cutting back old fruiting shoots after the fruit has been harvested to a point where a new shoot is emerging. You then train the new shoot along the wire. In early summer, prune off unwanted wood.

HARVESTING Pick the peaches as and when they are ripe; this is when they turn yellow and give very slightly under pressure.

Oranges, lemons, and mandarins

SOIL TREATMENT Citrus trees need well-drained soil, which should be a mixture of sand, compost, loam, and peat—ideally in roughly equal proportions.

PROPAGATION It is best to buy orange or lemon saplings from a nursery rather than try to grow your own from seed. Mandarins are especially good indoors, because they are small. Plant the trees just as you would outdoors (see p. 98) directly in greenhouse soil.

CARE WHILE GROWING Keep indoor citrus pruned small. Hose the foliage down on hot days. If you grow them in tubs, put them outdoors in the summer, but don't ever leave them out in a frost. In the winter, allow the temperature of the house to go down to 45°F (7°C).

One important point: you will have to fertilize the flowers of indoor citrus trees. You can easily see which are the male organs, the stamens, because they have pollen on them. Take some off with a small paintbrush and put it on the female stigma which sticks out beyond the petals.

HARVESTING Ripe fruit can be left on the tree for weeks, even months. Just pick it when you want it and don't worry about storing it.

Figs

SOIL TREATMENT It is best to grow figs in tubs that can be moved outside in summer. Fill the tubs with soil that contains plenty of compost, and mix in a little lime. Be sure the tubs have drainage holes.

PROPAGATION *Ficus carica* is the ideal variety for growing indoors in a tub, and the best thing is to buy a young one from a nursery. But, of course, like all figs, they can be propagated from cuttings (see p. 185). Plant them in their tubs as you would any other tree (see p. 98).

CARE WHILE GROWING Water enough to keep the soil damp, but not saturated, and spray the leaves from time to time. Move the tubs outside at the beginning of summer, and bring them in when the leaves have fallen or when the first mild frost comes in fall, whichever occurs first.

HARVESTING Eat figs fresh as soon as they are ripe. Otherwise dry them (see p. 216).

Grapes

SOIL TREATMENT The soil for grape vines must be well drained. Dig it deeply and apply rock phosphate and potash. Add lime if the pH is below 7.

PROPAGATION Buy year-old vines from a nursery or propagate from cuttings as you would if you were growing vines outdoors (see p. 188). Greenhouse vines can be planted inside or outside the greenhouse. If your greenhouse is heated it is best to plant the vines in the soil on the floor of the greenhouse. However, if you want to train vines inside an unheated greenhouse, it is a good idea to plant them in a well-prepared bed just outside the greenhouse, and train them through openings in the greenhouse wall. If you have more than one vine, plant them 10 feet (3 m) apart—indoors or outdoors.

CARE WHILE GROWING Vines should be trained up the south-facing wall of a greenhouse. Let each vine fan out to form six strong vertical branches, and tie these permanently in place against the wall.

Now, pretend these verticals are the ground and do your Guyot method pruning (see p. 190) using pairs of strong laterals that spring from them. In this way you will be able to cover your wall with fruiting spurs.

HARVESTING Cut bunches with pruners when the stems of the bunches begin to turn brown.

Strawberries

Strawberries are very easy to grow in the greenhouse. Treat them exactly as you would outside. You can plant them in good rich potting mix either on the ground, on a bench, or in pots.

CHAPTER NINE

Preserving Garden Produce

Containing instructions on the salting, drying, pickling, canning, and freezing of crops, and on the making of jams, jellies, chutneys, wines, and cider.

Preserving Garden Produce

Your aim as a self-sufficient gardener is to provide yourself and your family with a rich, varied, and high-quality diet throughout the year. This means that you must store a lot of your produce, and much of it will not store for long unless it is processed, or "preserved," in some way.

Grains, root vegetables, and potatoes are easy to store. Green vegetables are more difficult. However, if you live in a warm or mild climate, you do not even need to store them. I eat fresh green vegetables picked straight from the garden all year round. I have no desire to eat frozen peas in the winter or frozen Brussels sprouts in the summer, when, with much less trouble, I can eat them both fresh in their seasons. However, although I prefer to eat as much fresh food as I can, I still have to store and preserve a good deal.

Of course, in very cold climates, where snow covers the ground for months at a time, or where frost penetrates deep into the earth, you cannot simply go out into the garden and pick fresh vegetables in midwinter. Therefore all your winter food must come from your stores. And, even in warm climates, there is a strong case for storing certain crops. Tomatoes are a good example. There is no fresh equivalent of tomatoes at times of the year when you cannot pick them fresh. Nothing takes their place, and your cooking will be severely restricted if you do not have them in some form.

Food rotting agents

There are four main causes of food going bad:

ENZYMES These are natural chemicals within most plants. Over a period of time they can cause changes that will spoil food. They cannot function in freezing conditions and are destroyed by temperatures above 140°F (60°C).

MOLDS Molds can actually be seen—the white fluff on jellies or jams, the grayish dust on the rind of bacon, and so on. Some molds are not harmful, but many are, and in any case all molds—except those of blue cheeses—are best avoided, because they can weaken the food's resistance to more harmful organisms, especially bacteria. Molds won't spread at any temperature below freezing point, or above 120°F (50°C). They begin to die above 140°F (60°C). To be sure of killing them, you must heat food to 185°F (85°C).

YEASTS These function at about the same temperatures as molds. They cause fermentation, which turns sugars into alcohol. This has its uses in the making of wine, beer, and sauerkraut, but you don't want it to happen all the time.

BACTERIA Some bacteria are your worst enemy, although others actually assist in preserving processes. Harmful bacteria not only rot food, making it unpalatable, but some of them can even kill you. Bacteria vary as to the temperature that is needed to kill them. Two of the worst—the *Staphylococci* and, the most dangerous of all, *Clostridium botulinum*, which causes the deadly botulism—need 240°F (115°C) to kill them, their spores, and the poisons they leave. 240°F cannot be achieved by boiling water, which will only go to 212°F (100°C). So boiling—except for a very prolonged period, or in a pressure cooker—is not sufficient to rid food of these bacteria.

However, bacteria are not active in acidic food. All food that tends toward acidity—a pH of less than 4.5—is safe for canning without the use of a pressure cooker. These foods comprise all the fruits, including tomatoes, and rhubarb. Other vegetables must be pressure-cooked for safety.

Methods of preserving

There are six main ways of preserving food:

SALTING Salt draws some of the moisture out of vegetables, thereby inhibiting the activity of the rotting agents. It also keeps bacteria away, for the simple reason that they do not like salt.

DRYING This process removes the moisture that is necessary to the functioning of the various spoilage organisms. Vegetables should be dried so that they contain no more than ten percent water; fruits can contain up to 20 percent.

PICKLING AND CHUTNEYING With these methods you increase the acidity of whatever you are preserving by adding vinegar. Thus you do not need the prolonged boiling, or the superheating in a pressure cooker, or the very careful hermetic sealing, that you need with plain canning.

CANNING Canning works because all the living organisms and enzymes in the food are first destroyed by heat. Then the sterilized food is put into clean containers, which are sealed to stop the entry of any new organisms. The containers are then heated again to make sure of killing any organisms that may have gotten in by mistake.

JAM AND JELLY-MAKING You make use of both heat and sugar to preserve fruits that have a fairly high acid content.

FREEZING This works simply because spoilage organisms cannot operate at low temperatures. They are not necessarily killed, but they are prevented from multiplying and are unable to spread their beastly poisons.

SALTING

The art of salting is of great relevance to the gardener. Runner beans in particular are best preserved in this way: salted runners taste far nicer than frozen ones and in fact are almost indistinguishable from fresh beans. And salting them means that they are not taking up unnecessary room in the freezer. Salted runner beans are a great winter standby: a big crock or two, or a barrel, will give you good green food all winter, whatever happens to the garden outside; so it is well worth growing more than you can eat fresh and salting the remainder.

SALTING RUNNER BEANS

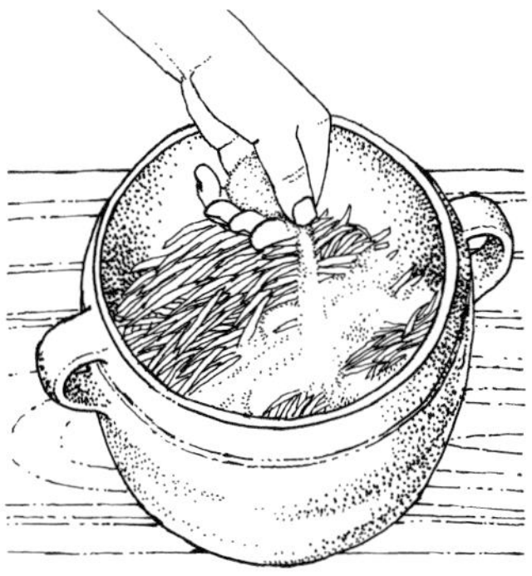

1 Take tender young beans and shred them. Put a layer of salt into a container and then a layer of beans on top of that. Add more salt.

2 Using a wooden pestle or a bottle, ram the beans down carefully but firmly. Continue to add salt and beans in alternate layers.

Beans

Only salt tender young beans, never old stringy ones. Put some dry salt in the bottom of a clean crock or barrel. If you can get "dairy" salt or block salt this is best, but you can make do with ordinary household salt. Absolute cleanliness is essential.

Put a layer of shredded beans in the container—it is useless to put them in unshredded—and ram them down gently but firmly with a wooden pestle or a bottle. Add more salt and more beans in alternate layers, as the bean-picking days go on, until the barrel is full. Then cover it with an airtight lid and leave it in a cool place until winter. A pickle will form that will drown the beans. Do not drain off the pickle, because it acts as the preserving agent, but every now and then skim off any scum that forms.

When winter comes, pull out a handful of beans when you want them. Rinse them well in cold water for five minutes, soak for no more than two hours, boil, and eat. I suggest you use a pound (500 g) of salt to three pounds (1.4 kg) of beans, but if you just use common sense, you should not go far wrong.

Other vegetables

You can salt other vegetables as well as runner beans—any vegetable, in fact, that is crisp and hard. Salting changes other vegetables more than it does runners, but they are still good to eat. The change occurs because of the formation of lactic acid, which is what turns milk sour. The lactic acid, created by bacterial action, is beneficial because it inhibits growth of harmful organisms, but it affects the taste of most vegetables slightly.

Take clean, fresh, undamaged vegetables. Wash them carefully, put them down in a crock, and cover with a ten percent brine solution. The brine is made of salt dissolved in water, which you boil as you stir in the salt. You then allow the brine to cool, keeping it covered. Generally speaking, your brine is strong enough if a potato will float on the surface. (Incidentally, this same brine can be used for preserving all kinds of meat.)

From time to time add more vegetables, more brine, and every now and then add more dry salt on top. This last is important, because the water in the vegetables will dilute your brine and you must keep its strength up. Cover the crock with a plate weighted down with a stone (never with metal) and keep it at ordinary room temperature for three weeks. Skim the surface of scum if necessary. Then move the crock to a cold pantry and pour half an inch (1 cm) of vegetable oil on top to seal the brine. Cover it again. When you want to use the vegetables, pull them out and rinse them for half an hour to get the salt out.

Sauerkraut

Sauerkraut is a fine substitute for clamping the hearting cabbages that you harvest in late fall.

MAKING SAUERKRAUT

1 Shred the hearts of white cabbage very finely, before packing them tightly into a barrel or crock in a series of layers.

2 Sprinkle dry salt between the layers. Cover the crock with a cabbage leaf and a cloth, before weighing the lid down with a stone.

You shred white cabbage hearts pretty finely, and pack them tightly into a barrel or crock, sprinkling dry salt between the layers of cabbage, at the rate of half an ounce (14 g) of salt to a pound (500 g) of cabbage. Cover the top of the crock with a big cabbage leaf and put a cloth over it. Then cover with a lid, weighted down with a stone. Store the barrel in a warmish place—about 70°F (21°C) is fine. Skim off scum from time to time. After two or three weeks, transfer the barrel to the coolest place you have (but not to a freezer), and keep it until you want it. Rinse the sauerkraut under a faucet before eating it. At all stages the salt must cover the sauerkraut. If it doesn't, add more salt. Additions of dill, ground celery seed, or ground caraway seed all help the flavor.

Dill crock

A great American institution is the dill crock. Pack a large number of cucumbers, along with a few onions, green peppers, cauliflower florets, green tomatoes, carrots, and chopped parsnips, into a pot with plenty of dill, and cover them with brine as described above. When you come to wash the salt out, the sharp taste of the dill remains.

DRYING

Drying is one of the simplest methods of storing vegetables, fruit, and herbs. Nothing is added: instead, just water is taken away. You could dry almost all your garden produce, but the process works best with all the herbs, many of the fruits, and just a few vegetables. If you live in a warm climate, seriously consider drying as a method of storage. Your produce will dry much more readily in a warm climate than in a cool one.

Fruit

Almost any fruit can be dried successfully, but apples, apricots, peaches, grapes, currants, plums, and figs dry most easily. To dry fruit, start by slicing it up. If you are using large fruit like apples, slice them into thin slices; smaller fruit, like peaches and apricots, should be halved; plums and anything smaller are best left intact. Fruit that is being dried whole benefits from being blanched before it is dried. To do this, either steam the fruit or plunge it into boiling water for a minute, and then put it into cold water to cool. Dry your fruit using one of the drying devices described below.

To preserve color, commercial driers burn sulfur under fruit. You can achieve this by dissolving two grams of ascorbic acid* in a quart (1 l) of water and dipping the fruit into it.

BLANCHING PEPPERS

1 Before drying whole peppers, blanch them by soaking them in boiling water for a minute.

2 Cool the peppers by soaking them in cold water. Drain immediately and dry the peppers.

Vegetables and herbs

All herbs, but only some vegetables—peas and beans, peppers, asparagus, and sweet corn—are easily dried. If you are going on a long Antarctic voyage, it might be sensible to dry all sorts of other vegetables too, like cabbages and members of the squash tribe. Blanch and dry vegetables just as you would fruit (see left). Herbs do not need to be blanched; dry them in any of the devices opposite.

SWEET CORN Sweet corn, especially off the cob, is an excellent thing to dry. It stores well in a small

DRYING SWEET CORN
Blanch the cobs in boiling water for ten minutes. Impale the cob on a nail sticking up at an angle from a piece of wood. Slice the kernels off the cob with a sharp knife. Dry in a slow oven and store in jars. To reconstitute the sweet corn, pour boiling water over it; then leave it until it has absorbed as much water as possible.

space and can be reconstituted easily. Begin by blanching the cobs in boiling water for ten minutes. If you just want to dry the kernels, impale the cob on a nail sticking up at an angle from a piece of wood. Then slice the corn off with a knife. Dry the kernels in a slow oven, and put them in jars. If you want to dry corn on the cob, strip off the husk and dry in a warm oven. When the cobs are dry you can either store them as they are, or get the corn off the cob by grabbing with both hands and twisting in opposite directions.

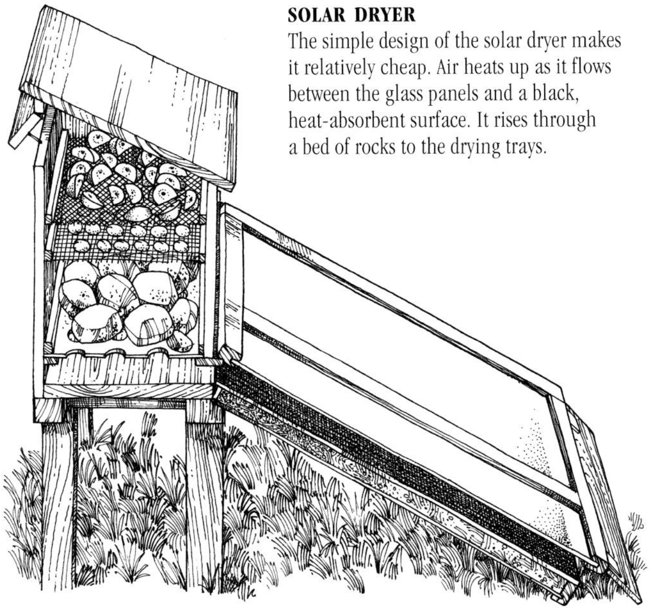

SOLAR DRYER
The simple design of the solar dryer makes it relatively cheap. Air heats up as it flows between the glass panels and a black, heat-absorbent surface. It rises through a bed of rocks to the drying trays.

Drying devices

TRAYS Any kind of tray that is perforated to let the air circulate can be used to dry fruit and vegetables. Put them either outdoors or in a warm place indoors.

CABINETS A drying cabinet can be simply a home-made structure that has slots to hold trays. An electric or kerosene heater can be placed underneath, if required. Alternatively, you can buy a drying cabinet, either metal or wood, which has an electric heater built in.

OVENS As long as you take great care, you can use your kitchen oven to dry most vegetables, fruit, and herbs. As a rule use low heat, especially for herbs. You may find it best to leave the oven door open. Use solid metal trays, instead of metal racks, to be sure that nothing catches fire.

SOLAR DRYERS These are catching on more and more, because they are an easy and effective way of using solar heat. Air is admitted through an adjustable flap and crosses over a blackened surface underneath glass panels, heating up as it goes. The hot air rises through a bed of rocks and then through a series of perforated trays, which hold the produce to be dried. The rock bed heats up slowly through the day, and retains some heat throughout the night, which prevents condensation from forming on the glass.

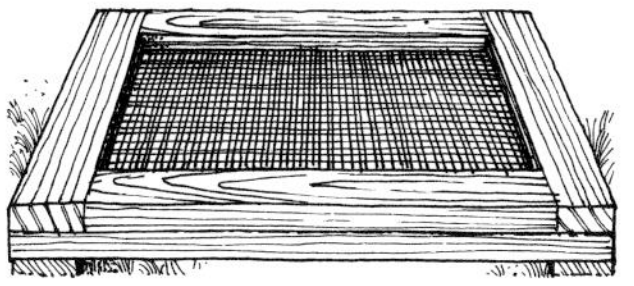

Wire mesh drying frame

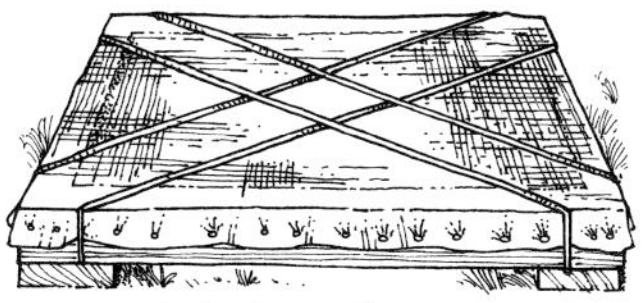

Cheesecloth drying frame

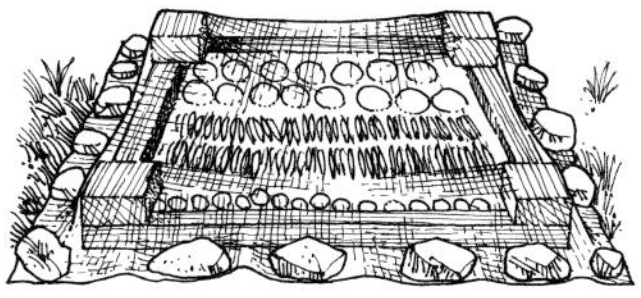

Cheesecloth protection for fruit

DRYING FRAMES
Frames can be made from wood and cheesecloth or wire mesh. Cover wire mesh with brown paper to protect the food.

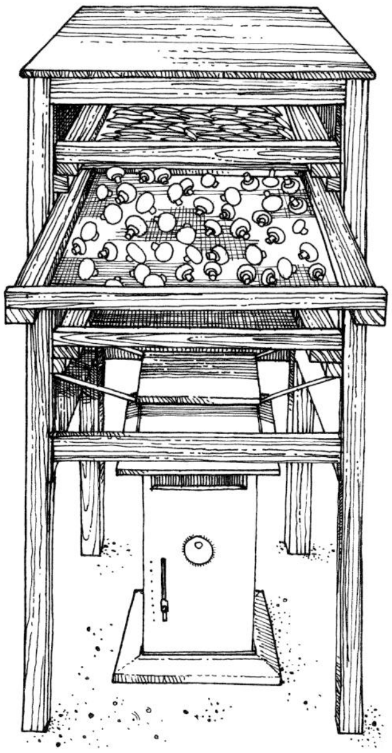

DRYING CABINET
A series of trays and a kerosene heater make up this cabinet.

DRYING IN BUNCHES
Dry herbs in bunches upside down, so that the volatile oils flow into the leaves.

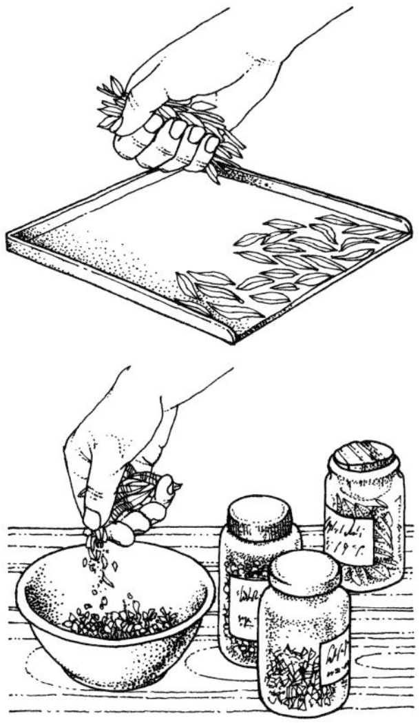

DRYING ON TRAYS
Alternatively, you can dry herbs on trays in your oven. Shred the leaves before storing them in jars.

HANGING BUNCHES You can dry herbs, apple slices, and mushrooms by stringing them and hanging them up. A temperature between 70 and 80°F (21–27°C) and a strong draft are ideal for herbs.

Reconstituting dried food

To reconstitute dried food, pour boiling water over it. Allow the food to absorb as much moisture as it can. Many fruits cannot be reconstituted: currants and raisins are a case in point.

MAKING PICKLES AND CHUTNEYS

Pickles and chutneys are both means of preserving produce and enhancing its flavor at the same time. Both processes involve flavoring fruit and vegetables with spices and then storing them in vinegar. To pickle something, you store it in cold, spiced vinegar, whereas to chutney something you cook it in vinegar until its consistency becomes thick and syrupy.

If you buy—rather than make—your vinegar, remember that there is a wide range of strengths and flavors. Distilled, or fortified vinegar is the very strongest; of the natural vinegars, wine vinegar is the strongest, and this also has a purer flavor than either cider or malt vinegar. These factors are worth bearing in mind: you will pay according to the quality of your vinegar. You may find it economical to use cider or malt vinegar in chutney—where the liquid is all boiled away—whereas in pickles you may well want to use a purer vinegar, since the flavor of the vinegar is more likely to affect the taste of the pickle.

Pickles

I believe in spicing the vinegar I use for pickling. You can add any spices you like, although I recommend that you use whole ones for an attractive end-product; ground spices will turn the vinegar cloudy. The very best way of making spiced vinegar is to steep all the spices in cold vinegar for two months; it will then be ready to use. For a quicker method, study the pictures below.

You can pickle virtually any vegetable or fruit, and you can use them whole or sliced up. Even okra, eggplants, and Jerusalem artichokes make good pickles. The principle of pickling is the same whatever the vegetable or fruit: draw out some of the excess water from moist vegetables by soaking them in brine or coating them with dry salt, and then put them in vinegar. Crisp vegetables can be put straight into the vinegar. Cold vinegar is usually perfectly adequate, but boiling your vinegar and sterilizing your jars is safer if you want to keep the pickle for a long time. As for spices and herbs—every person to his own recipe. Let imagination and experimentation reign supreme.

Remember, when you store away your jars of pickles, seal them tightly to prevent evaporation. Six months is probably the longest they will keep and still retain all their flavor. Take care that the vinegar is not in contact with the metal lid.

PICKLED ONIONS Soak small pickling onions in a brine of salt and water, using four ounces (114 g) of salt for each quart (1 l) of water. Leave them overnight and then skin them. Put them in fresh brine for three days; submerge them using a plate and a stone. Then drain them, pack them in jars, and fill the jars with cold spiced vinegar. Add a little sugar for a mellower flavor, and store for two months before eating.

PICKLED GREEN TOMATOES Slice the tomatoes and mix them with a few sliced onions. Sprinkle thickly with salt, and let the mixture stand overnight. Then rinse it well in water. Put it into hot sterilized jars and pour boiling spiced vinegar over it. You can pickle peppers in exactly the same way.

PICKLED CUCUMBERS Use a gallon (4.5 l) of spiced vinegar to six quarts (7 l) of cucumbers. The latter should have been salted (see p. 215), removed from the brine and desalted ready for pickling by being soaked in cold water for 12 hours. Boil the vinegar and then add the cucumbers. Boil for two minutes, and then leave them covered for three weeks. If you don't want to eat them immediately, drain the vinegar off and pack the cucumbers in sterilized jars. Pour fresh boiling vinegar over them, seal the jars, and immerse in boiling water for ten minutes. Treated in this way, pickled cucumbers should last indefinitely. To make a sweeter pickle, mix honey or sugar with the last addition of vinegar.

Chutneys

Most fruits and vegetables can be turned into chutney, but the best are tomatoes (green or red), eggplants, peppers, apples, squashes, pumpkins, rutabagas, plums, damsons, pears, grapefruit, oranges, lemons, and any other citrus fruits.

MAKING SPICED VINEGAR

1 Take a piece of cinnamon bark, some cloves, peppercorns, mustard seed, mace, allspice, garlic, and, if you like them, a chili pepper or two.

2 Tie the spices in a muslin bag, and pour in two pints of vinegar over them. Bring to a boil, and boil for a few minutes at most.

MAKING TOMATO CHUTNEY

1 Take two pounds (900 g) of tomatoes, two onions, a cooking apple, some raisins, garlic, brown sugar, salt, spice and half a pint of vinegar.

2 Skin and peel the onions, and peel and core the apple. Then chop them all up finely.

3 Simmer the onion in a small pan with a little water. Add the apple and raisins, and cook the mixture gently until it softens.

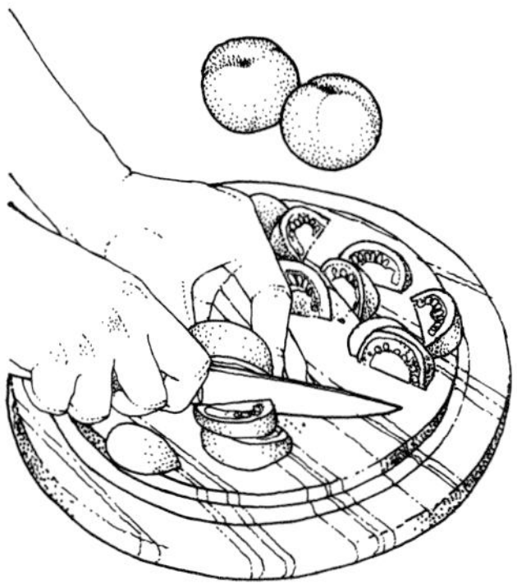

4 Skin the tomatoes (see p. 221), and then chop them up roughly into large chunks.

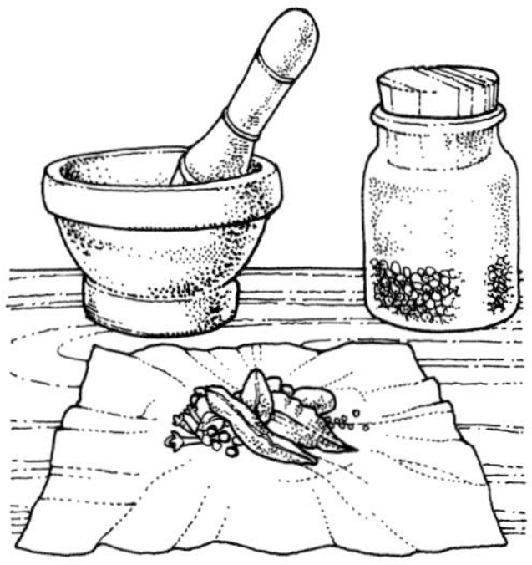

5 Crush the garlic and ginger with some salt in a mortar and pestle. Tie up a mixture of spices—say, some dried chili peppers, bay leaves, and cloves—in a little muslin bag.

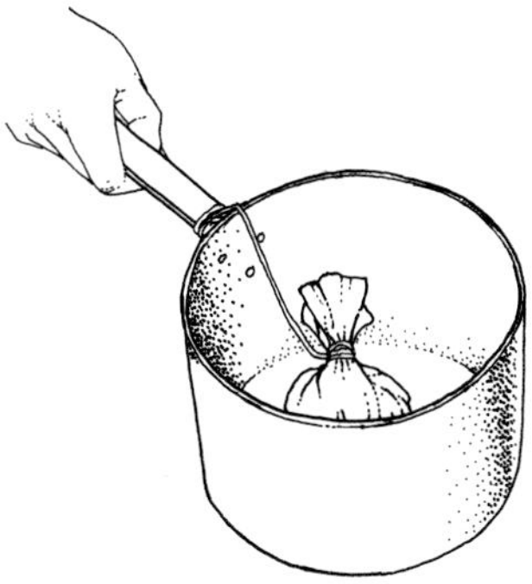

6 Put the muslin bag in a saucepan and tie it to the handle so as not to lose it in the chutney.

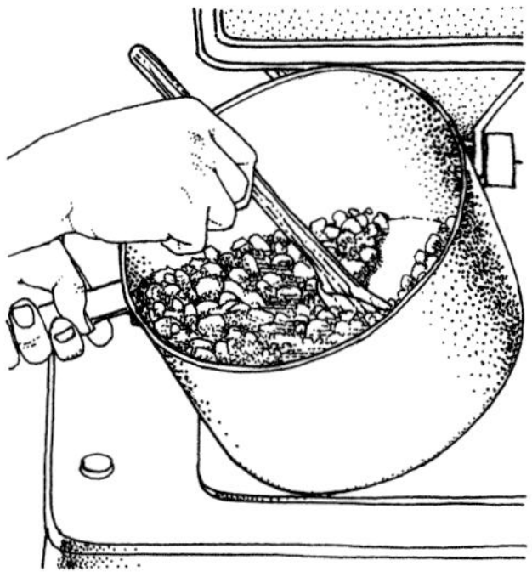

7 Pour in all the ingredients and cook over low heat for an hour, until the mixture thickens so that you can see the bottom of the pan when you draw a spoon through.

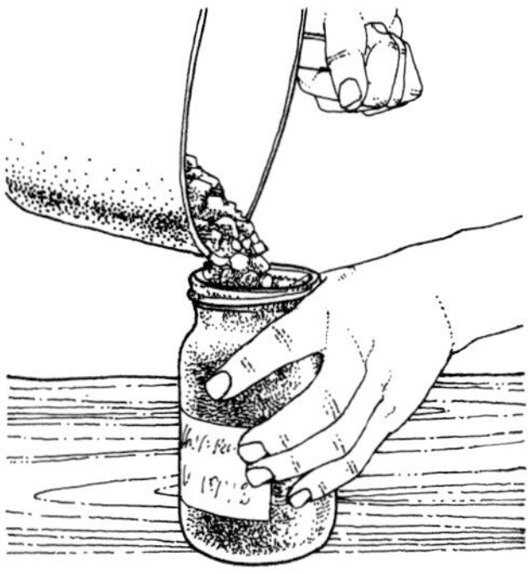

8 Pour the mixture into hot clean jars. Seal the jars to make them airtight, and label and store them.

All good chutneys are heavily spiced. You can use any herb or spice that you like or can get hold of. The following are commonly used: garlic, bay leaves, cayenne, chili, paprika, cumin, horseradish, coriander seed, mustard seed, cinnamon, peppercorns, cloves, ginger, and allspice. Salt and sugar also play an important part; most chutneys go dark as they cook, but if you want a really dark one, use brown sugar or even black molasses.

Never use copper, brass, or iron pans for boiling your chutneys. Unchipped enamel, stainless steel, or, in a pinch, aluminum* will do. You can make chutneys in earthenware crocks placed in the oven; I personally find this method is the best. If you are mixing hard produce like apples or onions with soft things like tomatoes or squash, simmer the hard ingredients first in water until they are soft. Soak any dried fruit you use. Put whole herbs and spices into a muslin bag, which I suggest you tie to the handle of the pan; otherwise, powder them first and then put them straight into the mixture. Before adding garlic or fresh ginger, crush them with a mortar and pestle.

Put all the ingredients into the container. Just cover them with vinegar, and boil very slowly until there is no free liquid left. Be very careful not to let the chutney burn on the bottom of the pan. Stir frequently during the final stage of boiling. Pack into sterilized jars, cover, and put them away. Make sure that the jars are tightly sealed; I recommend using twist-on metal caps from old jam or pickle jars. Then cover the lids with a cloth that has been dipped in melted wax.

CANNING

The principle of canning is very simple. You heat food in sterilized jars, close these hermetically, and then heat them again so as to kill any stray living organism that may have got in. Finally, the jars are allowed to cool. Provided there was no living thing inside the jar when it was sealed, and provided that the heating process was sufficient to kill all bacteria, mold, and viruses, there is no reason why the food should ever rot. Food canned 75 years ago—canning and tinning are identical processes—has been opened and found edible.

Canning most fruit—including plums and some soft fruit—is entirely satisfactory, completely safe, and well worth doing. Tomatoes, which are technically a fruit and not a vegetable, are easy to can and the results taste delicious. Tomatoes alone make canning a worthwhile process. However, I have reservations about the canning of vegetables, as opposed to fruit. Heating by boiling is not sufficient to sterilize them, so you need to use a pressure cooker, which affects the taste. And remember that vegetables can be stored easily and safely in many other ways.

COLD WATER BATH

1 Put the fruit into jars, along with a weak syrup of sugar and water.

2 Put the jars into a container with water. Place on the stove and heat very slowly (see chart, p. 221). Take the temperature with a thermometer.

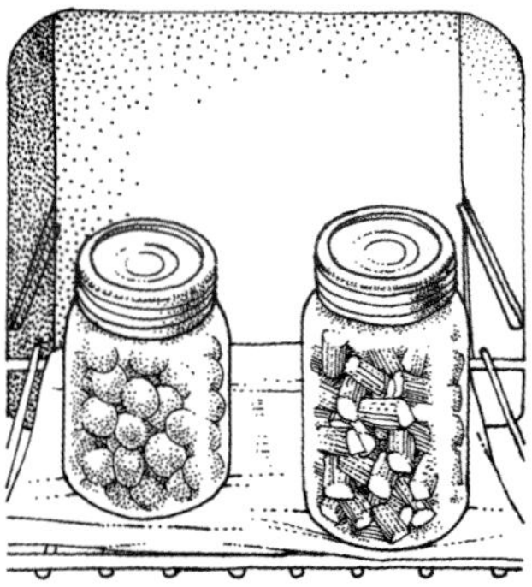

SLOW OVEN METHOD
Can the fruit without liquid. Cover with a loose lid and put into a low oven. Fill with boiling syrup after taking them out.

HOT WATER BATH
Pack your jars with fruit and pour in the boiling syrup. Put lids on loosely and lower into warm water; boil and simmer.

Jars for canning

Many kinds of commercial jars with tops that can be hermetically sealed are made especially for canning food. Most people use the very expensive but extremely good Mason jars, but it is worth considering ordinary screw-top jelly jars too. Some jams and jellies are also sold in suitable screw-top jars. Tops that have rubber rings stuck to the inside of the lid are best and will keep food for several years. Very good screw-top jars are imported to the west from Eastern Europe; these are far cheaper than the more complicated canning jars made in the west, and are perfectly good for canning purposes. Never let metal lids come in contact with the contents of your jars. Buy jars of a size you or your family can empty in a day, or at the most two days, because once you have opened a jar you must eat the contents before the bugs have time to get to work on it.

HEATING JARS You can heat the jars in any large pot filled with water on a fire or stove. Unless your pot has a false bottom, lay a towel, piece of wood, or piece of tin inside it—anything to put the jars on to stop them from touching the bottom. Otherwise they may crack.

Canning methods

There are three well-tried methods of canning fruit (including tomatoes).

COLD WATER BATH METHOD Put the fruit into jars. If you are canning tomatoes, fill the jars with brine. If you are canning other fruit, fill with a weak solution of sugar and water. Put the jars into a container of water on a stove; raise the temperature very slowly so that it takes an hour to reach 130°F (54°C), and then another half hour to reach the temperature given in the chart on p. 221. Use a thermometer to measure the temperature.

SLOW OVEN METHOD Put the fruit into jars, without adding any liquid. Cover each jar with a loose lid or saucer and put the jars into a low oven—about 250°F (120°C). Leave them in the oven for the time given in the chart on p. 221. Take them out and top off each jar with fruit from one of the jars (if there is any fruit left in that jar when all the other jars have been topped off, eat it for supper). Fill the jars with boiling brine (for tomatoes) or syrup (for fruit), screw on the tops, and leave them to cool. Do not delay for more than a few minutes between taking the jars out of the oven and filling them with boiling liquor.

HOT WATER BATH METHOD This method is for people who have neither oven nor thermometer. Again, use brine for tomatoes and other vegetables, and syrup—sugar and water—for fruit. Pack your jars with fruit and pour in boiling syrup or boiling brine—the jars must be hot first or they may crack. Put the lids on loosely—if they are tight the jars

SKINNING TOMATOES

1 Remove the stalks from your tomatoes, and use a knife to score the skins.

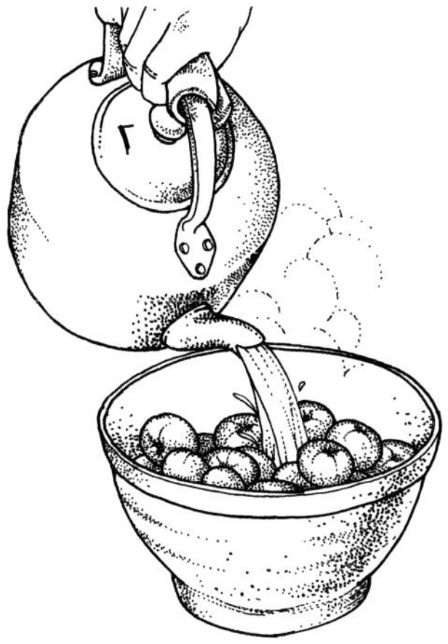

2 Put the tomatoes into a bowl and pour boiling water over them. Leave for a few minutes until the skins have loosened.

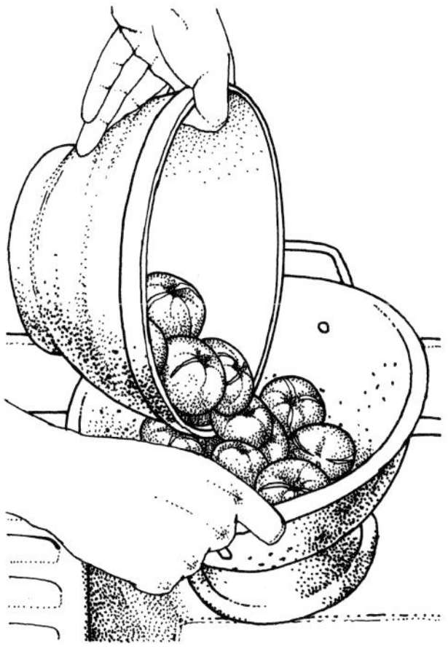

3 Drain the tomatoes; cover them with cold water, but don't leave them for very long or they will get soggy.

4 Using a sharp knife, peel off the skins carefully, so that the tomatoes keep their shape and don't lose any juices.

CANNING TOMATOES

1 Pack skinned tomatoes tightly into jars, using a wooden spoon to push large ones into position.

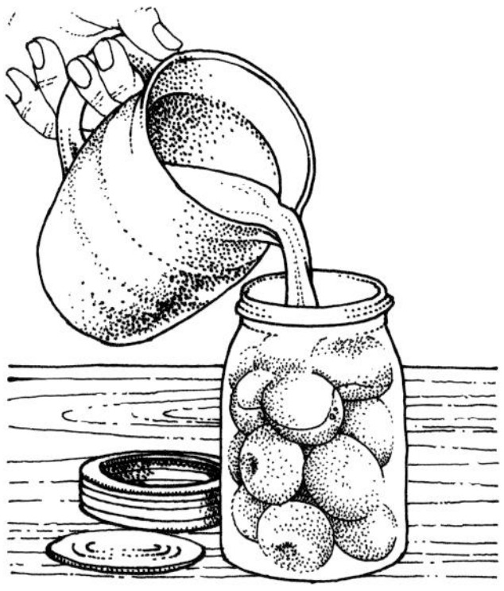

2 Pour brine into the jars beforehand if you are sterilizing in water; pour in afterward if you are using the oven. Put on airtight lids.

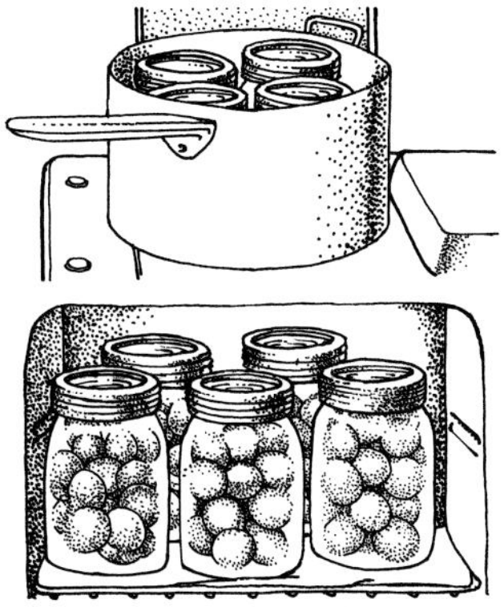

3 Put the jars into a pan of water or stand them on newspaper in the oven. Heat slowly in both cases (see below).

4 When the jars are cool, check that they are vacuum-sealed by lifting them up by the lids alone.

TIMES AND TEMPERATURES FOR CANNING

	COLD WATER BATH		HOT WATER BATH		SLOW OVEN	
Basic method	Take 90 minutes to bring water from cold to the required temperature. Then follow instructions given below.		Start at 100°F (39°C), taking 25–30 minutes to reach required temperature of 190°F (88°C). Follow instructions.		Preheat to 250°F (121°C). Leave jars according to the times given below.	
Liquid in canning jars	Put cold syrup or water in before processing.		Put hot liquid at 140°F (60°C) in before processing. For tomatoes, liquid is optional.		Add boiling liquid at end of processing.	
	TEMPERATURE	TIME	TEMPERATURE	TIME	TEMPERATURE	TIME
Soft fruits including apple slices	165°F (74°C)	10 min	190°F (88°C)	2 min	250°F (121°C)	45–55 min
Stone fruit and citrus fruit	180°F (83°C)	15 min	190°F (88°C)	10 min	Heat oven to 300°F (149°C) and put hot syrup in before processing them.	40–50 min
Tomatoes	190°F (88°C)	30 min	190°F (88°C)	40 min	250°F (121°C)	80–100 min
Purées and tight packs	Allow 5–10 min longer than times shown above and raise temperature a little.					

may explode—and lower them into warm water with the lids just above the surface. Bring to a boil and simmer for the length of time shown in the table on p. 221.

Skinning tomatoes

It is best to skin tomatoes before canning them. Do this by scalding them in boiling water and then plunging them into cold water. Score halfway a round each tomato first and the skins will come off easily.

Tomato juice

If you have plenty of tomatoes, a good way to store them is in the form of tomato juice. Cut tomatoes in halves and put them into a saucepan. Put the pan on very moderate heat until juice begins to flow. Keep pressing down with a pestle, and, as the juice flows, move the pan to a hotter source of heat and boil for half an hour. Then strain the juice through a fine sieve or colander, and return it to the stove. Add salt and pepper to taste (I like a crushed chili pepper as well), and boil for another half hour. Then pour the juice into hot sterilized jars and seal at once. The end result is infinitely better than store-bought tomato juice, which is packed with chemical preservative.

Canning rhubarb

Although rhubarb is not of course a fruit, it can be canned like one. Treat it as advised for "soft fruit" in the table on page 221. It is very good with elderberry juice added to it; this imparts a certain subtlety to the taste.

Opening jars

If you have trouble opening a jar, invert it in boiling water for half a minute and then unscrew. Jabbing a hole in the top renders the jar useless for further service, and the cost of special jars for canning is very high.

MAKING JAMS AND JELLIES

The secret substance that makes jam and jelly-making possible is pectin. This is released when the fruit is first cooked, and is what causes the jam or jelly to set.

There is plenty of pectin in apples, black currants, red currants, gooseberries, and damsons; there is less in plums (including greengages), apricots, peaches, and raspberries; and there is hardly any in blackberries, strawberries, cherries, pears, and rhubarb. When you make jam with fruit from the last category, you will need to add some pectin. To add lemon juice is one answer and it does sharpen the flavor; a more common method is to combine high-pectin fruits, like apples, with low-pectin fruits like blackberries. The third alternative is to go about making a form of fairly concentrated pectin yourself.

MAKING PECTIN

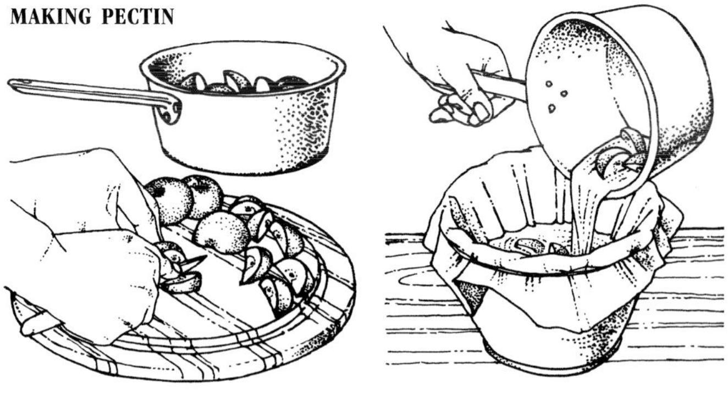

1 Peel and core some apples and cut them up into chunks. Put them into a pan and simmer them gently until they are thoroughly softened.

2 Strain them through a jelly bag and pour the juice that comes through into sterilized hot jars. This is pectin.

PECTIN To make pectin, cut up apples and boil them until they are really soft; strain them through a jelly bag, pour the juice that comes through into sterilized hot jars, and cover the jars. For jam or jelly of any sort, use this pectin in the proportion of a pound (500 g) of fruit, 12 ounces (300 g) of sugar, and half a pint (300 ml) of the pectin or apple juice. The jam or jelly will set.

Jams

Jam-making simply involves cooking fruit with sugar. After you have softened the fruit by slow cooking so that the pectin is released, you add sugar and boil the fruit rapidly. Weigh the fruit first, so that you know how much sugar to add. Use preserving sugar if you can because this dissolves fastest. Boil the fruit until setting point is reached. The setting point is critical. You can be sure when it has been reached by doing a simple test. After the mixture has boiled merrily for a while, take a little out in a wooden spoon and put it on a cold plate. If it is at setting point, a skin will form over the jam that will crinkle if you push it with your finger. The jam is now ready. If a skin does not form, allow the mixture to boil some more and try again.

STRAWBERRY JAM Take ten pounds (4.5 kg) of strawberries, eight pounds (3.5 kg) of sugar, and the juice of four lemons. Put the fruit and the lemon juice in the preserving pan and heat slowly, stirring gently. Add the sugar and boil until setting point is reached. Then immediately take the pot

TESTING FOR SET

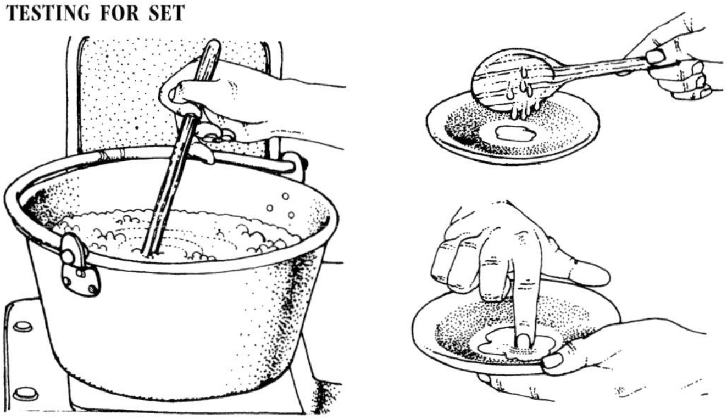

1 Boil the fruit for some time in a large container on the stove, until setting point has been reached. You can tell when this is by doing a simple test.

2 Put a little of the mixture into a wooden spoon and place it on a cold plate. If the jam has reached setting point, a skin will form over it; it will crinkle when you push it with your finger.

off the flame, skim off the scum, stir the jam once, and pour it into hot sterilized jars. Cover the jars and store them.

RHUBARB JAM This is much better than it sounds, particularly if you mix ginger with it. Take two pounds (900 g) of rhubarb, two pounds (900 g) of sugar, two lemons, and one ounce (28 g) of bruised ginger. Cut the rhubarb up small, put it in a bowl with sugar and lemon juice, stir it well and allow the juices to be drawn out. Then pour it into a saucepan, together with the ginger tied up in a muslin bag; boil until setting point is reached, and store the jam in sterilized jars.

GOOSEBERRY JAM Use slightly underripe gooseberries and boil them alone in a little water until they are soft; then add an equal weight of sugar and boil until setting point is reached.

Jellies

Jellies are made in the same way as jams, except that they are passed through muslin or a fine strainer, so that the solids are left behind. It is necessary to boil hard fruit for quite a long time to make jelly; soft fruit like raspberries or strawberries do not need quite so long. Use plenty of water for hard fruit and just enough to prevent burning for soft. Begin by boiling slowly to release the pectin. Add lemon juice if you are using low-pectin fruit. Next comes the part that makes it jelly: you strain the juice out of the fruit through a muslin jelly bag or a fine-meshed strainer. It is the juice that makes the jelly.

With gooseberries, currants, blackberries, and raspberries, you can often get two extractions of jelly-juice. After the first straining, put the pulp back in the saucepan with just enough water to make it a sloppy mess, boil it, and strain it again. In this way you will get more jelly. Your chickens will enjoy the residual pulp.

Now you have your fruit juice. Measure its volume and weigh out a pound (500 g) of sugar for every pint (600 ml) of juice. Bring the juice to a boil, pour in the sugar, and stir the mixture. Boil until setting point is reached, exactly as for jam. Then skim the brew, pour the hot jelly into hot sterilized jars, and cover.

BLACK CURRANT JELLY Add half a pint (300 ml) of water to every pound (500 g) of black currants. Boil and strain the fruit, then add a pound (500 g) of sugar to every pint (600 ml) of juice. Boil the mixture until setting point is reached.

BLACKBERRY JELLY Add the juice of a lemon to every pound (500 g) of blackberries, and follow the recipe for black currant jelly.

MAKING WINE

Making and consuming wine are two of the great joys of the self-sufficient way of life. And they are also extremely efficient methods of storing and ingesting goodness. I strongly urge you to try wine-making if you haven't already, for wine can be made from almost any vegetable or fruit that you have in surplus. However, there is no denying that grapes make the very best wine, and that certain vegetables and fruits—parsnips and rhubarb in particular—make a much more wholesome brew than others that I won't bother to mention.

Wine is made with sugar, in the form of fructose, which comes from fruit, or sucrose, which comes from sugar cane or sugar beets. This sugar is turned by yeast into carbon dioxide and alcohol. The carbon dioxide either escapes into the air or is trapped in the bottle, where it gives the fizz to champagne and other sparkling wines.

Really ripe grapes will make wine by themselves without the addition of extra sugar or yeast. In fact, pure grape juice, with nothing added to it at all, will of its own accord turn into wine. There is enough suitable yeast in the bloom and fructose in the flesh of ripe grapes. The yeast simply turns the sugar into alcohol.

But most of our "garden," or "country," wines are made from vegetables or fruits that do not contain sufficient sugar or suitable yeast. Thus sugar and yeast must be added.

Country wine

Many country wines are nothing more than cane or beet sugar dissolved in water, flavored with

MAKING PARSNIP WINE

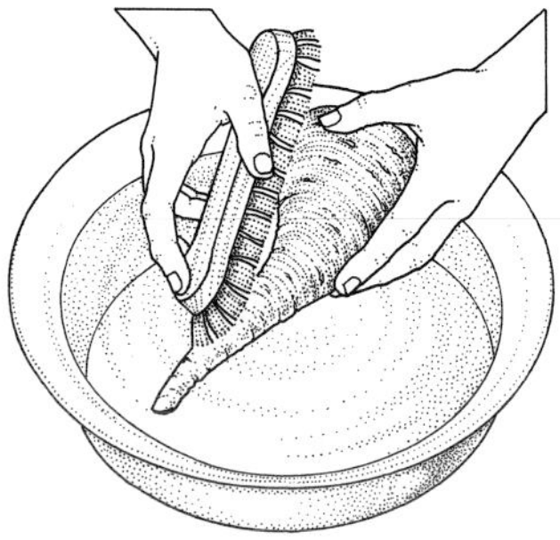

1 To make four gallons (18 l) of wine, you need 20 pounds (9 kg) of parsnips. Scrub them well, but do not peel them.

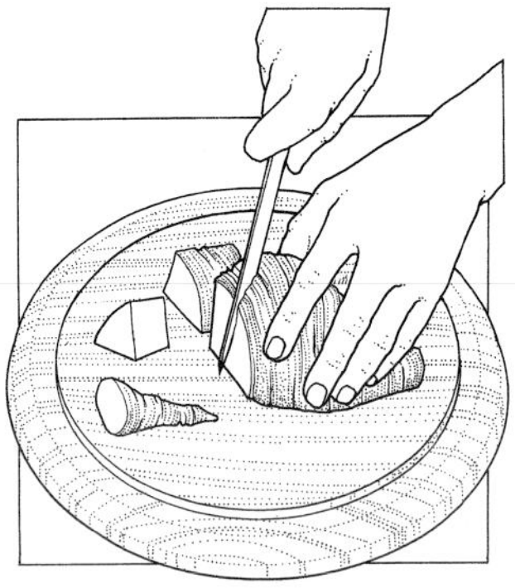

2 Chop the parsnips into cubes about two inches (5 cm) across, and put them into four gallons (18 l) of boiling water.

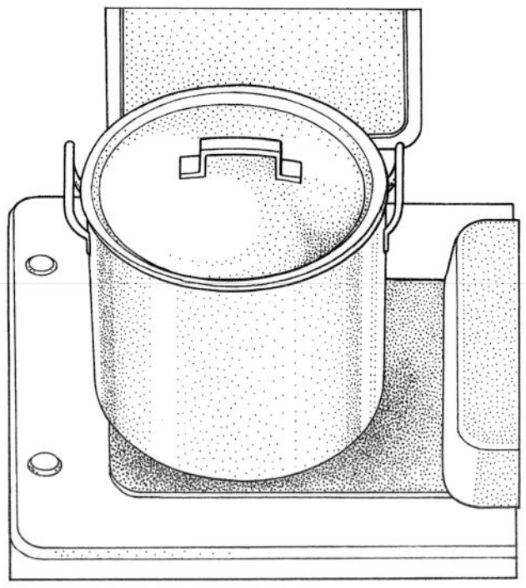

3 Boil the parsnips until a fork will penetrate them easily. Use a jug to scoop out half a pint (300 ml) of boiling liquor. Keep this to one side. Later you will add yeast to it, and it will be your "starter."

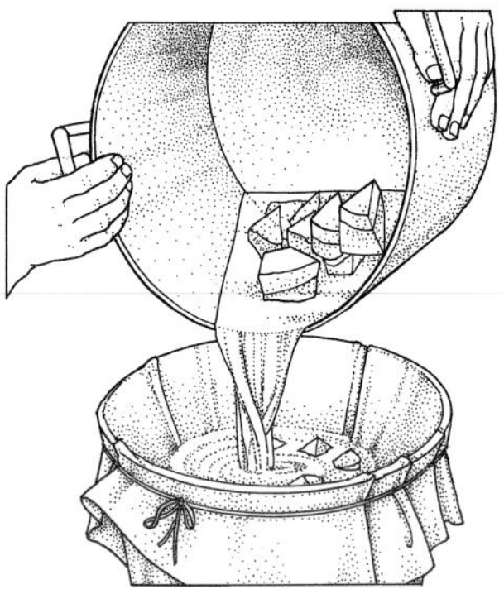

4 While the bulk of the liquor is still warm, strain it into another vessel through muslin or a fine strainer.

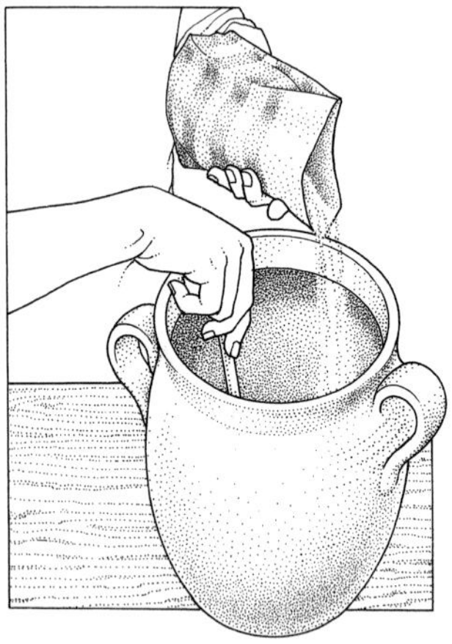

5 Stir in 12 pounds (5.5 kg) of sugar and two teaspoons of lemon juice or citric acid.

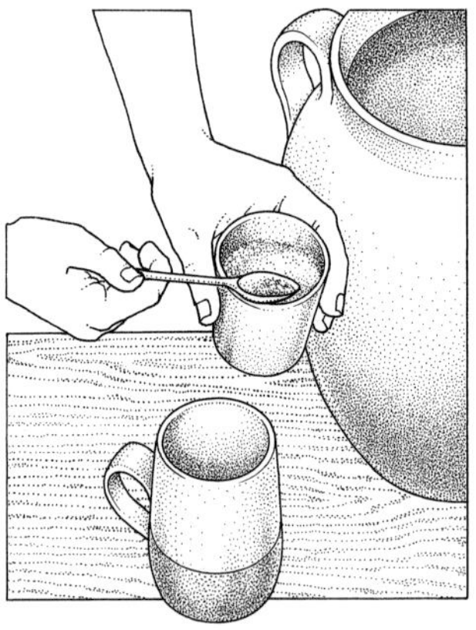

6 Stir two tablespoons of sugar into your half-pint jug. Cool this mixture to blood heat by standing the jug in cold water. Add yeast and cover with a cloth.

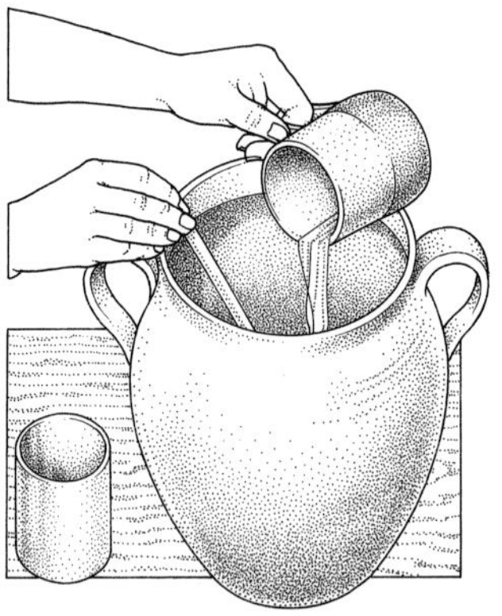

7 When the bulk of your liquor has cooled to blood heat, add the starter, which should be frothing well. Stir the mixture and cover it with a cloth.

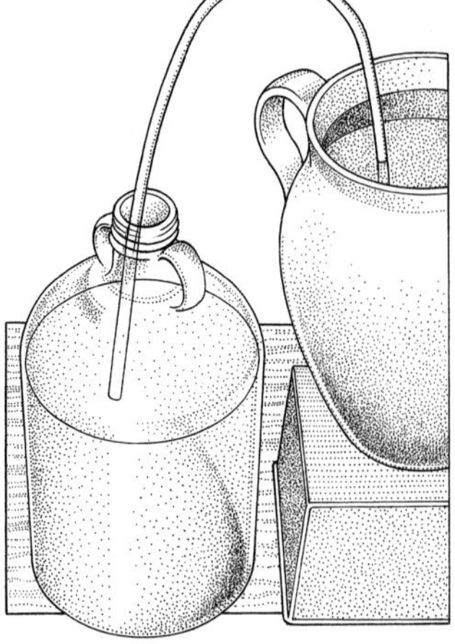

8 The next day move the vessel to a warm place and skim off the scum. Siphon the liquor into narrow-necked containers through a plastic tube.

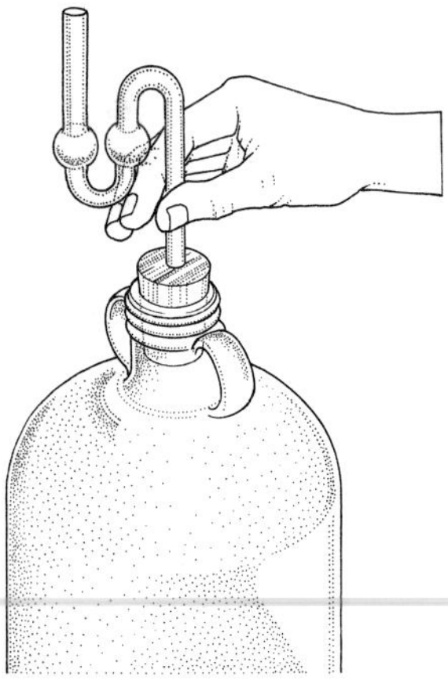

9 Seal the containers with fermentation locks, or cotton balls. When the bubbles cease to rise, rack the liquor into bottles, leaving the sediment behind.

CORKING WINE BOTTLES

1 Cork wine bottles as soon as you have filled them. Put your corks in boiling water and take them out as you need them.

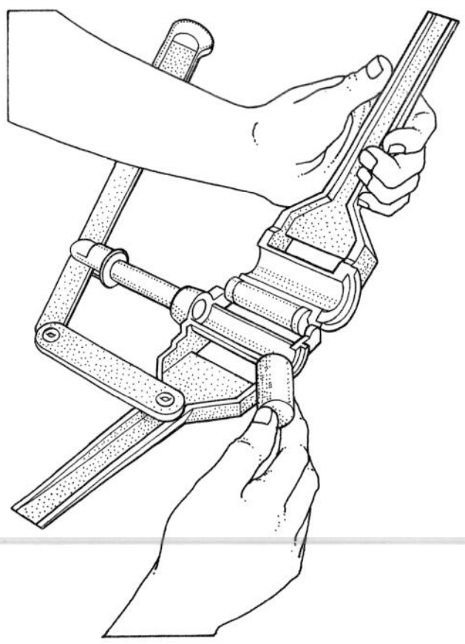

2 You can drive a cork in with a wooden mallet, but it is easier to use a corking tool. Open it up and put in a cork.

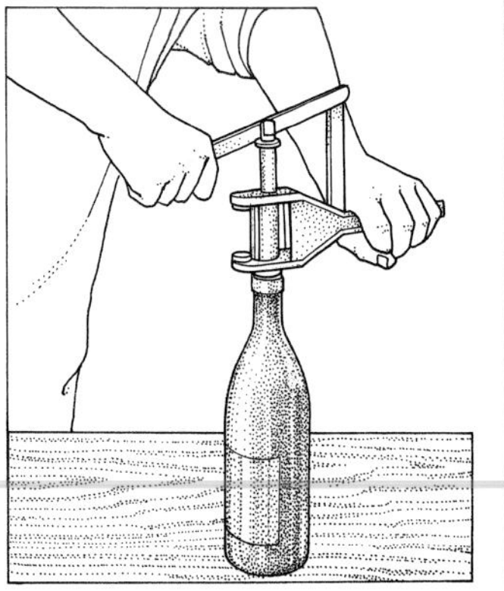

3 Rest the loaded tool on the neck of a bottle. Push down firmly on the lever to drive the cork in.

the fruit or vegetable in question, and fermented with yeast. But the best country wines, and in my view the only ones that are worthwhile, are those in which the garden produce itself makes the wine, apart from the addition of a relatively small amount of sugar to assist the sugar contained in the vegetables or fruit. The good country wines, in order, as I see it, of quality and usefulness, are the following: parsnip, rhubarb, black currant, raspberry, blackberry, strawberry, gooseberry, apple, pear, cherry, plum, damson, and carrot.

Once you have learned to make one fruit wine, you can make them all, and once you have learned to make one root wine, you can make all of them too. It is all common sense once you have learned the principles. To explain the principles I have written detailed directions for making two country wines: parsnip and black currant. As long as you adjust the amounts of water, you can substitute another root vegetable for parsnips, or another fruit for black currants.

PARSNIP WINE Of root wines, parsnip is by far the best. Why it should be that this rather earthy root should make such fine wine I cannot say: I just know that it does. You need: 20 pounds (9 kg) of parsnips; 12 pounds (5.5 kg) of white sugar, four gallons (18 l) of water; two teaspoons of citric acid or lemon juice; some yeast, ideally white wine yeast, available from home-brewing suppliers.

Scrub, but do not peel, the parsnips. Slice them into cubes two inches (5 cm) across, and boil them in the water until a fork will penetrate easily. If you boil them for too long, you will end up with a mush that will never clear. Strain off the liquor while it is still warm and stir the sugar into it. Pour in the lemon juice or citric acid. Wait until the mixture cools to blood heat and then add your yeast to it.

The best way to add the yeast is as follows. While the liquor is still boiling, scoop out half a pint (300 ml), stir two tablespoons of sugar into it, and cool it quickly by standing the jug in cold water. As soon as it reaches blood heat, add the yeast and cover with a cloth. By the time the bulk of your liquor is cool, your "starter," as this lesser amount is called, will be frothing merrily and can be added to the bulk. Stir it in with a wooden spoon, cover the container with a clean cloth, and leave. The reason for using this starter method is that it gets the yeast working more quickly in the bulk of the wine and there is less chance of alien organisms getting a hold.

The following day, after the first rapid fermentation is over, put the vessel in a warmish place—room temperature. Skim off the scum, pour the wine into narrow-necked containers. This is most easily done by siphoning through a rubber tube. Close the containers either with fermentation locks or with cotton balls. This is to allow the carbon dioxide gas to escape, but to stop harmful organisms from getting in. When all fermentation has ceased, "rack" the wine, which means pour it gently into bottles without disturbing the sediment. Cork the bottles and store.

If you want sparkling wine, put a teaspoonful of sugar and a couple of raisins into each bottle before corking. A secondary fermentation will then start in the bottles and form more gas to make the wine sparkle.

BLACK CURRANT WINE This is the best of the fruit wines, except of course grape wine. You need: twelve pounds (5.5 kg) of black currants; ten pounds (4.5 kg) of sugar; four gallons (18 l) of water; and yeast.

Crush the currants; don't bother to top and tail them and, if some have short stalks attached, leave them on as well. Boil the water and pour it on, cover well, and leave to soak. Stir once or twice daily. After three days, strain into another bowl, and add the sugar and the yeast. Pour the mixture into fermentation jars and leave it in a warm place until the fermentation has stopped. Then move them to a cool cellar or store room for three weeks, rack into another container, and store this in the cold for six months. Then rack into bottles and leave to mature for a year—if you can.

Grape wine

RED WINE Fourteen pounds (6.5 kg) of grapes should yield a gallon (4.5 1) of wine. Crush the fruit—an easy way is to pass it through an old-fashioned mangle set on its side—but don't crush the seeds. Put this, which is called the "must," in a big tub and rake most of the stalks and seeds out—a few don't matter.

Unless your grapes are really ripe, in which case the sugar content will be sufficiently high, measure the specific gravity (S.G.) at this point. It is best to measure it with a hydrometer. If the S.G. is less than 1.075, which is called 75°, add sugar to bring it up to at least 1.075, or even more—1.100 is sensible. To raise the S.G. by 5° you have to add three ounces (85 g) of sugar to a gallon of must.

Raise the temperature to 65°F (19°C) by taking a bucket or two out, heating the contents, and returning it to the bulk. It is a good idea at this stage to add a yeast starter. Stir the whole mass occasionally and constantly break the "cap" of skins and floating debris on the surface and submerge it. Do this several times a day. Keep the vat covered with a blanket or sheet at all times when you are not actually working on it. When the S.G. has dropped to 1.010, draw the wine off from the bottom of the vat. If you want some really fine wine, leaving the rest inferior, keep this wine separate from that which you get from subsequent pressings. Next press the "marc," as the mass of grape tissue left in the vat is called. Wrap it in muslin and press it in a press. If you don't have a press, you can improvise with a car jack. Get every drop of juice you can out of it.

Now put all the wine into a cask or casks for the second stage of fermentation, and make sure that from now on the wine is protected by a fermentation lock; otherwise it will turn to vinegar. Here is a difficulty—you have to keep topping off all vessels so as to keep them absolutely full. Air spells danger. You may find it best to keep some wine in small containers and top off the big ones when necessary.

When the S.G. has fallen to 1.000 (that of water) you can replace the fermentation lock with a solid bung. You will still have to top off at intervals. After a few weeks, rack the wine off its "lees" or sediment and bung it up again. Still top it off occasionally. After three months, rack the wine again. You can now store it as long as you like—six months or a year—before you rack it into bottles.

WHITE WINE Follow the instructions for red wine, but press the grapes as quickly as possible and skim off the grape skins as soon as you can. Do not allow the must to ferment with the grape skins in it. Black grapes make white wine, just as white ones do: it is fermenting with the skins that gives the color to red wine.

Hygiene

Only with perfect hygiene can you make fine wine. To sterilize your vessels, wash them to remove any solid matter, scald with boiling water or heat in a hot oven, and turn upside down to drain and store. Wooden casks are especially difficult to keep clean. Steam them thoroughly by inverting them and allowing a pipe from the spout of a tea kettle to blow steam upward through the open bung. The condensed water will run out of the bung. Then fill them with a solution of half a pound (200 g) of washing soda in 20 gallons (90 l) of boiling water and let them stand for 24 hours. Empty them and rinse them out thoroughly in cold water.

MAKING CIDER

If you crush apples, put the juice into a vessel and leave the vessel covered against unwanted organisms, the juice will eventually turn into cider. But the cider will be terribly sharp, "rough" enough to make your hair stand on end, and only a hardened drinker will be able to tolerate it. To make cider that will be more palatable to your neighbors, you will need to add sugar. If they are impatient and you want to speed up the fermentation process, you will need to add yeast. Wine yeast works faster than wild yeasts.

It is not always possible to estimate how much cider you will get from a given number of apples because apples vary considerably in their juice content. But as a rough guide you should get a gallon (4.5 l) of cider from 10 to 14 lb (4.5–6.3 kg) of apples. The best cider is made from a combination of very sweet and very sour apples. This means that the mixture is rich in sugar and acid. If crab apples are added, the mixture will also be rich in tannin, and this improves it.

Don't hurry to pick your apples. Wait until they are really ripe—ideally, pick them ripe and then leave them to soften in piles for two or three days. You can add windfalls and bruised or damaged apples to your pile as well—they don't seem to affect the quality of the cider at all. Then crush them. For this, a cider mill is ideal, but it is an expensive item to buy. Alternatively, you can use any hard object, such as a mallet, as long as it is not metal. Crushing by hand is very arduous, however; you might try using an old-fashioned horizontal mangle, which I have seen prove just as effective.

When the apples have been reduced to pulp, put the juice into a fermenting vat—a wooden barrel or an earthenware crock will do—and wrap the pulp in coarse cloth or burlap to form "cheeses." Then pile the "cheeses" on top of each other in a press, press two or three times to extract the juice, and pour the juice into your fermenting vat. Add a culture of yeast if you wish.

If you want a sweet cider, rack the fermenting cider off its lees (siphon it off without disturbing the sediment), and for every 10 gallons (45 l) add approximately six pounds (2.7 kg) of sugar. Allow the cider to ferment another week, then rack it again.

If you don't have the space or the means—or the apples—to make cider on this scale, there is a simpler method of making it on a small scale. Cut up your soft ripe apples very small, pulp them if you can, put the pulp in a crock, and cover it with boiling water. Leave the crock covered for ten days, then strain off the liquid, add one pound (500 g) of sugar to each gallon (4.5 l), bottle, and fit airlocks or cotton balls to the bottles. Cork after two weeks. The cider will improve with keeping.

If you want a sparkling cider, start by bottling a small quantity. Half-fill a screw-topped flagon, screw it up, and leave it in a warm place. Six hours later, open the flagon. If the cider has thrown a heavy deposit and if the flagon is filled with gas, the cider is not ready for bottling. Wait until there is no heavy sediment and the cider just gives off a little fizz of gas.

You can make an apple wine from crab apples that is sharper and sweeter than cider. Add about ten pounds (4.5 kg) of sliced crab apples to a gallon (4.5 l) of water, cover, and leave to soak for a week. Then strain and add three pounds (1.4 kg) of sugar to every gallon (4.5 l) of liquor. Leave to ferment for three days, skim off the scum, and rack into another vessel. After about two weeks, when the fermentation has ceased, rack into bottles and cork.

MAKING MEAD

To make mead you need about three pounds (1.4 kg) of honey to a gallon (4.5 l) of water. If you cannot spare this much pure honey, you can make up part of the required amount with comb cappings, bits of broken comb, and other honey oddments. Melt the honey in the water—don't boil it—and ferment. But as honey is deficient in acid and tannin—both necessary for proper fermentation—you will need to add these. The juice of three or four lemons will provide the acid; half a pound (200 g) of crushed crab apples will supply the tannin. You could add tea—I have heard of this but never tasted the result.

When the honey has dissolved, add your yeast starter. Leave it to ferment, then rack and bottle. But be prepared to wait a long time for your mead to be ready. It will take at least six months to ferment: if you can wait longer—two or three years—the mead will be even better.

FREEZING

If you have a large freezer, allow plenty of space for your meat and fish because this cannot easily be stored in any other way. Vegetables can be stored in other ways and should therefore take second place. However, as long as you have ample freezer space—six cubic feet (0.22 cu m) per member of the household is pretty adequate—you may like to freeze the following vegetables, which freeze easily and do not suffer unduly from the process: globe artichokes; asparagus; all kinds of beans; Brussels sprouts; cauliflower; sweet corn; peas; sweet peppers. Pumpkins and tomatoes freeze well if you purée them, and you can freeze tomato juice, if you don't can it (see p. 221).

Unless you have meat to freeze, or intend to buy cheap meat wholesale and store it by freezing, think hard before you buy a freezer. The money you spend on it initially, and on maintaining it, replacing it, and powering it, would buy an awful lot of food. I personally do not think it's worth buying a freezer for vegetables alone. If you do buy one, I would strongly recommend you get a chest freezer, rather than an upright one. Upright freezers lose all their cold air every time you open the door; because cold air is heavier than warm air, it just flops out.

Freezing vegetables

All fruit and vegetables should go to the freezer as soon as they are harvested. To leave them sitting around allows the sugars to start turning into

FREEZING AND THAWING VEGETABLES

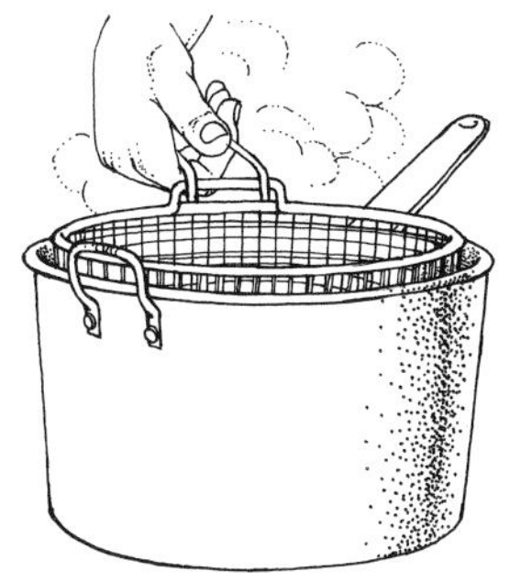

1 Before you freeze vegetables you must blanch them. Use a wire basket to immerse them in boiling water for a short time—between two and four minutes.

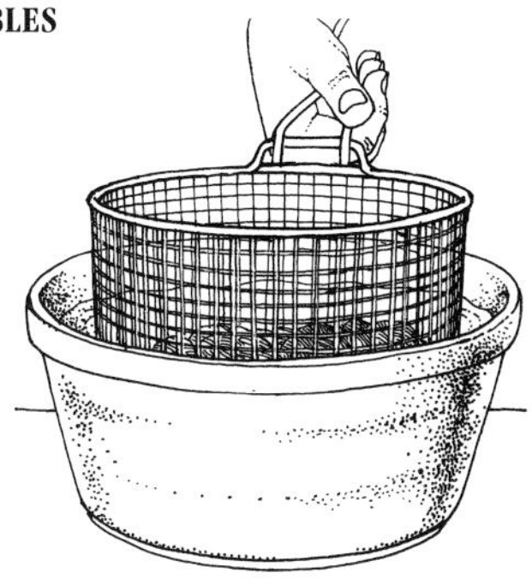

2 Remove the vegetables from the water and immediately plunge them into cold water for the same number of minutes.

3 Let the vegetables drain. Then put them in a plastic bag, and suck the air out through a straw. Put the bag in your freezer.

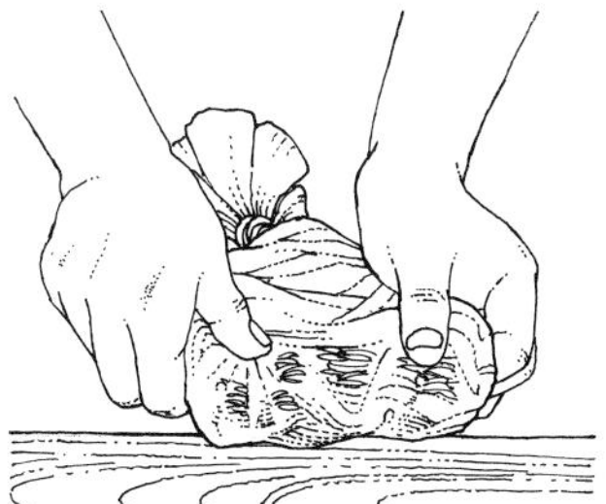

4 To thaw frozen vegetables quickly, bend and flex the bag with both hands. This will break up the ice and separate the vegetables.

FREEZING SOUP

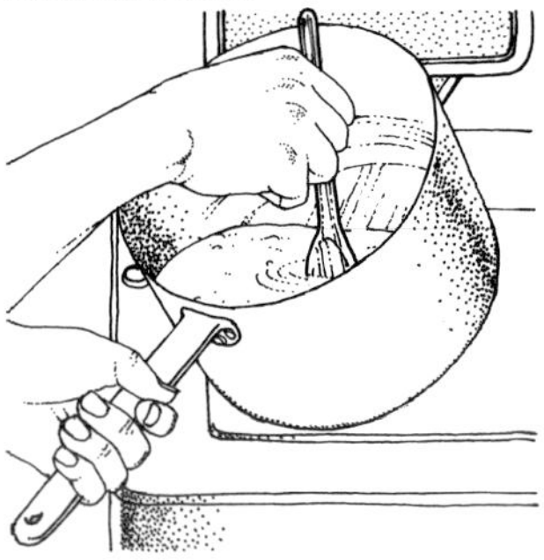

1 To freeze soup, first prepare the soup and allow it to get cold in a cool place; don't put it in the refrigerator.

2 Put a plastic bag inside the saucepan you will use to reheat the soup. Pour the soup into this bag, which should be a special freezer bag.

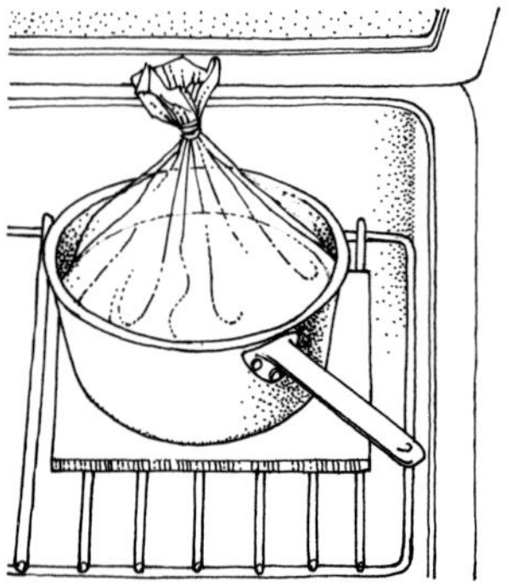

3 Put the saucepan containing the plastic bag and the soup inside the freezer. When the soup has frozen solid, remove the saucepan.

4 When you want to reheat the soup, remove the plastic bag and drop the frozen lump of soup into the saucepan. It will heat up very quickly.

starches and thereby the flavor is lost. If you don't have the time to prepare the food for freezing immediately, put it in the freezer as it is for a few minutes to chill it, but make sure you don't let it get frozen solid.

You should blanch most vegetables before you freeze them to kill any bugs that may be on them. This means plunging them into boiling water. Two or three minutes in the water is adequate for most vegetables, but a big solid thing, like a globe artichoke, should have four minutes. Blanching is easy if you use a wire fryer basket. When you remove the vegetables from the boiling water, plunge them straight into cold water with ice floating around in it. If you can, chill some water in your refrigerator the night before, preferably in large containers like cake pans. Dip the blanched vegetables, in the wire container, straight into the cold water, and pull them out after the same number of minutes that they were in the boiling water. Let them drain thoroughly and pack them in containers. They must be thoroughly dry before packing.

If you pack them in a plastic bag, expel all the air by sucking it through a straw before you put the twist-tie on to stop it from getting in again. It is a good idea to stuff the filled plastic bag into a carton. This gives it a square shape that stows more easily and makes better use of freezer space.

Sweet corn responds very well to freezing, and when it is unfrozen it is still fresh and sweet. It is best to freeze the kernels only; it is a waste of space to freeze whole cobs. Cook the corn on the cob first for ten minutes. Let it cool, and strip the corn off the cob with a knife. Let it drain, then pack it in containers and freeze. When you want to eat it, just simmer in water or milk for two to three minutes.

Freezing soup

When you freeze soups, put the plastic bag containing the soup inside the cooking pot that you intend eventually to reheat it in. The top of the pot must be wider than its diameter lower down. Put the pot into the freezer and, when the soup has frozen, pull it out. Knock the plastic bag with its block of frozen soup out of the pot, and pack the block in the freezer. When you come to heat the soup, just remove the plastic bag, and drop the frozen soup into the pot.

Freezing soft fruit

Soft fruit can be frozen in syrup, or in dry sugar, or just by itself. I always freeze it by itself because I can add what I want when it is unfrozen. You can just put soft fruit into containers and pack them in the freezer without more ado. Some people like to wash soft fruit in ice water before freezing it. If you do this you must dry it thoroughly before putting it in the freezer.

Containers for freezing

Everything you put in your freezer must be wrapped up in a fairly airtight container; otherwise it just dries up. You can just use plastic bags, but there are more sophisticated containers made from waxed cardboard, plastic, glass, and aluminum foil. Coffee cans with plastic tops are good for storing things in freezers, and so are glass jars. Beware of freezing food in any container that curves in toward the top, so that you can't get the food out before it thaws. Food will thaw much more quickly once it is out of its container. If you use plastic bags, make sure they are the heavier freezer bags: thin bags are not really suitable, although you can use them in a pinch. Once filled the container must be sealed tightly.

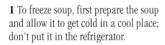

CHAPTER TEN

Miscellany

Containing the keeping of poultry, rabbits, and bees, the laying of paths and drains, the mending of fences, the maintenance of tools, the life of seeds, the variations of climate, and other topics.

Miscellany

CHICKENS

Chickens provide you with a constant supply of good fresh eggs, with chicken manure to activate your compost pile and with occasional table poultry. So chickens may well make the difference between just growing a few vegetables and true self-sufficient gardening. There is a great deal to learn about keeping chickens, and before starting, the beginner will be wise to seek advice from an experienced neighbor.

Anyone who understands organic gardening will appreciate that chickens must be allowed to scratch around outdoors as nature intended; to keep them otherwise is cruel and breaks the cycle of nature that is so beneficial to the garden. I know people who happily let hens run in their gardens most of the time. They do some damage, but also do good because they eat insects. But I personally haven't got the nerve to do this; a hen can scratch up a new seed-bed in half an hour flat.

The Balfour method

If you only have a small garden and do not want to let chickens run loose on it, you can still keep chickens, using the Balfour method—so named after Lady Eve Balfour, who invented it. With this method you will not need a separate compost pile, because the hen run is the compost pile. You use an ordinary hen house—that is, a good, solid, waterproof, draft-proof, well-ventilated wooden house with perches and nest boxes inside it. In front of it or around it you have a scratching pen, which ideally should be sheltered from wind. This becomes your "compost pile." You throw into this

THE BALFOUR METHOD
Build your hen house solidly out of wood, so that it is waterproof and draft-proof, yet well-ventilated. Next to it put a scratching pen, sheltered from wind. Throw in all the vegetable matter you can spare; the hens will turn the pen into a compost pile.

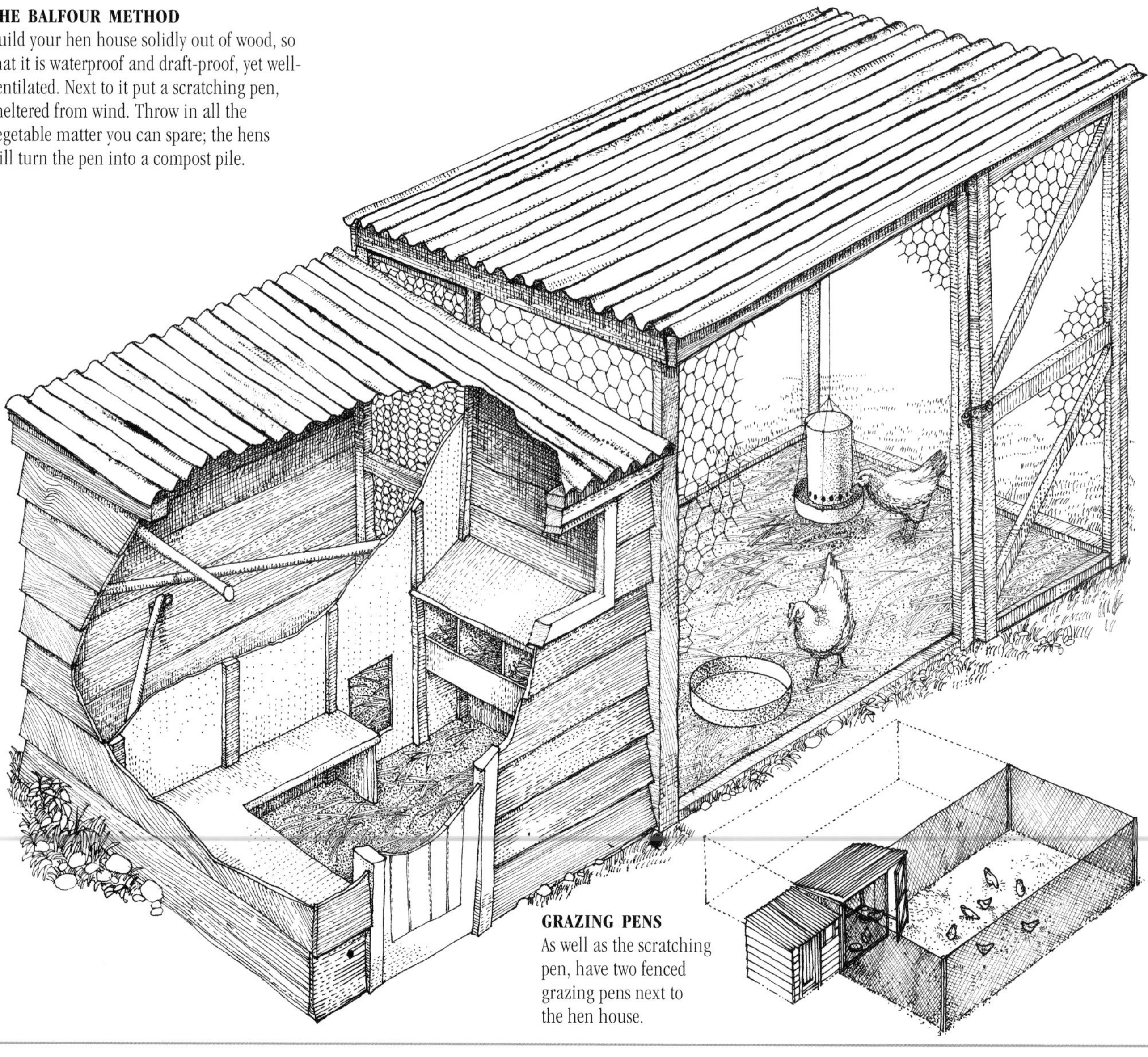

GRAZING PENS
As well as the scratching pen, have two fenced grazing pens next to the hen house.

area all the vegetable matter you can get; the more, the better. All the kitchen scraps, all the waste material from the garden, plenty of straw, bracken, spoiled hay, grass clippings, lawn mowings, everything you can lay your hands on goes into it. Your hens spend hours scrapping around in this material, because worms, earwigs, and other insects abound in it.

Apart from the scratching pen, you should have two grazing pens—or three if you can afford the space. These are just fenced pens, with gates arranged in such a way that the hens can be admitted into one of the pens while being denied access to the other.

The two pens should be put down to a grass, clover, and herb mixture. You allow the hens to run in one pen for two or three weeks, until the grass is eaten right down; then you admit them to the other. Because the hens are doing most of their scratching in the scratching pen, they should not tear up the grazing pens too severely. If you find that they do, you can limit their access to only a few hours a day.

The Balfour method has several advantages. Even though your acreage may be very small, your birds have access to herbage; at the same time the herbage is not lethally damaged by the hens' scratching; the main thing is that the scratching pen provides a quantity of magnificent compost. Every few weeks you empty the scratching pen completely and build it into a proper compost pile. You need add no extra nitrogen to activate it.

A refinement of the Balfour method is to arrange things so that after a year or two you can pull the fences down from the scratching pen and the two grazing pens, and reerect them on the other side of the hen house. Open the pop-hole on that side so the hens can use the fresh ground while you dig up their former pens and bring them into cultivation as part of your garden. You will thus regain the very considerable fertility built up by the hens.

Another possibility is to give your hens access to your soft fruit patch, and to your tree fruit orchard, during the winter. Hens can roam among fruit trees all summer as well, and they do a lot of good by killing many of the insects that would otherwise harm the trees, and by fertilizing the soil around the trees. They do a good job among soft fruit bushes too, except that obviously you cannot leave them there when the fruit is ripe or they will eat it. There is also a danger that they will eat the buds in spring. But certainly in winter it can do nothing but good to let them run among your bushes and trees. You will be sparing your Balfour grazing pens, so that they are more productive when spring comes and the hens have to be moved away from your soft fruit. If hens are run temporarily over any bit of land, they will do it good: and the more changes they get the better.

TRADITIONAL "ARK"
Made of weatherboarding and sawn lumber, this hen house has handles at each end and is easily moved.

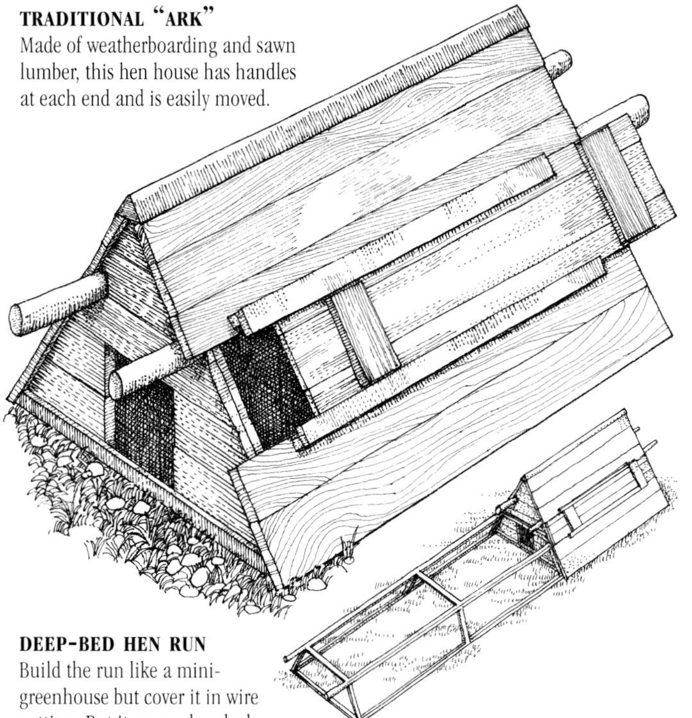

DEEP-BED HEN RUN
Build the run like a mini-greenhouse but cover it in wire netting. Put it over a deep bed.

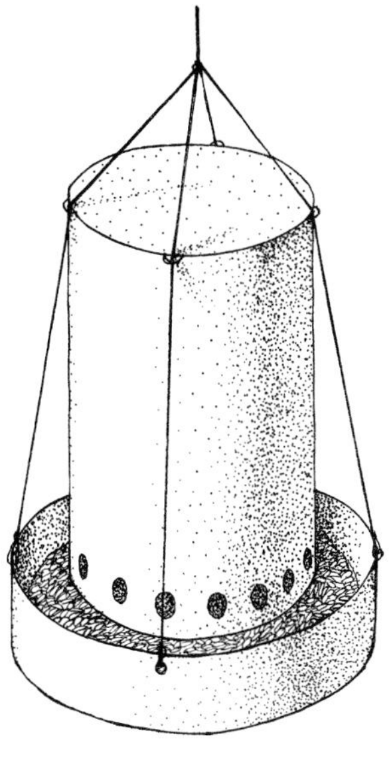

SELF-FEED HOPPER
This hopper, made from an oil drum with holes drilled around the base, is hung up for hens to peck at. The base of a larger drum catches spillage.

WORTHINGTON HOPPER
This trough is placed in the hen house away from birds and out of the reach of rats. The hens open the lid with the weight of their own bodies.

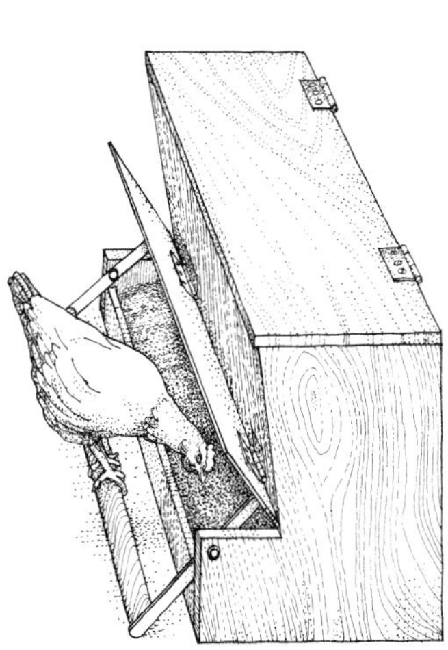

Chicken tractor

Another way of keeping hens in a small garden is the amusingly named "chicken tractor," developed at the Santa Barbara Urban Farm Project in California. This is still at an experimental stage and the exact effects it will have on your soil are not known. But used in the right way, it should considerably improve the fertility of your garden.

The tractor is simply a hen run, with a sleeping shelter attached to one end of it. The shelter should contain nesting boxes and have a floor made of spaced wooden dowels. The run itself can be built to fit exactly over the standard 20-foot-by-five-foot (6-x-1.5-m) deep bed (see p. 106). If you don't use deep beds, then build the run to the dimensions of your vegetable beds. Construct it in exactly the same way as the "mini-greenhouse" described on p. 111; the only difference is that you cover it with wire netting instead of plastic. It need have no floor.

The run can be separated from the night shelter so that it can be moved easily by two people, and then reattached. The night shelter can be carried, even with the birds in it.

Eight hens is the ideal number for a chicken-tractor. You place the run over one deep bed, which should have had rye or some other quick-growing grazing crop sown in it a month or two beforehand, to provide the hens with something to eat. The hens manure the bed, scratch it over deeply and eat all the insects they can find. They also destroy all the weeds. When another bed is ready to receive them, you just move the "tractor" onto it. The old bed will be well-manured and weed-free; dig it up and plant vegetables or fruit.

Feeding chickens

There are several views about feeding chickens: most orthodox methods recommend feeding a dry mash prepared in controlled quantities. I personally favor the Worthington method, developed by Jim Worthington. This is simply that you allow hens ample greenstuff—this means a lot of greenstuff, not just cabbage. You also let them have access to whole grain—wheat is best—and some high-protein food such as fish meal.

Let them eat as much as they want of whatever they want, and you will find that they eat a balanced diet, do not overeat, and lay plenty of eggs. What they eat will, in practice, average out at about four and a half ounces (128 g) of wheat per day each, and well under half an ounce (14 g) of fish meal. For fish meal you can substitute any other high protein food: soy meal; chick peas;

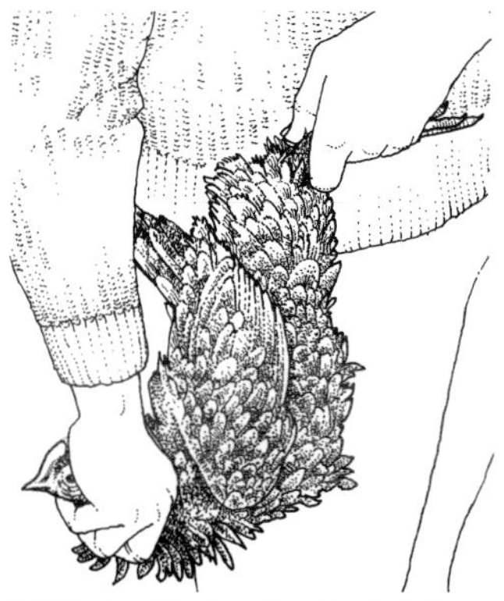

KILLING A CHICKEN

Take the legs in your left hand; hold the neck in your right hand so that it protrudes through your two middle fingers, with the head cupped in your palm. Push your right hand down and turn it so that the head bends back. Stop when you feel the backbone break. If you can, pluck the bird while it is still warm; the skin is far less likely to tear.

PREPARING THE CHICKEN

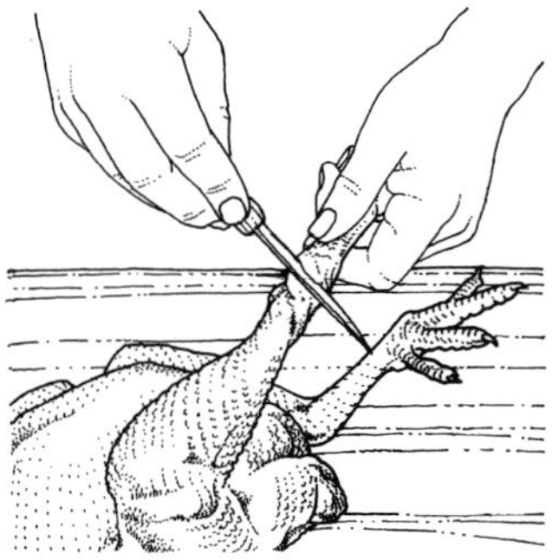

1 After you have plucked the bird, cut around its legs with a sharp knife and proceed to draw out the tendons.

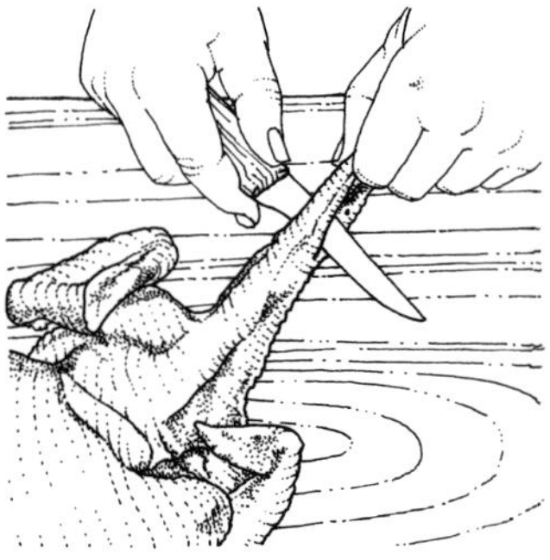

2 Cut the head off, and slide the knife down the bird's neck to leave a flap of skin.

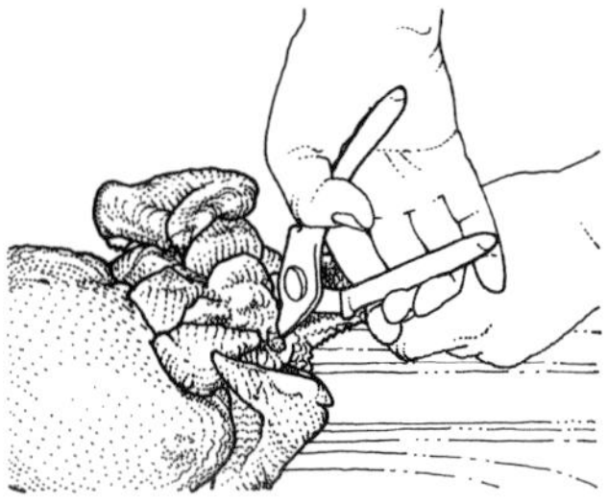

3 Break the neck off farther down than the flap, using a pair of pruners. You can then use it along with the giblets.

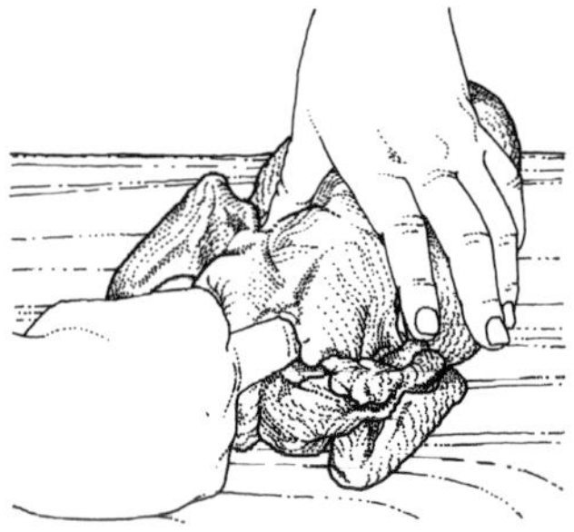

4 Put your finger into the hole left by the neck, and twist it around to break the ligaments that hold the innards.

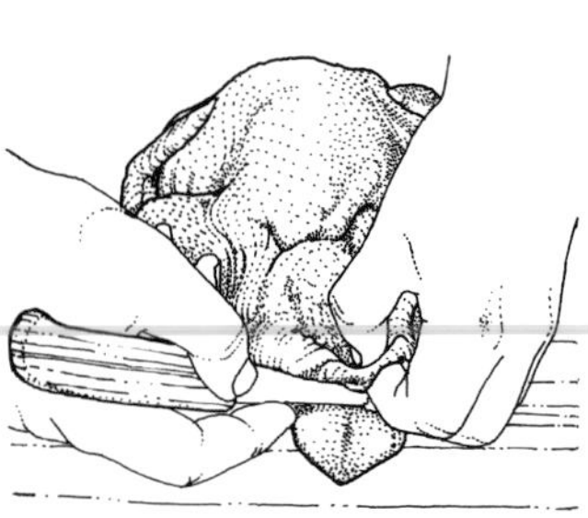

5 Cut the flesh all the way around the anus, taking care not to make a mess of the guts.

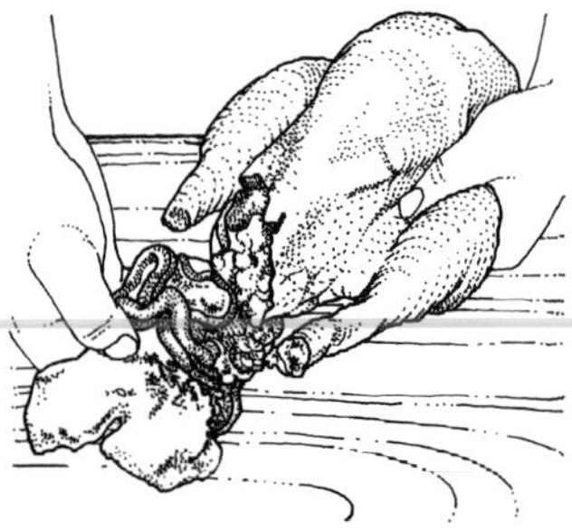

6 Put your hand in and pull out all the innards, including the lungs. Then wash the bird well inside and out.

cracked beans; meat meal or flakes. I knew a man who used to get fish heads and guts from his fishmonger, boil them up, and offer them to his hens. They laid superb eggs. As for the whole grain, use good oats or barley if these are more plentiful than wheat in your area.

Other garden produce that contains protein and is good for hens includes: sunflower seed, especially if you husk and grind them; lupine seed, ground or whole; peanut seed; alfalfa meal; crushed or ground peas and beans. If you grow too many potatoes, it is a kindness on cold winter evenings to boil some up with a little skim milk, or milk gone sour, or water the fish was boiled in, and feed them to the hens before it gets too dark.

If you feed your hens according to the Worthington scheme, you must feed them on this diet from an early age and you must feed them from hoppers. These should be in the hen house away from sparrows, and hung up or set up in such a way that the rats can't get at them. The Worthington hopper is a most excellent device and will save pounds of food over the years. The hens open the lid to the trough by the weight of their own bodies.

All poultry must at all times have access to dust baths, fresh water, and sharp grit—they use grit instead of teeth. Also, lime-rich material, such as crushed seashells, is very good for them.

If you are kind to your birds, they will be kind to you. But kindness should not go so far as keeping on non-laying birds for months. Cull the too fat, the too thin, the sick, lame, and lazy, and they will do you the final favor of making you some excellent soup.

Breeding chickens

You won't get any more eggs if you keep a cockerel among your hens, but your eggs will be fertile. Then, if a hen goes broody, you can let her sit on her eggs and soon you will have some young chickens to increase your flock, to eat, or to sell. Nine times out of ten, if you leave a broody hen alone and at peace to sit on her eggs, she will bring them off and care for the chicks, with no trouble at all.

Feed the chicks the diet recommended for ducks for a few days. If you wish to eat the chicks, feed them freely. It is better to feed them mash (a mixture of grain meal and protein), and a commercial mash is quite adequate. Either kill them at ten weeks as "broilers" weighing two or three pounds, or keep them until they are 14 or 15 weeks old and kill them as "fryers."

DUCKS

If you have a pond, or a stream running through your land, consider keeping ducks. They will provide you with eggs and meat. Duck eggs are delicious as long as they have not been lying in dirty water or mud; and some breeds such as Khaki Campbells lay more eggs than chickens do. It is cruel and unnatural to keep ducks away from water on which they can swim. Pond water must be clean and changed from time to time.

Breeding ducks

If you keep a friendly drake and up to six ducks at the bottom of your garden, the ducks will lay eggs and hatch them. If you manage things well, you can have a constant supply of eggs and meat. However, ducks make very bad mothers. Always confine a mother duck, even if you leave her ducklings free to wander away and come back to her, otherwise she will drag her brood through mud and wet grass, weakening them and even possibly killing them.

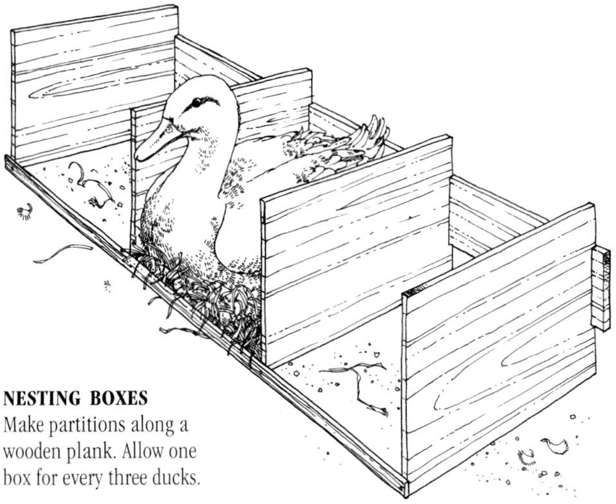

NESTING BOXES
Make partitions along a wooden plank. Allow one box for every three ducks.

Chickens are much better at hatching out duck eggs than ducks are, so it is best to put fertile duck eggs under a broody hen if you have one. Otherwise, treat ducklings in just the same way as you would treat young chicks.

You can also hatch duck eggs in an incubator. In this case, brood the ducklings at 94°F (34°C) for the first week, lower the temperature gradually to 50°F (10°C) the second week, and never let the ducklings get chilled. After two weeks, let them run out, but provide warm shelter, as well as giving them access to shelter from the sun.

Duck eggs take 28 days to hatch. When the ducklings are first hatched, feed them four or five times a day on a rich mash made of grain meal, preferably barley, with milk and a mashed hard-boiled egg in it. After three or four days you can reduce the number of times you feed them, but not the total amount of food. Leave out the egg, and, if you want to, you can give them bought pellets, and a little grain instead of, or as well as, the mash. They will do well on boiled-up kitchen scraps, or boiled potatoes and greens. Ducks are omnivorous, so you can give them meat or fish scraps.

Older birds should be fed exactly the same diet as chickens (see p. 232). But don't let them get too fat or they won't breed. Ducks need plenty of clean drinking water at all times.

You should kill table ducks when they are eight and a half to ten weeks old—no sooner and no later. A duck should have eaten about 20 lb (9 kg) of feed by this time and should weigh about three-and-a-quarter pounds (1.5 kg).

GEESE

You should only keep geese if you have surplus grass, because they are essentially grazing birds. So if you have a big garden or orchard, it will pay you to buy some young geese in the early summer and fatten them for eating in winter and, of course, at Christmas. As well as grass, feed them any surplus lettuces or other greens you happen to have, and about two weeks before you kill them, feed them heavily on barley meal or boiled potatoes.

Breeding geese

If you wish to breed geese, rather than just fatten young geese, you must of course have a gander, although you can get started by buying some goose eggs and putting them under a broody hen. She will hatch the eggs and look after the young goslings. People commonly keep two or three geese per gander. After 20 years of keeping geese, I have come to the conclusion that it is better to have just a pair. Geese are naturally monogamous. If you allow a goose to sit on her eggs in her own good time, helped and guarded by the gander, you will end up by rearing more live geese over the years than if you have one or more importunate auntie geese flapping around trying to sit on the eggs, or laying fresh eggs in a clutch already started.

Another factor to consider is that geese do not take to each other as readily as ducks and chickens do. A pair will probably have to be kept together for a minimum of six weeks before they will mate at all, and it may take two or three years before a goose is really breeding well. But once a pair is successfully established, you can expect a long, productive life from them. There are records—dubious, admittedly—of geese breeding for more than 70 years, but if you estimate an average of about ten years for a goose and five for a gander you will be about right. On the whole, geese make good sitters and good mothers, and ganders are usually attentive to the young as well. Eggs take about 28 days to hatch.

Make sure that a sitting goose gets enough to eat, as she will frequently be unwilling to leave the eggs even to feed. When the goslings are ready to leave the nest, they may be allowed to run out with their mother. If they are given the chance they will start grazing before they are a day old, and will thrive on good grass. For the first three weeks of their lives, feed them well on bread soaked in milk. Goslings grow very fast in the first 12 weeks of life, by which time they may already weigh as much as two-thirds of their eventual adult weight.

PIGEONS

If you have a family of four and you want to eat "squab," young pigeon, once a week, you should keep five pairs of breeding pigeons, because a couple can be expected to hatch out ten a year. When the undersides of the wings are fully feathered they are ready for eating. This should be when they weigh about a pound (500 g) and are four-and-a-half weeks old. Do not let them get much older than this, as they lose weight and also inhibit their parents from breeding. Kill, pluck, and truss them just like chickens.

Housing for pigeons

Pigeons are strictly monogamous; you should allow four square feet (0.37 sq m) of house-space per pair. Thus a pigeon house five feet (1.5 m) by four feet (1.2 m) is ideal for 5 pairs. You need more nesting boxes than you have pairs of pigeons: seven is about right for five pairs. Orange crates make fine nesting boxes. It is important that the pigeon house is rat-proof. If you build one specially, raise it on legs with inverted dishes under them to stop rats from climbing up. The house must also be draft-proof.

Feeding pigeons

Pigeons on free range need a little grain thrown to them every day, as well as some chick peas or other high protein seed. Give them as much as they can clear up in twenty minutes. They will forage for food, but they will probably do very little harm to your vegetables and fruit.

RABBITS

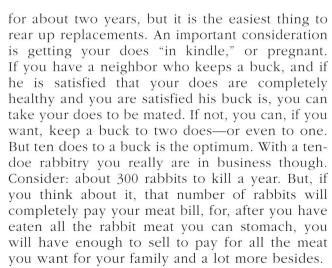

If you aim to be self-sufficient in food and you have a small garden, few things will contribute so much to your ideal as rabbits. In the first place, just one rabbit will provide three cubic feet (0.84 cu m) of droppings a year, which is enough to activate a big compost pile. Together with the soiled litter cleaned from your rabbit run, this equals a fertile garden. Plants and animals evolved to coexist and to support each other's life systems. Rabbits play their part especially well.

The other thing about rabbits is that they will give you excellent meat. Rabbit meat is very nutritious, free of fat, comes in convenient-sized parcels (one rabbit makes a splendid meal for a small family), and rabbits are easy to kill and to process. Any family can have one rabbit a week to eat, simply by keeping two does and rearing their litters for food.

A doe should have from four to five litters a year, each of about six babies. A doe should live for about two years, but it is the easiest thing to rear up replacements. An important consideration is getting your does "in kindle," or pregnant. If you have a neighbor who keeps a buck, and if he is satisfied that your does are completely healthy and you are satisfied his buck is, you can take your does to be mated. If not, you can, if you want, keep a buck to two does—or even to one. But ten does to a buck is the optimum. With a ten-doe rabbitry you really are in business though. Consider: about 300 rabbits to kill a year. But, if you think about it, that number of rabbits will completely pay your meat bill, for, after you have eaten all the rabbit meat you can stomach, you will have enough to sell to pay for all the meat you want for your family and a lot more besides.

Another thing to remember is that, if you do have to buy in food for your rabbits—and you likely will if you have ten does—this food not only feeds the rabbits, it feeds your garden too and in the end feeds you. The same applies to straw or bracken that you bring in for litter. Even if you have to buy the straw, you are not just buying rabbit bedding, you are buying fertility for your garden. Therefore there is every reason for keeping rabbits.

RAISED HUTCHES
This is a space-saving method of keeping rabbits. The hutches are one on top of the other, and raised a few inches off the ground. Keep them outside, up against a wall, with a lean-to roof positioned carefully above them to keep the rain off. The floor of each hutch is made of galvanized wire mesh, and a zinc tray slides in between each hutch to catch droppings.

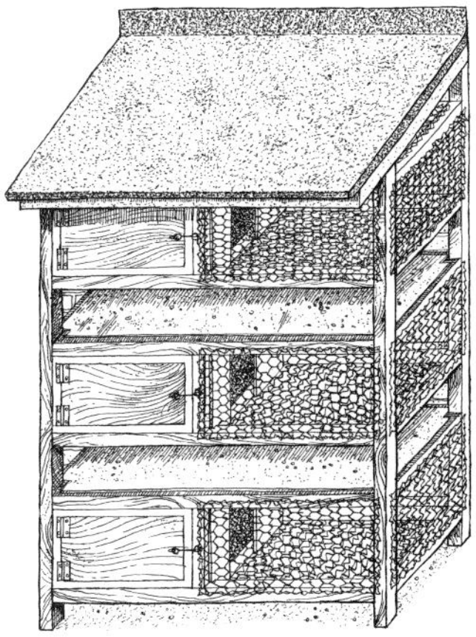

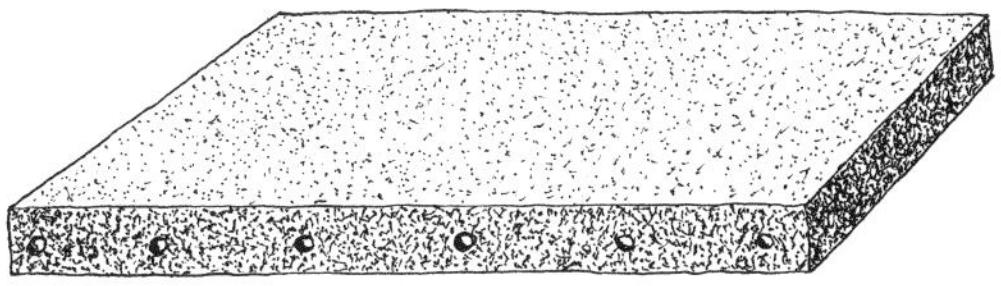

Caring for rabbits

There is work involved in keeping rabbits of course—particularly when you are getting started, but the thing to do is start small—say, with two does and a buck—and build up gradually as you gain experience. Once your rabbitry is established there is not so much work to do: minutes a day rather than hours.

You must consider the various needs and the instincts of rabbits and try to allow them to satisfy them. To keep a rabbit in a wire cage and feed it on nothing but pellets and dry hay is cruel. To allow it access to the ground in summer, so that it can nibble fresh grass and scratch the earth, is kind. And to keep it in a warm dry place in winter

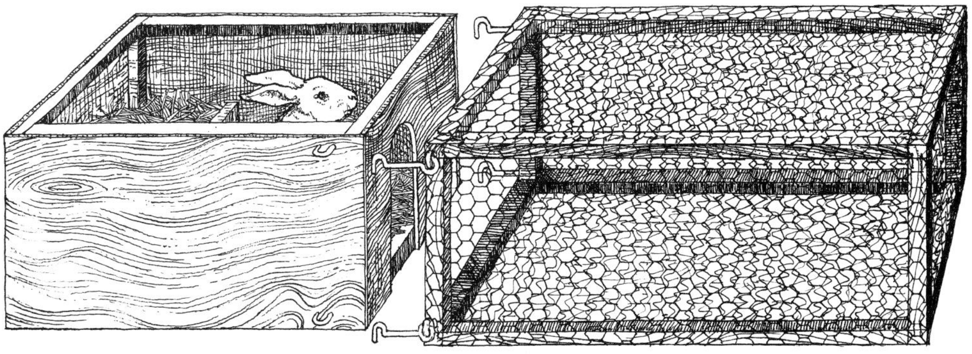

MOVABLE HUTCH
This hutch, or ark, consists of an enclosed area three feet by two feet (90 x 60 cm) and 18 inches (45 cm) high; if it contains a doe, it should have a nesting box inside it. Attached to the hutch should be a wire netting pen at least four feet by two feet (1.2 m x 60 cm).

and feed it on a variety of fresh green things is kind too. You can keep rabbits outdoors in movable hutches all winter as well, but this may not be ideal in cold climates.

Another requirement for rabbits is privacy. Their wild ancestors lived in holes, so you must give them the equivalent of holes to retire into. It is cruel to keep them out in the light all the time, or under the gaze of other animals.

Housing for rabbits

There are two basic forms of housing for rabbits. One, the kindest if it is done properly, is to have movable hutches, or arks, outdoors on grass. These can consist of an enclosed hutch three feet (90 cm) by two feet (60 cm) by 18 inches (45 cm) high and, if it contains a doe, put a nesting box inside it. Attached to this should be a wire netting pen at least four feet (1.2 m) long and two feet (60 cm) wide.

The other form of housing is raised hutches, either against a wall outside or, better still, in a shed. If you have them outside, I strongly recommend you have a simple lean-to roof over the hutches just to keep the rain off both you and the rabbits. Few things can be more unpleasant than cleaning out stinking wet rabbit litter from under stinking wet rabbits. If rabbits are kept dry and warm, they will be healthy and will not stink at all. And remember that rain can come horizontally as well as straight down, so be sure to arrange things so that, whatever the weather, neither litter nor rabbits get wet.

If your hutches are indoors, the alley for you to walk in front of them should be at least three feet six inches (105 cm) wide. The ceiling of an indoor rabbit house should be between eight and ten feet (2.4–3 m) high. If it is too low, it will be stuffy; if it is too high, it will be too cold.

If a raised hutch has wire floors, the galvanized mesh should be 14- to 16-gauge $^3/_4$-inch-by-$^3/_4$-inch (2-x-2-cm) mesh. You should only have a wire mesh floor under the outer eating and dunging pen. The sleeping quarters, which should be private and dark, should have a solid floor. Wood is the best material for building hutches, because it is warm. If you do not use tongue-and-groove boarding, coat the hutches with coal-tar epoxy.

Interior fixtures should be kept simple. A rack for hay is very important—you will waste far less hay this way. A bottle drinker—or else piped water running to automatic drinkers—will save endless time and be better for the rabbits. Rabbits must always have plenty of clean water. This is essential for their health, and don't believe any wiseacre who tells you anything else. If you feed them pellets, a hopper is very useful, or, failing that, some sort of dish which prevents the rabbits from scrapping the pellets out and wasting them.

Mating and breeding

Don't buy old or supposedly "in-kindle" does to start off with: buy young does, let them get used to their new homes, and put them to the buck when they are twenty weeks old. At twelve weeks a doe should have her own home. Handle your breeding stock gently—you wouldn't like to be carried about by the ears—and as often as you can. Get them and yourself used to it. If you are rough, they may become rough too, and scratch and bite you. Be gentle with them and they will be gentle.

Take the doe to the buck. If he mounts her within five minutes, well and good; if he doesn't, remove her and bring her back to him six hours later. After she has mated, carry her back to her own hutch, give her some food, and let her be quiet. When she gives birth, leave her alone in a nice warm nest box with plenty of hay for bedding, but inspect the litter next day and remove any dead, misshapen, or undersized infants—they will not do any good. After about four weeks you can wean the litter—place them in a pen of their own. Mate the doe again immediately. Never allow a buck access to his young children: he may well eat them.

With intensive feeding, meat rabbits are ready for killing when they are ten weeks old. I prefer to feed rabbits less intensively and keep them longer than ten weeks—say, for as long as four months. I let them grow slowly, mostly on greenstuff and hay with a little oats and boiled potatoes. This way they are bigger and better flavored when you come to kill them. The skins are then good for curing; at ten weeks they are almost useless because all the hair comes out. Starve a rabbit, but give it plenty of water, for 12 hours before killing it.

Feeding rabbits

You can feed rabbits entirely on commercial pellets and hay, but this does not come very close to meeting their natural requirements. You can feed them on nothing but greenstuff and hay, but if you do this you must provide a great deal of greenstuff of a wide variety and very good hay, and you take the risk of having small litters and small rabbits.

KILLING A RABBIT
To kill a rabbit, catch the hind legs in your left hand, put your right hand over the back of its neck, and pull suddenly, bending the head up and back as you do so. Death is absolutely painless and instantaneous.

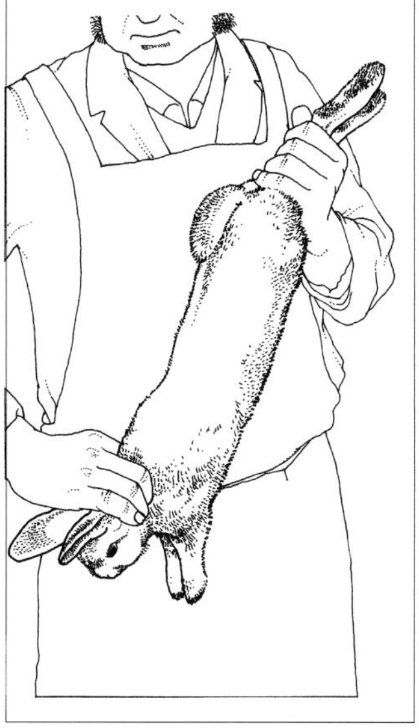

SKINNING A RABBIT

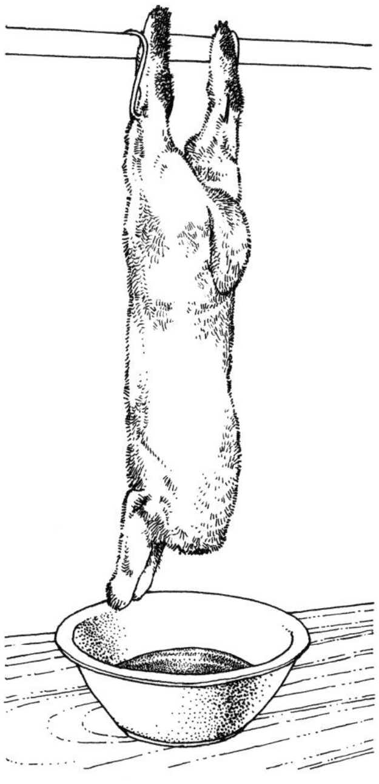

1 Hang the rabbit on two hooks passed through its hamstrings. Cut off its head and drain the blood into a bowl.

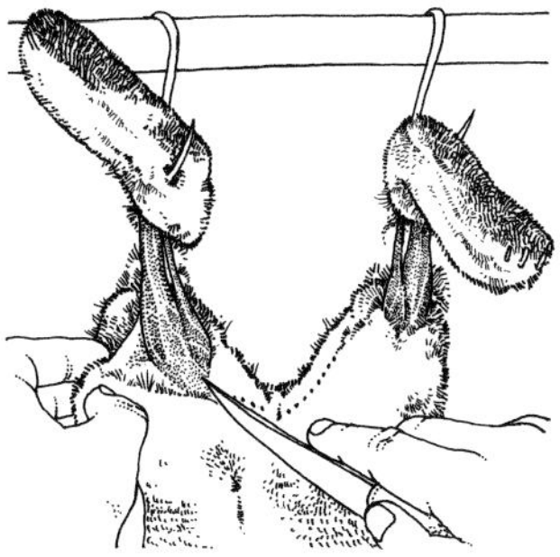

2 Slit the skin around the rabbit's hocks and down the inside of the thighs to the "vent."

3 Hold the rabbit by its legs and "skin-a-rabbit" in the same way as you would pull a sweater off a child.

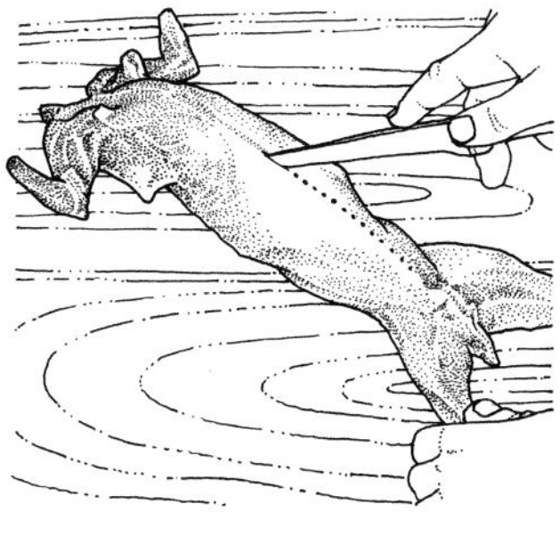

4 Cut off the rabbit's paws and slit it down the belly, taking great care not to puncture the innards.

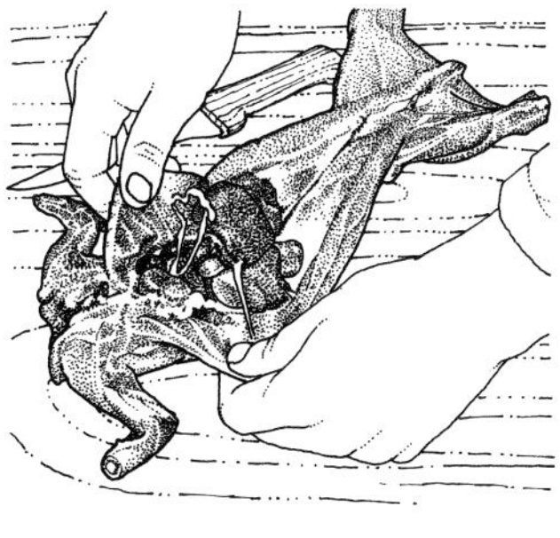

5 Take out the guts, cutting off the gall bladder, the heart, and the kidneys. Keep the liver, and bury the rest deep in a compost pile. Wash the rabbit thoroughly.

The best approach is to strike a happy medium and feed some grain—oats are best—mixed with hay and greenstuff, and, if you can get them, with bran and/or middlings. Rabbits also like potato peelings, boiled potatoes, and all the root crops: turnips, mangolds (after Christmas only), parsnips, carrots, kohlrabi, radishes. Sugar beet is fine for fattening young rabbits, but not for breeding stock; it makes them too fat.

Four ounces (114 g) of pellets, or six ounces (170 g) of oats, per day with as much hay and greenstuff as they want is a good ration for adult rabbits, but this amount must be increased for a doe in kindle to eight ounces (230 g) until the young are weaned, so that she has enough milk.

These rations are all ideal quantities for ideal rabbits. I have kept rabbits on and off for many years, and never weighed anything. I give them as much concentrate—grain or pellets—as they will eat up quickly, and then I give them as much hay and greenstuff as they want. Any food they leave should be cleared up before you give them more. They don't want rotting vegetation.

As for greenstuffs: grass is very good for rabbits, but not lawn-mowings because these ferment too quickly. Greenstuff must always be fresh or else made into hay. Dried nettles are marvelous—high in protein and very good for rabbits. Nearly all garden waste is good for them: the outer leaves of cabbages and the other *brassica*; all root tops except those of potatoes, which are poisonous; edible herbs; raspberry and blackberry leaves (particularly good if they get scours—diarrhea); shepherd's purse; sow thistle; dandelion, but not too much; sheep's parsley; coltsfoot; bindweed; sorrel; daisies; clover; and vetches. Grow kale for feeding to your rabbits in late winter and spring, and they will really thrive.

Silage is good for fattening rabbits for eating, and for feeding to milking does as well. The best way to make it is to cram lawn-clippings into plastic fertilizer bags and seal these so as to exclude all air. Stack the bags upside down on top of each other so the air cannot get in. Leave them until the grass clippings have fermented. This will add to the fertility of your soil once it has been through the rabbits. Don't feed them silage alone, though; give dry hay as well.

Never underestimate the appetites of your rabbits. They really need huge amounts of greenstuff, as much as they can eat. It keeps them fit and happy. Give them twigs to chew on: ash, thorn, apple, and rose prunings are all good for them and keep them amused.

BEES

The wonderful thing about bees is that they make use of food that does not cost anything—food that cannot, without their aid, be used at all.

Bees in small gardens

Many people are afraid to keep bees in small gardens in urban or suburban situations because of the one thing that all worker bees have in common: their sting. However, you and your neighbors will be quite safe as long as you keep them high up. An older solution is to keep them behind a hedge, which forces them to rise before flying away from the hive so that their flight path is above the human head. This works too, but you need room in your garden not just for the hives, but for a hedge. Keeping bees in urban situations actually has an advantage: you do not risk the appalling massacres that are caused by the spraying of crops in the countryside.

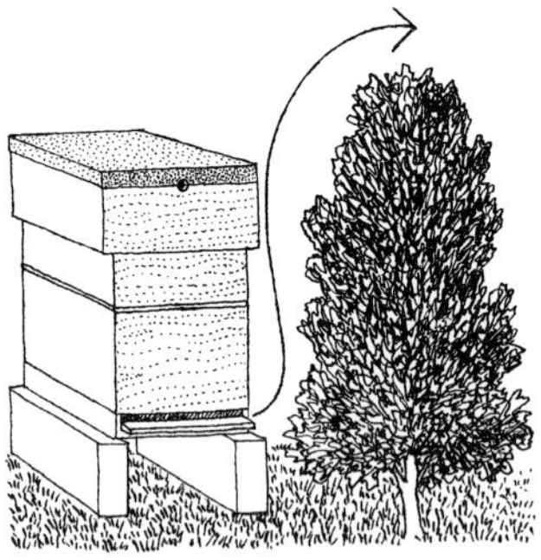

HIVES BEHIND A HEDGE
If you keep your hives behind a hedge, the bees are forced to rise before flying away, so that their flight path is higher than the human head.

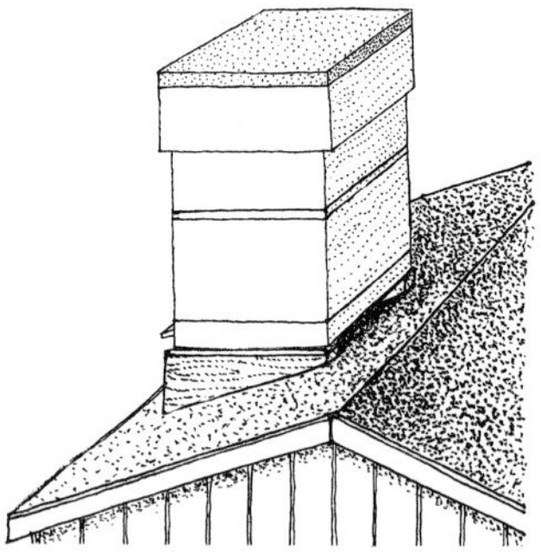

HIVES HIGH UP
Another safe place for hives is high up—on a platform or a roof. This uses up less space than keeping them behind a hedge.

Equipment for keeping bees

HIVES The modern method of keeping bees, which is quite rightly the only legal method in many countries, is to use wooden hives, which contain removable frames on which the honey is made. The frames contain wax sheets, which are printed with the beautiful hexagonal pattern of bee cells. The worker bees draw out the wax to form cells in which they hope their queen will come and lay her eggs. Sadly, the beekeeper fools the worker bees. He interposes a "queen excluder" between the queen, who is down in her royal apartment, called the "brood chamber," and the frames above. The workers fill the cells above the queen excluder with honey, but the queen never lays eggs in them. Therefore, the beekeeper does not kill unborn bees when he takes the honey.

Your beehives should contain the following parts, working from the bottom upward: a base, which is a flat piece of wood supported on legs, or struts; a brood chamber, which is a deep box, with no top or bottom, filled with deep frames; a queen excluder, a flat board that fits over the brood chamber with a hole in it big enough to allow worker bees through but not the queen; some supers, which are like the brood chamber, only shallower, and containing shallower frames; and a lid, which is a box with a top but no bottom.

CLOTHING As well as your hives, you will need various pieces of equipment for your apiary as a whole. Most important of all is protective clothing, and every time you put it on, take great care that you leave no openings through which a bee might be able to crawl.

CLEARING BOARD A clearing board is very useful. It is a flat board containing a "bee-valve," a device that will allow bees to go down but not up.

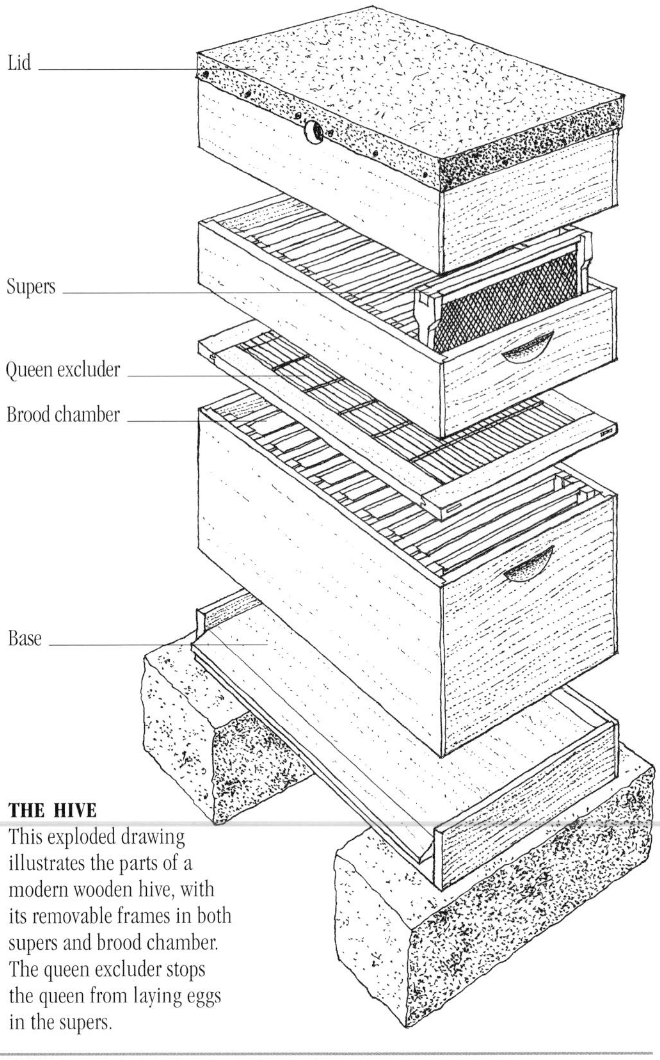

THE HIVE
This exploded drawing illustrates the parts of a modern wooden hive, with its removable frames in both supers and brood chamber. The queen excluder stops the queen from laying eggs in the supers.

You place it between the brood chamber and the supers so as to clear the bees out of the latter, when you want to take the frames out to extract the honey. You can do without a clearing board, in which case you will have to clear the bees out of the supers either by brushing them off with a soft brush, or by banging the supers on the ground. Both these techniques annoy the bees, but if you wear protective clothing you will be safe.

SMOKER A smoker is a container in which you burn corrugated paper, or cloth, to stupefy your bees so that you can work them without getting them too angry. This works because when bees smell smoke they think there is a forest fire; they therefore fill themselves up with honey, ready to move house. When bees are full of honey they cannot sting, but be careful not to give them too much smoke.

EXTRACTOR Once a frame is filled with honey, you can use a honey extractor to get it out. This is a centrifuge that spins the honey out, leaving the combs more or less intact so that they can be replaced in the hive, and used by the bees again. This saves the bees a lot of work and allows them to concentrate on making honey. Before you use an extractor you must decap the combs with a hot knife. Keep the knife hot in boiling water.

FEEDER If you are going to feed your bees in winter—and this will keep your bees healthy and give you more honey—you will need a feeder. This is a simple container made of plastic, wood, or metal, which you can fill with sugar water, and place above the top super just underneath the lid of the hive. The sugar water should be two parts by weight of sugar to one part of water. Boil the mixture and let it cool. Make sure that every colony has at least 35 lb (16 kg) of honey or sugar water to last it the winter. Nothing less than this will keep the bees happy, strong, and ready to make the best of the nectar flow next spring.

APPROACHING THE HIVE
Protective clothing is a vital part of beekeeping. Protect your hands and wrists with gloves and your face with a special bee-veil. Wear light-colored clothing, with your pants tucked into your socks. Use a hive tool or a screwdriver to pry open the hive, from the side or rear if possible. Have a smoker ready-lit, filled with corrugated paper, rags, or any material that will make smoke. The smoke will stupefy the bees; they will fill themselves up with honey, and be unable to sting.

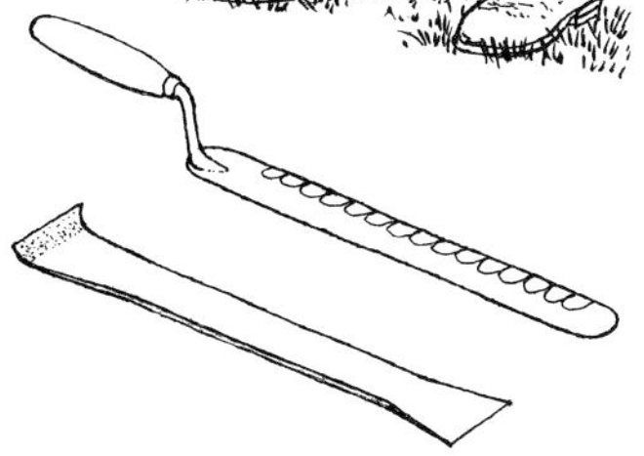

BEEKEEPING TOOLS
You can use a special hive tool to pry open the different parts of the hive, left. Before using an extractor, right, to obtain the honey, you must decap the combs with a hot knife.

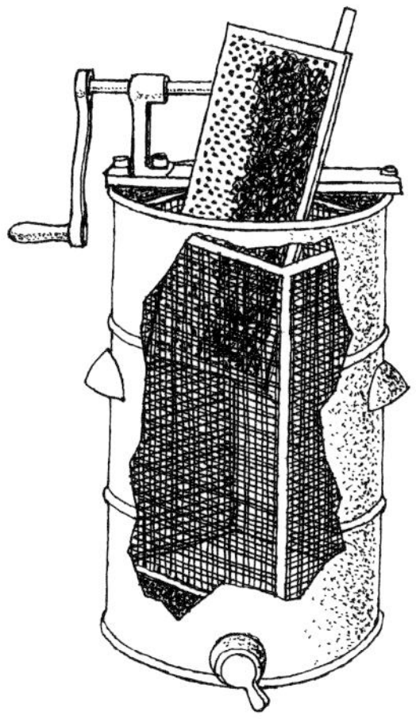

HONEY EXTRACTOR
This device acts as a centrifuge for spinning out honey, leaving the combs intact and reusable.

Establishing a bee colony

To start keeping bees you must either buy a colony in an existing hive, buy a nucleus, or hive a swarm. A nucleus is a queen and a few hundred workers in a box. These you must carefully feed with sugar water until they have established themselves sufficiently to survive unaided. Do not add a super to the nucleus brood chamber until all the frames in the latter are filled with honey.

To hive a swarm, you must first find one. Swarms are the children of bee colonies. They consist of a queen and several thousand workers. Their habit is to hang on a tree branch, like a huge football of bees, and stay there while they send out scouts to find a suitable home. When the scouts return, the whole swarm flies away and enters the new home. If you are lucky enough to find a swarm, hive it by shaking the branch hard, or cutting it off, so that the whole mass of bees falls into a box. Turn the box upside down and leave it until evening, with a stick under it to leave a gap through which the scouts can return to the swarm. Then carry the box to your empty hive. Lay a white sheet on the ground in front of the door to the hive, and shake the swarm out on to the sheet. As bees always tend to crawl upward, they will crawl into the new hive.

A final few words of advice: before you embark on beekeeping, join your local beekeeping society or group, or at least make friends with an experienced beekeeper. Buy, or borrow, a good book on the subject; there is much more to know about bees than I have managed to fit in here.

DRAINAGE

Wet land is, for nearly all land plants, bad land. The plants grow up late in the spring, grow slowly and badly, and fungus diseases flourish. And wet land is sour and acidic. You just have to drain it.

The water table

When water falls onto pervious ground, it sinks in and goes on sinking until it hits an impervious layer. If this layer is sloping, the water may continue downhill underground until it outcrops—that is, comes to the surface somewhere. If it outcrops in your garden, you will have a spring, and a very wet garden indeed. If the country is low-lying anyway, the water will not outcrop, but just stay where it is. The level to which water rises underground is called the water table. If it is only a few inches or even feet below the surface of the soil, you have a drainage problem. If it coincides with the surface, you have a swamp. If it is above the surface, you have a lake.

You also have a drainage problem if the surface of your soil is heavy clay because no water will percolate through it, either up or down. If your soil is sand or gravel, water will percolate easily, but if there is clay underneath, water can accumulate in the sand or gravel and you still have a drainage problem.

Ditches and land drains

You can lower your water table a certain amount by digging ditches around the edges of your land. If the problem is more acute, you can dig land drains, and these can go across the middle of your land—a herringbone pattern is ideal. Dig trenches and lay perforated pipes or open-jointed drains in the bottom. Cover over with gravel or other coarse material, and cover this with soil. Dig the trenches so that they lead the water to a stream or other watercourse.

Soakaways

Your garden may be wet because you have heavy clay topsoil above pervious subsoil. If so, you may well be able to cure the problem by digging a soakaway. Dig a pit in the lowest part of your garden; dig through the topsoil and well into the pervious subsoil. Fill the pit with big stones and dig land drains leading to the pit.

A soakaway can also be used to deal with damp soil caused by an impervious layer below the topsoil. Simply dig through the impervious layer when digging your soakaway. Lay land drains through the pervious topsoil, so that they catch water before it reaches the impervious layer.

If you can't dig through the impervious layer, or you can't lower the water table by any of the

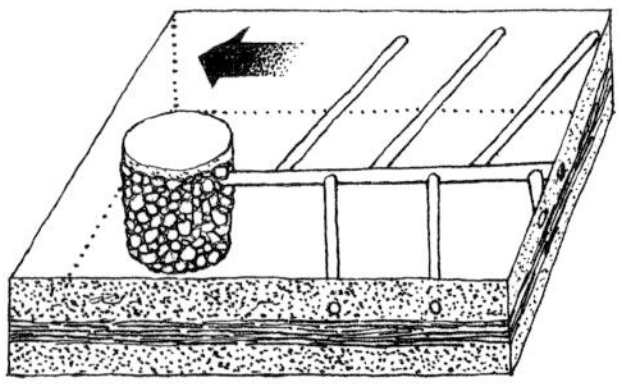

LAND DRAINS AND SOAKAWAY
If your garden has heavy clay topsoil on top of a pervious subsoil, you should be able to drain it by installing a soakaway. This is a hole that you need to dig deep enough—say, about four feet (1.2 m)—to carry the water down into the subsoil. Fill the pit with porous material, such as gravel or broken bricks. Dig land drains that run down to the soakaway at a gentle slope. Because the ground is porous at the bottom of the soakaway, the water will drain away. In a large garden, dig land drains in a herringbone—that is, with branch drains running into the main drain leading to the soakaway.

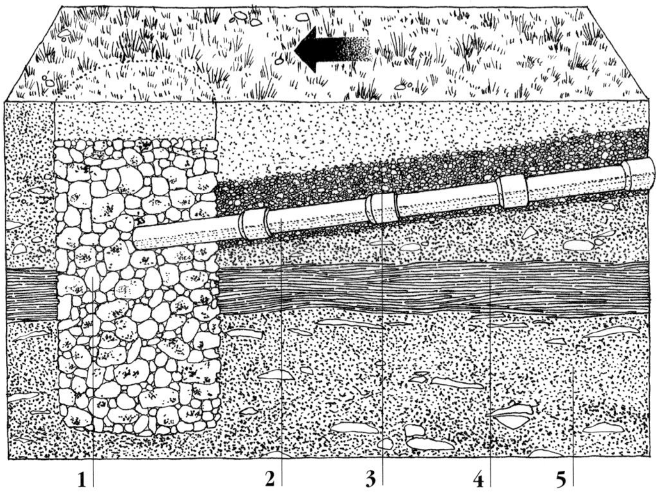

THE SOAKAWAY PIT
The cross-section above shows the soakaway **1** with land drains **2** running down to it. The drains are packed with small stones **3**. The layer of impervious material **4** is bypassed by the soakaway, so the water can reach the porous layer **5**.

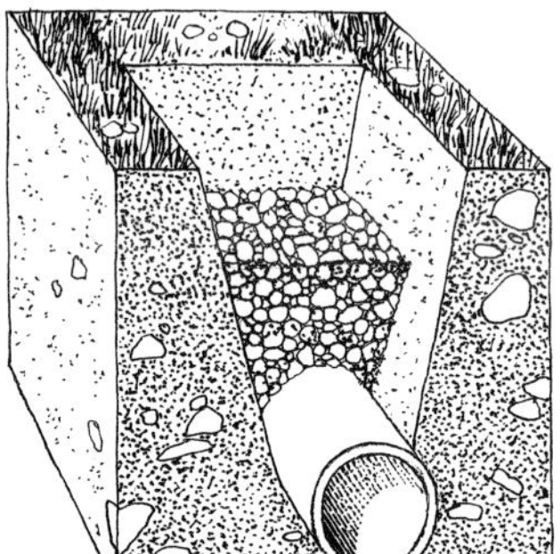

DRAIN CROSS-SECTION
When you lay a drain, cover it with small stones or gravel before you replace the soil.

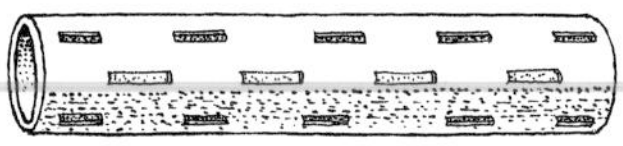

PERFORATED PIPES
This type of pipe is plastic and comes in long lengths; the perforations allow water to enter the drain all along the line.

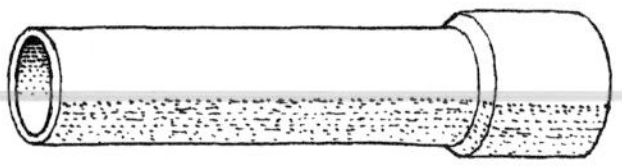

OPEN-JOINTED PIPES
These are short pipes laid end to end; they are not cemented together, so water can seep through the joints and flow down to the soakaway.

methods outlined above, you may find that you need a drain not just across your own yard, but across one, or even several, of your neighbors' yards. Such a ditch will benefit everyone, so it is worth getting together with your neighbors and digging it as a joint effort.

TERRACES

The terracing of steep hillsides has gone on since antiquity, and in many parts of the world it is the only way of farming or gardening on a permanent basis. Cultivation of unterraced hillsides inevitably leads to soil erosion and eventually a complete loss of soil cover. A properly terraced hillside will last forever.

If you have a steep, unproductive slope, it is well worth terracing it. If you acquire land with terraces on it, it is likely to be very productive land. This may be to do with the fact that the initial labor involved in terracing is very high, the cost of terraced land is therefore correspondingly high, and so terraced land is cherished and lovingly cared for.

Constructing a terrace

Your retaining walls can be of stone, brick, concrete blocks or, on not too steep a slope, sod. Masonry retaining walls—and remember they have to be very high—are too expensive nowadays for most people to contemplate. However, you can build sod terrace walls at minimum expense, but at a high cost of labor.

Mark out the width of the proposed terrace and dig a level foundation for the retaining wall at the base of it right along that contour. Peel the sod off the side of the hill to the width of the intended bed. Use the strips of sod to build a wall on your level foundation and give the wall a slight batter in toward the terrace. The wall needs to be half as high as the vertical height of the stretch of slope you are terracing. Level the terrace by throwing the soil from the uphill half of the proposed terrace down to the downhill half that stretches to your wall. The sod that forms the wall will put out fresh grass on its exposed vertical side and have a new lease on life. It will stand up as well as a stone wall.

Whatever material you use for your walls, you must arrange some drainage. Water building up behind a wall can burst it. Below the topsoil, close to the terrace wall, you should build up some permeable material—stones or pebbles. Insert short drainpipes through the wall and into this permeable fill at intervals of about ten yards.

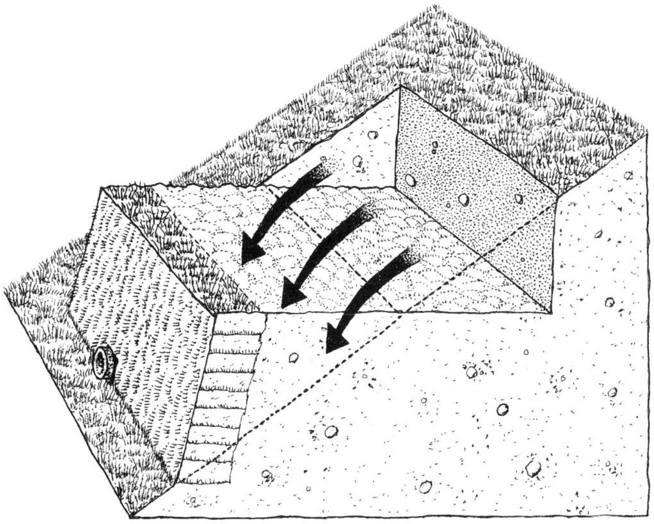

BUILDING A TERRACE

Mark out the width of the proposed terrace, and peel off the sod from the side of the hill to this width. Dig a level foundation at the base of the terrace, and build a retaining wall on top of it with the sod. Give the wall a slight batter in toward the terrace. Build in some porous material close to the wall; then insert short drain pipes through the wall and into this porous fill. Finish by moving the soil from the top half of the terrace to the bottom half. The sod in the wall will put out fresh grass on its exposed side.

GARDEN PATHS

A good paved path running down the backbone of your garden is a great boon. The main highway of a vegetable garden gets too much traffic for a grass path to cope with. A dirt path becomes mud, except on very light sandy soil, and pushing a wheelbarrow is much easier on a paved surface. Now that I am going over to deep-bed gardening (see p. 106), I am making paved paths between each five foot (1.5 m) wide deep bed by simply throwing all the stones and pebbles that I dig up onto the narrow strips between the beds. I strongly recommend this practice: with a paved path running down the middle of a small garden and deep beds running off at right angles with stone paths between them, it is never necessary to get your feet dirty at all! This makes quite a difference to the state of your living room carpet.

Types of paths

There are various ways of making a paved path. Concrete is permanent, but is very ugly. And after all, you garden for pleasure as well as for food, and therefore aesthetics should be given a fairly high priority. A brick path, possibly with a zigzag pattern of bricks, is very attractive, and if well made, will last a lifetime. But if it is made badly it will fall to pieces in no time. Crazy paving is fine

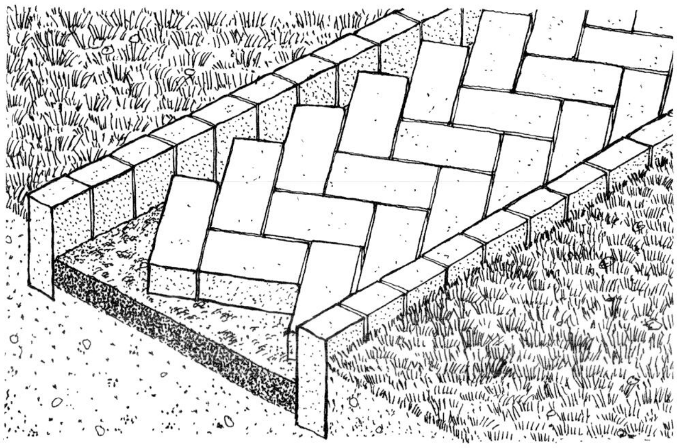

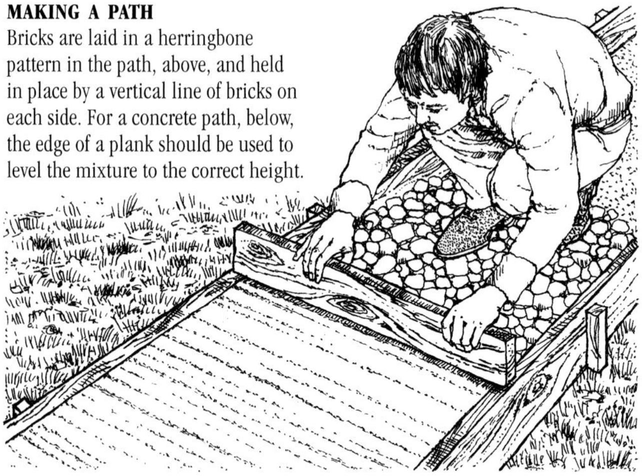

MAKING A PATH
Bricks are laid in a herringbone pattern in the path, above, and held in place by a vertical line of bricks on each side. For a concrete path, below, the edge of a plank should be used to level the mixture to the correct height.

if you like it: personally I think the end result is not worth all the trouble of intricately fitting it together. But flat stone slabs laid properly can be excellent, or a mixture of stone slabs and bricks is serviceable and attractive. You can use old bricks and you don't even have to clean them. Gravel, granite chips, or other small loose pebbles are fairly good: perfectly all right if you have a wheelbarrow with a pneumatic tire, but laborious if your wheelbarrow has a bare steel rim; the good old wide wooden wheel, with an iron tire on it, goes quite sweetly over gravel.

Laying a path

GRAVEL Dig a trench about four inches (10 cm) deep, and to the width you want the path. Fill with plenty of gravel. Never skimp on the gravel: use a lot or it will get trampled away to nothing.

BRICKS AND SLABS Dig a trench four inches (10 cm) deep and put two inches (5 cm) of sand in it. Flatten the sand with a plank to make it level and firm. Lay your bricks to a herringbone pattern on the sand. Hit them with a mallet to make the surface level, being careful not to chip off corners. Sprinkle sand over the bricks and sweep over the surface so the sand falls into the cracks. Hose the path down so that the sand binds together.

The edges are a problem, for they can break away. One method of preventing this is to lay a line of bricks along each edge sloping down away from the path and embedded quite deeply in the soil at the side. Bricks set vertically in the soil will fulfill the same purpose.

A slab path can be set in place using the same method. The edges will not be a problem because the slabs are so large.

CONCRETE If your soil is soft and muddy, you will have to dig quite a deep trench—say, a foot (30 cm) deep—and fill this with crushed stone or rubble. If your soil is hard and well drained, you need only dig a trench about two inches (5 cm) deep. Now lay planks on their edges for shuttering along each side of the path. Keep the planks securely in position by driving pegs into the soil beside your trench.

Your concrete should be a mixture of one part cement, two parts sand, and four parts small aggregate. If you can borrow or rent a cement mixer to do the mixing, do so, for it will save you an awful lot of sweat. If you can't do this, mix your concrete this way: make a pyramidal pile of your aggregate on a hard surface such as concrete. Throw your sand over this pile so that it covers the pile evenly. Then throw on your cement in the same way. Now start shoveling from one side, sliding your shovel in along the floor, and build, shovelful by shovelful, another conical pile nearby. When you have completed this pile, repeat the process, putting the material back where it was before.

Move it a third time, and then make a hole in the top of your cone like the crater in the top of a volcano. Throw water with a dash of emulsifier in it—if you can't get emulsifier, use detergent—into the hole. Now splodge around in the hole with your shovel so as to blend the dry mix with the water. Beware a break in the circular wall at the bottom of the pile, or you will lose water, cement, and all. When you have mixed as much as you can, quickly shovel the outside of the pile into the middle. Slice it repeatedly with your shovel and sprinkle on more water as you need it. You don't want the mixture too sloppy: just make it all thoroughly wet.

Pour the mixture into your trench using a bucket. Level the surface to the height of the retaining planks, and firm the top by pressing down with the edge of a plank. This should leave a pattern of small furrows and ridges that makes for a nonslip surface.

If you can give the surface a slight camber, this is an advantage, for the path will shed water.

You may well find it easier to make a long path in sections, say, ten feet (3 m) long. If your path is more than 40 feet (12 m) long, unbroken, put in expansion joints roughly every 30 feet (9 m). These can be thin planks standing up on edge. Just leave the planks there after the concrete has set. They allow for expansion of the concrete; otherwise it might crack in hot weather. After you have made your path, keep the concrete wet for three days. Hose it occasionally or cover it with wet sacks. Do not let the frost get at it.

HEDGES AND FENCES

Good fences make good neighbors and bad fences make the other kind. Certainly no one should put up a boundary fence, or wall or hedge, without consulting his neighbor, and it is far better to compromise if there is a disagreement. You will be far happier with the wrong sort of fence and a friend on the other side of it, than with the fence you wanted and a mortal enemy.

Hedges and fences each have advantages over the other. Hedges look nice, and they are alive and therefore foster other living things, like small animals, birds, moths, and butterflies. They add to the biological richness of your garden, and they can be productive. Consider the merits of a crab apple or hazel hedge, for example.

However, hedges sterilize a lot of ground: that is to say, they shade a wide stretch, but more drastically their roots spread out and draw all the moisture and nourishment out of a wide area. In a big garden this may not matter, but in a small one it does. Hedges can also harbor weeds and harmful insects. Personally I would not grumble about this, because they harbor useful insects—predators—as well, and to an organic gardener anything that increases the richness of life in his garden is to be encouraged. And it is easy enough to stop weeds from seeding in the soil below a hedge. Creeping perennial weeds like couch grass are a problem, though, and in a small garden they represent a very strong argument against planting a hedge at all.

Fences and walls take up far less ground than hedges and do not send out roots to deprive the garden soil. Nor do they harbor insects and weeds. A major advantage is that they are quickly established: an instant solution. Most important of all though, is that they can be used for training cordon, espalier, or fan-trained fruit trees or any food-bearing climbers, such as runner beans, tomatoes, cucumbers, squashes, or melons. This, to my mind, gives them a very big advantage over hedges.

There are not many plants that make a good hedge and at the same time provide food, while a fence or wall will support a great deal of food-bearing life, and you can therefore increase the effective area of your garden by the area of the side of the fence. You cannot do this with a hedge.

However, bear in mind that fences can look ugly, though of course they needn't. Choose one that appeals to you visually—as long as it is strong—use it to support fruit and vegetables, and you will have a beautiful fence. Also remember that fences are generally more expensive than hedges, especially if you gather the material for planting a hedge yourself. And fences don't last as long as hedges. Indeed, there are many hedges that are known to be hundreds of years old. Walls are, of course, more expensive than hedges, but they do last.

Building a hedge

For my hedges I always plump for some simple indigenous plant like hawthorn or beech, which is attractive and keeps its brown leaves all winter. I lay, or plash, my hedges every three or four years so as to keep them dense and prevent them from opening out at the bottom. Plashing is the art of cutting almost through the upper stems of the hedge, laying them over at a slight angle to the horizontal without breaking them off, and weaving them together.

The many varieties of trees and bushes normally used for garden hedges, like privet and boxwood, can be trained by clipping with shears. Personally, I could live quite happily without ever seeing another privet hedge, and would only grow boxwood for proper trees; they produce marvelous hard lumber that is suitable for making carvings, and for "priests," the small clubs that are so useful for administering the last rites to unlawfully caught salmon.

Building a fence

Some gardeners have the fencing they inherited with their gardens. But when you have to put up your own fences, there are many considerations. The first is probably cost. In most parts of North America, wood is relatively cheap and is probably the best material. In much of Europe the price of wood is prohibitive. It is as cheap, and certainly more permanent, to use brick. Plain brick walls, such as surround many a city garden, can mellow and look attractive, and will last for centuries.

Fences can also be claustrophobic, but they have the advantages of halting the progress of rabbits and storing the heat of the sun. Whatever you use, consider the fence as part of the garden, for growing things on, rather than merely a boundary marker.

If you want wire fences, the sunny side of your garden is an ideal place to have them. You can train sun-loving plants up it and get the benefit of your neighbor's sunshine as well as your own. A fence intended to break the force of the wind should not be solid. A solid fence causes strong and damaging wind eddies in its lee. A fence through which a little of the wind can percolate—what engineers call spoiling eddies—is a much better windbreak.

MAKING A WATTLE FENCE
To build a wattle fence, take split hazel or willow withies. Weave them into the uprights.

A wattle fence, which you can build yourself, is ideal for this purpose. A solid wall, particularly if it is painted black, on the north side of a garden, will warm in the sun and force on any tender trees that you train up it.

GARDEN SHEDS

A tiny garden shed that will just hold your tools is better than nothing, but if you can afford the money and space, or the labor and space, a good, big shed that you can use as a potting shed, and that has room for a workbench for mending tools, is very much better. And, if the shed is big enough to store potatoes, roots, seed, dried beans, and even strings of onions, better still. I know a man who keeps two sixty-gallon cider casks in his tool shed: one for parsnip wine and one for rhubarb wine. He is a wise and happy man.

Interior layout

My own belief about the insides of sheds is that anything that can be hung up should be hung up. That way, you can find it when you want it, and it doesn't get forgotten in some damp corner where it will go rusty or rot. Personally, I like to paint the outline of every tool on the wall where it is hung. Then I know if the tool is out of place and I go and look for it.

Inside and to one side of the door, out of any rain that may drive in, you should keep your sandbox (see p. 246). It only takes a second to jab your spade, fork, hoe, or whatever into this as you bring it in, and if you always do this, you will always have clean, shiny, well-oiled tools. A "man" (see p. 246) should be hung on a string, also near the door. You can use this to scrape the worst mud off your tools before you bring them in. I think it is also a good idea to have a boot jack on the floor for pulling off your boots. You then leave them in their appointed place in the shed and put on your indoor shoes before going into the house. Calico bags of beans and other large seeds should be hung up overhead. Paper seed packets should be put in a drawer. Herbs and vegetables can be hung up to dry. A small bookshelf for books like this one and for seed catalogs is a good idea.

Building a shed

Before you decide to buy an expensive factory-built shed, consider making one yourself out of scrap material. A well-built shed, of second-hand timber framing and weatherboarding, creosoted* inside and well tarred outside, looks extremely good. The roof can be of planking with tarred felt or other fabric treated with coal-tar epoxy over it. A roll of tarred felt costs a lot less than new corrugated iron.

Another method of roofing that I find works very well is to begin by laying on a covering of old corrugated iron with holes drilled in it. Above this lay on plastic fertilizer bags overlapping like tiles so that they shed any water that comes from above. Cover this with another layer of old corrugated iron and nail right through it. Paint the top layer with coal-tar epoxy. This form of roof is well insulated, quite weatherproof, and very long-lasting, though I cannot claim that it is a delight to the eye.

The most important thing with all wooden buildings is that they should have dry feet. No timber should be set in the ground. It is a good idea to build the shed on piers of brick or concrete, and put in some form of damp coursing between the masonry and the timber. The very best thing to do is to lay a concrete floor below the shed and build on this.

Insulating a shed

It is a good idea to have your shed insulated. If you wish to store potatoes or other crops in the shed in very cold climates this is essential. Fiberglass insulation, like that used in attics and around hot water tanks, is ideal, but it is expensive. A method that should cost nothing is to use plastic fertilizer bags to build a false ceiling and false walls inside the shed. This creates a cavity of air between the outer walls and the plastic layer; as long as there are no gaps in the plastic, insulation of this kind can be very effective.

EQUIPPING A GARDEN SHED
This shed is made of timber framing and weatherboarding; it is creosoted* inside and tarred outside, with a roof made of tarred felt treated with coal-tar epoxy. The floor is concrete, so that the wood has no chance of getting damp. Inside the shed, all the items are arranged for greatest convenience. The tools are hung up on the wall, each one with its outline in whitewash, to make its absence conspicuous. A sandbox and a "man," along with a bootjack, are placed beside the door, so that tools can be cleaned and muddy boots removed first thing upon entering the shed. Books, netting, animal feed, and fruit canes are among the many and various things that can be kept in the shed.

CARING FOR GARDEN TOOLS

Tools, like boots, are paradoxically pricy if they are cheap. A cheap pair of boots will last a hard-working gardener a year; I have an expensive pair that is just wearing out after fifteen years. The very same principle applies to tools, and my advice is, as long as you can afford them, buy good-quality, well-made tools. Examine the joint between the handle and the metal working part. Are the two held together with one or two nails, or is there a well-crafted snugly fitting joint?

Replacing handles

The handles on good tools wear out faster than the tools themselves. But you can increase the life of wooden handles enormously by doing two things: oiling them with boiled linseed oil once a year, and keeping them indoors out of the rain.

Handles are notoriously expensive nowadays, but remember that every time a farmer lays a hedge, he cuts out and burns scores of good potential handles. It pays, on visits to the country, to keep a lookout for lengths of clean, straight-grained, ash (or curved ash for particular jobs). Bring them home, hang them up in the tool shed to season, and then shape them with a bill-hook and a draw-knife to fit into your tools as handles.

Sharpening tools

Tools that are meant to be sharp should be kept sharp. Hoes particularly should have an edge: not a razor-sharp one that will crumple on hitting the first stone, of course, but a slightly rounded obtuse-angled edge. Spades, too, although the best ones are made to be pretty well self-sharpening, should be kept fairly sharp.

This does not mean that you should grind away at your hoes and spades until they are all worn to stumps, but that, when you find an edge badly worn away, you should use a file. A file is better for such tools than a grindstone.

All cutting tools, like axes, hatchets, and pruning knives, should be kept sharp at all costs. It is a complete waste of time to work with blunt tools. Cutting tools can be sharpened on a grindstone, with a hand stone, also known as a "whetstone," whether of carborundum or millstone grit, or with a file. If you use a circular grindstone, be sure to keep it wet, otherwise the heat generated may take the temper out of the steel at the very edge of your blade. Be very wary of using a carborundum wheel too much on any one tool at one time. They can easily take the temper out of an edge.

Before sharpening any blade, examine the shape of the edge carefully so that you know how to sharpen it. Always keep very strictly to the original angle of the edge of the blade. The blade must always be ground right down again to bring its cutting edge to the original angle.

Some blades, like those of axes and some knives, are ground away on both sides so as to leave a symmetrical section. Sharpen both sides of these. Other blades, like those of chisels, draw-knives, planes, and some pruning knives, are only ground on one side, so as to leave an asymmetrical section. With these, do all your grinding and whetting on one side, and then simply pass the other side, laid quite flat, a few times over the grinding device, just to take off the burr.

USING A MAN
A man is a wedge-shaped piece of hard wood with a sharpened edge. Hang it on a string by the door of your tool shed and use it to scrape the worst of the mud off spades, forks, shovels and all other steel blades before you dip them into your sandbox.

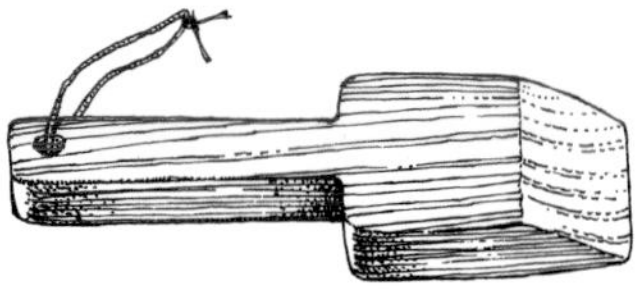

Caring for tools

MAN A man is a wedge-shaped piece of hard wood with a sharpened edge. Tie a piece of string to the handle and hang it on a hook by the door of your tool shed. You can use it to scrape the mud off your spades, forks, shovels, hoes, and everything else before you take them into your shed and plunge them into the sandbox.

USING A SANDBOX
Fill a box with a mixture of sand and sump-oil. The box should be big enough to take the blade of your largest spade. Every time you finish working with a steel tool, dip the blade into the sandbox. Keep the box out of the rain and it will last years.

SANDBOX There is one terribly simple method of doubling the life, and effectiveness, of steel tools, and that is to have a sandbox. Find a box deep enough to hold the blade of your biggest spade. Fill it with a mixture of sand and motor oil, which you can get free when you, or someone else, changes the oil in your car. Every time you come in from working with any steel tool, just plunge the blade into the sandbox. This both cleans and oils the blade. If you keep your sandbox out of the rain, it will last for many years without needing a change of contents.

SEED LIVES AND VEGETABLE YIELDS

	Percentage germination in years			Average yield per 10-ft row
	Up to 100%	75% +	50% +	
Asparagus	2	3–4	4–5	10 lb (4.5 kg)
Beans, fava	2	4	6	8 lb (3.6 kg)
green	2	3	–	8 lb (3.6 kg)
Lima	1	3	–	2.5 lb (1.2 kg)
runner	2	3	5	17–30 lb (8–14 kg)
soybeans	1	2	6	1.5 lb (0.7 kg)
Beets	3	6	10	15 lb (6.8 kg)
Broccoli	3	4	6	12 lb (5.4 kg)
Brussels sprouts	2	4	5–6	11 lb (5 kg)
Cabbage, spring	3	5	10	5–8 heads
summer	3	5	10	5–8 heads
winter	3	5	10	5–8 heads
Chinese	3	5	7	10 heads
Carrots	2	3	5	8 lb (3.6 kg)
Cauliflower	3	4	5	5–8 heads
Celeriac	3	4	5–6	12–20 lb (5.4–9 kg)
Celery	3	4	5–6	12–14 lb (5.4–6.4 kg)
Chicory	3	4	5–6	20 heads or 40 roots
Cress and land cress	3	5	9	Lots
Cucumbers	1	4	6	50 cucumbers
Dandelions	1	2	5	Lots
Eggplants	4	5	–	20 lb (9 kg)
Endives	5–6	7–8	10	10-12 plants
Florence fennel	3	4	7	20 bulbs
Hamburg parsley	1	2	4	10–15 lb (4.5–6.8 kg)
Kale	3–4	4–5	6–7	12 lb (5.4 kg)
Kohlrabi	3	4–5	6–7	12 lb (5.4 kg)
Leeks	1	2	3–4	20–30 leeks
Lettuce	3	4	6	15 lettuce
Melons	1	2	4	12–16 melons
Okra	4	5	–	16–20 pods
Onions	2	2	4	8–10 lb (3.6–4.5 kg)
Parsnips	1	1–2	4	15–20 lb (6.8–9 kg)
Peanuts	1	–	–	2–5 lb unshelled (0.9–2.3 kg)
Peas	3	4	9	20 lb (9 kg) pods
Peppers	2	2–4	7–8	7.5–10 lb (3.4–4.5 kg)
Potatoes	Plant seed potatoes			25 lb (11.3 kg)
Radishes	4	5	–	Lots
Rhubarb	Plant crowns			30–60 stalks
Rutabagas and turnips	2	2–3	5–6	8–14 lb (3.6-6.4 kg)
Salsify	1	1–2	3	6 lb (2.7 kg)
Seakale	1	1–2	3	6 lb (2.7 kg)
Spinach	2	3–4	5–6	8–10 lb (3.6–4.5 kg)
Spinach beet	1	2–4	5–6	10–15 lb (4.5–6.8 kg)
Squashes	4	6	–	8–15 squashes
Sweet corn	1	2	3	30–50 cobs
Swiss chard	2	6	10	8 lb (3.6 kg)
Tomatoes	3	6	10	20 lb (9 kg)
Watercress	2	5	9	Lots
Watermelons	3	6	10	5–10 melons

USEFUL ADDRESSES

Information about organic gardening

Garden Organic
Ryton Organic Gardens
Coventry
Warwickshire, UK
CV8 3LG
www.gardenorganic.org.uk
(Garden Organic is the working name for The Henry Doubleday Research Association)

The Soil Association
South Plaza
Marlborough Street
Bristol, UK
BS1 3NX
www.soilassociation.org

Information about trees, shrubs and perennial crops

Agroforestry Research Trust
46 Hunters Moon
Dartington
Totnes
Devon, UK
TQ9 6JT
www.agroforestry.co.uk

Information about vegetable seeds

Real Seeds
Brithdir Mawr
Newport
Pembrokeshire, UK
SA42 0QJ
www.realseeds.co.uk

Information about smallholding

www.acountrylife.com

Information about alternative technologies

Centre for Alternative Technology
Machynlleth
Powys, UK
SY20 9AZ
www.cat.org.uk

Information about community gardens

American Community Garden Association
www.communitygarden.org

CLIMATE ZONES

If you grow only the crops that grow indigenously in your climate zone, you will not suffer many losses due to climate, nor will you have to protect your plants from the weather. But of course gardeners don't do this. Quite rightly, they like a varied diet, so they push their luck, and try to grow the more succulent and tasty crops farther north than the plants really want to grow. This is why an understanding of climate is important to the gardener. It tells him when to plant, when to harvest and, perhaps most important, when to protect his plants artificially.

Cities are always warmer than the open countryside. The waste heat from houses and all those people contribute toward this. So, if you live in a city or its suburbs, you can plant a little earlier, and enjoy a longer growing season, than the gardeners in the countryside nearby.

Frost

The period that elapses between the last freeze of the spring and the first freeze of the fall is a crucial time for vegetable gardeners. If you want to grow outdoors those plants that are tender—and they include all the crops that are indigenous to warm climates, such as tomatoes, cucumbers, peppers, and squashes—you can grow them outside only during this period. Except for root crops and brassicas, all your vegetables should be harvested before the first fall frost.

The growing period of food plants can be measured against the growing period of grass. When the grass in your lawn starts to grow, after the dormant winter period, you can start putting in seed. Grass begins to grow when the soil temperature reaches 43°F (6°C) in the spring. When the soil temperature falls below that temperature in the fall, grass stops growing.

There are certain factors that affect the dates of the first and last freezes. For example, proximity to the sea, or any deep water, tends to warm the air and prevent frost, while altitude generally increases the cold.

Rainfall

Lucky is the gardener whose land gets just the right amount of water naturally from rainfall. Too much winter rain washes the nutrients out of the soil, erodes the soil itself, and prevents the gardener from getting out onto the land as early as he would like. Planting green manure crops on

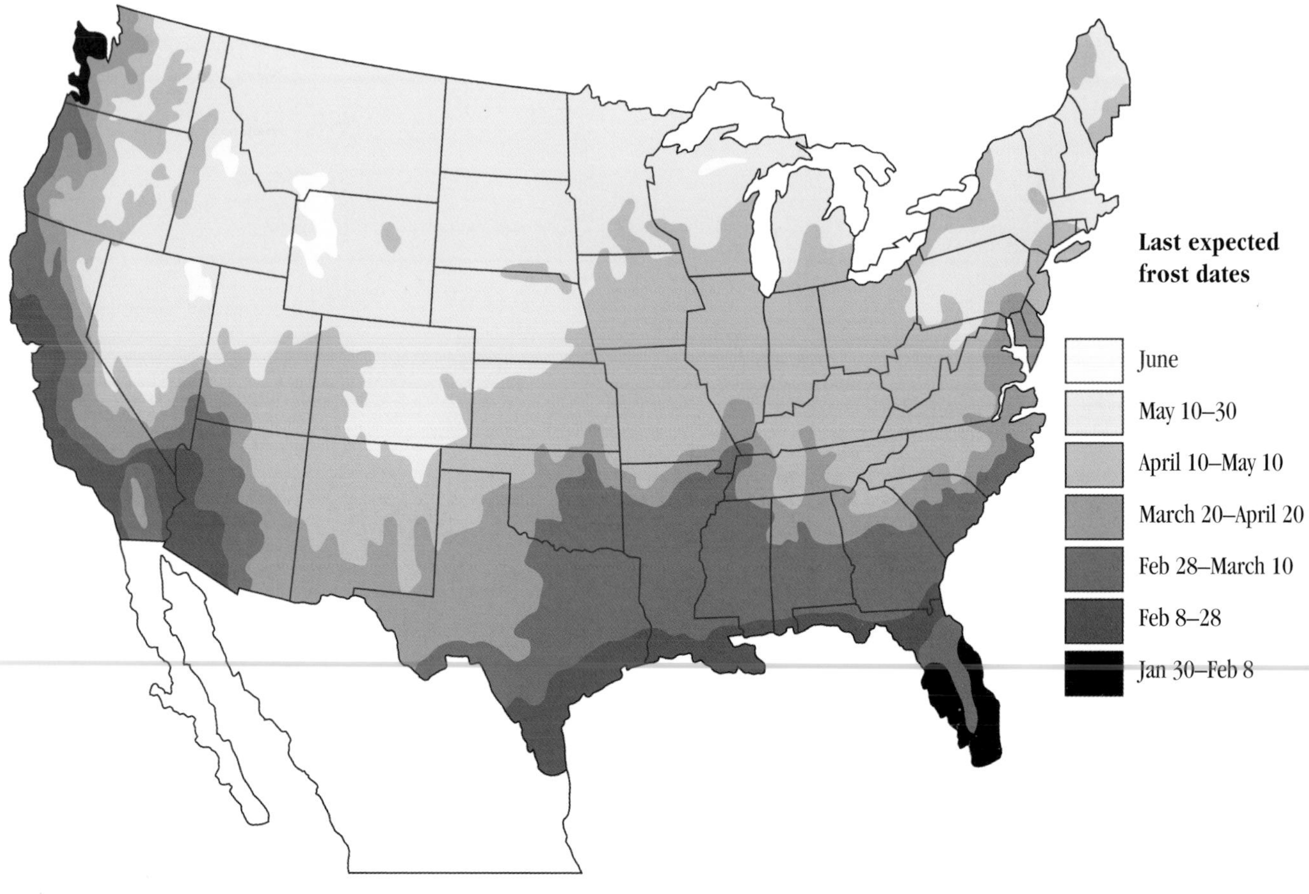

vacant vegetable beds in winter prevents erosion and keeps the nutrients in the soil.

The garden needs rain in the spring and early summer, and not too much rain in midsummer. Most of us have to give nature a hand and either dig irrigation furrows or get out the hose or watering can.

Sunshine

Some sunshine is important for all food-producing plants except mushrooms. Crops like sweet corn, eggplants, peppers, melons, tomatoes, cucumbers, peaches, and grapes will not ripen without plenty of sun. In areas that get little sun you can, of course, grow these crops under glass or plastic and, if necessary, provide artificial heat. On the other hand, some crops cannot take too much sun; lettuce plants, for instance, go to seed and die when the summer heat sets in.

Wind

Many plants suffer badly in windy positions. As well as the strength of the wind itself, there is the problem that wind exaggerates the effect of frost. In an area of high winds, try to utilize sheltered spots, or build windbreaks.

The USDA regions

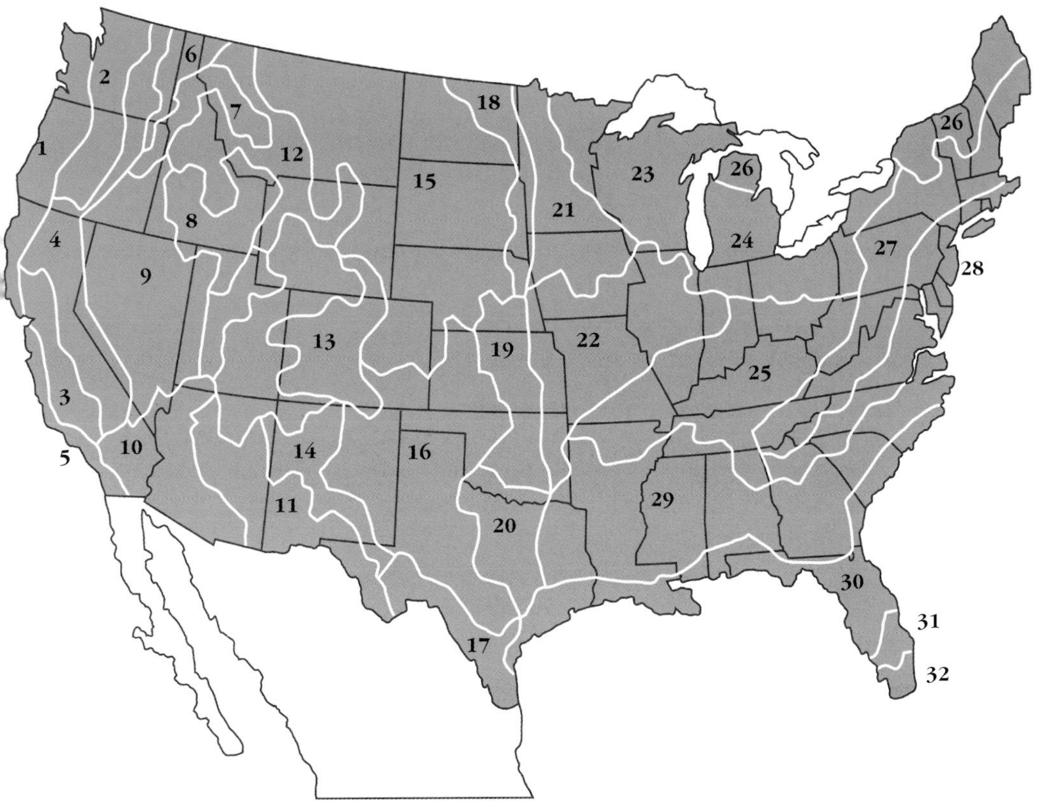

CLIMATE REGIONS

Region 1 Cool, dry summers with frequent fogs. Heavy winter rainfall.
Region 2 Summers drier and warmer than Region 1. Average low temperatures 10°F to 20°F.
Region 3 Hot, dry summers. Mild winters with 8 to 10 inches of rain. Low temperatures: 22°F to 24°F.
Region 4 A very mountainous region. Conditions vary greatly according to elevation.
Region 5 Summers on coast dry and cool, warmer inland. 10 inches of rainfall in valleys, 30 inches in the mountains.
Region 6 Summers warm. Winter temperatures average 10°F to 15°F.
Region 7 Summers warm. Winter temperatures range from 0°F to 15°F.
Region 8 Semiarid—hot summers and cold winters with temperatures between 0°F and –10°F.
Region 9 Cold winters. Hot summer days with frosty nights.
Region 10 Scorching hot with a rainfall of 3 to 10 inches.
Region 11 Same as Region 9 but hotter.
Region 12 Elevation and exposure variations mean big differences in rainfall and temperature.
Region 13 Similar to Region 12 but temperatures at same elevations average 7°F hotter.
Region 14 Similar to Regions 12 and 13 but warmer.
Region 15 Moderately warm summers, very cold winters.
Region 16 Dry farming area, warmer than Region 15. Rainfall 12 to 22 inches.
Region 17 Dry and hot with 12 to 22 inches of rainfall but excessive evaporation.
Region 18 Fairly humid with cold dry winters.
Region 19 Sudden variations in winter temperature. Hot winds in summer.
Region 20 Transition between dry farming regions to the west and humid climate of eastern Texas.
Region 21 Cold winters with drying winds. Rain (20 to 30 inches) comes mostly in summer.
Region 22 Prairie country with cold drying winter winds. Rainfall: 30 to 40 inches.
Region 23 East warmer and more humid than west, which is similar to Region 21.
Region 24 Moisture-laden atmosphere, 30 to 40 inches of rainfall spread over the year.
Region 25 Warm summer with risk of drought. Moderate winter. Rainfall: 40 to 50 inches.
Region 26 Long days and cool nights in summer. Heavy snowfalls in winter.
Region 27 Abundant rain (35 to 50 inches) through the year. Heavy snow in colder areas.
Region 28 Warmer than Region 27 with possibility of drought at end of summer. Moderate winters.
Region 29 Warm summers and abundant rain (45 to 60 inches).
Region 30 Hot in summer. Short winter with much rain.
Region 31 Warm summers, but killing annual frosts. Rainfall: about 50 inches.
Region 32 No killing frosts and only slight temperature variations. Rainfall: 50 to 60 inches.

Index

Figures in **bold** indicate a major reference. Figures in *italics* indicate illustrations or illustrated text.

Acknowledgments

Author's Acknowledgments to the original edition

I would like to express my gratitude to the many people who have helped in the making of this book, particularly the staff of Dorling Kindersley, who have worked with great devotion and good will for many months—far beyond the call of ordinary duty. I would also like to thank the artists for doing such excellent illustrations, and the people who live and work at my farm for their help in carrying out the many experiments and trials that have provided information for the book. Lastly I would like to thank everyone who helped with advice, demonstrations and information, with special thanks to Lawrence D. Hills, Dr. Shewell-Cooper, the East Malling Research Station, the Covelo Garden Trust California, Ecology Action California, and many private gardeners in Britain, the United States, Italy, and France.

John Seymour

Dorling Kindersley Limited would like to thank the following for their special contributions to this edition of the book:

Annelise Evans and Andrew Halstead.

Dorling Kindersley Limited would like to thank the following for their special contributions to the original edition of the book:

Michael Carlo, Georgina D'Angelo, East Malling Research Station, Katherine Fenlaugh, Fred Ford, Ramona Ann Gale, Lesley Gilbert, K. Holmes of Cramers Ltd., Sally Seymour, Dr. W. E. Shewell-Cooper, Martin Solomons, Mick Leahy, John Rudzitis and the staff of Vantage, Alan Lynch, Geoff Smith and the staff of Cowells.

Artists

David Ashby, David Bryant, Brian Craker, Julian Holland, Peter Kesteven, Robert Micklewright, Peter Morter, Nigel Osborne, Jim Robins, Malcolm Smythe, Eric Thomas.

Notes and cautions

The following notes and cautions are the result of an evaluation of the author's original text in the light of changes to nomenclature, horticultural advice, and legislation regarding the use of chemicals as home garden remedies.

NOTES

Actinomycetes – these are fungi-bacteria, which includes *Actinomyces* and other genera such as *Streptomyces*, that give the soil its smell.

Aphids – the new Latin name for the aphids called phylloxera is *Daktulosphaira vitifoliae* (previously they were *Phylloxera vastatrix*).

Cucumbers – as well as the strain called 'Burpless,' other smooth-skinned cucumbers are now available for growing in the open garden in temperate regions.

Family names – some plant family names have been changed in the book, to conform to current botanical nomenclature, as follows:

Old name	Current name
Compositae	*Asteraceae*
Cruciferae	*Brassicaceae*
Gramineae	*Poaceae*
Labiatae	*Lamiaceae*
Leguminosae	*Fabaceae*
Umbelliferae	*Apiaceae*

Fish meal – as a feed for poultry, fish meal is acceptable as part of a balanced diet but is not adequate on its own.

Garden geography – in Chapter Two: Gardening through the Year, a number of terms, such as north and south and those relating plant growth to the months of the year, refer to the northern hemisphere.

Molybdenum deficiency – this is not common in gardens.

Parsnip rust – parsnips may develop a rusty discoloration on the roots that is more usually referred to as parsnip canker.

Peat – commercial harvesting of peat destroys wetland habitats; there are now many good peat alternatives available to gardeners.

Peat bags – plastic bags bought ready-filled with peat are now known as grow bags.

Tarred felt collars – commercial or homemade collars with a smear of kerosene for use on brassicas against cabbage root fly are unlikely to be of any value.

Wart disease – this is now extremely rare on potatoes. As most potato cultivars in use today have full resistance to wart disease, it is unlikely that any gardener would need to take steps to combat it.

CAUTIONS

Ascorbic acid – dipping fruit in ascorbic acid solution after drying may add to the nutrient status of the fruit. However, if it is to control pests and diseases, it is illegal in some countries.

Aluminum – aluminum pans are no longer recommended for cooking acidic foodstuffs such as chutneys.

Bitumen – modern wound sealants are available that do not contain bitumen.

Bordeaux mixture – as with any pesticide, this is illegal to make at home in some countries. It may still be available from garden centers and can be useful for potato blight.

Bullfinches – these birds are are now red-listed (globally threatened) by the IUCN (the World Conservation Union).

Burgundy mixture – as with any pesticide, this is illegal to make at home in some countries. It is not readily available commercially.

Calomine – the calomine mentioned on p. 123 for use against clubroot was probably Calomel, which contained a mercury salt (mercurous chloride), but is no longer available and is illegal in some countries.

Copper sulfate – spraying this solution (copper sulfate, lime and water) on grapes is illegal in some countries.

Creosote is banned for domestic and amateur horticultural use and as a wood preservative in some countries.

Denatured alcohol – painting denatured alcohol on apple trees is illegal in some countries.

Formaldehyde solution – this is possibly acceptable for cleaning mushroom boxes if it is being used as a disinfectant. However, formalin (also known as formaldehyde) is a potentially dangerous chemical that would be difficult for a gardener to obtain.

Formalin – as a seed dip against leaf spot and as a spray against onion smut, formalin is illegal in some countries.

Human manure – putting human manure on the compost heap is questionable as there may be health concerns, including transmission of heavy metals and parasites. Check local and national legislation before using human manure in your garden.

Kerosene – the use of kerosene is illegal in some countries and cannot be recommended for garden applications.

Kudzu – this plant is rampantly invasive and is illegal to buy, sell, or plant in most states because of the economic and ecological damage it causes.

Lime-sulfur spray – this spray is no longer on sale and is illegal in some countries.

Liquid manure – as a fertilizer on celery, liquid manure is acceptable. However, if the intention is to kill a pest (e.g., leaf miner), it is illegal. As celery is usually eaten raw, there could be health problems if the manure is of animal origin.

Nicotine mixture – the use of nicotine as a pesticide for domestic and amateur horticultural use is now banned in some countries.

Quassia – this is no longer on sale and is illegal in some countries.

Quicklime – quicklime gives out a great deal of heat when in contact with water and would need to be handled with great care. It cannot be recommended for any garden use on health and safety grounds.

Soft soap – this is illegal as a pesticide in some countries but acceptable in the form of commercial pesticides containing fatty acids.

Soot and lime – if the intention is to control caterpillars, this remedy use would be illegal in some countries; if the treatment is a deterrent, it might be legal.

Sulfur spray – powdery mildew on strawberries is now treated with sulfur dust, not spray.

Washing soda – washing soda solution sprayed on gooseberry bushes against mildew is illegal in some countries.

Winter wash/tar wash – tar-oil winter washes are now banned for amateur horticultural use in some countries; washes based on plant oils are available.